Seabird Biodiversity and Human Activities

Books in the Series

Seabird Biodiversity and Human Activities
Jaime A. Ramos and Leonel Pereira (eds.), 2022

Series: Aquatic Sciences

Series Editor: Leonel Pereira

Volume 1

Seabird Biodiversity and Human Activities

Editors

Jaime A. Ramos
MARE – Marine and Environmental Research Centre
University of Coimbra
Department of Life Sciences
Coimbra, Portugal

Leonel Pereira
MARE – Marine and Environmental Research Centre
University of Coimbra
Department of Life Sciences
Coimbra, Portugal

CRC Press
Taylor & Francis Group
Boca Raton London New York

CRC Press is an imprint of the
Taylor & Francis Group, an **informa** business
A SCIENCE PUBLISHERS BOOK

Cover credit: The front cover is a Dolphin Gull (*Larus scoresbii*) in the city of Punta Arenas, Chile (photo of Lucas Krüger). The back cover shows a Southern Fulmar (*Fulmarus glacialoides*) in the Magellan Strait (photo of Lucas Krüger), a Little Tern adult (*Sternula albifrons*) incubating in the Algarve, Portugal, and a White-tailed Tropicbird (*Phaethon lepturus*) chick in the Seychelles (photos of Jaime A. Ramos).

First edition published 2022
by CRC Press
6000 Broken Sound Parkway NW, Suite 300, Boca Raton, FL 33487-2742

and by CRC Press
2 Park Square, Milton Park, Abingdon, Oxon, OX14 4RN

CRC Press is an imprint of Taylor & Francis Group, LLC

Library of Congress Cataloging-in-Publication Data

Names: Ramos, Jaime A., 1965- editor. | Pereira, Leonel, editor.
Title: Seabird biodiversity and human activities / editors, Jaime A. Ramos, MARE - Marine and Environmental Research Centre, University of Coimbra, Department of Life Sciences, Coimbra, Portugal, Leonel Pereira, MARE - Marine and Environmental Research Centre, University of Coimbra, Department of Life Sciences, Coimbra, Portugal.
Description: First edition. | Boca Raton : CRC Press, 2022. | Series: Aquatic sciences : Volume 1 | Includes bibliographical references and index.
Identifiers: LCCN 2021047056 | ISBN 9780367498276 (hardcover)
Subjects: LCSH: Sea birds. | Sea birds--Ecology. | Sea birds--Conservation. | Sea birds--Effect of human beings on. | Marine biodiversity conservation.
Classification: LCC QL678.52 .S425 2022 | DDC 598.177--dc23/eng/20211021
LC record available at https://lccn.loc.gov/2021047056

ISBN: 978-0-367-49827-6 (hbk)
ISBN: 978-0-367-49828-3 (pbk)
ISBN: 978-1-003-04752-0 (ebk)

DOI: 10.1201/9781003047520

Typeset in Times New Roman
by Radiant Productions

Preface to the Series

Aquatic Science is the multidisciplinary study of aquatic ecosystems, both freshwater and marine. Scientific investigations range in scale from the molecular level of contaminants to the stresses on entire ecosystems.

Some of the major fields of study within the this aquatic sciences include: limnology (study of lakes, rivers, wetlands and groundwater); biogeochemistry; aquatic ecology; oceanography; Marine biology; ecology applied to marine macroalgae; shrimp farming; ecophysiology and behavior of fish; microalgae ecology; microbial aquatic ecology, nutrient cycling and limnology; ecology of marine benthos and cephalopods; zooplankton ecology; aquatic ecotoxicology; seabirds biodiversity; dynamics of populations of aquatic beings; fisheries biology; physical chemistry, marine pollution and coastal management; malacology; crustacean nutrition; oceanography; pathology of aquatic organisms; continental fish farming; aquaculture technology, mariculture and coastal management. In each of these fields, different volumes edited by renowned experts will be published.

The Series Editor
Leonel Pereira

Preface

Seabirds are very charismatic for birdwatchers, ornithologists, conservationists and the larger public. They have been and are popular in ornithology, and, more recently, they have been used as important model species for theoretical and applied ecological studies. Seabirds are very good indicators of the relationships established between humans and nature, and can act as models in the long quest of achieving environmental sustainability by modern human societies. Seabirds are long-lived top predators, reflecting changes occurring at lower trophic levels, and their populations are largely influenced by a vast array of human activities that may interfere with their breeding, foraging and migration patterns. Seabirds have for long provided cues to important human activities such as navigation and the location of fishing grounds. However, the overexploitation of marine resources, linked to the continuous rise of the human population, is pushing many seabird populations to the verge of extinction. The once safe islands where breeding could take place are being ravaged by introduced mammal predators, the vast and abundant fish stocks that supported enormous seabird populations have been decimated, the once calm waters are now covered by ship navigation routes and energy-producing platforms (either fossil or renewable). On top of this, the once pristine and healthy ocean waters are now littered with plastic and other pollutants, and coastal seashores are now urbanized, lighted up day and night, and packed with recreational and touristic attractions. Given such widespread environmental changes, it is not a surprise that seabirds are classified as one of the most threatened bird group. Ultimately the fate of seabirds will reflect the fate of our own societies.

There are quite a lot of academic and other publications addressing seabirds in the market, but we believe this book fills a gap that has been so far neglected: it starts by covering the main known seabird research methods and then focuses on a selection of pragmatic issues related with interactions between seabirds and humans, the use of seabirds as ecological indicators and the conservation and management of seabird populations. We believe this is an approach that would appeal to a wide range of professionals and students, including wildlife researchers, practitioners, managers and other professionals who are actively engaged in the areas of ecological research and environmental sustainability. This book summarizes our current knowledge on the relationships between seabirds and human activities, with a view to stimulate research, conservation, and the development of tools and policies to achieve effective mitigation measures benefitting both humans and seabird populations. The relationships between seabirds and human activities are now popular topics of research, due to the application of recent technology and analytical methods. Most seabird researchers are now familiar with the use of biologging, such as GPS loggers, to study their spatial and behavioural ecology. Recent developments in the use of stable isotope analysis or DNA metabarcoding are now widely applied to study seabirds' diet and feeding ecology. In parallel, new analytical methods that help us to understand and summarize vast quantities of data, such as those generated by biologging, are now available to us. These modern methods are enabling us to solve many mysteries of the life of seabirds, but, at the same time, pose a challenge to those who are at the start of their research or are in conservation careers.

Therefore this book is split into four broad sections (from A to D). Section A begins with an introduction to seabirds and their study and includes three more chapters: (1) conventional and modern techniques to study seabird diet and feeding ecology, (2) the conservation physiology toolbox that provides a mechanistic understanding of how environmental changes and human activities impact physiological responses of seabirds, and (3) the use of tracking technology to study relationships between seabirds and environmental stressors. In Section B, we focus on the interactions between seabirds and human activities, either positive or negative. This section includes the following chapters: (1) seabirds and fisheries interactions, (2) the ecology of urban gulls, (3) seabirds and renewable oceanic energy sources, and (4) seabirds and biotoxins. Section C tackles the issue of seabirds as ecological indicators, for: (1) forage fish, (2) oceanographic changes, (3) metal and plastic pollutants, and (4) climate change. Finally, Section D is dedicated to specific actions and tools for seabird conservation, and three main topics are highlighted: (1) light pollution as a seabirds' conservation threat, (2) eradication and control of invasive mammal species as a tool for the conservation of seabirds, and (3) the use of seabirds to identify and establish Marine Protected Areas. This book not only provides an updated insight on all these topics but also highlights gaps that would require updated efforts in terms of research, conservation and management measures needed to understand the relationships between seabirds and human activities. We hope it provides a source of inspiration for a new generation of seabird students and conservationists.

This book benefitted from a wide range of contributors and experts coming from diverse backgrounds, and ranging from professional scientists, late graduate and post-graduate students and practitioners from Non-Governmental Organizations. The entire book owes a great deal to the engagement of researchers and students of the research group Ecology and Conservation of Top Predators (EcoTop) from the Marine and Environmental Research Center (MARE), of the University of Coimbra, Portugal. We would like to express our gratitude to the EcoTop members, to the panel of authors that accepted to write the chapters, and finally to all the reviewers for their comments and suggestions.

Coimbra, June 2021
Jaime A. Ramos
Leonel Pereira

Contents

Part A
Techniques to Study Seabirds

CHAPTER 1

An Introduction to Seabirds and Their Study

Marie Claire Gatt,[1,*] *José Pedro Granadeiro*[2] and *Paulo Catry*[3]

What is a seabird?

The lives of humans and seabirds have long intertwined, from the exploitation of seabirds for food and feathers by coastal and island communities, to the importance of guano and its trade in and from South America, and to the seabird's strong significance as an indicator for fishermen and sailors, culminating in great works of prose and poetry. The term "seabird" is used to refer to species from several taxonomic orders of birds which feed largely or entirely on marine organisms and are adapted to make extensive use of the marine environment for at least part of the annual cycle (Furness and Monaghan 1987, Croxall et al. 2012). The definition is not a fixed one, and certain avian groups—such as waders, seaducks, divers and phalaropes—are included or left out depending on the definition adopted. However, all the Sphenisciformes (penguins) and Procellariiformes ("tubenoses"), the Suliformes (gannets, boobies, cormorants, frigatebirds) excluding darters, the Charadriiformes (auks, larids, skuas) excluding waders and other families, and some of the Pelecanidae (pelicans) most often classify as seabirds. This selection encompasses around 350 extant species around the world (Handbook of the Birds of the World and BirdLife International 2019). The relative dependence on the marine environment varies significantly within these groups. Among the cormorants, gulls, terns and pelicans, some species associate very strongly with terrestrial and/or freshwater environments, rarely wandering far offshore. Penguins, tubenoses, auks, skuas and most sulid groups, on the other hand, depend solely on the sea for food and only come to land in order to breed.

Seabirds constitute a polyphyletic unit, with many unrelated taxa displaying convergent evolution to adapt to the marine environment. Functioning within a marine environment requires seabirds to be able to effectively manage salt balance. While some species may make use of available fresh water, many do not, and much of their hydration comes from food. In any case, saltwater

[1] Dipartimento di Scienze e Politiche Ambientali, Università degli Studi di Milano via Celoria 26, Milano, I-20133, Italy.
[2] Centro de Estudos do Ambiente e do Mar (CESAM), Departamento de Biologia Animal, Faculdade de Ciências, Universidade de Lisboa, 1749-016 Lisboa, Portugal.
Email: jpgranadeiro@fc.ul.pt
[3] Marine and Environmental Sciences Centre (MARE), ISPA-Instituto Universitário, Rua Jardim do Tabaco n.º 34, 1149-041 Lisboa, Portugal.
Email: paulo_catry@ispa.pt
* Corresponding author: marie.gatt@unimi.it

ingestion cannot be completely avoided. To control osmolality, marine birds have supra-orbital salt glands which secrete concentrated saline solution from their orbital or nasal cavity, excreting excess salt. These glands are the major pathway of salt excretion, in contrast with other vertebrates in which this role is taken up by the kidneys (Gerstberger and Gray 1993).

Among seabird species that pursue prey in dives, there is a requirement to see their prey sharply underwater. Birds appear to be the only vertebrate group that have excellent amphibious vision—that is, their eyes are adapted to focus both in air as well as in water. This is done by reducing the refractive contribution of the cornea, relying almost entirely on the flexible refractive ability of the lens. Among seabird species studied, there appears to be more than one evolutionary strategy to amphibian vision (Glasser and Howland 1996, Cronin 2012): Albatrosses and penguins have a flatter cornea and a near-spherical lens, the shape of which is controlled by the ciliary muscle. A ring of intraocular bones, the scleral ossicles, provide further rigidity to the ocular structure to stabilise the focal length against external forces (Sivak and Millodot 1977, Martin 1998, Fischer and Schoenemann 2019). Cormorants and probably gannets have a more unique ocular structure, using the ciliary muscle to pull the lens forward through the rigid iris, increasing refractive ability (Katzir and Howland 2003, Machovsky-Capuska et al. 2012).

Flight and wing-propelled diving require different wing forms—efficient locomotion in one medium is traded off with that in the other. Flying seabird species specialising in wing-propelled pursuit diving, such as the auks and diving petrels, have smaller, rounder wings, high weight-to-wing area ratios (wing loading) and high flight costs as they are largely flapping flyers—the Brünnich's Guillemot (*Uria lumvia*) has the highest wing-loading of any other volant bird. Moreover, while many bird bones are typically pneumatised (i.e., hollow, filled with air and connected to the respiratory system), the breast bone and long bones in auks are not, reducing their buoyancy underwater (Gaston and Jones 1998). The resulting high flight costs have led to flightlessness in penguins, Flightless Cormorants (*Phalacrocorax harrisi*), and some extinct species, notably the Great Auk (*Pinguinus impennis*) (Elliott et al. 2013). Being liberated from the energetic constraints of flapping flight, many of these species could grow to large body masses compared to their flying counterparts, greatly improving their diving capacities (Schreer and Kovacs 1997).

However, some of the largest volant avian species are among seabirds, and therefore they must be capable of energy efficient flight at sea. The Wandering Albatross (*Diomedea exulans*) is a classical example, weighing up to 12 kg and with a wingspan of up to 3.5 metres—the largest in any extant bird, it maintains remarkably low mass-specific field metabolic rates during gliding flight, with heart rates during flight only ca 50% above those when resting (Weimerskirch et al. 2000). This is achieved through dynamic soaring, whereby albatrosses and other Procellariiforme seabirds gain energy from shear wind by cyclically climbing in headwinds, curving leewards, descending, and curving again into headwinds, sustaining long-distance non-flapping flight at very little energetic expense (Weimerskirch et al. 2000, Shaffer et al. 2001, Sachs et al. 2013). As an adaptation to this means of travel, albatrosses, petrels and shearwaters have long, narrow wings and a few species even possess a locking elbow joint which allows them to keep their wings open without muscular effort (Pennycuick 1982). Frigatebirds use a different flying technique, owing to their inability to take off from the water's surface despite their dependence on the marine environment for foraging (Weimerskirch et al. 2016). They have the lowest wing loading of any bird (Pennycuick 1983) and travel by rollercoaster flight, soaring up thermals formed below cumulus clouds to altitudes generally up to 700 m, then gliding down, covering long distances at very little energetic cost, sometimes remaining on the wing for weeks on end (Weimerskirch et al. 2016). Many other seabirds flap-glide throughout powered flight, with the extent of flapping to gliding varying with their morphology (Pennycuick 1986).

Distribution and geographical niche

No family of seabirds is restricted to a single ocean, and those restricted to a particular hemisphere are exceptional. Seabird distribution around the globe is tightly linked to foraging ecology and morphology. Seabirds with a strong association to the coast even outside the breeding period experience additional spatial constraints to those affecting pelagic seabirds.

Penguins, auks and diving petrels, best suited to dive-pursuit, tend to occur at higher latitudes where marine productivity is high (Bunt 1975). The energetic demands of wing-propelled dive-pursuit coupled with low flying efficiency or flightlessness generally make it unfeasible for these species to expand into or across the tropics, where dive-pursuit and short-distance flight are not suited to low productivity and patchy distribution of fast-moving prey (Brooke 2002).

Procellariiformes also show extra-tropical peaks in species richness and are asymmetrically distributed, with far higher diversity in southern high latitudes than in northern ones. The reduced energy expenditure of dynamic soaring in many procellariiforme species allows them to capitalise on patchily distributed resources over large distances (Lack 1968, Ashmole 1971). Wind speeds are highest at high latitudes, but are also higher and more constant in the southern hemisphere, where there is more contiguous ocean with fewer land barriers compared to the northern hemisphere. Both the available wind energy and the expanse of open ocean increase the potential for long-distance foraging trips to highly productive areas (Weimerskirch et al. 2000, Shaffer et al. 2001), and may partially explain the spatial distribution of species richness in this order (Davies et al. 2010).

The tropics are characterised by less productive waters and more patchily distributed food resources (Longhurst and Pauly 1987). Deep-diving seabirds which are common at higher latitudes are largely absent from the tropics. Seabirds here have adapted to foraging in surface waters, often in association with abundant underwater top-predators, such as tuna and dolphins, which concentrate prey items at the surface in feeding frenzies (Ballance and Pitman 1999). This patchiness and unpredictability of food resources, dependent on inter-specific feeding associations, results in a lower foraging site fidelity among tropical seabirds compared to populations studied at higher latitudes (Weimerskirch 2007).

The population sizes of seabirds, particularly those on oceanic islands in resource-poor zones, appear to be strongly limited by the amount of food available to a given species under the restrictions of central-place foraging. This phenomenon follows a common trend: pelagic species, which have a larger area of foraging ground available to them, are more abundant than inshore species, which experience higher competition in a restricted foraging area immediately surrounding the colony. Among both classifications, migrants are also more abundant than residents, possibly a result of their exploitation of food resources elsewhere (Diamond 1978).

A common feature of the vast majority of seabirds is coloniality. Over 95% of seabird species are colonial, higher than any other avian group (Coulson 2002). Colonial breeding is in fact rather rare among terrestrial species of birds. Seabird colonies vary in density but nesting sites cluster closer together than would be expected by chance alone, and imply a high level of interaction between the individuals that form part of the colony. The factors promoting colonial breeding are still under investigation. While protection against predators is a common argument (Lack 1968), colonies come with density-dependent costs, such as prevalence of contagious disease and agonistic interindividual interactions (Ashbrook et al. 2010). Another theory revolves around the dynamic distribution of food, making it futile to defend a feeding territory and rewarding to take up cues on the location of resources from the rest of the population at the colony (information transfer) or at sea due to the conspicuousness of foraging aggregations (local enhancement) (Ward and Zahavi 1973, Buckley 1997). Work on Australasian Gannets (*Morus serrator*), Guanay Cormorants (*Phalacrocorax bougainvillii*) and

Peruvian Boobies (*Sula variegata*) uncovered evidence of both information transfer at the colony and local enhancement of foraging, with strong ties to species foraging ecology (Weimerskirch et al. 2010, Jones et al. 2018, 2020). Today, the importance of intraspecific communication to reach patchy resources under poor foraging conditions could increase the threat to survival in shrinking colonies of endangered seabird populations (Boyd et al. 2016).

The conspicuousness of breeding colonies has long made them the focus of attention of naturalists, which sought to quantify the size of seabird populations for long-term monitoring. Continuous census data for populations of some species go back to previous centuries, and many long-term monitoring programmes started as early as the mid-20th century, increasing in their number and scope (Wooller et al. 1992). From the early 20th century, the capacity of bird monitoring was increased by the advent of bird ringing (or "banding" in the US)—fitting uniquely numbered metal rings to the legs of birds. This allowed for information on life history and movements to be associated with individuals of known origins, greatly improving the level of knowledge obtained. Among the first bird species to be ringed were gulls (Preuss 2001), and the diversity of ringed seabirds spread with the practice of ringing itself, now a ubiquitous tool in monitoring studies. More recently, the use of plastic rings with alphanumeric codes or colour combinations legible from a distance has increased the potential of obtaining information from individuals without physical capture, particularly in coastal seabird species which are more likely to be observed.

Ecologists' curiosity has long dwelled away from the confines of the colony and on the at-sea distribution and habitat use of seabirds. As economic interest in offshore infrastructure increased in the last century, it also became apparent that the at-sea activities of seabirds needed to be quantitatively understood to assess potential environmental risks of human exploitation of the marine environment, and these impact studies offered a financial means to carry out such observations. However, quantifying seabirds at sea in an accurate and comparable manner, e.g., via vessel-based distance sampling along transect lines, is an arduous task and there are many biases that need to be taken into consideration. The most basic is the difference in visibility between species as a function of their size, colour and the conspicuousness of their at-sea behaviour, all of which are further confounded by the weather conditions during observation, transect direction and speed. These considerations merited the recommendation of common monitoring methods allowing comparisons between sites and species (Tasker et al. 1984), leading to the European Seabirds at Sea (ESAS) methodology. The general methodology is based on distance sampling at sea on board a vessel travelling at a fixed speed along planned transect lines to obtain comparable measures of bird densities. While this can be carried out on specifically organised vessel trips, it is also often carried out on ferries and other "vessels of opportunity". As the scope of the collection of such data evolved from mapping the at-sea distribution of seabirds to obtaining more information on seabird ecology, the standardised collection of behavioural data was also integrated (Camphuysen and Garthe 2004). Another cost-effective surveying method for seabirds is strip-transect aerial surveys. Aerial surveys are usually limited to within a few hundred kilometres of the coast, but allow observations over large scales in a shorter period of time than would be possible from a ship, and usually at lower cost (Certain and Bretagnolle 2008).

Movement ecology

The at-sea movements of seabirds have until recently been poorly understood, particularly in pelagic and migratory species due to the remoteness of their at-sea distribution. However, advances in remote tracking technologies have allowed us to track in great detail individuals' foraging trips and year-round movements (Pereira et al. 2022). Before the wider-scale use of remote tracking, information on the life of seabirds at sea was obtained from land-based studies on breeding phenology, colony attendance cycles, delivery rates of food loads to offspring, studies on diet and limited observations at sea. Ring recoveries and resightings, both of live and dead birds, tend to be biased towards coastal seabird species which are more likely to be observed or washed ashore, leaving many mysteries regarding pelagic species. Nevertheless, these early studies revealed the migratory nature of some species, the existence of a pre-laying exodus in some seabird groups, attendance cycles and the durations of foraging trips during incubation and chick-provisioning and the location of some foraging hotspots offshore.

To the human eye, it is spectacular how seabirds traverse the open ocean through a seemingly featureless seascape. The mechanisms of bird orientation and, particularly, navigation are still not fully understood; there is some evidence for the use of sun-compass orientation in seabirds (Padget et al. 2018) and magnetic cues seem to guide natal philopatry in at least one species (Pollonara et al. 2015, Wynn et al. 2020). Most notably, however, is the apparent importance of olfactory cues both for navigation and for finding food (Nevitt 2008, Wikelski et al. 2015, Pollonara et al. 2015). One candidate chemical for olfactory cues is dimethyl sulphide (DMS), a strong-smelling substance which is emitted by phytoplankton when they are grazed upon by zooplankton (Nevitt et al. 1995, Nevitt 2008).

Foraging movements

The nature of seabird movements depends on the scale being considered. At larger spatial scales, birds often appear to leave the colony in a non-random direction and head towards features of high productivity, such as shelf break areas, eddies and upwelling currents, demonstrating an awareness of the spatial distribution of available resources (Weimerskirch 2007). Foraging movements are very rarely observed to follow a random walk, but typically take the form of long, directional commuting trips to a foraging spot, or long loop or figure-of-eight trips with foraging stops along the route. On entering an area of potentially higher prey density, the scale of which is determined by the oceanography and foraging strategy, seabirds carry out nested Area Restricted Searching (ARS), flying with an increased turning rate to locate food patches or individual items (Kareiva and Odell 1987, Viswanathan et al. 1996, Hamer et al. 2009). At the individual level, foraging site fidelity is recorded at various temporal scales, with evidence of some consistency even between years, probably a result of individual biases and site familiarity obtained in early life (Patrick et al. 2014, Wakefield et al. 2015).

During the breeding season, most seabirds are constrained in the distance and duration of foraging trips by the need to regularly return to the nest to carry out parental duties, such as incubation or food-provisioning of chicks. This results in central-place foraging strategies. Foraging radii differ greatly among taxa according to flight capacity, diet, and breeding stage (Thaxter et al. 2012). Owing to their efficient means of travel and the ability to regurgitate food (and often stomach oil) to chicks, petrels, shearwaters and albatrosses are capable of traversing a few hundred kilometres on foraging trips of a few days. Spectacularly, the foraging trips of Desertas Petrel (*Pterodroma deserta*) span the North Atlantic Ocean, reaching total distances of 12000 km (Ventura et al. 2020), while Murphy's Petrel (*Pterodroma ultima*) in the South Pacific carry out foraging trips extending over 4000 km from the colony on round-trips that can be well over 13000 km during incubation (Clay et al. 2019). Auks typically remain closer, generally within 100 km of the colony, due to high flight costs and their inability to regurgitate which limits them to delivering freshly caught prey to chicks. Most gulls and terns also occur within these ranges. Other coastal and near-shore foraging seabirds, such as cormorants and shags, are constrained to the area in close proximity of the colony, but they efficiently explore the water column, which can result in accentuated prey depletion over the course of the breeding season (Birt et al. 1987).

Migration

For many seabird species the end of the breeding season marks the end of the central-place constraint, and allows migrations into more suitable foraging grounds. As seabird tracking data increased, it became evident how some of the longest, fastest and most wide-ranging migrations in the animal kingdom are carried out by seabird species. Among the shearwaters, individual Sooty Shearwaters (*Ardenna grisea*) breeding in New Zealand travel from the Southern Ocean to the Bering Sea and span from Japanese waters to the coast of Chile (Shaffer et al. 2006). Arctic Terns (*Sterna paradisaea*) hold the record for the longest round-trip migrations, sometimes in excess of 90 000 km from Northern hemisphere temperate or Arctic breeding grounds to the Antarctic Ocean and back (Egevang et al. 2010, Fijn et al. 2013). The daily distances attained by seabirds on migration are significantly higher than those of most terrestrial avian migrants, reaching several hundred kilometres per day, and are particularly high in dynamic soaring procellariiformes which can afford spending longer periods of the day in flight compared to land birds (Croxall 2005, Shaffer et al. 2006, Hedd et al. 2012, Dias et al. 2012). The ability to employ a "fly and forage" strategy—being able to carry out short foraging bouts *en route*—also reduces the overall energetic costs of migration in seabirds (Dias et al. 2012).

Among migratory seabirds are several examples of differential migration—intraspecific variability in migratory phenology, routes or destinations, commonly among individuals of different age groups, social dominance status or sex (Briedis and Bauer 2018). Within the same seabird population, different individuals have been found to spend the non-breeding season in distinct and distant oceanic zones, some even exhibiting partial migration, often with significant individual consistency between years (Phillips et al. 2005, Shaffer et al. 2006, González-Solís et al. 2007, Kopp et al. 2011, Ramírez et al. 2013, McFarlane Tranquilla et al. 2014, Fifield et al. 2014, Müller et al. 2014, Perez et al. 2014, Davis et al. 2016, Fayet et al. 2016a). These variations have been linked to age, previous reproductive outcome and sex, particularly in dimorphic seabird species (Phillips et al. 2017).

Foraging behaviour and trophic niche

Studying the diet of seabirds allows us to recognise their role in the marine ecosystem. Through their diet, we can identify the trophic niche they occupy, the existence of interspecific competition and ecological specialisations. Moreover, seabird diet reflects the marine assemblages of available prey, allowing seabirds to act as biomarkers of the health of fish stocks (Velarde et al. 1994). Most diet studies report frequency of occurrence and numerical abundance of prey items per bird in order to produce comparable data between studies (Duffy and Jackson 1986).

Among the earliest attempts to uncover seabird diet are methods still employed today: the collection of pellets (regurgitations of indigestible material by gulls and cormorants), analysis of identifiable items in faeces, recording prey items found at the colony and direct observation of prey items in the beak (Duffy and Jackson 1986, Barrett et al. 2007). A general limitation across many of these methods is the representativeness of the sample obtained. Pellets comprise of hard parts, such as larger skeletal elements, shells, and cephalopod beaks, and thus underrepresent soft or more digestible prey items (Hartley 1948). Otoliths from smaller fish may also be too digested for identification or absent (Duffy and Laurenson 1983). Analyses of faeces come with similar limitations. These limitations are partially addressed by the complementary use of molecular methods, such as DNA barcoding (Ceia et al. 2022).

The foraging ecology of seabirds is diverse, reflecting their taxonomic diversity and global distribution. They occupy almost all consumer trophic levels, from planktivorous diets, to hunters of fish and squid, to scavengers and predators of other seabirds, with an array of foraging specialisations in morphology and behaviour. Most seabirds are opportunistic feeders and will perform more than one foraging behaviour. However, there are behaviours that are more typical of particular groups and species than others.

One of the behaviours most commonly employed by seabird species is pursuit diving from the water's surface, such as that employed by penguins, the auks and diving petrels. Diving depths differ greatly between seabird groups, determined by their morphology, as previously discussed. The use of capillary-tube depth gauges (Burger and Wilson 1988) first exposed the extent of seabird diving capacities. For instance, while shearwaters were considered largely surface foragers and shallow divers, dive depth investigations in Sooty Shearwaters and Manx Shearwaters (*Puffinus puffinus*) have recorded occasional dives well below 40 m (Weimerskirch 1996, Weimerskirch and Cherel 1998, Shoji et al. 2016), similar to the diving abilities of diving petrels. The greatest diving depths recorded were carried out by Emperor Penguins (*Aptenodytes forsteri*), greater than 400 m and lasting up to 21 minutes (Wienecke et al. 2007).

Plunge diving—whereby birds dive into the water from the air—is a foraging mechanism most notably employed by the Sulidae (gannets and boobies), Sternidae (terns), Phaethontidae (tropicbirds) and the Brown Pelican (*Pelecanus occidentalis*) (Wodzicki and Robertson 1955, Dunn 1971, Ropert-Coudert et al. 2004). Gannets in particular are morphologically adapted to plunge diving at high speeds (over 20 m/s) from heights frequently up to 45 m above the sea surface, withstanding the force of impact against the water surface (Wodzicki and Robertson 1955, Garthe et al. 2000, Ropert-Coudert et al. 2004, Chang et al. 2016).

Among surface-foraging seabirds, larids, frigatebirds, storm-petrels and larger petrels forage by picking prey from the water's surface. A highly specialised foraging mode is surface filtering. This is carried out by krill-eating and planktivorous petrels and prions. Some prions are particularly adapted to this form of foraging in that their upper mandibles are lined with lamellae, much like the baleen of a whale.

Many predatory seabirds scavenge facultatively. Giant petrels often feed on carcasses of seabirds and seals, while large albatrosses often scavenge squid, for example. Scavenging has been increasingly observed as an interaction with anthropogenic food sources, such as landfills and fisheries discards. The ubiquity and increased intensity of fisheries in the world's oceans over the past half a century have shaped seabird foraging behaviour, as exposed through dietary analysis and observations of their spatial and temporal overlap (Le Bot et al. 2018). The high density and multi-species interactions with fisheries often give rise to kleptoparasitism, particularly in skuas and large gulls (Garthe and Hüppop 1998). Kleptoparasitism is also commonly seen in other contexts among larids, skuas and frigatebirds in particular, either directed towards individuals of their own species or interspecifically, both by aerial pursuit and harassment at sea or at the colony during chick provisioning (Brockmann and Barnard 1979).

Mesopelagic prey, which occupy waters deeper than 200 m during the day, are present in the diet of many seabirds. While this depth is available to some penguins, mesopelagic fish and squid also form part of the diet of surface foraging or shallow-diving seabirds. Mesopelagic organisms rise to epipelagic waters during dark nights in their diel vertical migrations. Their increased availability makes nocturnal foraging a relatively common trait among seabirds, to differing extents (Cruz et al. 2013, Waap et al. 2017).

Life history

Most of what we know about seabird life history and demography is a result of long-term population studies. Such studies require the identification of individuals through bird ringing in order to elucidate demographic trends. The early stages of long-term studies serve for the collection of information on the species biology, such as annual phenology and reproductive output estimates. With more years of data, the dataset allows for the exploration of themes such as population dynamics, environmental correlates with temporal trends, age effects and inter-annual variation (Wooller et al. 1992, Jenouvrier et al. 2005, Frederiksen et al. 2008, Bauch et al. 2020).

Seabirds are K-strategists, with high adult survival, slow maturation and low annual reproductive output, investing in parental care but not at the cost of future survival. The oldest known wild bird is a female Laysan Albatross (*Phoebastria immutabilis*) affectionately named *Wisdom*, still breeding at 69 years of age in 2020 on Midway Atoll, Hawaii. While Wisdom's is a spectacular age for a breeding bird, a long-life expectancy is not an uncommon trait among even the smallest of seabirds. Seabird annual adult survival rates lie at or above 90% for many groups, living long enough to experience the fitness declines of senility (Coulson and Wooller 1976, Reed et al. 2008, Catry et al. 2011). The highest post-fledging mortality occurs in the first years of life, among immature non-breeding birds (Fay et al. 2015, Szostek and Becker 2015).

In many species, the breeding season extends for a substantial part of the annual cycle, with pair bonding playing an important role in largely monogamous species. Pair fidelity is high across seabird taxa, with higher reproductive success in birds that have been paired together for longer. In fact, divorce appears to elicit costs in terms of reproductive success, and birds that lose a partner may not easily find a replacement (Bried et al. 2003, Naves et al. 2007, Ismar et al. 2010).

While some near-shore species raise multiple chicks and may lay a replacement clutch, pelagic seabirds invest in a single, large egg, and are unable to replace it within the same season. Chick rearing also lasts longer in offshore feeding species, typically over two months (Furness and Monaghan 1987). In procellariiformes, penguins, some auk species and tropic birds, young fledge independently, and the parents have no input in provisioning once they leave the nest. However, in boobies, terns, larids, skuas and frigatebirds parents feed the fledged young for a longer period before full independence. In some auks, the flightless precocial chicks leave the colony and follow the male parent offshore, where food resources are more predictable and predation risk is lower. There, they form large nurseries and overlap the chick provisioning period with intense post-breeding moult during which the adult birds are also flightless (Kuletz and Piatt 1999). Across different seabird taxa, the recruitment of young into the breeding population may take two years since fledging in some species, but typically takes much longer, with some offshore feeding seabirds having maturation periods spanning a decade or longer (Furness and Monaghan 1987, Weimerskirch 2002). During this time, young birds acquire flight and foraging skills, as well as knowledge of foraging areas, in order to be fully equipped for the high energetic demands of breeding and improve their chances of success (Yoda et al. 2004, Riotte-Lambert and Weimerskirch 2013, Mendez et al. 2017).

The costs of reproduction in seabirds are evidenced in trade-offs between breeding and self-maintenance—the results of reproductive investment are often phenological shifts in migration, moult or future timing of return to the breeding colony, potentially leading to future reproductive consequences (Berzins et al. 2011, Catry et al. 2013, Fayet et al. 2016b, Gatt et al. 2021). Skipped breeding attempts (sabbatical years) are a life history feature which may result from the requirement to replenish body condition, increasing the probability of success the subsequent breeding attempt (Mougin et al. 1997, Weimerskirch et al. 2015). This is particularly evident in some albatrosses, such as the Wandering Albatross, which generally breeds biennially owing to the length of the breeding season (Tickell 1968). Contrastingly, seabirds in the tropics can exhibit sub-annual breeding cycles, resulting in some populations breeding year round. This phenomenon does not appear to be a phylogenetic trait since it is recorded in terns, gulls, boobies, frigatebirds and shearwaters. Instead, it is likely an adaptation to life outside predictable seasonal variation in the availability of resources. Increasing the number of potential breeding attempts across a lifetime may improve chances to reproduce successfully in the low marine productivity of the tropics (Reynolds et al. 2014).

Seabirds are typically considered highly philopatric; new breeders have a high probability of establishing in their natal colony, often returning within metres of their own natal nest (Huyvaert and Anderson 2004, Steiner and Gaston 2005, Campioni et al. 2017). However, long-term monitoring of seabird colonies and large-scale ringing of chicks have revealed the importance of natal dispersal—

the permanent movement of new recruits from their natal colony to their first breeding site—for population dynamics (Frederiksen and Petersen 2000, Lavers et al. 2007). Adult birds have also been observed to change breeding colonies (breeding dispersal). In Audouin's Gulls (*Ichthyaetus audouinii*) in the western Mediterranean, the propensity to emigrate is highest in colonies of poor habitat quality and increases after periods of high environmental perturbations (Fernández-Chacón et al. 2013). Dispersing adults of Audouin's Gull (Fernández-Chacón et al. 2013), Roseate Tern (*Sterna dougallii*) (Spendelow et al. 1995), and Great Cormorant (*Phalacrocorax carbo*) (Hénaux et al. 2007) appear to base their decisions on where to settle on the population density of conspecifics in the potential sites, possibly as a proxy for habitat quality, availability of mates and resource availability. In meta-populations, the importance of distance from the source colony appears to be species-dependent, having little effect in Audouin's Gulls and Roseate Terns (Spendelow et al. 1995, Fernández-Chacón et al. 2013).

Main threats

Seabirds are relatively well-monitored, with 19% of all seabird populations having been monitored regularly since 1950 (Wooller et al. 1992). This wealth of information allows for a long-term perspective into their population dynamics. Unfortunately, the results are dire—the losses experienced by the global seabird population places them among the most threatened groups of birds (Croxall et al. 2012).

An assessment of the threats seabirds face, both on land and at sea, identified bycatch, non-native alien species, and/or the adverse effects of climate change as the most important challenges currently faced worldwide (Dias et al. 2019). At the colonies, invasive alien species are the most prolific threat to seabirds, impacting nearly half of all seabird species (Dias et al. 2019). Rats and cats are the major predators, responsible for the depredation of eggs, chicks and even adult birds, directly and strongly impacting the reproductive output of entire colonies. The presence of goats and rabbits can also have negative impacts on burrow-nesting seabirds through the indirect effect of soil erosion. Harvesting is the next major threat on land (Chen et al. 2009, Merkel et al. 2014), with some populations still recovering from extensive harvesting for feathers, fat and meat in the last century (Spennemann 1998, Granadeiro et al. 2006). The persistence of seabird colonies in increasingly urbanised and disturbed coastal environments, such as the Mediterranean, suggests that the current distribution is limited to refugia where the pressures are less intense (Ruffino et al. 2009). The proximity to light-polluted areas is a significant threat to fledglings globally, particularly for Procellariiformes, which are attracted to and grounded by lights and likely perish if not encountered and released (Rodríguez et al. 2017).

Fisheries provide seabirds with relatively easy and predictable food sources, concentrating marine organisms and providing resources through baiting and discards. While this may be advantageous to generalist seabirds (Votier et al. 2004), fisheries interactions risk acting as an ecological trap as they increase the probability of bycatch and incidental mortality (Oliveira et al. 2022a). Longline fisheries are notorious for the large numbers of bycaught Procellariiformes attracted by bait, and have particularly devastating effects on endangered albatrosses and other large petrels (Tuck et al. 2001, Phillips et al. 2016). Alcids and other diving seabirds are often caught in gill nets. Warp cables used in trawling are also responsible for seabird deaths as a result of collision. Overall, almost 30% of all seabird species are affected by bycatch risk (Dias et al. 2019), with hundreds of thousands of birds estimated to die annually as a result of fisheries bycatch (Anderson et al. 2011, Žydelis et al. 2013). Additionally, intense fishing effort has caused declines in fish stocks, resulting in direct competition between fisheries and seabirds. This is particularly pertinent in the case of industrial fleets targeting forage fish and euphausiids (Grémillet et al. 2018).

Finally, climate change is possibly one of the most ubiquitous and multifaceted challenges to be faced by marine communities, including seabirds. We anticipate shifts in oceanographic systems (namely surface winds, currents and sea surface temperature, amongst other critical phenomena and parameters), nesting habitat loss due to sea level rise, an increase in the frequency and intensity of extreme weather events, spread of disease, and melting sea ice, all of which already are having consequences on the global seabird community. The strongest climate change effects are expected to come from long-term bottom-up mechanisms affecting productivity and food resources. At the poles, sea ice phenology controls primary production, with implications throughout the trophic chain. As the extent of sea ice decreases and the timing of spring melting shifts, bottom-up consequences on arctic seabird communities are starting to be seen (Ramírez et al. 2017). Above-average temperatures experienced in the oceans—termed "marine heatwaves"—have increased in frequency and intensity over the last century and are expected to cause irreversible changes to the marine ecosystem. Marine heatwaves result in food shortages as prey moves deeper or to higher latitude, cooler waters and/or have diminished body condition and nutritional value (Frölicher et al. 2018). Apart from directly causing morbidity, such prey shortages may also cause breeding deferral or nest abandonment in seabirds, resulting in low reproductive outputs (Piatt et al. 2020). Black-legged kittiwakes GPS-tracked before, during, and after a severe marine heatwave in the northeast Pacific Ocean in 2014 were shown to greatly increase their foraging ranges during the event, but also after sea surface temperatures returned to normal, suggesting that the effects of temperature anomalies are long-lasting (Osborne et al. 2020). Systematic citizen-science surveys along the West Coast of the United States by The Coastal Observation and Seabird Survey Team (COASST; https://coasst.org/) revealed how this marine heatwave was related to mass die-offs, extreme reproductive failure and behavioural shifts in Common Guillemot (*Uria aalge*) the dominant piscivorous seabird in the North Pacific. The estimated at-sea mortality was around 10–20% of the total Common Guillemot population in western North America. While offshore seabird die-offs do occur sporadically, the magnitude, duration and spatial extent was globally unprecedented (Piatt et al. 2020).

While we are unlikely to mitigate climate change effects, we can attempt to manage other co-occurring threats; most species affected by climate change also experience more manageable threats at land and at sea such as alien invasive predators and bycatch (Dias et al. 2019). Reducing such pressures, through alien species control/eradications (Oliveira et al. 2022b), bycatch regulation, and the creation of safe zones that promote biodiversity, could potentially buffer some of the negative consequences of a changing climate and improve seabird population resilience. By now, alien species eradication from many high-priority areas has been completed—the largest attempt to date has been the eradication of rats and mice from South Georgia, over an area of 3500 km^2 (Martin and Richardson 2019)—and a range of measures to mitigate bycatch have been shown to be effective in a number of fisheries, although efforts targetting gillnet bycatch have been less advanced (Bull 2007). Critically, the spatial distribution of many seabird populations, particularly in highly pelagic species, means that effective marine conservation management must link several countries and regional fisheries management organisations in their collective responsibility to protect marine biodiversity (Beal et al. 2021).

Acknowledgments

The authors would like to thank Mark Bolton for helpful comments on an earlier version of this chapter. This work was funded by the Fundação para a Ciência e a Tecnologia (FCT, Portugal) through the strategic projects UIDB/04292/2020 and UIDP/04292/2020 granted to MARE, and UIDB/50017/2020 and UIDP/50017/2020 granted to CESAM.

References cited

Anderson, O.R.J., C.J. Small, J.P. Croxall, E.K. Dunn, B.J. Sullivan, O. Yates et al. 2011. Global seabird bycatch in longline fisheries. Endanger. Species Res. 14: 91–106. doi: 10.3354/esr00347.

Ashbrook, K., S. Wanless, M.P. Harris and K.C. Hamer. 2010. Impacts of poor food availability on positive density dependence in a highly colonial seabird. Proc. R. Soc. B Biol. Sci. 277: 2355–2360. doi: 10.1098/rspb.2010.0352.

Ashmole, N.P. 1971. Seabird ecology and the marine environment. pp. 224–286. *In*: Farmer, D.S. and J.F. King [eds.]. Avian Biology Volume 1. Academic Press Inc., New York, USA.

Ballance, L.T. and R.L. Pitman. 1999. Foraging ecology of tropical seabirds. BirdLife South Africa. Proceedings of the 22nd International Ornithological Congress, Durban. Natural History Book Service. 2057–2071.

Barrett, R.T., K. Camphuysen, T. Anker-Nilssen et al. 2007. Diet studies of seabirds: A review and recommendations. ICES J. Mar. Sci. 64: 1675–1691. doi: 10.1093/icesjms/fsm152.

Bauch, C., M.C. Gatt, J.P. Granadeiro, S. Verhulst and P. Catry. 2020. Sex-specific telomere length and dynamics in relation to age and reproductive success in Cory's shearwaters. Mol. Ecol. 29: 1344–1357. doi: 10.1111/mec.15399.

Beal, M., M.P. Dias, R.A. Phillips et al. 2021. Global political responsibility for the conservation of albatrosses and large petrels. Sci. Adv. 7: eabd7225. doi: 10.1126/sciadv.abd7225.

Berzins, L.L., H.G. Gilchrist, K.D. Matson and G. Burness. 2011. Sex-specific effects of increased incubation demand on innate immunity in Black Guillemots. Physiol. Biochem. Zool. 84: 222–229. doi: 10.1086/658373.

Birt, V., T. Birt, D. Goulet, D. Cairns and W. Montevecchi. 1987. Ashmole's halo: direct evidence for prey depletion by a seabird. Mar. Ecol. Prog. Ser. 40: 205–208. doi: 10.3354/meps040205.

Boyd, C., D. Grünbaum, G.L. Hunt, A.D.S.E. Punt, H. Weimerskirch and S. Bertrand. 2016. Effectiveness of social information used by seabirds searching for unpredictable and ephemeral prey. Behav. Ecol. 27: 1223–1234. doi: 10.1093/beheco/arw039.

Bried, J., D. Pontier and P. Jouventin. 2003. Mate fidelity in monogamous birds: A re-examination of the Procellariiformes. Anim. Behav. 65: 235–246. doi: 10.1006/anbe.2002.2045.

Briedis, M. and S. Bauer. 2018. Migratory connectivity in the context of differential migration. Biol. Lett. 14: 20180679. doi: 10.1098/rsbl.2018.0679.

Brockmann, H.J. and C.J. Barnard. 1979. Kleptoparasitism in birds. Anim. Behav. 27: 487–514.

Brooke, M. de L. 2002. Seabird systematics and distribution: A review of current knowledge. pp. 57–85. *In*: Schreiber, E.A. and J. Burger [eds.]. Biology of Marine Birds. CRC Press, Boca Raton, USA.

Buckley, N.J. 1997. Spatial-concentration effects and the importance of local enhancement in the evolution of colonial breeding in seabirds. Am. Nat. 149: 1091–1112. doi: 10.1086/286040.

Bull, L.S. 2007. Reducing seabird bycatch in longline, trawl and gillnet fisheries. Fish Fish. 8: 31–56. doi: 10.1111/j.1467-2979.2007.00234.x.

Bunt, J.S. 1975. Primary productivity of marine ecosystems. pp. 169–183. *In*: Lieth, H. and R.H. Whittaker [eds.]. Primary Productivity of the Biosphere. Ecological Studies (Analysis and Synthesis), vol. 14. Springer-Verlag, Berlin.

Burger, A.E. and R.P. Wilson. 1988. Capillary-tube depth gauges for diving animals: An assessment of their accuracy and applicability. J. F. Ornithol. 59: 345–354.

Camphuysen, K.C.J. and S. Garthe. 2004. Recording foraging seabirds at sea: Standardised recording and coding of foraging behaviour and multi-species foraging associations. Atl. Seabirds 6: 1–32.

Campioni, L., J.P. Granadeiro and P. Catry. 2017. Albatrosses prospect before choosing a home: intrinsic and extrinsic sources of variability in visit rates. Anim. Behav. 128: 85–93. doi: 10.1016/j.anbehav.2017.04.008.

Catry, P., J.P. Granadeiro, J. Ramos, R.A. Phillips and P. Oliveira. 2011. Either taking it easy or feeling too tired: Old Cory's Shearwaters display reduced activity levels while at sea. J. Ornithol. 152: 549–555. doi: 10.1007/s10336-010-0616-7.

Catry, P., M.P. Dias, R.A. Phillips and J.P. Granadeiro. 2013. Carry-over effects from breeding modulate the annual cycle of a long-distance migrant: an experimental demonstration. Ecology 94: 1230–1235. doi: 10.1890/12-2177.1.

Ceia, F.R., J.C. Xavier, A.R. Carreiro, I. dos Santos and Y. Cherel. 2022. Conventional and modern approaches to study seabird trophic ecology and diet. pp. 19–35. *In*: Ramos, J.A. and L. Pereira [eds.]. Seabird Biodiversity and Human Activities. CRC Press, Boca Raton, FL.

Certain, G. and V. Bretagnolle. 2008. Monitoring seabirds population in marine ecosystem: The use of strip-transect aerial surveys. Remote Sens. Environ. 112: 3314–3322. doi: 10.1016/j.rse.2008.01.019.

Chang, B., M. Croson, L. Straker, S. Gart, C. Dove, J. Gerwin et al. 2016. How seabirds plunge-dive without injuries. PNAS 113: 12006–12011. doi: 10.1073/pnas.1608628113.

Chen, S., S. Chang, Y. Liu, S. Chan, Z. Fan, C. Chen et al. 2009. A small population and severe threats: status of the Critically Endangered Chinese crested tern *Sterna bernsteini*. Oryx 43: 209. doi: 10.1017/S0030605308001142.

Clay, T.A., S. Oppel, J.L. Lavers, R.A. Phillips and M. de L. Brooke. 2019. Divergent foraging strategies during incubation of an unusually wide-ranging seabird, the Murphy's petrel. Mar. Biol. 166: 1–14. doi: 10.1007/s00227-018-3451-7.

Coulson, J.C. and R.D. Wooller. 1976. Differential survival rates among breeding kittiwake gulls *Rissa tridactyla* (L.). J. Anim. Ecol. 45: 205–213.

Coulson, J.C. 2002. Colonial breeding in seabirds. pp. 87–114. *In*: Schreiber, E.A. and J. Burger [eds.]. Biology of Marine Birds. CRC Press, Boca Raton, USA.

Cronin, T.W. 2012. Visual optics: Accommodation in a splash. Curr. Biol. 22: 871–873. doi: 10.1016/j.cub.2012.08.053.

Croxall, J.P. 2005. Global circumnavigations: Tracking year-round ranges of nonbreeding albatrosses. Science 307: 249–250. doi: 10.1126/science.1106042.

Croxall, J.P., S.H.M. Butchart, B. Lascelles, A.J. Stattersfield, B. Sullivan, A. Symes et al. 2012. Seabird conservation status, threats and priority actions: A global assessment. Bird Conserv. Int. 22: 1–34. doi: 10.1017/S0959270912000020.

Cruz, S.M., M. Hooten, K.P. Huyvaert, C.B. Proaño, D.J. Anderson, V. Afanasyev et al. 2013. At-sea behavior varies with lunar phase in a nocturnal pelagic seabird, the swallow-tailed gull. PLoS One 8: e56889. doi: 10.1371/journal.pone.0056889.

Davies, R.G., U.M. Irlich, S.L. Chown and K.J. Gaston. 2010. Ambient, productive and wind energy and ocean extent predict global species richness of procellariiform seabirds. Glob. Ecol. Biogeogr. 19: 98–110. doi: 10.1111/j.1466-8238.2009.00498.x.

Davis, S.E., M. Maftei and M.L. Mallory. 2016. Migratory connectivity at high latitudes: Sabine's Gulls (*Xema sabini*) from a colony in the Canadian High Arctic migrate to different oceans. PLoS One 11: e0166043. doi: 10.1371/journal.pone.0166043.

Diamond, A.W. 1978. Feeding strategies and population size in tropical seabirds. Am. Nat. 112: 215–223. doi: 10.1086/283261.

Dias, M.P., R. Martin, E.J. Pearmain et al. 2019. Threats to seabirds: A global assessment. Biol. Conserv. 237: 525–537. doi: 10.1016/j.biocon.2019.06.033.

Dias, M.P., J.P. Granadeiro and P. Catry. 2012. Do seabirds differ from other migrants in their travel arrangements? On route strategies of Cory's Shearwater during its trans-equatorial journey D. Hyrenbach [ed.]. PLoS One 7: e49376. doi: 10.1371/journal.pone.0049376.

Duffy, D.C. and L.J.B. Laurenson. 1983. Pellets of Cape cormorants as indicators of diet. Condor 85: 305–307. doi: 10.2307/1367066.

Duffy, D.C. and S. Jackson. 1986. Diet studies of seabirds: A review of methods. Colon. Waterbirds 9: 1–17. doi: 10.2307/1521138.

Dunn, E.K. 1971. Effect of age on the fishing ability of Sandwich Terns *Sterna sandvicensis*. Ibis 114: 360–366. doi: 10.1111/j.1474-919X.1972.tb00833.x.

Egevang, C., I.J. Stenhouse, R.A. Phillips, A. Petersen, J.W. Fox and J.R.D. Silk. 2010. Tracking of Arctic terns *Sterna paradisaea* reveals longest animal migration. PNAS 107: 2078–2081. doi: 10.1073/pnas.0909493107.

Elliott, K.H., R.E. Ricklefs, A.J. Gaston, S.A. Hatch, J.R. Speakman and G.K. Davoren. 2013. High flight costs, but low dive costs, in auks support the biomechanical hypothesis for flightlessness in penguins. Proc. Natl. Acad. Sci. 110: 9380–9384. doi: 10.1073/pnas.1304838110.

Fay, R., H. Weimerskirch, K. Delord and C. Barbraud. 2015. Population density and climate shape early-life survival and recruitment in a long-lived pelagic seabird. J. Anim. Ecol. 84: 1423–1433. doi: 10.1111/1365-2656.12390.

Fayet, A.L., R. Freeman, A. Shoji, D. Boyle, H.L. Kirk, B.J. Dean et al. 2016a. Drivers and fitness consequences of dispersive migration in a pelagic seabird. Behav. Ecol. 27: 1061–1072. doi: 10.1093/beheco/arw013.

Fayet, A.L., R. Freeman, A. Shoji, H.L. Kirk, O. Padget, C.M. Perrins et al. 2016b. Carry-over effects on the annual cycle of a migratory seabird: an experimental study. J. Anim. Ecol. 85: 1516–1527. doi: 10.1111/1365-2656.12580.

Fernández-Chacón, A., M. Genovart, R. Pradel et al. 2013. When to stay, when to disperse and where to go: Survival and dispersal patterns in a spatially structured seabird population. Ecography 36: 1117–1126. doi: 10.1111/j.1600-0587.2013.00246.x.

Fifield, D.A., W.A. Montevecchi, S. Garthe, G.J. Robertson, U. Kubetzki and J. Rail. 2014. Migratory tactics and wintering areas of Northern Gannets (*Morus bassanus*) breeding in North America. Ornithological Monographs 79: 1–63. doi: 10.1642/aoum.79-1.

Fijn, R.C., D. Hiemstra, R.A. Phillips and J. Van Der Winden. 2013. Arctic Terns *Sterna paradisaea* from the Netherlands migrate record distances across three oceans to Wilkes Land, East Antarctica. Ardea 101: 3–12. doi: 10.5253/078.101.0102.

Fischer, O. and B. Schoenemann. 2019. Why are bones in vertebrate eyes? Morphology, development and function of scleral ossicles in vertebrate eyes—a comparative study. J. Anat. Physiol. Stud. 3: 1–26.

Frederiksen, M. and E. Petersen. 2000. The importance of natal dispersal in a colonial seabird, the Black Guillemot *Cepphus grylle*. Ibis. 142: 48–57. doi: 10.1111/j.1474-919X.2000.tb07683.x.

Frederiksen, M., F. Daunt, M.P. Harris and S. Wanless. 2008. The demographic impact of extreme events: stochastic weather drives survival and population dynamics in a long-lived seabird. J. Anim. Ecol. 77: 1020–1029. doi: 10.1111/j.1365-2656.2008.01422.x.

Frölicher, T.L., E.M. Fischer and N. Gruber. 2018. Marine heatwaves under global warming. Nature 560: 360–364. doi: 10.1038/s41586-018-0383-9.

Furness, R.W. and P. Monaghan. 1987. Seabird Ecology. Blackie, Glasgow, UK.

Garthe, S. and O. Hüppop. 1998. Foraging success, kleptoparasitism and feeding techniques in scavenging seabirds: Does crime pay? Helgolander Meeresuntersuchungen 52: 187–196. doi: 10.1007/BF02908747.

Garthe, S., S. Benvenuti and W.A. Montevecchi. 2000. Pursuit plunging by northern gannets (*Sula bassana*) feeding on capelin (*Mallotus villosus*). Proc. R. Soc. B Biol. Sci. 267: 1717–1722. doi: 10.1098/rspb.2000.1200.

Gaston, A.J. and I.L. Jones. 1998. The Auks. Oxford University Press, Oxford, UK.

Gatt, M.C., M. Versteegh, C. Bauch, B.I. Tieleman, J.P. Granadeiro and P. Catry. 2021. Costs of reproduction and migration are paid in later return to the colony, not in physical condition, in a long-lived seabird. Oecologia 195: 287–297. doi: 10.1007/s00442-020-04775-w.

Gerstberger, R. and D.A. Gray. 1993. Fine Structure, Innervation and Functional Control of Avian Salt Glands. Int. Rev. Cytol. 144: 129–215. doi: 10.1016/S0074-7696(08)61515-6.

Glasser, A. and H.C. Howland. 1996. A history of studies of visual accommodation in birds. Q. Rev. Biol. 71: 475–509.

González-Solís, J., J.P. Croxall, D. Oro and X. Ruiz. 2007. Trans-equatorial migration and mixing in the wintering areas of a pelagic seabird. Front. Ecol. Environ. 5: 297–301. doi: 10.1890/1540-9295(2007)5[297:TMAMIT]2.0.CO;2.

Granadeiro, J.P., M.P. Dias, R. Rebelo, C.D. Santos and P. Catry. 2006. Numbers and population trends of Cory's Shearwater *Calonectris diomedea* at Selvagem Grande, Northeast Atlantic. Waterbirds 29: 56–60. doi: 10.1675/1524-4695(2006)29[56:naptoc]2.0.co;2.

Grémillet, D., A. Ponchon, M. Paleczny, M.L.D. Palomares, V. Karpouzi and D. Pauly. 2018. Persisting worldwide seabird-fishery competition despite seabird community decline. Curr. Biol. 28: 4009–4013.e2. doi: 10.1016/j.cub.2018.10.051.

Hamer, K.C., E.M. Humphreys, M.C. Magalhães, S. Garthe, J. Hennicke, G. Peters et al. 2009. Fine-scale foraging behaviour of a medium-ranging marine predator. J. Anim. Ecol. 78: 880–889. doi: 10.1111/j.1365-2656.2009.01549.x.

Hartley, P.H.T. 1948. The assessment of the food of birds. Ibis 90: 361–381. doi: 10.1111/j.1474-919X.1948.tb01700.x.

Handbook of the Birds of the World and BirdLife International. 2019. Handbook of the Birds of the World and BirdLife International digital checklist of the birds of the world. Version 4. Available at: http://datazone.birdlife.org/userfiles/file/Species/Taxonomy/HBW-BirdLife_Checklist_v4_Dec19.zip.

Hedd, A., W. Montevecchi, H. Otley, R. Phillips and D. Fifield. 2012. Trans-equatorial migration and habitat use by Sooty Shearwaters *Puffinus griseus* from the South Atlantic during the nonbreeding season. Mar. Ecol. Prog. Ser. 449: 277–290. doi: 10.3354/meps09538.

Hénaux, V., T. Bregnballe and J.D. Lebreton. 2007. Dispersal and recruitment during population growth in a colonial bird, the great cormorant *Phalacrocorax carbo sinensis*. J. Avian Biol. 38: 44–57. doi: 10.1111/j.2006.0908-8857.03712.x.

Huyvaert, K.P. and D.J. Anderson. 2004. Limited dispersal by Nazca boobies *Sula granti*. J. Avian Biol. 35: 46–53. doi: 10.1111/j.0908-8857.2004.03131.x.

Ismar, S.M.H., C. Daniel, B.M. Stephenson and M.E. Hauber. 2010. Mate replacement entails a fitness cost for a socially monogamous seabird. Naturwissenschaften 97: 109–113. doi: 10.1007/s00114-009-0618-6.

Jenouvrier, S., C. Barbraud, B. Cazelles and H. Weimerskirch. 2005. Modelling population dynamics of seabirds: Importance of the effects of climate fluctuations on breeding proportions. Oikos 108: 511–522. doi: 10.1111/j.0030-1299.2005.13351.x.

Jones, T.B., S.C. Patrick, J.P.Y. Arnould, M.A. Rodriguez-Malagon, M.R. Wells and J.A. Green. 2018. Evidence of sociality in the timing and location of foraging in a colonial seabird. Biol. Lett. 14: 9–12. doi: 10.1098/rsbl.2018.0214.

Jones, T.B., J.A. Green, S.C. Patrick, J.C. Evans, M.R. Wells, M.A. Rodríguez-Malagón et al. 2020. Consistent sociality but flexible social associations across temporal and spatial foraging contexts in a colonial breeder. Ecol. Lett. 23: 1085–1096. doi: 10.1111/ele.13507.

Kareiva, P. and G. Odell. 1987. Swarms of predators exhibit "preytaxis" if individual predators use area-restricted search. Am. Nat. 130: 233–270. doi: 10.1086/284707.

Katzir, G. and H.C. Howland. 2003. Corneal power and underwater accommodation in Great Cormorants (*Phalacrocorax carbo sinensis*). J. Exp. Biol. 206: 833–841. doi: 10.1242/jeb.00142.

Kopp, M., H. Peter, O. Mustafa, S. Lisovski, M.S. Ritz, R.A. Phillips et al. 2011. South polar skuas from a single breeding population overwinter in different oceans though show similar migration patterns. Mar. Ecol. Prog. Ser. 435: 263–267. doi: 10.3354/meps09229.

Kuletz, K.J. and J.F. Piatt. 1999. Juvenile Marbled Murrelet nurseries and the productivity index. Wilson Bull. 111: 257–261.

Lack, D. 1968. Ecological Adaptations for Breeding in Birds. Methuen, London, UK.

Lavers, J.L., I.L. Jones and A.W. Diamond. 2007. Natal and breeding dispersal of Razorbills (*Alca torda*) in eastern North America. Waterbirds 30: 588–594. doi: 10.1675/1524-4695(2007)030[0588:NABDOR]2.0.CO;2.

Le Bot, T., A. Lescroël and D. Grémillet. 2018. A toolkit to study seabird-fishery interactions. ICES J. Mar. Sci. 75: 1513–1525. doi: 10.1093/icesjms/fsy038.

Longhurst, A.R. and D. Pauly. 1987. Ecology of Tropical Oceans. Academic Press, New York, USA.

Machovsky-Capuska, G.E., H.C. Howland, D. Raubenheimer, R. Vaughn-Hirshorn, B. Würsig, M.E. Hauber and G. Katzir. 2012. Visual accommodation and active pursuit of prey underwater in a plunge-diving bird: The Australasian gannet. Proc. R. Soc. B Biol. Sci. 279: 4118–4125. doi: 10.1098/rspb.2012.1519.

Martin, A.R. and M.G. Richardson. 2019. Rodent eradication scaled up: Clearing rats and mice from South Georgia. Oryx 53: 27–35. doi: 10.1017/S003060531700028X.

Martin, G.R. 1998. Eye structure and amphibious foraging in albatrosses. Proc. R. Soc. London. Ser. B Biol. Sci. 265: 665–671. doi: 10.1098/rspb.1998.0345.

McFarlane Tranquilla, L.A., W.A. Montevecchi, D.A. Fifield, A. Hedd, A.J. Gaston, G.J. Robertson et al. 2014. Individual winter movement strategies in two species of murre (*Uria* spp.) in the Northwest Atlantic. PLoS One 9: e90583. doi: 10.1371/journal.pone.0090583.

Mendez, L., A. Prudor and H. Weimerskirch. 2017. Ontogeny of foraging behaviour in juvenile red-footed boobies (*Sula sula*). Sci. Rep. 7: 13886. doi: 10.1038/s41598-017-14478-7.

Merkel, F., A.L. Labansen, D. Boertmann, A. Mosbech, C. Egevang, K. Folk et al. 2014. Declining trends in the majority of Greenland's thick-billed murre (*Uria lomvia*) colonies 1981–2011. Polar Biol. 37: 1061–1071. doi: 10.1007/s00300-014-1500-3.

Mougin, J.-L., C. Jouanin and F. Roux. 1997. Intermittent breeding in Cory's Shearwater *Calonectris diomedea* of Selvagem Grande, North Atlantic. Ibis. 139: 40–44. doi: 10.1111/j.1474-919X.1997.tb04502.x.

Müller, M.S., B. Massa, R.A. Phillips and G. Dell'omo. 2014. Individual consistency and sex differences in migration strategies of Scopoli's shearwaters *Calonectris diomedea* despite year differences. Curr. Zool. 60: 631–641. doi: 10.1093/czoolo/60.5.631.

Naves, L.C., E. Cam and J.Y. Monnat. 2007. Pair duration, breeding success and divorce in a long-lived seabird: benefits of mate familiarity? Anim. Behav. 73: 433–444. doi: 10.1016/j.anbehav.2006.10.004.

Nevitt, G.A., R.R. Veit and P. Kareiva. 1995. Dimethyl sulphide as a foraging cue for Antarctic Procellariiforme seabirds. Nature 376: 680–682.

Nevitt, G.A. 2008. Sensory ecology on the high seas: The odor world of the procellariiform seabirds. J. Exp. Biol. 211: 1706–1713. doi: 10.1242/jeb.015412.

Osborne, O.E., P.D.O. Hara, S. Whelan, P. Zandbergen, S.A. Hatch and K.H. Elliott. 2020. Breeding seabirds increase foraging range in response to an extreme marine heatwave. Mar. Ecol. Prog. Ser. 646: 161–173. doi: 10.3354/meps13392.

Oliveira, N., J.A. Ramos, J.G. Calado and J.M. Arcos. 2022a. Seabirds and fisheries interactions. pp. 77–89. *In*: Ramos, J.A. and L. Pereira [eds.]. Seabird Biodiversity and Human Activities. CRC Press, Boca Raton, FL.

Oliveira, P., F.M. Medina, M. Nogales and P.L. Geraldes. 2022b. Eradication and control of invasive mammal species as a seabird conservation tool. pp. 224–242. *In*: Ramos, J.A. and L. Pereira [eds.]. Seabird Biodiversity and Human Activities. CRC Press, Boca Raton, FL.

Padget, O., S.L. Bond, M.M. Kavelaars, E. van Loon, M. Bolton, A.L. Fayet et al. 2018. *In situ* clock shift reveals that the sun compass contributes to orientation in a pelagic seabird. Curr. Biol. 28: 275–279.e2. doi: 10.1016/j.cub.2017.11.062.

Patrick, S.C., S. Bearhop, D. Grémillet, A. Lescroël, W.J. Grecian, T.W. Bodey et al. 2014. Individual differences in searching behaviour and spatial foraging consistency in a central place marine predator. Oikos 123: 33–40. doi: 10.1111/j.1600-0706.2013.00406.x.

Pennycuick, C.J. 1982. The flight of petrels and albatrosses (Procellariiformes), observed in South Georgia and its vicinity. Philos. Trans. R. Soc. Lond. B. Biol. Sci. 300: 75–106.

Pennycuick, C.J. 1983. Thermal soaring compared in three dissimilar tropical bird species, *Fregata magnificens*, *Pelecanus occidentals* and *Coragyps atratus*. J. Exp. Biol. 102: 307–325.

Pennycuick, C.J. 1986. Flight of seabirds. pp. 43–62. *In*: J.P. Croxall [ed.]. Seabirds: Feeding Ecology and Role in Marine Ecosystems. Cambridge University Press, Cambridge, UK.

Pereira, J.M., V.H. Paiva, L. Kruger and S.C. Votier. 2022. Tracking seabirds for conservation and marine spatial planning. pp. 59–74. *In*: Ramos, J.A. and L. Pereira [eds.]. Seabird Biodiversity and Human Activities. CRC Press, Boca Raton, FL.

Perez, C., J.P. Granadeiro, M.P. Dias, H. Alonso and P. Catry. 2014. When males are more inclined to stay at home: insights into the partial migration of a pelagic seabird provided by geolocators and isotopes. Behav. Ecol. 25: 313–319. doi: 10.1093/beheco/art119.

Phillips, R.A., J.R.D. Silk, J.P. Croxall, V. Afanasyev and V.J. Bennett. 2005. Summer distribution and migration of nonbreeding albatrosses: Individual consistencies and implications for conservation. Ecology 86: 2386–2396. doi: 10.1890/04-1885.

Phillips, R.A., R. Gales, G.B. Baker et al. 2016. The conservation status and priorities for albatrosses and large petrels. Biol. Conserv. 201: 169–183. doi: 10.1016/j.biocon.2016.06.017.

Phillips, R.A., S. Lewis, J. González-Solís and F. Daunt. 2017. Causes and consequences of individual variability and specialization in foraging and migration strategies of seabirds. Mar. Ecol. Prog. Ser. 578: 117–150. doi: 10.3354/meps12217.

Piatt, J.F., J.K. Parrish, H.M. Renner, S.K. Schoen, T.T. Jones, M.L. Arimitsu et al. 2020. Extreme mortality and reproductive failure of common murres resulting from the northeast Pacific marine heatwave of 2014–2016. Plos One 15: e0226087. doi: 10.1371/journal.pone.0226087.

Pollonara, E., P. Luschi, T. Guilford, M. Wikelski, F. Bonadonna and A. Gagliardo. 2015. Olfaction and topography, but not magnetic cues, control navigation in a pelagic seabird: Displacements with shearwaters in the Mediterranean Sea. Sci. Rep. 5: 1–10. doi: 10.1038/srep16486.

Preuss, N.O. 2001. Hans Christian Cornelius Mortensen: Aspects of his life and of the history of bird ringing. Ardea 89: 1–6.

Ramírez, I., V. Paiva, D. Menezes, I. Silva, R.A. Phillips, J. Ramos et al. 2013. Year-round distribution and habitat preferences of the Bugio Petrel. Mar. Ecol. Prog. Ser. 476: 269–284. doi: 10.3354/meps10083.

Ramírez, F., A. Tarroux, J. Hovinen, J. Navarro, I. Afán, M.G. Forero et al. 2017. Sea ice phenology and primary productivity pulses shape breeding success in Arctic seabirds. Sci. Rep. 7: 1–9. doi: 10.1038/s41598-017-04775-6.

Reed, T.E., L.E.B. Kruuk, S. Wanless, M. Frederiksen, E.J.A. Cunningham and M.P. Harris. 2008. Reproductive senescence in a long-lived seabird: Rates of decline in late-life performance are associated with varying costs of early reproduction. Am. Nat. 171: E89–E101. doi: 10.1086/524957.

Reynolds, S.J., G.R. Martin, A. Dawson, C.P. Wearn and B.J. Hughes. 2014. The sub-annual breeding cycle of a tropical seabird. PLoS One 9: e93582. doi: 10.1371/journal.pone.0093582.

Riotte-Lambert, L. and H. Weimerskirch. 2013. Do naive juvenile seabirds forage differently from adults? Proc. R. Soc. B Biol. Sci. 280: 20131434. doi: 10.1098/rspb.2013.1434.

Rodríguez, A., N.D. Holmes, P.G. Ryan et al. 2017. Seabird mortality induced by land-based artificial lights. Conserv. Biol. 31: 986–1001. doi: 10.1111/cobi.12900.

Ropert-Coudert, Y., D. Grémillet, P. Ryan, A. Kato, Y. Naito and Y. Le Maho. 2004. Between air and water: The plunge dive of the cape gannet *Morus capensis*. Ibis 146: 281–290. doi: 10.1111/j.1474-919x.2003.00250.x.

Ruffino, L., K. Bourgeois, E. Vidal, C. Duhem, M. Paracuellos, F. Escribano et al. 2009. Invasive rats and seabirds after 2,000 years of an unwanted coexistence on Mediterranean islands. Biol. Invasions 11: 1631–1651. doi: 10.1007/s10530-008-9394-z.

Sachs, G., J. Traugott, A.P. Nesterova and F. Bonadonna. 2013. Experimental verification of dynamic soaring in albatrosses. J. Exp. Biol. 216: 4222–4232. doi: 10.1242/jeb.085209.

Schreer, J.F. and K.M. Kovacs. 1997. Allometry of diving capacity in air-breathing vertebrates. Can. J. Zool. 75: 339–358.

Shaffer, S.A., D.P. Costa and H. Weimerskirch. 2001. Behavioural factors affecting foraging effort of breeding wandering albatrosses. J. Anim. Ecol. 70: 864–874. doi: 10.1046/j.0021-8790.2001.00548.x.

Shaffer, S.A., Y. Tremblay, H. Weimerskirch, D. Scott, D.R. Thompson, P.M. Sager et al. 2006. Migratory shearwaters integrate oceanic resources across the Pacific Ocean in an endless summer. PNAS 103: 12799–12802. doi: 10.1073/pnas.0603715103.

Shoji, A., B. Dean, H. Kirk, R. Freeman, C.M. Perrins and T. Guilford. 2016. The diving behaviour of the Manx Shearwater *Puffinus puffinus*. Ibis 158: 598–606. doi: 10.1111/ibi.12381.

Sivak, J.G. and M. Millodot. 1977. Optical performance of the penguin eye in air and water. J. Comp. Physiol. A 119: 241–247. doi: 10.1007/BF00656636.

Spendelow, J.A., J.D. Nichols, I.C. Nisbet, H. Hays, G.C. Cormons, J. Burger et al. 1995. Estimating annual survival and movement rates of adults within a metapopulation of roseate terns. Ecology 76: 2415–2428. doi: 10.2307/2265817.

Spennemann, D.H.R. 1998. Excessive exploitation of Central Pacific seabird populations at the turn of the 20th Century. Mar. Ornithol. 26: 49–57.

Steiner, U.K. and A.J. Gaston. 2005. Reproductive consequences of natal dispersal in a highly philopatric seabird. Behav. Ecol. 16: 634–639. doi: 10.1093/beheco/ari035.

Szostek, K.L. and P.H. Becker. 2015. Survival and local recruitment are driven by environmental carry-over effects from the wintering area in a migratory seabird. Oecologia 178: 643–657. doi: 10.1007/s00442-015-3298-2.

Tasker, M.L., P.H. Jones, T. Dixon and B.F. Blake. 1984. Counting seabirds at sea from ships: A review of methods employed and a suggestion for a standardized approach. Auk 101: 567–577.

Thaxter, C.B., B. Lascelles, K. Sugar, A.S.C.P. Cook, S. Roos, M. Bolton et al. 2012. Seabird foraging ranges as a preliminary tool for identifying candidate Marine Protected Areas. Biol. Conserv. 156: 53–61. doi: 10.1016/j.biocon.2011.12.009.

Tickell, W.L.N. 1968. The biology of the great albatrosses Diomedea exulans and Diomedea epomophora. Antarct. Res. Ser. 12: 1–55.

Tuck, G.N., T. Polacheck, J.P. Croxall and H. Weimerskirch. 2001. Modelling the impact of fishery by-catches on albatross populations. J. Appl. Ecol. 38: 1182–1196. doi: 10.1046/j.0021-8901.2001.00661.x.

Velarde, E., M. Tordesillas, L. Vieyra and R. Esquivel. 1994. Seabirds as indicators of important fish populations in the Gulf of California. Calif. Coop. Ocean. Fish. Investig. Reports 35: 137–143.

Ventura, F., J.P. Granadeiro, O. Padget and P. Catry. 2020. Gadfly petrels use knowledge of the windscape, not memorized foraging patches, to optimize foraging trips on ocean-wide scales. Proc. R. Soc. B Biol. Sci. 287: 20191775. doi: 10.1098/rspb.2019.1775.

Viswanathan, G.M., V. Afanasyev, S.V. Buldyrev, E.J. Murphy, P.A. Prince and H.E. Stanley. 1996. Lévy flight search patterns of wandering albatrosses. Nature 381: 413–415. doi: 10.1038/381413a0.

Votier, S.C., R.W. Furness, S. Bearhop, J.E. Crane, R.W.G. Caldow, P. Catry et al. 2004. Changes in fisheries discard rates and seabird communities. Nature 427: 727–730. doi:10.1038/nature02315.

Waap, S., W.O.C. Symondson, J.P. Granadeiro, H. Alonso, C. Serra-Gonçalves, M.P. Dias et al. 2017. The diet of a nocturnal pelagic predator, the Bulwer's petrel, across the lunar cycle. Sci. Rep. 7: 1384. doi: 10.1038/s41598-017-01312-3.

Wakefield, E.D., I.R. Cleasby, S. Bearhop, T.W. Bodey, R.D. Davies, P.I. Miller et al. 2015. Long-term individual foraging site fidelity-why some gannets don't change their spots. Ecology 96: 3058–3074. doi: 10.1890/14-1300.1.sm.

Ward, P. and A. Zahavi. 1973. The importance of certain assemblages of birds as "information-centres" for food-finding. Ibis. 115: 517–534. doi: 10.1111/j.1474-919X.1973.tb01990.x.

Weimerskirch, H. 1996. Diving depths of Sooty Shearwaters *Puffinus griseus*. Ibis 138: 786–794.

Weimerskirch, H. and Y. Cherel. 1998. Feeding ecology of short-tailed shearwaters: breeding in Tasmania and foraging in the Antarctic? Mar. Ecol. Prog. Ser. 167: 261–274. doi: 10.3354/meps167261.

Weimerskirch, H. 2002. Seabird demography and its relationship with the marine environment. pp. 115–135. *In*: Schreiber, E.A. and J. Burger [eds.]. Biology of Marine Birds. CRC Press, Boca Raton, USA.

Weimerskirch, H., T. Guionnet, J. Martin, S.A. Shaffer and D.P. Costa. 2000. Fast and fuel efficient? Optimal use of wind by flying albatrosses. Proc. R. Soc. B Biol. Sci. 267: 1869–1874. doi: 10.1098/rspb.2000.1223.

Weimerskirch, H. 2007. Are seabirds foraging for unpredictable resources? Deep. Res. Part II 54: 211–223. doi: 10.1016/j.dsr2.2006.11.013.

Weimerskirch, H., S. Bertrand, J. Silva, J.C. Marques and E. Goya. 2010. Use of social information in seabirds: Compass rafts indicate the heading of food patches. PLoS One 5: e9928. doi: 10.1371/journal.pone.0009928.

Weimerskirch, H., K. Delord, A. Guitteaud, R.A. Phillips and P. Pinet. 2015. Extreme variation in migration strategies between and within wandering albatross populations during their sabbatical year and their fitness consequences. Sci. Rep. 5: 8853. doi: 10.1038/srep08853.

Weimerskirch, H., C. Bishop, T. Jeanniard-du-Dot, A. Prudor and G. Sachs. 2016. Frigate birds track atmospheric conditions over months-long transoceanic flights. Science 353: 74–78. doi: 10.1126/science.aaf4374.

Wienecke, B., G. Robertson, R. Kirkwood and K. Lawton. 2007. Extreme dives by free-ranging emperor penguins. Polar Biol. 30: 133–142. doi: 10.1007/s00300-006-0168-8.

Wikelski, M., E. Arriero, A. Gagliardo, R.A. Holland, M.J. Huttunen, R. Juvaste et al. 2015. True navigation in migrating gulls requires intact olfactory nerves. Sci. Rep. 5: 1–11. doi: 10.1038/srep17061.

Wodzicki, K. and F. Robertson. 1955. Observations on diving of Australasian gannet. Notornis 6: 72–76.

Wooller, R.D., J.S. Bradley and J.P. Croxall. 1992. Long-term population studies of seabirds. Trends Ecol. Evol. 7: 111–114. doi: 10.1016/0169-5347(92)90143-Y.

Wynn, J., O. Padget, H. Mouritsen, C. Perrins and T. Guilford. 2020. Natal imprinting to the Earth's magnetic field in a pelagic seabird. Curr. Biol. 30: 1–5. doi: 10.1016/j.cub.2020.05.039.

Yoda, K., H. Kohno and Y. Naito. 2004. Development of flight performance in the brown booby. Biol. Lett. 271 Suppl: S240–S242. doi: 10.1098/rsbl.2003.0157.

Žydelis, R., C. Small and G. French. 2013. The incidental catch of seabirds in gillnet fisheries: A global review. Biol. Conserv. 162: 76–88. doi: 10.1016/j.biocon.2013.04.002.

Chapter 2

Conventional and Modern Approaches to Study Seabird Trophic Ecology and Diet

Filipe R. Ceia,[1,*] *José C. Xavier,*[1,2] *Ana R. Carreiro,*[1] *Ivo dos Santos*[1] and *Yves Cherel*[3]

Introduction

The study of diet and trophic ecology is essential in ecology because it is related with the survival, speciation and evolution of species. Diet variability is very common in natural communities, both among individuals and species, and may occur at any trophic level within a food web (Svanbäck et al. 2015). In this sense, diet studies of marine top predators, such as seabirds, are of great importance in several key areas of applied ecology. For instance, as sentinel species of overall marine ecosystem variability, the diet of seabirds can provide early indications of fluctuations in fish stocks (Velarde et al. 2013, Ramos and Furness 2022). It is well known that dietary shifts of seabirds indicate changes in the trophic structure of marine ecosystems, especially in pelagic food webs, because most seabirds rely on small mid-trophic level species living in the upper-water column (Montevecchi and Myers 1996). On the other hand, and due to diet variability, the conservation of seabirds is highly dependent on the knowledge of their trophic ecology at both the population and individual levels (Ceia and Ramos 2015). The breeding failure and population decline in several species of seabirds have been attributed to abundance, quality and availability of their prey (Kowalczyk et al. 2014). Searching for the true diet of seabirds (i.e., an accurate and precise estimation of the prey consumed) is essential for their conservation and to understand mechanisms underlying the structure of pelagic food webs (Chiaradia et al. 2014). However, diet studies of seabirds are often biased concerning the limitations of techniques (Barrett et al. 2007).

[1] University of Coimbra, Marine and Environmental Sciences Centre, Department of Life Sciences, 3000-456 Coimbra, Portugal.
Emails: carreiro.ar92@gmail.com; ivonobredossantos@gmail.com

[2] British Antarctic Survey, Natural Environment Research Council, Cambridge CB3 0ET, UK.
Email: jccx@cantab.net

[3] Centre d'Etudes Biologiques de Chizé (CEBC), UMR 7372 du CNRS- La Rochelle Université, 79360 Villiers-en-Bois, France.
Email: cherel@cebc.cnrs.fr

* Corresponding author: ceiafilipe@zoo.uc.pt

Up to the early 1980's, killing of some individuals was the primary source of dietary information from seabirds (Duffy and Jackson 1986). This archaic technique was then substituted by emetics and stomach pumps to obtain complete stomach contents, but these are still intrusive methods, which may cause trauma or injury to birds in a few cases (Wilson 1984). Most ornithologists and ecologists are now reluctant to use stomach sampling due to their limitations and ethical issues (Hunt and Wilson 2012), but they still form the only way to collectively quantify the diet by mass and estimate prey size. Hence, it complements a panoply of conventional methods that do not require any disturbance of individuals such as field observations, collection of regurgitated pellets (i.e., a compact ball of undigested hard parts of prey) or analyses of faeces, as these techniques are often time-consuming or limited in their outcomes. Hopefully, modern techniques significantly increase our knowledge of seabirds' trophic ecology, even during periods in which scientists do not have access to them, with minimal detriment to seabirds. Some of the most used and powerful methods to assess seabird diets are presently the analyses of stable isotopes and lipids, and DNA metabarcoding (Barrett et al. 2007, Karnovsky et al. 2012, McInnes et al. 2016). The main limitations of conventional methods to assess seabird diets in the past, can be now complemented by these modern techniques to gain further and crucial information on seabirds' diets and trophic ecology (Fig. A2.1). It is important to note that such new approaches did not replace and/or surpass the need of collecting conventional diet samples, because no method is free of bias or error (Chiaradia et al. 2014). Indeed, each technique has its own limitations, and it is generally recommended to use at least two complementary methods to both appropriately identify and quantify the food items (Fig. A2.1). Due to the challenges and limitations described in studying seabirds' diets, several knowledge gaps remain across most species. For example, little is known about the trophic ecology of immature and non-breeding birds, and of birds at their wintering sites (Barrett et al. 2007).

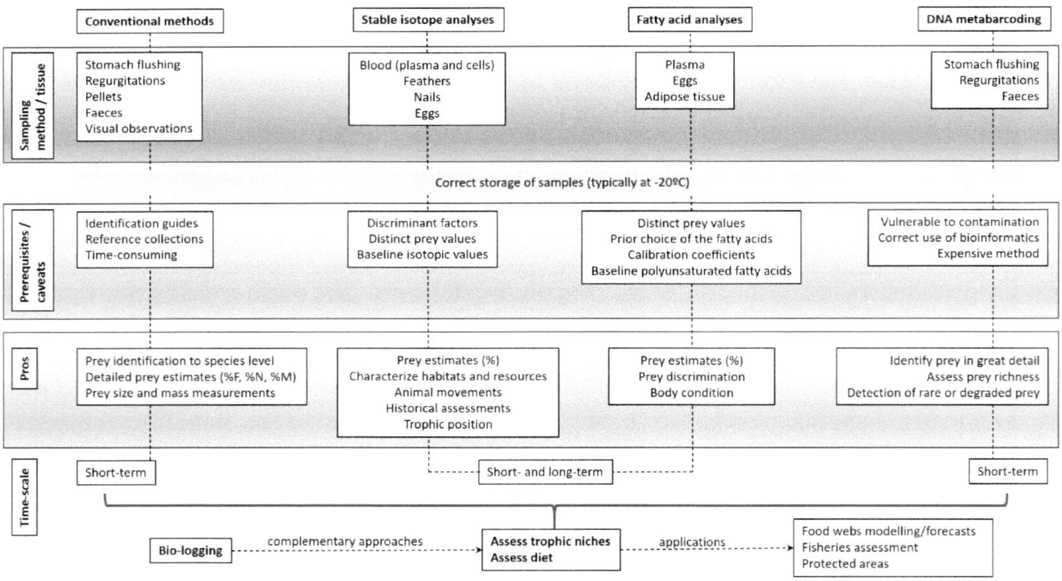

Figure A2.1. Schematic diagram showing the pros and cons (and prerequisites) of conventional and modern methods to study the prey and the feeding ecology of seabirds. A good reference database is required for every method.

Conventional methods to assess diet

The diet of marine predators through the analyses of their stomach contents provides valuable information to understand trophic interactions within food webs, and for the assessment of ocean resources (McInnes et al. 2016, Thompson et al. 2020). The conventional methods to assess the diet of seabirds (here considered the morphological analyses of prey from traditional techniques such as stomach contents/regurgitations, faeces, pellets/boluses and visual observations of seabirds carrying prey in their bills) have been the most used methods until very recently (Duffy and Jackson 1986, Barrett et al. 2007, McInnes et al. 2016), namely in various long-term monitoring programs, providing us with invaluable information (Gaston et al. 2009, Constable 2011, Mills et al. 2020). In the early 2000's, there was a pronounced shift in the preferred method of characterizing the diet of seabirds, from the morphological examination of prey remains to stable isotopic analyses of tissues, reducing the ability to identify prey consumed and detect changes in prey groups (McInnes et al. 2016). When compared to other modern techniques, the most important aspect of using conventional methods is that it allows a direct prey identification, sometimes to species level, as well as a range of parameters on their biology (e.g., size, mass, maturity stage), and consumption rates of the predators at a specific time period. With such information, it is possible to study not only the diet of seabirds (e.g., prey switching and requirements, overlap with fisheries, foraging behaviour) but also to obtain clues on the biology of their prey (distribution, availability, growth/reproduction rates, spawning and post-mortality areas, and migration) (Cherel and Weimerskirch 1999, Alvito et al. 2015, Xavier et al. 2016). Besides its invasive nature, a major limitation, however, is that most oceanic seabirds carry food in their stomach during the chick-rearing period only. Hence, in many cases, the method is useless during the pre-laying period, incubation and during the long inter-breeding period.

The analyses of stomach contents/regurgitations and pellets/boluses are probably the most used conventional methods to assess seabirds' diet. When fresh prey is collected, the analyses should be done as soon as possible after collection to avoid deterioration or, alternatively, samples must be kept at –20ºC. When the prey is not in good condition, undigested parts (e.g., cephalopod beaks, fish otoliths and bones, crustacean exoskeletons, polychaete jaws, mollusc shells) can allow their identification but not mass quantification (Hecht 1987, Xavier and Cherel 2009, Xavier et al. 2020). After sorting stomach content samples into different fresh components (e.g., fish, cephalopods, crustaceans, others) that are weighed, prey items are identified and measured (either directly if the prey is complete, or indirectly via undigested items). The most used indices to quantify the ingested prey include the Frequency of Occurrence (%F), Number (%N) and Reconstituted Mass (%M) (Xavier et al. 2015, Petta et al. 2020), which provide valuable information about seabird diets, particularly if combined.

The conventional methods have a range of advantages and disadvantages that have been documented (Duffy and Jackson 1986, Votier et al. 2003, Barrett et al. 2007, Karnovsky et al. 2012). Some of the key issues are:

(1) A physical contact with the seabirds is needed (e.g., spontaneous regurgitations or using stomach flushing (Wilson 1984, Xavier et al. 2017), which stress individuals and may cause mortality in rare cases (Duffy and Jackson 1986, Phillips 2006).

(2) Different prey present different digestion rates, with a bias towards prey with hard parts that can remain in the predators' stomachs for weeks and months (e.g., cephalopod beaks). Hence the importance to not pool fresh and accumulated items in the analysis.

(3) It represents the most recent feeding events (Phillips 2006), so it should not be used for long-term dietary habits.

(4) It is dependent of high-quality identification guides of prey, as well as good allometric equations to estimate prey size. Even for well-known prey from the Southern Ocean, there is a lack of good allometric equations of cephalopod and crustacean species (Xavier and Cherel 2009, Xavier et al. 2020).

(5) Sorting and identifying prey items is time consuming and difficult for inexperienced scientists (Jackson et al. 2007).

There is a need to standardise the use of this conventional method in seabirds so that studies are comparable (Xavier et al. 2015, McInnes et al. 2016). Indeed, using %F, %N and %M would facilitate future comparative studies, taking advantage of an emerging use of ecological data from the digital world (Jarić et al. 2020). The knowledge obtained from conventional methods can provide essential information for the management of seabirds, and for monitoring broader changes in marine ecosystems (McInnes et al. 2016). However, modern techniques should complement conventional methods to evaluate the diet of seabirds, increasing our ability to understand this enigmatic group of vertebrates.

Stable isotope analysis

Stable Isotope Analysis (SIA) is an indirect biochemical method of determining seabird diet, and probably the most popular of the new approaches to study trophic niches of seabirds (Hunt and Wilson 2012). SIA integrates diet information in the tissues of consumers, and it is a very useful complement to diet sampling by conventional methods. While conventional dietary analyses often indicate a 'snapshot' of what the individual has just eaten, SIA can provide short- and long-term insights into the seabird diet because different tissues have different turnover rates (see below) and, therefore, can reflect temporal changes in trophic niche (Ceia et al. 2012). This is a valuable tool to estimate diets using mixing models. The stable isotope method can provide information on both seabirds' habitat (scenopoetic) and diet (bionomic) over time and space independently of the time of sampling (Newsome et al. 2007). The technique allows investigating the foraging, migratory and trophic ecology of seabirds through the collection of specific tissues during one single capture (Ceia et al. 2018). SIA represents a sensitive, relatively easy and inexpensive approach to assess trophic relationships within marine food webs, being also successfully used for monitoring and evaluating relevant changes in marine productivity patterns using seabirds as bioindicators (Ramírez et al. 2014, Chiu-Werner et al. 2019).

SIA has been used increasingly in ecological studies since the 1980's (De Lange and Van den Brink 2006, Hobson 2011), with a tremendous growth since the 2000's (Phillips et al. 2014). Stable isotopes of carbon ($\delta^{13}C$) and nitrogen ($\delta^{15}N$) are the most frequently used in marine ecosystems, particularly in the study of seabirds (Ceia and Ramos 2015), although sulphur stable isotopes ($\delta^{34}S$) are also useful to investigate the diet of coastal generalist seabirds such as gulls (Moreno et al. 2010). $\delta^{13}C$ values indicate sources of dietary carbon, thus reflecting the seabird foraging habitat, while $\delta^{34}S$ values allow discriminating marine, estuarine and terrestrial foraging areas. In contrast to $\delta^{13}C$ and $\delta^{34}S$, $\delta^{15}N$ shows a marked stepwise increase between diet and the consumer, being thus extremely useful to infer consumers' trophic position. Based on these stable isotopes, and using isotopic mixing models (e.g., SIBER; Jackson et al. 2011), the resulting isotopic niche provides standardised means of characterising seabirds' ecological niches (Newsome et al. 2007), a crucial concept in theoretical and applied ecological studies. Moreover, the advance of Compound-Specific Isotopic Analyses (CSIA) of Amino Acids (AAs) provides a new tool to track isotopic baseline levels and can be used to improve our knowledge on the trophic position of seabirds. By combining the results of AA-CSIA with bulk (mainly protein) SIA, seabird ecologists can examine whether variation in bulk isotope values is due to differences in baseline isotope values or differences in trophic position (Lorrain et al. 2009, Wallace et al. 2014). Despite a powerful method to estimate trophic niches, SIA does not give information on what specifically individuals consume to the species level; such information must be validated by methods such as conventional diet analyses or DNA metabarcoding. However, once aware of the main potential food sources, mixing models (e.g., SIMMR; Parnell 2020) can be used to estimate the relative proportion of different dietary sources, acknowledging that a sensitivity analysis (i.e., isotope mixing models need to fully account for uncertainty, see limitations below) should be used in order to accurately estimate source contributions (Moore and Semmens 2008, Chiaradia et al. 2014, Phillips et al. 2014). It is also important to follow the best practices in reporting results of SIA by using consistent terminology and proper expression of results and methods applied (Bond and Hobson 2012).

Currently, the stable isotope method is used routinely at low costs, which allows to measure isotopic ratios of a large number of samples, and it has a great potential for tracing food web structure and function in pelagic and coastal systems. However, some isotopic basic information on trophic relationships between prey and consumers are essential to reconstruct diets, assess variability in isotopic values, estimate trophic positions and investigate food web structures. In particular:

(1) The isotopic values of the potential food sources must be distinct.

(2) The baseline isotope values should be steady or change predictably.

(3) Trophic isotopic discrimination factors between food and consumers' tissues must be known.

(4) Correction factors must be used when comparing the isotopic values of different tissues due to tissue-specific isotopic discrimination factors, mainly resulting from tissue-specific protein (and hence amino acid) composition (e.g., Cherel et al. 2014b).

These criteria are not always met because accurate information cannot be gathered for all species in all environments, and the sources of uncertainty are difficult to assess in field studies. For that reason, it is recommended (i) to adopt species-specific discrimination factors available in the literature, (ii) to determine isotopic values in multiple tissues to provide a temporal assessment of diet, (iii) to characterise baseline isotope values, (iv) to use conventional methods (e.g., stomach contents), and (v) to integrate life history of study animals in the system (Jardine et al. 2006).

SIA bypassed some limitations of the conventional methods of food analysis. Tissues with different isotopic turnover rates allowed investigating the food and feeding ecology of seabirds over the short- to the long-term, including the poorly known inter-breeding and moulting periods (Ceia et al. 2012, Cherel et al. 2016). Several tissues can be sampled non-intrusively during the breeding season yet providing information throughout the whole annual cycle. The most used tissues collected from living birds are blood and feathers, and to a lesser extend nails and eggs (Inger and Bearhop 2008). Whole blood is a metabolically active tissue, which includes plasma and blood cells that have different isotopic turnovers. While turnover rate of plasma is high (ca. 12–15 days), with a half-life of ~ 3 days, the complete turnover rate of blood cells is much slower (ca. 120–150 days), with a half-life of ~ 30 days (Hobson and Clark 1993). Thus, with a single blood sample, scientists can collect information on trophic ecology from two distinct short- to medium-time periods (although a small overlap exists in the synthesis of the two tissues). Another added-value of blood sampling is that plasma isotopic values frequently depict seabirds' GPS or satellite-tracking periods (e.g., Jaeger et al. 2010, Ceia et al. 2014, Mendes et al. 2018), thus greatly improving our understanding on spatial and trophic use of resources by seabirds. Unlike blood, fully-grown feathers are metabolically inactive and their $\delta^{13}C$ and $\delta^{15}N$ values reflect seabirds' habitat and diet, respectively, during their synthesis (Cherel et al. 2000, Cherel et al. 2016). However, it is essential to be aware of moult patterns to know the sequence and timing of moult in the studied species, because moult chronology may vary substantially between species and even between individuals (Ramos et al. 2009, Carravieri et al. 2014, Grecian et al. 2015). Feathers are also an excellent tool to reconstruct historical diet of seabirds by measuring isotopic values from museum specimens (e.g., Cherel et al. 2014a), gathering information on foraging ecology from decades or even centuries ago. Nails, another keratinous tissue, are also inert after synthesis and can provide a continuous short- to long-term record of stable isotope signal (Barquete et al. 2013, Dimitrijević et al. 2018). Egg components, on the other hand, can be useful to investigate the females' feeding ecology during the specific period of egg formation, which in some Sphenisciforms (penguins) and Procellariforms corresponds to the enigmatic pre-laying exodus at sea (Connan et al. 2019).

The application of SIA to policy developments on conservation of seabirds is noticeable. It can provide insights into (i) the impact of introduced species, such as rats, on seabirds' breeding populations in insular ecosystems (e.g., Hervías et al. 2014), (ii) responses of seabirds to commercial fisheries and their vulnerability to bycatch (e.g., Votier et al. 2010), (iii) prey ecology (e.g., Guerreiro et al. 2015), and (iv) responses of endangered species to changes in environmental conditions (e.g., Chiu-Werner et al. 2019). The growing number of studies using stable isotope techniques considerably increased our knowledge on the prey and the feeding ecology of seabirds, but one should be aware of the main limitations of the method to avoid misinterpretations of the outcomes.

Lipids as trophic markers

Lipids are hydrophobic molecules that play an essential role in cell membrane structure, signalling and energy storage. Mainly represented by Fatty Acids (FA) (around 65% of total lipids; Harvey 2006), lipids are ubiquitous in ecosystems, from terrestrial to marine environments, and from low to high trophic levels (Iverson 2009, Parrish 2009). Seabirds store fat in white adipose tissue as triacyglycerols (TAG), each containing three FA, while most of their prey store neutral lipids either as triacyglycerols or Wax Esters (WE). Each wax ester contains one FA and one fatty alcohol (FAlc), thus explaining why fatty alcohols have to be taken into account in some dietary investigations, with the main prerequisite that both TAG and WE have to be sorted from total lipids to determine their FA pattern, and FA and FAlc patterns, respectively (Connan et al. 2007). In marine food webs, FA have been indicated as trophic markers (Dalsgaard et al. 2003, Feiner et al. 2018) and considered an indirect tool to depict predator-prey relationships (Iverson et al. 2007, Raclot et al. 1998). Ideally, a trophic marker should be of unique origin, non-modified, quantitatively stable throughout successive trophic levels and with the potential to provide insights about trophic links and food web baseline (Dalsgaard et al. 2003, Iverson 2009). However, this is not always possible as FA are of a limited number, ubiquitous, and they can be modified by the metabolism of consumers (Dalsgaard et al. 2003). Nevertheless, FA present several characteristics that make them valuable trophic markers to study marine food web dynamics: (i) unlike proteins or carbohydrates, FA undergo little modifications during digestion, being absorbed, transported and deposited almost intact (Dalsgaard et al. 2003, Iverson 2009); (ii) most animals are not able to biosynthesize some FA, having a reduced ability to modify or introduce double bonds to FA chain (Iverson 2009, Parrish 2009); (iii) for monogastric animals, and particularly in seabirds, the lipid-rich diets reduce the biosynthesis and bioconversion of FA, which creates the opportunity to evaluate the bioaccumulation of dietary FA (Williams and Buck 2010, Karnovsky et al. 2012). Thus, FA have been used as ecological tracers to study energy flows and nutrient transfer from the lower to upper trophic levels, with minimal invasiveness (Hebert et al. 2006, Iverson 2009). Obviously, the method is most useful when seabirds are in an anabolic state and feed on fatty prey, while it is useless when they feed on lean prey, with the resulting data more difficult to interpret when seabirds are in a catabolic state (e.g., fasting).

Preliminary studies on the FA composition of seabird tissues started in the 1960/70's by analysing the stomach oil of several Procellariforms (Lewis 1969, Cheah and Hansen 1970, Clarke and Prince 1976). Stomach oils are FA (and Falc) rich oils with a complex and diverse lipid composition (TAG,WE, free FA, cholesterol, polar lipids), which is formed in the proventriculus (Roby et al. 1989, Roby et al. 1992). They act as an alternative source of energy when seabirds experience periods of fasting (Roby et al. 1997), thus reducing the energetic costs of fat mobilization (Roby et al. 1997, Roby et al. 1989). Stomach oils have the advantage that their origin come exclusively from the prey (Lewis 1969, Cheah and Hansen 1970, Wang et al. 2007), giving an excellent opportunity to study predators' diets without injuring the bird, and to identify the presence/absence of some prey groups based on FA and FAlc compositions (Connan et al. 2005, Connan et al. 2007, Connan et al. 2010). Although stomach oil FA and FAlc were not yet metabolized, resembling more closely prey FA and FAlc compositions (Wang et al. 2007, Richoux et al. 2010), it only reflects a 'snapshot' of the last meals (Wang et al. 2007), providing no evidence of what is being incorporated in seabird tissues. To overcome this issue it is recommended to analyse plasma (Käkelä et al. 2009), eggs (Wang et al. 2014) or adipose tissue FA profile (i.e., FA signature) (Iverson et al. 2007). Plasma and adipose tissue have been extensively used in the field of seabird trophic ecology, because they can be sampled with little detriment to individuals (Käkelä et al. 2009, Rocha et al. 2016). Eggs are less commonly used, because only freshly laid eggs are suitable for analysis, due to preferential incorporation of some yolk FA during embryo development (Groscolas et al. 2003). Thus, as for SIA, different tissues reflect different time-windows of FA assimilation, and consequently, reflect short- to long-term periods of diet (Iverson et al. 2007, Williams and Buck 2010).

Since the 2000's, several studies have focused on the qualitative/semi-quantitative use of FA signatures as potential indicators of seabird trophic ecology (Käkelä et al. 2006, Williams et al. 2008). The validation of the qualitative use of FA in marine systems relies on the fact that FA can work as indicators of different food web baselines (Dahl et al. 2003), different food resources (Sargent and Falk-Petersen 1981), and that FA are transferred in a conservative way throughout the successive trophic levels (Dalsgaard et al. 2003, Iverson et al. 2004). For instance, Dahl et al. (2003) found that Common Eiders (*Somateria mollissima*), Black-legged Kittiwakes (*Rissa tridactyla*) and Northern Fulmars (*Fulmarus glacialis*) exhibited different fat, liver and muscle FA compositions; these differences were the direct consequence of distinct diets and reliance on different food webs (benthic vs. pelagic). Later, a captive experiment conducted with Herring Gulls (*Larus argentatus*) revealed a clear influence of prey type on plasma FA pattern (Käkelä et al. 2005), which was different between gulls feeding on demersal vs. pelagic fish. The authors proposed that some FA, particularly some FA ratios (e.g., 20:4n-6/(18:3n-3+18:4n-3+20:5n-3)), can be indicators of food web baselines (Käkelä et al. 2007, Wold et al. 2011), and may help to confirm the presence/absence of some prey groups in predators' diet. Thus, FA can be used qualitatively/semi-quantitatively in two different ways:

(1) by studying only individual or populational FA signature variations during a given timeframe, making inferences about predators' diet shifts (Iverson et al. 1997, Karnovsky et al. 2008), prey spatial distribution and availability;
(2) by evaluating specific FA, or ratios among FA, which are characteristic of a single prey species or group of prey (Käkelä et al. 2006, Käkelä et al. 2007).

However, the most important advance on the use of FA as a tool to study seabirds' trophic ecology is the quantitative estimation of diets (Budge et al. 2006). Currently, there are several methodological models that allow the quantitative assessment of diet, but the most promising is the Quantitative Fatty Acid Signature Analysis (QFASA) developed by Iverson et al. (2004). This algorithm uses predator and potential prey FA signatures to calculate the weighted mixture of prey that most closely represents predator signatures, by minimizing the model distance and assigning a proportion to each diet item included in the model. The prior choice of the FA subset is required (Budge et al. 2006, Iverson et al. 2004), and it is recommended to add prey total lipid content to improve model accuracy (Iverson et al. 2004, Bromaghin 2017). QFASA also requires the calculation of calibration coefficients, i.e., constant values that account for predators' lipid metabolism (Iverson et al. 2004, Rosen and Tollit 2012). These values are obtained through captive experiments where the researchers are in complete control over what the animals are eating (Iverson et al. 2004, Williams et al. 2009). Despite the improvements of QFASA during the last decade, other modelling approaches arose (Galloway et al. 2014, Galloway et al. 2015, Neubauer and Jensen 2015), attempting to overcome the limited assumptions required by QFASA. These new approaches are based on Bayesian measurements (similar to isotopic mixing models), now applied to predator FA (Galloway et al. 2014, Galloway et al. 2015), but with little alterations to account for differences between SIA and FA data structure (e.g., FA are usually expressed in proportions, while SIA are not) (Neubauer and Jensen 2015). Moreover, the recent advances on compound-specific isotope analysis extended to FA (FA-CSIA), with $\delta^{13}C$ values of some FA differentiating carbon sources (Budge et al. 2011, Mohan et al. 2016). For instance, the analysis of FA-$\delta^{13}C$ denoted that only the isotopic values of polyunsaturated FA (PUFA) remained stable throughout trophic levels, highlighting their potential as dietary tracers (Burian et al. 2020). PUFA have been used as biomarkers of marine primary producers (e.g., diatoms, dinoflagellates; Budge et al. 2008, De Troch et al. 2012), because consumers cannot biosynthesize them. Additionally, the long double-bonded chains of PUFA guarantee their high preservation throughout trophic levels, enhancing their use as trophic tracers of energy and nutrient flows (Gladyshev et al. 2009, Galloway and Budge 2020). Yet, the method merits further investigations, because knowledge of FA-$\delta^{13}C$ trophic fractionation factors and the quantification of the impact of FA metabolism on carbon isotopic composition are not well known (Burian et al. 2020, Galloway and Budge 2020).

The new methodologies, such as Bayesian FA mixing models (Galloway et al. 2014, Neubauer and Jensen 2015), the continual improvements of the current models (Bromaghin et al. 2017), and the combined use of multiple approaches have enabled accurate assessments of predator-prey relationships using FA (Nielsen et al. 2017). The understanding of how metabolic activity affect predator's FA composition, and how different life-stages, body condition and physiological requirements may influence individual FA must be deeply investigated in various consumers within and across food webs, to allow comparing FA characteristics (synthesis, abundance) between ecosystems.

DNA metabarcoding

DNA-based techniques have been more recently developed as a conservation and management tool in dietary studies, helping to understand the trophic links within the marine food webs (e.g., Mcinnes et al. 2017b). This technique relies on identifying DNA sequences unique to particular prey taxa in predators' diet samples, namely from their regurgitations, guts or faeces (Symondson 2002). Several approaches have been developed throughout the years (King et al. 2008, Pompanon et al. 2012), and it is possible to target a specific DNA region and to amplify it into millions of copies using Polymerase Chain Reaction (PCR) (Jarman et al. 2002, Parsons et al. 2005). With the increasing use of DNA-based techniques, large databases of animal DNA sequences from several genes have been developed. This has been referred to as "DNA barcoding", and such databases can be freely accessed and directly applied to identify prey DNA isolated in diet studies.

In the last few years, a series of High-Throughput Sequencing (HTS) devices have been commercially developed. These are able to produce thousands or millions of sequences at once, detecting even rarer and short sequences, and lowering the cost of each DNA sequencing (King et al. 2015). Such technologies have revolutionised the number of high-quality reads that can be obtained from samples containing a mixture of unidentified species. This so-called "DNA metabarcoding" approach enhanced the use of molecular techniques on dietary studies, contributing greatly to the growth of international databases (Sheppard and Harwood 2005). Many studies have used metabarcoding to successfully gain knowledge on organisms' diets in great details, allowing to study ecosystems' condition, status of conservation strategies and responses to environmental changes (Murray et al. 2011).

The experimental design of a metabarcoding project must take several aspects into consideration, including sampling, selection of primers, DNA extraction, DNA sequencing, bioinformatics analysis and the associated costs. The final goal should be the storage and sharing of the molecular data, even for studies that do not produce scientific papers. This allows the growth of databases such as the National Institutes of Health Genetic sequence database (GenBank) and the Barcode of Life Data systems (BOLD), which comprise millions of barcodes for a large array of genes and species.

The sampling—and its processing—is one of the most crucial steps of the DNA metabarcoding approach. DNA from diet samples must be of good quality, and these can be sampled from the animals' stomach. However, as discussed earlier, stomach flushing might be invasive, especially for small seabird species. Prey remains can be collected from natural regurgitation of pellets and defensive regurgitations of undigested or semi-digested meals. Seabirds, such as Procellariforms and gulls, tend to regurgitate when captured or handled (Alvito et al. 2015, Calado et al. 2018), and such samples potentially provide less invasive means to acquire samples for DNA metabarcoding analyses. Faecal sampling is the least invasive method to obtain diet samples from seabirds and the most indicated for small species, but faeces may present degraded DNA and some PCR inhibitors (McInnes et al. 2017a, Ando et al. 2020).

Extracted DNA is amplified using PCR by choosing different primers to identify target prey groups, e.g., fish, cephalopods or amphipods. Then, PCR products are run in a HTS sequencer. The different available HTS technologies use diverse mechanisms and incorporation/detection tools (Glenn 2011, Zhang et al. 2011), and such differences should be taken into consideration when comparing results among studies using different platforms. Other factors, such as biological differences in digestion (Leray et al. 2013) and the bioinformatics analysis pipeline, may also affect interpretation of dietary composition or richness (Clare 2014, Salipante et al. 2014). Therefore, it is important to know the correct use of bioinformatics pipelines to remove sequences that were most likely a PCR or sequencing error. For inexperienced users, it can be difficult to choose the best bioinformatical tools with the proper parameters, so time and resources can be spent in analysing the results.

In addition to the bias between platforms and bioinformatic algorithms, there are some limitations of DNA-based methods to study seabird diets. The intent of many dietary studies is to obtain quantitative data on the relative amounts of the different prey consumed by a species. However, quantification of the relative proportion of a given prey through HTS is still not possible in any cases, and is considered the main limitation of metabarcoding (Pompanon et al. 2012, but see de Sousa et al. 2019). This happens not only because different predator taxa can differ markedly in their DNA digestion rates (e.g., Chen et al. 2000), but also due to potential bias during PCR when target DNA is exponentially amplified (Pompanon et al. 2012). Moreover, DNA-based techniques cannot distinguish between seabird prey from prey of seabird prey (the so-called secondary ingestion). This can be an issue for studies which use gut contents or regurgitation samples. However, for studies using faecal samples it is not a main concern, because the secondary prey is likely to have been thoroughly digested by the time it reaches the faeces of study species (Sheppard and Harwood 2005). Nonetheless, when studying seabird diets, it is important to consider the accidental ingestion of non-prey organisms, because large numbers of small planktonic organisms may be ingested in seawater together with the intended prey. The contamination of the samples with the environment or surrounding DNA (in the field or the laboratory) is a very important consideration since it can lead to misleading results.

Despite these limitations, DNA-based studies have the major advantage of identifying components of diet that are not apparent through physical examination (Ando et al. 2020, Deagle et al. 2018, de Sousa et al. 2019). They allow to study the diet of animals in a non-invasive way through analysis of their spontaneous regurgitates and faeces. Such techniques can be particularly useful for smaller seabird species for which invasive methods, such as stomach flushing, should be avoided. As these DNA-based techniques become more widely applied to ecological studies, more DNA sequences from different genes are becoming available, providing many opportunities for research. Metabarcoding is still an expensive method when compared to most traditional methods, but such costs are steadily decreasing with the improvement of the new generation of sequencing technology. Also, when compared to SIA or lipids, the genetic approach offers a more diverse and complete data for assessing prey richness, since it relies on identifying DNA sequences that are unique to particular prey (Jarman et al. 2013). It is also faster than traditional approaches of prey identification in faeces and stomach contents, making it highly cost-effective for large dietary studies (Pompanon et al. 2012, de Sousa et al. 2019). These molecular techniques allow detection of rare or already degraded prey species that were underestimated by traditional methods (e.g., gelatinous organisms). Furthermore, when combined with ecological tools such as SIA or tracking devices, metabarcoding is a powerful tool to evaluate resource partitioning between consumers by providing high resolution data on seabird prey (Carreon-Martinez et al. 2011).

The use of different methods to study the diet and trophic ecology of storm petrels

There are 27 species of storm petrels, and although most of them are categorized as Least Concern in the IUCN Red List, they are sensitive to human disturbance and the majority are threatened by invasive and/or opportunistic species (IUCN 2020). Due to their small size and elusive behaviour, the diet and feeding ecology of storm petrels is one of the least known of all seabirds. Here, we detail a case study showing how well complementary approaches can help to decipher the trophic ecology of this cryptic group of seabirds (Fig. A2.2).

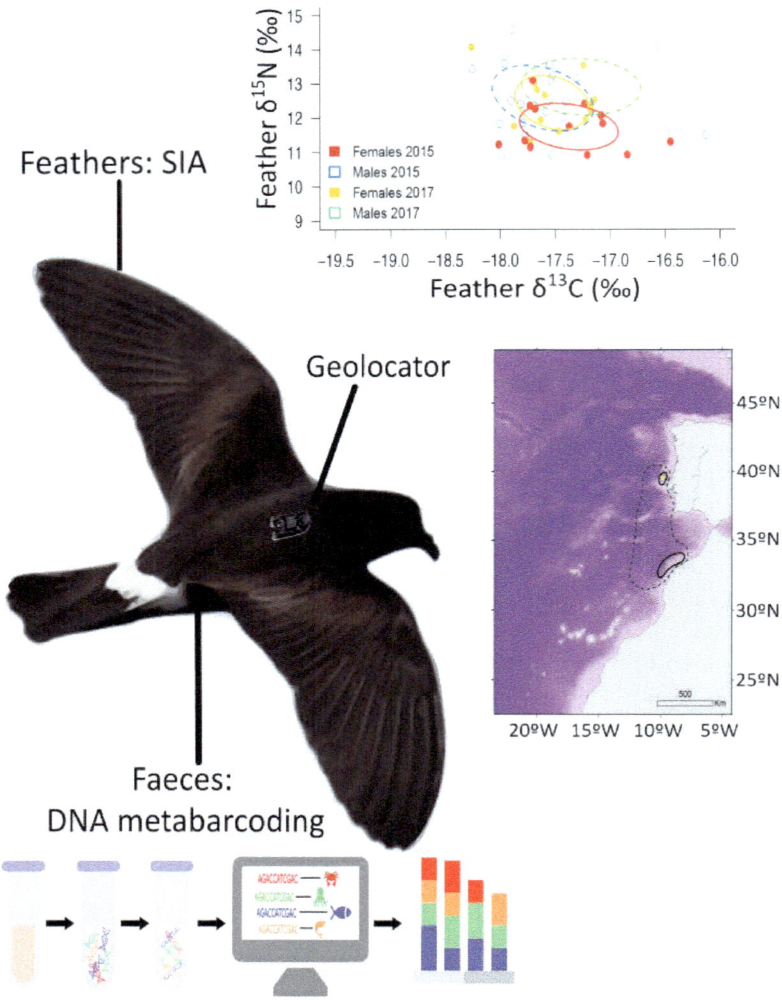

Figure A2.2. Combined diagram showing the use of complementary methods to study the trophic ecology of a small seabird species, the Madeiran Storm Petrel (*Hydrobates castro*) (adapted from Carreiro et al. 2020). Bio-logging (geolocators) enables to describe foraging areas, Stable Isotope Analyses (SIA) provide important information on the trophic niche, and DNA metabarcoding detailed identification of prey items.

Through behavioural observations, it was thought that storm petrel's diet was largely based on zooplankton and small mesopelagic fish. However, regurgitations of chicks of Leach's Storm Petrel (*Hydrobates leucorhous*) showed mostly mesopelagic fish (Myctophidae and Gadidae), crustaceans and even cephalopods (Hedd and Montevecchi 2006, Hedd et al. 2009, Frith et al. 2020). Analysis of food samples from the Mediterranean Storm Petrel (*H. pelagicus melitensis*) revealed two feeding strategies, thus showing the adaptive behaviour of this species—birds performed traditional long trips during the day in which they feed essentially on pelagic fish, and short trips close to the colony at night and prey upon opossum shrimps (Mysidacea) (Albores-Barajas et al. 2011). A combination of SIA and FA analyses indicated that Mediterranean Storm Petrels rely on ichthyoplankton and zooplankton (Aguado-Giménez et al. 2016). However, the lipid source could be the pelagic baitfish that is used to feed tunas in pelagic fish farms, because many Mediterranean Storm Petrels were observed feeding on the oily plume present in such farms (Aguado-Giménez et al. 2016). Because regurgitates presumably represent the chick diet only, the use of molecular techniques arose as an alternative to study the feeding ecology of storm petrels (Medeiros et al. 2012), which was successfully applied to the Madeiran Storm Petrel (*H. castro*) (Carreiro et al. 2020).

The Madeiran Storm Petrel is a monomorphic medium size storm petrel (Monteiro et al. 1996), breeding in oceanic islands from equatorial to subtropical latitudes, mostly in winter (Monteiro and Furness 1998). Like other storm petrels, the Madeiran Storm Petrel is a K-strategist, returning to nesting sites at night but avoiding bright nights of full moon (Bolton et al. 2004). A study on this species breeding in Berlengas archipelago, Portugal, incorporated sexual, seasonal (breeding and non-breeding period) and temporal (two years) information on trophic variability, to assess its foraging ecology (Carreiro et al. 2020). The molecular techniques used in this study allowed identifying many prey taxa to the genus and species levels, some of which would have been unlikely to be identified through traditional methods. A predominance of Gadiformes (codfishes) DNA in the samples was detected. Moreover, feather $\delta^{15}N$ values were similar to those reported for Cory's Shearwaters (*Calonectris borealis*) and the Macaronesian Shearwaters (*Puffinus baroli*) (Paiva et al. 2010, Paiva et al. 2016), thus suggesting that the Madeiran Storm Petrel feed on higher trophic level prey such as mesopelagic fish and fishery discards rather than being exclusively zooplanktivorous. In addition, the limited tracking data suggests that Madeiran Storm Petrels breeding in Berlengas adopt a dual foraging strategy (i.e., alternating between short and long trips), as already described in other Procellariforms (Weimerskirch 1998). Overall, the combination of different techniques described the food and feeding ecology of the Madeiran Storm Petrel for the first time (Carreiro et al. 2020), a rare example of winter breeder storm petrel. The complementarity between tracking, SIA and metabarcoding is an example of the great framework that can be used to study the foraging ecology of seabirds.

Acknowledgments

This study benefited from the strategic program of MARE (MARE - UID/MAR/04292/2020) and the transitory norm contract DL57/2016 to FRC (DL57/2016/CP1370/CT90) at the University of Coimbra, financed by the Foundation for Science and Technology (FCT; Portugal). ARC acknowledges her doctoral grant (SFRH/BD/139019/2018) attributed by FCT and the European Social Fund (POPH, EU). We acknowledge Ricardo Ceia and Renata Medeiros for helpful comments in the pre-final draft of this manuscript.

References cited

Aguado-Giménez, F., A. Sallent-Sánchez, S. Eguía-Martínez, J. Martínez-Ródenas, M.D. Hernández-Llorente, C. Palanca-Maresca et al. 2016. Aggregation of European storm-petrel (*Hydrobates pelagicus* ssp. *melitensis*) around cage fish farms. Do they benefit from the farmś resources? Mar. Environ. Res. 122: 46–58. doi: 10.1016/j.marenvres.2016.09.006.

Albores-Barajas, Y.V., F. Riccato, R. Fiorin, B. Massa, P. Torricelli and C. Soldatini. 2011. Diet and diving behaviour of European Storm Petrels *Hydrobates pelagicus* in the Mediterranean (ssp. *melitensis*). Bird Study 58: 208–212. doi: 10.1080/00063657.2011.560244.

Alvito, P.M., R. Rosa, R.A. Phillips, Y. Cherel, F. Ceia, M. Guerreiro et al. 2015. Cephalopods in the diet of nonbreeding black-browed and grey-headed albatrosses from South Georgia. Polar Biol. 38: 631–641. doi: 10.1007/s00300–014–1626–3.

Ando, H., H. Mukai, T. Komura, T. Dewi, M. Ando and Y. Isagi. 2020. Methodological trends and perspectives of animal dietary studies by noninvasive fecal DNA metabarcoding. Environ. DNA 2: 391–406. doi: 10.1002/edn3.117.

Barquete, V., V. Strauss and P.G. Ryan. 2013. Stable isotope turnover in blood and claws: A case study in captive African Penguins. J. Exp. Mar. Bio. Ecol. 448: 121–127. doi: 10.1016/j.jembe.2013.06.021.

Barrett, R.T., K. Camphuysen, T. Anker-Nilssen, J.W. Chardine, R.W. Furness, S. Garthe et al. 2007. Diet studies of seabirds: A review and recommendations. ICES J. Mar. Sci. 64: 1675–1691. doi: 10.1093/icesjms/fsm152.

Bolton, M., R. Medeiros, B. Hothersall and A. Campos. 2004. The use of artificial breeding chambers as a conservation measure for cavity-nesting procellariiform seabirds: A case study of the Madeiran storm petrel (*Oceanodroma castro*). Biol. Conserv. 116: 73–80. doi: 10.1016/s0006–3207(03)00178–2.

Bond, A. and K. Hobson. 2012. Reporting stable-isotope ratios in ecology: Recommended terminology, guidelines and best practices. Waterbirds 35: 324–331. doi: 10.1675/063.035.0213.

Bromaghin, J.F. 2017. QFASAR: Quantitative fatty acid signature analysis with R. Methods Ecol. Evol. 8: 1158–1162. doi: 10.1111/2041–210X.12740.

Bromaghin, J.F., S.M. Budge, G.W. Thiemann and K.D. Rode. 2017. Simultaneous estimation of diet composition and calibration coefficients with fatty acid signature data. Ecol. Evol. 7: 6103–6113. doi: 10.1002/ece3.3179.

Budge, S.M., S.J. Iverson and H.N. Koopman. 2006. Studying trophic ecology in marine ecosystems using fatty acids: A primer on analysis and interpretation. Mar. Mammal Sci. 22: 759–801. doi: 10.1111/j.1748–7692.2006.00079.x.

Budge, S.M., A.M. Springer, S.J. Iverson, G. Sheffield and C. Rosa. 2008. Blubber fatty acid composition of bowhead whales, *Balaena mysticetus*: Implications for diet assessment and ecosystem monitoring. J. Exp. Mar. Bio. Ecol. 359: 40–46. doi: 10.1016/j.jembe.2008.02.014.

Budge, S.M., S.W. Wang, T.E. Hollmen and M.J. Wooller. 2011. Carbon isotopic fractionation in eider adipose tissue varies with fatty acid structure: Implications for trophic studies. J. Exp. Biol. 214: 3790–3800. doi: 10.1242/jeb.057596.

Burian, A., J.M. Nielsen, T. Hansen, R. Bermudez and M. Winder. 2020. The potential of fatty acid isotopes to trace trophic transfer in aquatic food-webs. Philos. Trans. R. Soc. Lond. B375: 20190652. doi: 10.1098/rstb.2019.0652.

Calado, J.G., D.M. Matos, J.A. Ramos, F. Moniz, F.R. Ceia, J.P. Granadeiro et al. 2018. Seasonal and annual differences in the foraging ecology of two gull species breeding in sympatry and their use of fishery discards. J. Avian Biol. 1–12. doi: 10.1111/jav.01463.

Carravieri, A., P. Bustamante, C. Churlaud, A. Fromant and Y. Cherel. 2014. Moulting patterns drive within-individual variations of stable isotopes and mercury in seabird body feathers: Implications for monitoring of the marine environment. Mar. Biol. 161: 963–968. doi: 10.1007/s00227–014–2394–x.

Carreiro, A.R., V.H. Paiva, R. Medeiros, K.A. Franklin, N. Oliveira, A.I. Fagundes et al. 2020. Metabarcoding, stables isotopes, and tracking: Unraveling the trophic ecology of a winter-breeding storm petrel (*Hydrobates castro*) with a multimethod approach. Mar. Biol. 167: 1–13. doi: 10.1007/s00227–019–3626–x.

Carreon-Martinez, L., T.B. Johnson, S.A. Ludsin and D.D. Heath. 2011. Utilization of stomach content DNA to determine diet diversity in piscivorous fishes. J. Fish Biol. 78: 1170–1182. doi: 10.1111/j.1095–8649.2011.02925.x.

Ceia, F.R., R.A. Phillips, J.A. Ramos, Y. Cherel, R.P. Vieira, P. Richard et al. 2012. Short- and long-term consistency in the foraging niche of wandering albatrosses. Mar. Biol. 159: 1581–1591. doi: 10.1007/s00227–012–1946–1.

Ceia, F.R., V.H. Paiva, S. Garthe, J.C. Marques and J.A. Ramos. 2014. Can variations in the spatial distribution at sea and isotopic niche width be associated with consistency in the isotopic niche of a pelagic seabird species? Mar. Biol. 161: 1861–1872. doi: 10.1007/s00227–014–2468–9.

Ceia, F.R. and J.A. Ramos. 2015. Individual specialization in the foraging and feeding strategies of seabirds: A review. Mar. Biol. 162: 1923–1938. doi: 10.1007/s00227–015–2735–4.

Ceia, F.R., Y. Cherel, V.H. Paiva and J.A. Ramos. 2018. Stable isotope dynamics (δ13C and δ15N) in neritic and oceanic waters of the North Atlantic inferred from GPS-tracked Cory's shearwaters. Front. Mar. Sci. 5: 377. doi: 10.3389/fmars.2018.00377.

Cheah, C.C. and I.A. Hansen. 1970. Stomach oil and tissue lipids of the petrels *Puffinus pacificus* and *Pterodroma Macroptera*. Int. J. Biochem. 1: 203–208. doi: 10.1016/0020–711X(70)90095–9.

Chen, Y., K.L. Giles, M.E. Payton and M.H. Greenstone. 2000. Identifying key cereal aphid predators by molecular gut analysis. Mol. Ecol. 9: 1887–1898. doi: 10.1046/j.1365–294X.2000.01100.x.

Cherel, Y. and H. Weimerskirch. 1999. Spawning cycle of onychoteuthid squids in the southern Indian Ocean: New information from seabird predators. Mar. Ecol. Prog. Ser. 188: 93–104. doi: 10.3354/meps188093.

Cherel, Y., K.A. Hobson and H. Weimerskirch. 2000. Using stable-isotope analysis of feathers to distinguish moulting and breeding origins of seabirds. Oecologia 122: 155–162. doi: 10.1007/PL00008843.

Cherel, Y., M. Connan, A. Jaeger and P. Richard. 2014a. Seabird year-round and historical feeding ecology: Blood and feather δ13C and δ15N values document foraging plasticity of small sympatric petrels. Mar. Ecol. Prog. Ser. 505: 267–280. doi: 10.3354/meps10795.

Cherel, Y., S. Jaquemet, A. Maglio and A. Jaeger. 2014b. Differences in δ13C and δ15N values between feathers and blood of seabird chicks: Implications for non-invasive isotopic investigations. Mar. Biol. 161: 229–237. doi: 0.1007/s00227–013–2314–5.

Cherel, Y., P. Quillfeldt, K. Delord and H. Weimerskirch. 2016. Combination of at-sea activity, geolocation and feather stable isotopes documents where and when seabirds molt. Front. Ecol. Evol. 4: 3. doi: 10.3389/fevo.2016.00003.

Chiaradia, A., M.G. Forero, J.C. McInnes and F. Ramírez. 2014. Searching for the true diet of marine predators: Incorporating bayesian priors into stable isotope mixing models. PLoS One 9: e92665. doi: 10.1371/journal.pone.0092665.

Chiu-Werner, A., F.R. Ceia, S. Cárdenas-Alayza, M. Cardeña-Mormontoy, M. Adkesson and J.C. Xavier. 2019. Inter-annual isotopic niche segregation of wild humboldt penguins through years of different El Niño intensities. Mar. Environ. Res. 150: 104755. doi: 10.1016/j.marenvres.2019.104755.

Clare, E.L. 2014. Molecular detection of trophic interactions: Emerging trends, distinct advantages, significant considerations and conservation applications. Evol. Appl. 7: 1144–1157. doi: 10.1111/eva.12225.

Clarke, A. and P.A. Prince. 1976. The origin of stomach oil in marine birds: Analyses of the stomach oil from six species of subantarctic procellariiform birds. J. Exp. Mar. Bio. Ecol. 23: 15–30. doi: 10.1016/0022–0981(76)90082–4.

Connan, M., P. Mayzaud, M. Boutoute, H. Weimerskirch and Y. Cherel. 2005. Lipid composition of stomach oil in a procellariiform seabird *Puffinus tenuirostris*: Implications for food web studies. Mar. Ecol. Prog. Ser. 290: 277–290. doi: 10.3354/meps290277.

Connan, M., Y. Cherel and P. Mayzaud. 2007. Lipids from stomach oil of procellariiform seabirds document the importance of myctophid fish in the Southern Ocean. Limnol. Oceanogr. 52: 2445–2455. doi: 10.4319/lo.2007.52.6.2445.

Connan, M., P. Mayzaud, K. Hobson, H. Weimerskirch and Y. Cherel. 2010. Food and feeding ecology of the Tasmanian short-tailed shearwater (*Puffinus tenuirostris*, Temminck): Insights from three complementary methods. J. Oceanogr. Res. Data 3: 19–32.

Connan, M., B.J. Dilley, T.O. Whitehead, D. Davies, C.D. Mcquaid and P.G. Ryan. 2019. Multidimensional stable isotope analysis illuminates resource partitioning in a sub-Antarctic island bird community. Ecography 42: 1948–1959. doi: 10.1111/ecog.04560.

Constable, A.J. 2011. Lessons from CCAMLR on the implementation of the ecosystem approach to managing fisheries. Fish Fish. 12: 138–151. doi: 10.1111/j.1467–2979.2011.00410.x.

Dahl, T.M., S. Falk-Petersen, G.W. Gabrielsen, J.R. Sargent, H. Hop and R.M. Millar. 2003. Lipids and stable isotopes in common eider, black-legged kittiwake and northern fulmar: Atrophic study from an Arctic fjord. Mar. Ecol. Prog. Ser. 256: 257–269. doi: 10.3354/meps256257.

Dalsgaard, J., M. St. John, G. Kattner, D. Müller-Navarra and W. Hagen. 2003. Fatty acid trophic markers in the pelagic marine environment. Adv. Mar. Biol. 46: 225–340. doi: 10.1016/S0065–2881(03)46005–7.

Deagle, B.E., A.C. Thomas, J.C. Mcinnes, L.J. Clarke, E.J. Vesterinen, E.L. Clare et al. 2018. Counting with DNA in metabarcoding studies: How should we convert sequence reads to dietary data? Mol. Ecol. 28: 391–406. doi: 10.1111/mec.14734.

De Lange, H. and N. Van den Brink. 2006. Literature review of available techniques to characterize marine and estuarine food webs. With emphasis for application in the model OMEGA. Alterra-Rapport 1372. Wageningen, The Netherlands.

De Sousa, L.L., S.M. Silva and R. Xavier. 2019. DNA metabarcoding in diet studies: Unveiling ecological aspects in aquatic and terrestrial ecosystems. Environ. DNA 1: 199–214. doi: 10.1002/edn3.27.

De Troch, M., P. Boeckx, C. Cnudde, D. Van Gansbeke, A. Vanreusel, M. Vincx et al. 2012. Bioconversion of fatty acids at the basis of marine food webs: Insights from a compound-specific stable isotope analysis. Mar. Ecol. Prog. Ser. 465: 53–67. doi: 10.3354/meps09920.

Dimitrijević, D., V.H. Paiva, J.A. Ramos, J. Seco, F.R. Ceia, N. Chipev et al. 2018. Isotopic niches of sympatric Gentoo and Chinstrap Penguins: Evidence of competition for Antarctic krill? Polar Biol. 41: 1655–1669. doi: 10.1007/s00300–018–2306–5.

Duffy, D.C. and S. Jackson. 1986. Diet studies of seabirds: A review of methods. colon. Waterbirds 9: 1–17. doi: 10.2307/1521138.

Feiner, Z.S., C.J. Foley, H.A. Bootsma, S.J. Czesny, J. Janssen, J. Rinchard et al. 2018. Species identity matters when interpreting trophic markers in aquatic food webs. PLoS One 13: e0204767. doi: 10.1371/journal.pone.0204767.

Frith, R., D. Krug, R.A. Ronconi, S.N.P. Wong, M.L. Mallory, L.A.M. Tranquilla et al. 2020. Diet of Leach's storm-petrels (*Hydrobates leucorhous*) among three colonies in Atlantic Canada. Northeast Nat. 27: 612–630.doi: 10.1656/045.027.0402.

Galloway, A., M. Eisenlord, M. Dethier, G. Holtgrieve and M. Brett. 2014. Quantitative estimates of isopod resource utilization using a Bayesian fatty acid mixing model. Mar. Ecol. Prog. Ser. 507: 219–232. doi: 10.3354/meps10860.

Galloway, A.W.E., M.T. Brett, G.W. Holtgrieve, E.J. Ward, A.P. Ballantyne, C.W. Burns et al. 2015. A fatty acid based bayesian approach for inferring diet in aquatic consumers. PLoS One 10: e0129723. doi: 10.1371/journal.pone.0129723.

Galloway, A.W.E. and S.M. Budge. 2020. The critical importance of experimentation in biomarker-based trophic ecology. Philos. Trans. R. Soc. Lond. B. 375: 20190638. doi: 10.1098/rstb.2019.0638.

Gaston, A.J., D.F. Bertram, A.W. Boyne, J.W. Chardine, G. Davoren, A.W. Diamond et al. 2009. Changes in Canadian seabird populations and ecology since 1970 in relation to changes in oceanography and food webs. Environ. Rev. 17: 267–286. doi: 10.1139/A09–013.

Gladyshev, M.I., N.N. Sushchik, O.V. Anishchenko, O.N. Makhutova, G.S. Kalacheva, A.A. Kolmakova et al. 2009. Efficiency of transfer of essential polyunsaturated fatty acids along trophic chains in aquatic ecosystems. Oecologia 165: 521–531. doi: 10.1134/S1607672909030089.

Glenn, T.C. 2011. Field guide to next-generation DNA sequencers. Mol. Ecol. Resour. 11: 759–769.doi: 10.1111/j.1755–0998.2011.03024.x.

Grecian, W.J., R.A.R. McGill, R.A. Phillips, P.G. Ryan and R.W. Furness. 2015. Quantifying variation in δ13C and δ15N isotopes within and between feathers and individuals: Is one sample enough? Mar. Biol. 162: 733–741. doi: 10.1007/s00227–015–2618–8.

Groscolas, R., F. Fréchard, F. Decrock and B.K. Speake. 2003. Metabolic fate of yolk fatty acids in the developing king penguin embryo. Am. J. Physiol. – Regul. Integr. Comp. Physiol. 285: 850–861. doi: 10.1152/ajpregu.00105.2003.

Guerreiro, M., R.A. Phillips, Y. Cherel, F.R. Ceia, P. Alvito, R. Rosa et al. 2015. Habitat and trophic ecology of Southern Ocean cephalopods from stable isotope analyses. Mar. Ecol. Prog. Ser. 530: 119–134. doi: 10.3354/meps11266.

Harvey, H.R. 2006. Sources and cycling of organic matter in the marine water column. pp. 1–25. *In*: Volkman, J.K. [ed.]. Marine Organic Matter: Biomarkers, Isotopes and DNA. Springer, Berlin. doi: 10.1007/698_2_001.

Hebert, C.E., M.T. Arts and D.V.C. Weseloh. 2006. Ecological tracers can quantify food web structure and change. Environ. Sci. Technol. 40: 5618–5623. doi: 10.1021/es0520619.

Hecht, T. 1987. A guide to the identification of otoliths of Southern Ocean fishes. S. Afr. J. Antarct. Res. 17: 1–87.

Hedd, A. and W.A. Montevecchi. 2006. Diet and trophic position of Leach's storm-petrel *Oceanodroma leucorhoa* during breeding and moult, inferred from stable isotope analysis of feathers. Mar. Ecol. Prog. Ser. 322: 291–301. doi: 10.3354/meps322291.

Hedd, A., W.A. Montevecchi, G.K. Davoren and D.A. Fifield. 2009. Diets and distributions of Leach's storm-petrel (*Oceanodroma leucorhoa*) before and after an ecosystem shift in the Northwest Atlantic. Can. J. Zool. 87: 787–801. doi: 10.1139/z09–060.

Hervías, S., F.R. Ceia, T. Pipa, M. Nogales, R.R. De Ybáñez and J.A. Ramos. 2014. How important are seabirds in the diet of black rats on islands with a superpredator? Zoology 117: 171–178. doi: 10.1016/j.zool.2013.12.003.

Hobson, K.A. and R.G. Clark. 1993. Turnover of 13C in cellular and plasma fractions of blood: Implications for nondestructive sampling in avian dietary studies. Auk. 110: 638–641.

Hobson, K.A. 2011. Isotopic ornithology: A perspective. J. Ornithol. 152: 49–66. doi: 10.1007/s10336–011–0653–x.

Hunt, G.L. and R.P. Wilson. 2012. The coming of age of marine ornithology. Mar. Ecol. Prog. Ser. 451: 227–229. doi: 10.3354/meps09725.

Inger, R. and S. Bearhop. 2008. Applications of stable isotope analyses to avian ecology. Ibis 150: 447–461. doi: 10.1111/j.1474–919X.2008.00839.x.

IUCN. 2020. The IUCN Red List of Threatened Species. https://www.iucnredlist.org/.

Iverson, S.J., K.J. Frost and L.F. Lowry. 1997. Fatty acid signatures as indicators of foraging ecology and distribution of harbor seals in the Gulf of Alaska. Mar. Ecol. Prog. Ser. 151: 255–271.

Iverson, S.J., C. Field, W.D. Bowen and W. Blanchard. 2004. Quantitative fatty acid signature analysis: A new method of estimating predator diets. Ecol. Monogr. 74: 211–235. doi: 10.1890/02–4105.

Iverson, S., A. Springer and A. Kitaysky. 2007. Seabirds as indicators of food web structure and ecosystem variability: Qualitative and quantitative diet analyses using fatty acids. Mar. Ecol. Prog. Ser. 352: 235–244. doi: 10.3354/meps07073.

Iverson, S.J. 2009. Tracing aquatic food webs using fatty acids: From qualitative indicators to quantitative determination. pp. 281–308. *In*: Kainz, M., M.T. Brett and M.T. Arts [ed.]. Lipids in Aquatic Ecosystems. Springer, New York. doi: 10.1007/978–0–387–89366–2.

Jackson, G.D., P. Bustamante, Y. Cherel, E.A. Fulton, E.P.M. Grist, C.H. Jackson et al. 2007. Applying new tools to cephalopod trophic dynamics and ecology: Perspectives from the Southern Ocean Cephalopod Workshop, February 2–3, 2006. Rev. Fish Biol. Fish. 17: 79–99. doi: 10.1007/s11160–007–9055–9.

Jackson, A.L., R. Inger, A.C. Parnell and S. Bearhop. 2011. Comparing isotopic niche widths among and within communities: SIBER—Stable Isotope Bayesian Ellipses in R. J. Anim. Ecol. 80: 595–602. doi: 10.1111/j.1365–2656.2011.01806.x.

Jaeger, A., V.J. Lecomte, H. Weimerskirch, P. Richard and Y. Cherel. 2010. Seabird satellite tracking validates the use of latitudinal isoscapes to depict predators' foraging areas in the Southern Ocean. Rapid Commun. Mass Spectrom. 24: 3456–3460. doi: 10.1002/rcm.

Jardine, T.D., K.A. Kidd and A.T. Fisk. 2006. Applications, considerations, and sources of uncertainty when using stable isotope analysis in ecotoxicology. Environ. Sci. Technol. 40: 7501–7511. doi: 10.1021/es061263h.

Jarić, I., R.A. Correia, B.W. Brook, J.C. Buettel, F. Courchamp, E. Di Minin et al. 2020. iEcology: Harnessing large online resources to generate ecological insights. Trends Ecol. Evol. 35: 630–639. doi: 10.1016/j.tree.2020.03.003.

Jarman, S.N., N.J. Gales, M. Tierney, P.C. Gill and N.G. Elliott. 2002. A DNA-based method for identification of krill species and its application to analysing the diet of marine vertebrate predators. Mol. Ecol. 11: 2679–2690. doi: 10.1046/j.1365-294X.2002.01641.x.

Jarman, S.N., J.C. McInnes, C. Faux, A.M. Polanowski, J. Marthick, B.E. Deagle et al. 2013. Adelie penguin population diet monitoring by analysis of food DNA in scats. PLoS One 8: 11. doi: 10.1371/journal.pone.0082227.

Käkelä, R., A. Käkelä, S. Kahle, P. Becker, A. Kelly and R. Furness. 2005. Fatty acid signatures in plasma of captive herring gulls as indicators of demersal or pelagic fish diet. Mar. Ecol. Prog. Ser. 293: 191–200. doi: 10.3354/meps293191.

Käkelä, A., J. Crane, S. Votier, R. Furness and R. Käkelä. 2006. Fatty acid signatures as indicators of diet in great skuas *Stercorarius skua*, Shetland. Mar. Ecol. Prog. Ser. 319: 297–310. doi: 10.3354/meps319297.

Käkelä, A., R. Furness, A. Kelly, U. Strandberg, S. Waldron and R. Käkelä. 2007. Fatty acid signatures and stable isotopes as dietary indicators in North Sea seabirds. Mar. Ecol. Prog. Ser. 342: 291–301. doi: 10.3354/meps342291.

Käkelä, R., R.W. Furness, S. Kahle, P.H. Becker and A. Käkelä. 2009. Fatty acid signatures in seabird plasma are a complex function of diet composition: A captive feeding trial with herring gulls. Funct. Ecol. 23: 141–149. doi: 10.1111/j.1365-2435.2008.01475.x.

Karnovsky, N.J., K.A. Hobson, S. Iverson and G.L. Hunt, Jr. 2008. Seasonal changes in diets of seabirds in the North Water Polynya: A multiple-indicator approach. Mar. Ecol. Prog. Ser. 357: 291–299. doi: 10.3354/meps07295.

Karnovsky, N.J., K.A. Hobson and S.J. Iverson. 2012. From lavage to lipids: Estimating diets of seabirds. Mar. Ecol. Prog. Ser. 451: 263–284. doi: 10.3354/meps09713.

King, R.A., D.S. Read, M. Traugott and W.O.C. Symondson. 2008. Molecular analysis of predation: A review of best practice for DNA-based approaches. Mol. Ecol. 17: 947–963. doi: 10.1111/j.1365-294X.2007.03613.x.

King, R.A., W.O.C. Symondson and R.J. Thomas. 2015. Molecular analysis of faecal samples from birds to identify potential crop pests and useful biocontrol agents in natural areas. Bull. Entomol. 105: 261–272. Res. doi: 10.1017/S0007485314000935.

Kowalczyk, N.D., A. Chiaradia, T.J. Preston and R.D. Reina. 2014. Linking dietary shifts and reproductive failure in seabirds: A stable isotope approach. Funct. Ecol. 28: 755–765. doi: 10.1111/1365-2435.12216.

Leray, M., J.Y. Yang, C.P. Meyer, S.C. Mills, N. Agudelo, V. Ranwez et al. 2013. A new versatile primer set targeting a short fragment of the mitochondrial COI region for metabarcoding metazoan diversity: Application for characterizing coral reef fish gut contents. Front. Zool. 10: 34. doi: 10.1186/1742-9994-10-34.

Lewis, R.W. 1969. Studies on the stomach oils of marine animals–II. Oils of some procellariiform birds. Comp. Biochem. Physiol. 31: 725–731. doi: 10.1016/0010-406X(69)92072-6.

Lorrain, A., B. Graham, F. Ménard, B. Popp, S. Bouillon, P. van Breugel et al. 2009. Nitrogen and carbon isotope values of individual amino acids: A tool to study foraging ecology of penguins in the Southern Ocean. Mar. Ecol. Prog. Ser. 391: 293–306. doi: 10.3354/meps08215.

McInnes, J.C., B. Raymond, R.A. Phillips, S.N. Jarman, M.A. Lea and R. Alderman. 2016. A review of methods used to analyse albatross diets—Assessing priorities across their range. ICES J. Mar. Sci. 73: 2125–2137. doi: 10.1093/icesjms/fsw105.

McInnes, J.C., R. Alderman, B.E. Deagle, M.A. Lea, B. Raymond and S.N. Jarman. 2017a. Optimised scat collection protocols for dietary DNA metabarcoding in vertebrates. Methods Ecol. Evol. 8: 192–202. doi: 10.1111/2041-210X.12677.

McInnes, J.C., S.N. Jarman, M.A. Lea, B. Raymond, B.E. Deagle, R.A. Phillips et al. 2017b. DNA metabarcoding as a marine conservation and management tool: A circumpolar examination of fishery discards in the diet of threatened albatrosses. Front. Mar. Sci. 4: 1–22. doi: 10.3389/fmars.2017.00277.

Medeiros, R.J., R.A. King, W.O.C. Symondson, B. Cadiou, B. Zonfrillo, M. Bolton et al. 2012. Molecular evidence for gender differences in the migratory behaviour of a small seabird. PLoS One 7: e46330. doi: 10.1371/journal.pone.0046330.

Mendes, R.F., J.A. Ramos, V.H. Paiva, J.G. Calado, D.M. Matos and F.R. Ceia. 2018. Foraging strategies of a generalist seabird species, the yellow-legged gull, from GPS tracking and stable isotope analyses. Mar. Biol. 165: 168. doi: 10.1007/s00227-018-3421-0.

Mills, W.F., J.C. Xavier, S. Bearhop, Y. Cherel, S.C. Votier, C.M. Waluda et al. 2020. Long-term trends in albatross diets in relation to prey availability and breeding success. Mar. Biol. 167: 1–12. doi: 10.1007/s00227-019-3630-1.

Mohan, S.D., T.L. Connelly, C.M. Harris, K.H. Dunton and J.W. Mcclelland. 2016. Seasonal trophic linkages in Arctic marine invertebrates assessed via fatty acids and compound-specific stable isotopes. Ecosphere 7: 1–21. doi: 10.1002/ecs2.1429.

Monteiro, L.R., J.A. Ramos, R.W. Furness and A.J. del Nevo. 1996. Movements, morphology, breeding, moult, diet and feeding of seabirds in the Azores. Colon. Waterbirds 19: 82–97. doi: 10.2307/1521810.

Monteiro, L.R. and R.W. Furness. 1998. Speciation through temporal segregation of Madeiran storm petrel (*Oceanodroma castro*) populations in the Azores? Philos. Trans. R. Soc. London Ser. B. Biological Sci. 353: 945–953. doi: 10.1098/rstb.1998.0259.

Montevecchi, W. and R. Myers. 1996. Dietary changes of seabirds indicate shifts in pelagic food webs. Sarsia 80: 313–322.

Moore, J.W. and B.X. Semmens. 2008. Incorporating uncertainty and prior information into stable isotope mixing models. Ecol. Lett. 11: 470–480. doi: 10.1111/j.1461–0248.2008.01163.x.

Moreno, R., L. Jover, I. Munilla, A. Velando and C. Sanpera. 2010. A three-isotope approach to disentangling the diet of a generalist consumer: the yellow-legged gull in northwest Spain. Mar. Biol. 157: 545–553. doi: 10.1007/s00227–009–1340–9.

Murray, D.C., M. Bunce, B.L. Cannell, R. Oliver, J. Houston, N.E. White et al. 2011. DNA-Based faecal dietary analysis: a comparison of qpcr and high throughput sequencing approaches. PLoS One 6: e25776. doi: 10.1371/journal.pone.0025776.

Neubauer, P. and O.P. Jensen. 2015. Bayesian estimation of predator diet composition from fatty acids and stable isotopes. PeerJ 2015: 1–19. doi: 10.7717/peerj.920

Newsome, S.D., C.M. Rio, S. Bearhop and D.L. Phillips. 2007. A niche for isotopic ecology. Front. Ecol. Environ. 5: 429–436. doi: 10.1890/060150.01.

Nielsen, J.M., E.L. Clare, B. Hayden, M.T. Brett and P. Kratina. 2017. Diet tracing in ecology: Method comparison and selection. Methods Ecol. Evol. 9: 278–291. doi: 10.1111/2041–210X.12869.

Paiva, V.H., P. Geraldes, I. Ramírez, A. Meirinho, S. Garthe and J.A. Ramos. 2010. Foraging plasticity in a pelagic seabird species along a marine productivity gradient. Mar. Ecol. Prog. Ser. 398: 259–274. doi: 10.3354/meps08319.

Paiva, V.H., A.I. Fagundes, V. Romão, C. Gouveia and J.A. Ramos. 2016. Population-scale foraging segregation in an apex predator of the north Atlantic. PLoS One 11: 1–19. doi: 10.1371/journal.pone.0151340.

Parnell, A. 2020. Package 'simmr.' doi: https://github.com/andrewcparnell/simmr.

Parrish, C.C. 2009. Essential fatty acids in aquatic food webs. pp. 309–326. *In*: Kainz, M., M.T. Brett and M.T. Arts [eds.]. Lipids in Aquatic Ecosystems. Springer, New York. doi: 10.1007/978–0–387–89366–2.

Parsons, K.M., S.B. Piertney, S.J. Middlemas, P.S. Hammond and J.D. Armstrong. 2005. DNA-based identification of salmonid prey species in seal faeces. J. Zool. 266: 275–281doi: 10.1017/S0952836905006904.

Petta, J.C., O.N. Shipley, S.P. Wintner, G. Cliff, M.L. Dicken and N.E. Hussey. 2020. Are you really what you eat? Stomach content analysis and stable isotope ratios do not uniformly estimate dietary niche characteristics in three marine predators. Oecologia 192: 1111–1126. doi: 10.1007/s00442–020–04628–6.

Phillips, D.L., R. Inger, S. Bearhop, A.L. Jackson, J.W. Moore, A.C. Parnell et al. 2014. Best practices for use of stable isotope mixing models in food-web studies. Can. J. Zool. 92: 823–835. doi: 10.1139/cjz–2014–0127.

Phillips, R.A. 2006. Efficacy and effects of diet sampling of albatross chicks. Emu. 106: 305–308. doi: 10.1071/MU06035.

Pompanon, F., B.E. Deagle, W.O.C. Symondson, D.S. Brown, S.N. Jarman and P. Taberlet. 2012. Who is eating what: diet assessment using next generation sequencing. Mol. Ecol. 21: 1931–1950. doi: 10.1111/j.1365–294X.2011.05403.x.

Raclot, T., R. Groscolas and Y. Cherel. 1998. Fatty acid evidence for the importance of myctophid fishes in the diet of king penguins, *Aptenodytes patagonicus*. Mar. Biol. 132: 523–533. doi: 10.1007/s002270050418.

Ramírez, F., I. Afán, K.A. Hobson, M. Bertellotti, G. Blanco and M.G. Forero. 2014. Natural and anthropogenic factors affecting the feeding ecology of a top marine predator, the Magellanic penguin. Ecosphere 5: 38. doi: 10.1890/ES13–00297.1.

Ramos, R., J. González-Solís and X. Ruiz. 2009. Linking isotopic and migratory patterns in a pelagic seabird. Oecologia 160: 97–105. doi: 10.1007/s00442–008–1273–x.

Ramos, J.A. and R.W. Furness. 2022. Seabirds as indicators of forage fish stocks. pp. 137–148. *In*: Ramos, J.A. and L. Pereira [eds.]. Seabird Biodiversity and Human Activities. CRC Press, Boca Raton, FL.

Richoux, N.B., S. Jaquemet, B.T. Bonnevie, Y. Cherel and C.D. McQuaid. 2010. Trophic ecology of Grey-headed albatrosses from Marion Island, Southern Ocean: Insights from stomach contents and diet tracers. Mar. Biol. 157: 1755–1766. doi: 10.1007/s00227–010–1448–y.

Roby, D., K. Brink and A. Place. 1989. Relative passage rates of lipid and aqueous digesta in the formation of stomach oils. Auk. 106: 303–313. doi: 10.1093/auk/106.2.303.

Roby, D.D., J.R.E. Taylor and A.R. Place. 1992. Investigation of the adaptive role of stomach oils in seabird reproduction. Antarct. J. US. 7: 147–178.

Roby, D.D., J.R.E. Taylor and A.R. Place. 1997. Significance of stomach oil for reproduction in seabirds: An interspecies cross-fostering experiment. Auk. 114: 725–736. doi: 10.2307/4089292.

Rocha, A.D., P.M. Araújo, F.R. Martinho, J.A. Ramos and J.A. Masero. 2016. A non-lethal biopsy technique for sampling subcutaneous adipose tissue of small and medium-sized birds. J. F. Ornithol. 87: 213–221. doi: 10.1111/jofo.12145.

Rosen, D.A.S. and D.J. Tollit. 2012. Effects of phylogeny and prey type on fatty acid calibration coefficients in three pinniped species: Implications for the QFASA dietary quantification technique. Mar. Ecol. Prog. Ser. 467: 263–276. doi: 10.3354/meps09934.

Salipante, S.J., T. Kawashima, C. Rosenthal, D.R. Hoogestraat, L.A. Cummings, D.J. Sengupta et al. 2014. Performance comparison of Illumina and Ion Torrent next-generation sequencing platforms for 16S rRNA-based bacterial community profiling. Appl. Environ. Microbiol. 80: 7583–7591. doi: 10.1128/AEM.02206–14.

Sargent, J.R. and S. Falk-Petersen. 1981. Ecological investigations on the zooplankton community in balsfjorden, northern Norway: Lipids and fatty acids in *Meganyctiphanes norvegica*, *Thysanoessa raschi* and *T. inermis* during mid-winter. Mar. Biol. 62: 131–137. doi: 10.1007/BF00388175.

Sheppard, S.K. and J.D. Harwood. 2005. Advances in molecular ecology: Tracking trophic links through predator-prey food-webs. Funct. Ecol. 19: 751–762. doi: 10.1111/j.1365–2435.2005.01041.x.

Svanbäck, R., M. Quevedo, J. Olsson and P. Eklöv. 2015. Individuals in food webs: The relationships between trophic position, omnivory and among-individual diet variation. Oecologia 178: 103–114.doi: 10.1007/s00442–014–3203–4.

Symondson, W.O.C. 2002. Molecular identification of prey in predator diets. Mol. Ecol. 11: 627–641. doi: 10.1046/j.1365–294X.2002.01471.x.

Thompson, M.S.A., H. Pontalier, M.A. Spence, J.K. Pinnegar, S.P.R. Greenstreet, M. Moriarty et al. 2020. A feeding guild indicator to assess environmental change impacts on marine ecosystem structure and functioning. J. Appl. Ecol. 57: 1769–1781. doi: 10.1111/1365–2664.13662.

Velarde, E., E. Ezcurra and D.W. Anderson. 2013. Seabird diets provide early warning of sardine fishery declines in the Gulf of California. Sci. Rep. 3: 1332. doi: 10.1038/srep01332.

Votier, S.C., S. Bearhop, A. MacCormick, N. Ratcliffe and R.W. Furness. 2003. Assessing the diet of great skuas, *Catharacta skua*, using five different techniques. Polar Biol. 26: 20–26. doi: 10.1007/s00300–002–0446–z.

Votier, S.C., S. Bearhop, M.J. Witt, R. Inger, D. Thompson and J. Newton. 2010. Individual responses of seabirds to commercial fisheries revealed using GPS tracking, stable isotopes and vessel monitoring systems. J. Appl. Ecol. 47: 487–497. doi: 10.1111/j.1365–2664.2010.01790.x.

Wallace, B.P., J. Schumacher, J.A. Seminoff and M.C. James. 2014. Biological and environmental influences on the trophic ecology of leatherback turtles in the northwest Atlantic Ocean. Mar. Biol. 161: 1711–1724. doi: 10.1007/s00227–014–2467–x.

Wang, S.W., S.J. Iverson, A.M. Springer and S.A. Hatch. 2007. Fatty acid signatures of stomach oil and adipose tissue of northern fulmars (*Fulmarus glacialis*) in Alaska: Implications for diet analysis of Procellariiform birds. J. Comp. Physiol. B. 177: 893–903. doi: 10.1007/s00360–007–0187–y.

Wang, S.W., T.E. Hollmén and S.J. Iverson. 2014. Egg yolk fatty acids as a proxy to quantify diets of female Spectacled Eiders (*Somateria fischeri*). Can. J. Zool. 92: 453–461. doi: 10.1139/cjz–2013–0293.

Weimerskirch, H. 1998. How can a pelagic seabird provision its chick when relying on a distant food resource? Cyclic attendance at the colony, foraging decision and body condition in sooty shearwaters. J. Anim. Ecol. 67: 99–109. doi: 10.1046/j.1365–2656.1998.00180.x.

Williams, C.T., S.J. Iverson and C.L. Buck. 2008. Stable isotopes and fatty acid signatures reveal age- and stage-dependent foraging niches in tufted puffins. Mar. Ecol. Prog. Ser. 363: 287–298. doi: 10.3354/meps07477.

Williams, C.T., S.J. Iverson and C.L. Buck. 2009. The effects of diet and caloric restriction on adipose tissue fatty acid signatures of tufted puffin (*Fratercula cirrhata*) nestlings. J. Comp. Physiol. B 179: 711–720. doi: 10.1007/s00360–009–0354–4.

Williams, C.T. and C.L. Buck. 2010. Using fatty acids as dietary tracers in seabird trophic ecology: Theory, application and limitations. J. Ornithol. 151: 531–543. doi: 10.1007/s10336–010–0513–0.

Wilson, R.P. 1984. An improved stomach pump for penquins and other seabirds. J. F. Ornithol. 55: 109–112.

Wold, A., I. Jæger, H. Hop, G.W. Gabrielsen and S. Falk-Petersen. 2011. Arctic seabird food chains explored by fatty acid composition and stable isotopes in Kongsfjorden, Svalbard. Polar Biol. 34: 1147–1155. doi: 10.1007/s00300–011–0975–4.

Xavier, J.C. and Y. Cherel. 2009. Chephalopod beak guide for the Southern Ocean. British Antarctic Survey, Cambridge, UK.

Xavier, J.C., S.L. Hill, M. Belchier, T.J. Bracegirdle, E.J. Murphy and J.L. Dias. 2015. From ice to penguins: The role of mathematics in Antarctic research. *In*: Bourguignon, J.P., R. Jeltsch, A. Pinto and M. Viana [eds.]. Mathematics of Energy and Climate Change. CIM Series in Mathematical Sciences, vol 2. Springer, Cham, Switzerland. doi: 10.1007/978–3–319–16121–1_20.

Xavier, J.C., B. Raymond, D.C. Jones and H. Griffiths. 2016. Biogeography of cephalopods in the Southern Ocean using habitat suitability prediction models. Ecosystems 19: 220–247. doi: 10.1007/s10021–015–9926–1.

Xavier, J.C., P.N. Trathan, F.R. Ceia, G.A. Tarling, S. Adlard, D. Fox et al. 2017. Sexual and individual foraging segregation in gentoo penguins pygoscelis papua from the Southern Ocean during an abnormal winter. PLoS One 12: e0174850. doi: 10.1371/journal.pone.0174850.

Xavier, J.C., Y. Cherel, G. Boxshall, A. Brandt, T. Coffer, J. Forman et al. 2020. Crustacean guide for predator studies in the Southern Ocean. Scientific Committee on Antarctic Research, Cambridge, UK.

Zhang, J., R. Chiodini, A. Badr and G. Zhang. 2011. The impact of next-generation sequencing on genomics. J. Genet. Genomics 38: 95–109. doi: 10.1016/j.jgg.2011.02.003.

CHAPTER 3

A Physiological Toolbox to Explore the Relationships Between Seabirds and Their Changing Environments

Ana Cláudia Norte,[1,*] *Pedro Miguel M. Araújo*[1] and *David Costantini*[2]

Introduction

The goal of biomonitoring is to detect early signs of environmental change before it affects populations and communities. Ideally, the mechanism of response to an environmental pressure, and how this response relates to population trends, should be well understood to develop predictive models, which can be incorporated into conservation plans and monitoring programs. The capacity to detect an environmental change, and to link it to population trends and community/ecosystem changes, depends on the organizational level at which the response to a given stressor is evaluated. A biomarker, a term originally used in ecotoxicological studies (Oost et al. 2005), may be considered, in a broad sense, as a measurable environmental disruption-induced change in morphology, cellular or biochemical components or processes, structures and functions, assessed in a sample or biologic system. In this sense, an environmental disruption is considered a potential hazard, whether chemical, physical, or biological (World Health Organization 1993). When a general decrease in the breeding success of individuals belonging to a given population (i.e., an effect at a high organizational level) is observed, it is fairly straightforward to predict the subsequent effect on population trends, but it may be difficult to understand the causes (i.e., identify the environmental stressors) of such low breeding performance. This gap may, however, be filled by considering physiological factors. In so doing, we need to balance the limitations in understanding the molecular mode of action of an environmental stressor, with the possibility to extrapolate the consequences of the observed changes in a given biomarker to the population and community levels, i.e., the ecological significance of the observed biomarker response. For instance, changes in hormone levels in breeding parent birds (e.g., prolactin)

[1] University of Coimbra, MARE - Marine and Environmental Sciences Centre, Department of Life Sciences, Coimbra, Portugal.
Email: Pmaraujo.ecotop@gmail.com
[2] Unité Physiologie Moléculaire et Adaptation, UMR 7221, Muséum National d'Histoire Naturelle, CNRS, Paris, France.
Email: david.costantini@mnhn.fr
* Corresponding author: acgnorte@ci.uc.pt

can be the mechanistic link between potential environmental stressors (e.g., contaminants) that act as endocrine disruptors, and the lower breeding success observed at the population level (Verreault et al. 2008, Angelier and Chastel 2009, Tartu et al. 2015). Conservation physiology is a new discipline that relies on physiological measurements (conservation physiology toolkit; Seebacher and Franklin 2012, Cooke et al. 2013, Madliger et al. 2018) at the individual level to identify relevant stressors, to assess and predict consequences for individual fitness and population dynamics, and to implement this information into conservation programs. These physiological measurements differ in terms of specificity of the response, which is related to the mechanism of action of a given stressor upon a biological system on one side, and on its link to fitness (i.e., reproductive output and survival) on the other side. Often, researchers opt for an ample panel of physiological measurements so that the stressor in action can be more easily identified and linked to fitness (Mallory et al. 2010, Maceda-Veiga et al. 2015, Madliger et al. 2018).

Traditionally, assessing the impacts of a given environmental change on the health of seabird populations and species has been evaluated by the long-term monitoring of population demography in their breeding colonies (Jouventin and Weimerskirch 1990). However, long term studies need a large human, logistical and economic investment, and often show long temporal lags between environmental and population trend variations, which impairs the predictive, proactive or timely-response management actions (Sorenson et al. 2017). This limitation can be overcome by the use of physiological biomarkers. These biomarkers can be classified depending on physiological subdiscipline, organizational level, type of molecule, biological function or responsiveness to particular stressors (Madliger et al. 2018). Biochemical evaluations may encompass both nutrition-related metabolite assessments and xenobiotic-(i.e., a substance that does not occur naturally, or may occur naturally but at significantly lower concentrations) induced alterations of enzymatic activities. An endocrine biomarker can provide information on stress physiology (e.g., glucocorticoids and mineralocorticoids), reproductive status (e.g., prolactin, progestins, androgens and oestrogens), nutritional state and metabolic rate (e.g., thyroid hormones) and exposure to xenobiotics (Madliger et al. 2016). An often used haematological biomarker to assess the presence of a wide variety of stressors in wild bird populations is the heterophil: lymphocyte (H:L) ratio. This is a general biomarker that increases when an organism is being exposed to several stressful stimuli (Maxwell 1993). This increase in H:L is due to the release of corticosterone, the main glucocorticoid in birds, in response to the activation of the Hypothalamus-Pituitary-Adrenals (HPA) axis, after the perception of a stressor. In turn, corticosterone induces the release of heterophils into the blood stream from the bone marrow, and transmigration of lymphocytes from circulation into other tissues; this may happen in situations of predator sighting, competition, heat stress and food shortage, among others (Davis et al. 2008). In seabirds, H:L has been mainly used to assess energetic trade-offs during breeding (Vleck et al. 2000, Wojczulanis-Jakubas et al. 2015, Kulaszewicz et al. 2017) and exposure to contaminants. However, because the immune status may greatly affect the circulating white blood cells' profile, changes in H:L ratios may be influenced by infection and inflammation processes, which are intrinsically linked to immunity. More recently, there has been growing interest in dynamics of telomeres (i.e., changes in their length over time). Telomeres are nucleoprotein structures located at the terminal ends of chromosomes that play a major role in maintaining genome stability and in avoiding loss of genetic material (O'Sullivan and Karlseder 2010). A major pillar underlying the ecological research on telomeres is that their length might integrate the individual history of wear and tear, thus representing a valuable metric of individual phenotypic quality. For example, changes in telomere length were associated with senescence (Foote et al. 2011), glucocorticoid secretion (Herborn et al. 2014, Angelier et al. 2018), reproductive effort (Bauch et al. 2013, Bichet et al. 2020), environmental quality (Young et al. 2013), and chemical contamination (Blévin et al. 2017, Sebastiano et al. 2020) in seabirds.

In this chapter, we will focus on examples of physiological biomarkers, that may be measured in samples collected in a minimally invasive manner (Fig. A3.1a,b), and that have been successfully

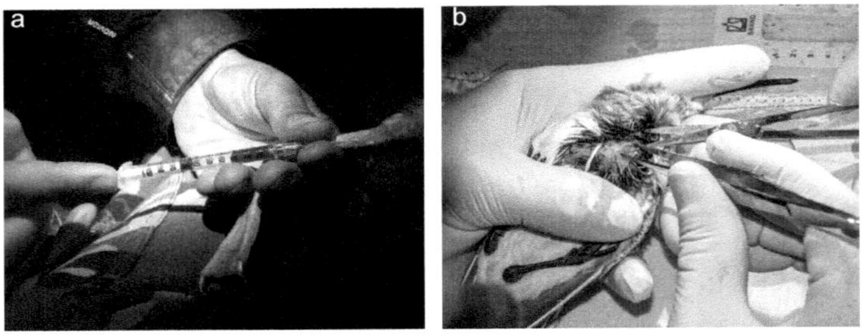

Figure A3.1. Minimally invasive biological samples for physiological assessments. (a) Blood collection from the metatarsal vein, and (b) non-invasive fat biopsy technique from the furcular fat deposits. Photo credits: (a) Jorge Pereira and (b) Pedro Miguel Araújo.

applied to seabird conservation to identify stressors acting upon populations, while explaining the mechanistic mode of action of environmental pressures on the observed responses. We will highlight physiological responses observed in seabirds in response to food availability and nutritional state, contamination, direct human disturbance, parasitism and disease and climate change-induced heat stress.

Food availability and nutritional state

Seabirds are central place foragers (Orians and Pearson 1979), and therefore very sensitive to changes in food availability in the colony surroundings, particularly when feeding chicks. Most species are also colonial breeders and often face high levels of intraspecific competition during reproduction (Cairns 1989). At the individual level, physiological biomarkers that may prove useful to reflect food availability include body mass/condition index, protein, lipid and carbohydrate metabolites in blood (Totzke et al. 1999, Quillfeldt et al. 2004, Bauch et al. 2010, Maceda-Veiga et al. 2015, Storey et al. 2017, Grilli et al. 2018; see below), haemoglobin levels (reviewed in Johnstone et al. 2015, Minias 2015), immunocompetence indicators (Norris and Evans 2000, Gasparini et al. 2006), oxidative stress biomarkers (Beaulieu and Costantini 2014, Costantini 2008), plasma thyroxin levels (Totzke et al. 1999) and glucocorticoids.

Glucocorticoids

Glucocorticoids are hormones released on stimulation of the HPA axis during stressful events (e.g., unpredictable environmental events such as food shortages and/or predictable life-history events such as wintering, migration and breeding), to help mediating energetic and resource allocation among body functions that regulate body maintenance, immunity, growth and reproduction (Siegel 1980, Wingfield and Kitaysky 2002, Romero 2004). Therefore, although they are not specific biomarkers of nutritional state and limited food resources, they are the most often used biomarker to assess pressure on energy balance in seabirds (e.g., in blood, feathers or faeces, Quillfeldt and Möstl 2003, Will et al. 2014, Fairhust et al. 2015). Corticosterone (CORT) is the main glucocorticoid in birds (Holmes and Phillips 1976). It may vary among seabird colonies and years differing in food availability and mediates trade-offs between current and future reproduction or survival (Quillfeldt and Möstl 2003, Storey et al. 2017), which is a delicate compromise for long lived species such as seabirds, given their slow reproducing rates and extended period of chick dependence. Kitaysky et al. (1999) compared CORT levels of Black-legged Kittiwakes (*Rissa trydactyla*) breeders, from egg laying to chick rearing, between two colonies located on two islands of Alaska. These islands showed contrasting food availabilities. Food shortages had been a recognised problem in the area, with some colonies of seabirds in Alaska showing chronic food shortages, and limited breeding success over 25 years (Hatch et al. 1993). The baseline and stress-induced CORT levels were evaluated for birds of the two colonies with a capture-restraint sampling protocol (Kitaysky et al. 1999, Romero 2004). Baseline CORT levels correlated negatively with body condition index, were higher for individuals from the colony with poorer feeding conditions, and increased more strongly during reproduction in this colony (Kitaysky et al. 1999). Kitaysky et al. also showed that kittiwakes breeding in the colony with poor food conditions had suppressed acute stress responses that allowed continued investment in reproduction, even during periods of limited food resources. However, this suppression likely impaired post-breeding survival of parent kittiwakes owing to potential metabolic and energetic costs associated with a low CORT production (Kitaysky et al. 1999). The link between nutritional stress during reproduction and population trends mediated by CORT secretion in seabirds was also confirmed by breeding population size estimations of Common Murres (*Uria algae*) in Alaska during two decades, and by assessing both breeding individuals' CORT levels and changes in food availability in the foraging areas during 5 years (Kitaysky et al. 2007, Kitaysky et al. 2010). This long time series of data showed that CORT variation was largely explained by variation in food supply, independently of colony and breeding stage. The authors also found a persistent negative relationship between CORT secretion and fecundity. Additionally, they found that in a declining colony, CORT levels were higher on average and individuals that had higher CORT levels were less likely to be re-sighted. These series of studies were able to identify changes in food abundance as a major factor affecting CORT, and provided direct support for the hypothesis that food-related stress during reproduction can contribute to the decreased fecundity of seabirds and potentially their survival as well.

A meta-analysis on the suitability of baseline CORT to predict fitness metrics in seabird species concluded that food availability and reproductive success show overall significant negative trends with an increase in baseline CORT (Sorenson et al. 2017). However, glucocorticoid biomarkers are complex and highly context dependent, and their use and interpretation as indicators of fitness should be careful and paired with other metrics of energetic state. In addition, there are several "downstream" physiological metrics in the HPA activation cascade that may be assessed, and may, inclusively, be more directly linked to challenging conditions: those include oxidative status biomarkers, immunological and inflammatory biomarkers, and reproductive hormones, among others (Breuner et al. 2013). For instance, antioxidant defences were shown to be highly correlated with population trends in colonies of breeding Gentoo (*Pygoscelis papua*) and Adélie penguins (*Pygoscelis adeliae*) in the Antarctic Peninsula, possibly reflecting the availability of krill in the colony surroundings (Beaulieu et al. 2013). The types of prey may affect circulating antioxidant defences or oxidative damage of seabirds (e.g., García-Tarrasón et al. 2014, Beaulieu et al. 2015). However, the relationships between dietary antioxidants and species-specific antioxidant status might not be always so straightforward to predict, possibly owing to the important contribution of endogenously synthesised antioxidants (Colominas-Ciuró et al. 2021).

Lipids

The analyses of particular dietary macromolecules could be complementary to those of glucocorticoids in order to have a better insight into nutritional stress. Lipids are among the most important energetic nutritional compounds in an individual's diet, as they are essential compounds for key physiological functions (Hebert et al. 2009, Parrish 2009, Colombo et al. 2017). Lipids are crucial in the function of all biological membranes that work as barriers for cells and cell organelles, and facilitate some basic processes such as respiration, signal transduction and motility (Palsdottir and Hunte 2004). Furthermore, neutral lipids in triglycerides are the main form of fatty acids' storage and constitute main energetic reserves in all animals (Wolins et al. 2006). The majority of fat is stored in subcutaneous deposits and the amount of fat stores and corresponding Fatty Acid (FA) composition can be related to changes in food supplies and diet (Price and Guglielmo 2009, Guglielmo 2010). Unlike mammals and other birds, in seabirds, foraging and offspring rearing involve endurance flights that are mainly fuelled by the oxidation of FA stored in adipose tissue, and delivered to muscles by the circulatory system (McWilliams et al. 2004). The ability to mobilize and catabolize extracellular fat is crucial to meet the energetic demands of muscles during such intense activity, such as offspring rearing or migration. Factors such as FA chain length and degrees of unsaturation can affect the rate of mobilization, transport and oxidation of lipid reserves during flight, thus affecting the foraging behaviour and overall fitness of seabird species (Maillet and Weber 2006). Nonetheless, little is known about the role of different fatty acids in the various phases of seabirds' annual cycle, leading to important gaps in knowledge about limitation and requirements of dietary sources on their physiology.

All terrestrial vertebrates, including birds, cannot synthesize *de novo* polyunsaturated fatty acids (PUFAs-essential FAs in vertebrate metabolism) by themselves, and the proportion of PUFAs in their tissues therefore reflects the PUFA content in the diet (Hennessy et al. 2021). There are two particularly important types of PUFA, omega-3 and omega-6. Vertebrates lack the ability to interconvert these two types of PUFA, and both must be acquired through diet. FA can be assessed quantitative or qualitatively, allowing inter- and intraspecific comparisons (Iverson 2009, Williams et al. 2009, Williams and Buck 2010). These analyses can be conducted in different tissues such as adipose tissue, blood and eggs, involving generally non-lethal and little-invasive sampling techniques (Williams and Buck 2010). Rocha et al. (2016) developed a non-lethal methodology to collect fat samples through a small incision in the furcular region (Fig. A3.1b), which allows to analyse the fat tissue composition without sacrificing birds. The main advantage in analysing FA through fat samples collected from live birds is to examine the dietary variation of the same individuals across their annual cycle. Therefore, FA profiles may provide important insights on the foraging patterns of predators, within or between species/populations, and be used as warning indicators of prey availability, population trends and changes in food web dynamics (Hebert et al. 2009). Using plasma and adipose tissue, Iverson et al. (2004) developed the Quantitative Fatty Acid Signature Analysis (QFASA, see Ceia et al. in this book) to quantify the proportions of different prey species in a predator's diet. With Nuclear Magnetic Resonance (NMR), FA profile can be estimated (in percentage) due to lipid species such as n-3 omega, monounsaturated fatty acids (MUFAs) and polyunsaturated fatty acids (PUFA) providing distinguishable peaks in specific regions of the NMR spectrum, which can be quantified by comparison with an internal pyrazine standard (Duarte et al. 2014). This methodology was used to evaluate the importance of fatty acid composition in the breeding success and migratory strategies of waders (Viegas et al. 2017, Araújo et al. 2019), and could theoretically be transposed to seabirds.

Integrating different metrics of food availability and nutritional state

Recently, the development of transcriptomic approaches and RNA sequencing technology allowed great advances in dietary studies, mainly because the quantification of gene expression may foster the evaluation of dietary preferences of predators, including seabirds (Kleinschmidt et al. 2019, Young et al. 2020). A comprehensive understanding of the molecular pathways controlling lipid deposition can be crucial to understand the genetic mechanisms regulating fuelling during breeding and pre-migration in seabirds when supported with other methodologies. By coupling analysis of traditional physiological indicators of energy metabolism (e.g., glucose, triacylglycerides, non-esterified fatty acids) with NMR, Gas Chromatography–Mass Spectrometry (GC-MS) and RNA sequencing technology to determine the composition of Fatty Acids Signatures (FAS), we can better understand a predators' energetic status and correlate it with several fitness parameters. Several studies have already combined these dietary methods to assess intrinsic and extrinsic factors that may influence diet specialisation in colonially breeding seabird species (Barrett et al. 2007, Connan et al. 2017, King et al. 2017, Young et al. 2020). King et al. (2017) used essential fatty acids, stable isotopes and regurgitates to detect subtle differences in the diet of Double-crested Cormorants

(*Phalacrocorax auritus*) foraging in different sites, which may underlie differences in exposure to contaminants and their effects. Connan et al. (2017) analysed stable isotopes and fatty acids to understand how the diet of Cape Gannets (*Morus capensis*) reflected intrinsic factors related with sex and the costs of egg production. Overall, comprehensive long-term diet studies encompassing these techniques will truly contribute to improve our understanding of how the apparent variation in food availability affects seabird populations. This is an important area of research given the strong declines of many seabird populations, which have been hypothesized to result from nutritional stress affecting reproduction and survival of individuals, and this could be exacerbated by climate change (Keogan et al. 2018).

Human disturbance

Direct human disturbance may be detrimental during the reproduction of seabirds, for instance for those species nesting in sandy beaches used for recreational purposes (Medeiros et al. 2007), or for penguins in colonies visited by tourists (Yorio et al. 2001). The assessment of how stressful human presence may be is a priority, especially in colonies characterized by a low reproductive performance (Watson et al. 2014, Ellenberg et al. 2007, Carey 2009).

Glucocorticoids

In most wild vertebrates, the perception of human presence is expected to activate the HPA leading to CORT release. This hormone has, therefore, been used in birds to evaluate stress response to human disturbance (resulting from research or touristic activities; Fowler 1999, Walker et al. 2006, Brewer et al. 2008, Poisbleau et al. 2009, Watson et al. 2014, Soldatini et al. 2015). Fowler (1999) showed that CORT levels of Magellanic Penguins (*Spheniscus magellanicus*) from highly visited breeding colonies did not increase when penguins were exposed to human presence. Conversely, in penguins that were not accustomed to human presence, even when no behavioural responses were detectable by observation (Yorio and Boersma 1992), blood CORT levels increased significantly (see also Walker et al. 2006). The time needed for this habituation varies, depending on the degree of human visitation pressure and its predictability, but may take years to develop in colonies with moderate levels of visitation (Yorio and Boersma 1992, Fowler 1999, Viblanc et al. 2012). However, in colonies where visitation is unregulated, Yellow-eyed Penguins (*Megadyptes antipodes*) breeding in a high tourist-exposed site show increased CORT levels, with no signs of habituation, and presented a reduced breeding success in comparison with an undisturbed site (Ellenberg et al. 2007). One consequence of chronic CORT production due to human disturbance may be a poor immune and general health condition, as shown in Magellanic Penguins breeding in touristic areas (Palacios et al. 2018), which may contribute to a poor reproductive performance.

Data-loggers

Halfway between changes in blood hormone levels and the visual observation of behavioural changes in response to human disturbance, lies the use of bio-loggers. This can be applied with different degrees of invasiveness to record subtle changes in physiology. One of the first studies to use such loggers applied safety-pin electrocardiogram electrodes and implantable radio-transmitters in Adélie Penguins (*Pygoscelis adeliae*) to register heart rate in response to the presence of other birds around the nest, and anthropogenic stimuli (Culik et al. 1990; but see Stout and Schwab 1980 where they evaluated Ring-billed Gull's (*Larus delawarensis*) heart rate responses to intentional dispersal procedures). Heart Rate (HR) loggers, similar to those used in sports, have also been tapped to the body of penguins (Viblanc et al. 2012). Ellenberg et al. (2013) used an even less invasive bio-logger to monitor HR of the endangered Yellow-eyed Penguin, using a dummy egg equipped with a microphone. Egg dummies had been previously used in penguins (Nimon et al. 1995, Nimon et al. 1996, Giese et al. 1999) and other seabirds (e.g., terns and gulls; Arnold et al. 2011) with slight modifications in the type of sensor (infrared sensor versus microphone). This fake egg is added to the clutch (or substitutes the real egg without any detrimental impact on incubation if the real egg is simultaneously artificially incubated and given back to parents before hatching) to register the incubating parent's HR (Fig. A3.2). With this device, Ellenberg et al. (2013) showed that the duration of disturbance and type of human behaviour were the most important factors explaining increased heart rates in Yellow-eyed Penguins. A human stopping for 1 minute elicited a HR increase (above that registered during nest maintenance) equivalent to a nest check by a researcher, and a photographer "crawling around" for 20 minutes increased the HR as high as that registered when the penguin was handled by a researcher. Depending on the sensitivity of the microphone and tailored pipelines for data conversion (e.g., through machine learning approaches), there are numerous behavioural traits that these devices can collect (e.g., respiration rates, gular fluttering behaviours, among others), and valuable information can be retrieved if linked with fitness measures.

Ocean-based human activities, such as recreation, ship traffic, offshore drilling and renewable energy sources, produce aerial and underwater noise that may impact foraging and resting seabirds. The effects of underwater noise have been addressed in marine mammals (Erbe et al. 2019), but there are almost no studies on seabirds (Hansen et al. 2020). Bio-loggers registering heart rate (Weimerskirch et al. 2018) and/or behavioural responses and satellite transmitters registering take-off/flight behaviour (Burger et al. 2019) could be useful to study the impacts of anthropogenic noise on seabirds. Thermal imaging has been used to evaluate thermoregulation in wild vertebrates (Veghte and Herreid 1965, Cilulko et al. 2013, Walcott et al 2020), and can also be applied to monitor physiological stress in birds, thus providing a non-invasive complement to physiological data gathering (McCafferty 2013, Jerem et al. 2019). As a consequence of the HPA-axis activation, rapid changes in the pattern of blood flow occur, increasing from the periphery to the core of the body via sympathetically mediated vasoconstriction and reducing skin temperature. This influx of peripheral blood to the core, along with stress-induced thermogenesis, increases core temperature. Then, a regulation of a new, higher thermoregulatory set-point for core body temperature raises the peripheral temperature of the bird (Herbornetal 2015, Jerem et al. 2019). A

Figure A3.2. Egg dummy (on the left) placed in a Yellow-legged Gull (*Larus michahellis*) nest located in a building's roof top. Photo credit: Ana Cláudia Norte.

study with Blue Tit (*Cyanistes caeruleus*) showed that baseline corticosterone predicted eye region surface temperature changes during acute stress, with no influence of environmental conditions (Jerem et al. 2019). This may be a promising technique to evaluate physiological stress also for coastal breeding species and grounded seabirds.

Chemical contamination

Industrialization and agriculture are responsible for the release of tons of toxic chemical compounds into the environment, which become imbedded into food webs worldwide. Aquatic and marine environments are particularly at risk of chemical contamination because of the rapid movement of contaminants in water. Biomagnification of many types of contaminants along the food webs can also lead to extremely high concentrations in piscivorous species at the top of the trophic networks. Therefore, marine birds are particularly exposed to a wide range of chemicals, including heavy metals, organochlorides, oil products and plastics and other forms of contamination because they spend most of their time in aquatic environments, where they can be exposed by direct external contact, inhalation, and particularly, ingestion of food and water (Schreiber and Burger 2001, Lopes et al. of this book). The classical example of the detrimental effects of biomagnifying contaminants on seabirds is that of organochlorine pesticides (DDT) that induced egg-shell thinning in raptors but also in fish-eating birds such as cormorants, gannets and terns (Rattner et al. 1984, Walker 1990, Burger and Gochfeld 2004). In addition, contaminants may also act as endocrine disruptors (Verreault et al. 2004, Jenssen 2006, Verreault et al. 2008, Tartu et al. 2015, Blévin et al. 2017, Sebastiano et al. 2021) or neurodisruptors (e.g., cholinesterases inhibition; Oropesa et al. 2007), cause metabolic rate imbalance and behavioural abnormalities, and detrimentally impact reproduction (see above; Bustnes et al. 2001). Some chemicals may cause anaemia: e.g., lead inhibits the activity of the delta-aminolevulinic acid dehydratase (ALAD), an enzyme responsible for the synthesis of haeme (a component of haemoglobin; Walton et al. 1997, Franson et al. 2000), and Polycyclic Aromatic Hydrocarbons (PAHs), present for instance in crude oil, cause haemolytic anaemia via oxidative attack of erythrocytes by PAH metabolites resulting in haemoglobin leakage and formation of Heinz bodies (Troisi et al. 2007). Contaminants may depress immunity (see below; Briggs et al. 1996, Grasman 2002) and increase oxidative stress through processes involved in their metabolism and detoxification, thereby potentially leading to accelerated senescence (Parkinson 1996, Hansen 2003). Porphyrin levels in excreta have been reported as a good indicator of exposure to persistent organic contaminants and metals, and indicate a disruption of the haeme synthesis and pro-oxidant status of tissue cells (Fox et al. 1988, Woods 1995, Jara-Carrasco et al. 2016). These, and other physiological consequences of contaminant exposure in wildlife have already been reviewed (Fossi et al. 1999, Schreiber and Burger 2001, Burger and Gochfeld 2004, Isaksson 2010, Maceda-Veiga et al. 2015).

Growing evidence points to the utility of oxidative status biomarkers as reliable bioindicators of exposure to several types of chemical contaminants (Isaksson 2010, Koivula and Eeva 2010, Einor et al. 2016). Generation of oxidative damage to biomolecules (e.g., lipids, proteins, nucleic acids), increased ROS production, decreased GSH:GSSG, or up/down-regulation of antioxidant enzymes all reflect a perturbation of the oxidative status, and each of them may have significant fitness consequences for the organism, such as a detrimental impact on growth, fertility or senescence (Costantini 2019). Molecular evidence provides mechanistic support for a direct role of some contaminants in generating perturbations of the oxidative status. For example, metals, such as cadmium, mercury, lead and arsenic, can induce cleavage of peroxides and organic hydroperoxides through the Fenton reaction, leading to the generation of new free radicals that can fuel oxidation processes (Halliwell and Gutteridge 2015). Metals can also bind to thiol groups, such as the endogenous antioxidant glutathione, inducing oxidation of the thiol group and depleting antioxidant defenses, which could promote an imbalance between pro- and antioxidants. In contrast, polychlorinated biphenyls (PCBs), such as PCB 77 or PCB 126, can increase the generation of free radicals by binding to the Aryl hydrocarbon Receptor (AhR), and activating the enzyme cytochrome P450 1A (Hennig et al. 1999, Hennig et al. 2002), a subfamily of enzymes that contribute to metabolize contaminants. These interactions between PCBs and receptors can uncouple the catalytic cycle of the enzyme CYP1A, allowing the heme iron within the active site of this enzyme complex to undergo cycles of oxidation and reduction, and act as a Fenton catalyst, generating free radicals (Schlezinger et al. 1999). Also, the pro-oxidant effects of pesticides represent the central mechanism that unifies their toxicity (Banerjee et al. 2001). Further to the direct pro-oxidant effects of certain types of contaminants, oxidative damage can also be generated indirectly through the reactions that occur as a part of the organism's detoxification response. A growing number of studies are showing an association between oxidative status biomarkers and fitness-related traits in seabirds. For instance, higher lipid oxidative damage was found to be negatively related to future breeding probabilities in male Wandering Albatrosses (*Diomedea exulans*) (Costantini et al. 2015). Similarly, Scopoli's Shearwaters (*Calonectris diomedea*) with higher levels of a biomarker of oxidative damage had a lower resight probability, and a lower number of chicks raised in the next seven years (Costantini and Dell'Omo 2015). Thus, oxidative status biomarkers might offer a good toolbox to address the sublethal effects of contaminant exposure on the reproduction and survival of seabirds.

A large number of studies found significant relationships among biomarkers of oxidative status and several types of chemical contaminants in seabirds. Moreover, these studies enabled to identify the contaminants that might have had a stronger effect on the bird physiology. For example, although the plasma concentration of lipid oxidative damage was not associated with concentrations of pesticides or PCBs, it increased with red blood cell mercury concentration in female but not in male Wandering Albatrosses (Costantini et al. 2014). In contrast, while lipid oxidative damage was not associated with mercury in Black-vented Shearwaters (*Puffinus opisthomelas*), individuals with higher mercury concentrations showed a lower activity of the antioxidant enzyme glutathione peroxidase, and a lower non-enzymatic antioxidant capacity in plasma (Soldatini et al. 2020). In female Scopoli's Shearwaters, the activity of the antioxidant enzyme superoxide dismutase was higher in those individuals having higher plasma concentrations of β-HCH or HCB, but was not associated with other persistent organic contaminants (Costantini et al. 2017). The assessment of oxidative status biomarkers in male Arctic Black-legged Kittiwakes (*Rissa tridactyla*) enabled to infer that the physiological effects of perfluoroalkyl and polyfluoroalkyl substances (PFASs) might be stronger for the long-chain congeners (Costantini et al. 2019): (a) higher protein oxidative damage was found in birds having higher concentrations of perfluorododecanoic acid (PFDoA), perfluorotridecanoic acid (PFTriA) and perfluorotetradecanoic acid (PFTeA), and (b) lower plasmatic non-enzymatic micro-molecular antioxidants were found in birds having higher concentrations of perfluoroundecanoic acid (PFUnA), PFDoA and PFTeA. Besides oxidative damage on proteins and lipids, free radicals can also damage DNA, structurally causing DNA strand breaks, forming DNA cross-links and DNA adducts, among others, and affect DNA repair and replication processes leading to mutation and epigenetic changes (Bonisoli-Alquati 2014). Genotoxicity is usually evaluated using molecular approaches that are laborious or that require expensive equipment that is not always available (e.g., Comet assay, quantitative PCR assays to evaluate telomere length, chromatography, flow cytometry; see Bonisoli-Alquati 2014 for a review). However, some authors have assessed Erythrocyte Nuclear Abnormalities (ENAs) by observing stained blood smears under the light or fluorescence microscope, to evaluate genomic damage caused by chromosome breakage and centromere or spindle dysfunction during cell division resulting from exposure to genotoxic substances or radiation (Müller et al. 1996, Quirós et al. 2008). This test is generally considered to be reliable and sensitive, especially if used in bird species with low levels of spontaneous micronuclei (Zúñiga-González et al. 2001, Bonisoli-Alquati 2014). Although this technique is time-consuming and requires an experienced observer, it has been useful to show the presence of genomic damage in several penguin species in colonies where high levels of heavy metals and persistent organic contaminants occur (De Mas et al. 2015, Barbosa et al. 2013). ENAs were also shown to differ between closely-related penguin species breeding in sympatry but with contrasting population trends, as they were found to be higher in Adélie Penguins than in sympatric Gentoo Penguins (*Pygoscelis papua*). These results suggest that Adélie Penguins are more sensitive to contamination, and ENAs were suggested to represent a good indicator of population trends and sensitivity to ongoing global changes (D'Amico et al. 2016). Micronuclei (a type of ENA) in Common Terns (*Sterna hirundo*) were also found to reflect the levels of industrialization around their colonies, and probably contaminant exposure in the Mediterranean coast of Tunisia. The chicks from colonies in more industrialized areas also showed depleted antioxidant defence mechanisms and higher lipid peroxidation, and their micronuclei levels were negatively correlated with their body mass (Oudi et al. 2019).

In conclusion, these studies show that it is important to rely on a carefully selected panel of oxidative status biomarkers tailored by study goals, in order to identify the types of chemical contaminants that have a larger impact on bird physiology. A multi-biomarker approach would also enable to better understand the interactions that occur among biomarkers, and why biomarkers do not respond consistently to a given contaminant. This approach enables to assess *a posteriori* the effects of contaminants on birds. It is, however, also important to explore the fitness consequences of changes in given oxidative status biomarkers. In doing so, it would be possible to evaluate *a priori* perspectives of reproduction or survival of individuals being exposed to different types or amounts of contaminants.

Parasitism, disease and their interactions with other environmental stressors

Seabirds have often been used as sentinels for pathogens. Because of their large range of movements, they may be exposed to a larger diversity of parasites and may transport infectious disease agents to remote areas such as the Antarctic (Woods et al. 2009, Grimaldi et al. 2010), contributing to what is known as "pathogen pollution". Disease emergence in wildlife is a growing problem in the current scenario of increased movements and translocation events of wild species and wildlife-related products, and encroachment between wildlife habitats and urbanised areas. Usually, disease is recognised as a problem affecting bird populations when mass mortality events are detected (Newman et al. 2007). Massive seabird die-offs are more likely to increase awareness of the general public to disease and related environmental stressors. However, it is more effective (and preventive) to perform disease surveillance through tests that look for the presence of antibodies to specific pathogens or for genetic material of given pathogens. Such approaches would improve our understanding about disease epidemiology (Mörner et al. 2002), and parasite pathogenicity in seabirds.

Humans and their domestic animals can be a source of infectious agents as suggested in studies with terrestrial mammals and some species of birds that explore urban environments (Daszak 2000, Dobson and Foufopoulos 2001, Smith et al. 2009, Watson et al. 2017). Therefore, species suffering from higher parasite exposure are those exploring coastal habitats and with opportunistic habits, rather than pelagic species that spend long periods at sea and breed in remote oceanic islands. Charadriiformes and Procellariiformes are the most studied orders regarding their parasites (Khan et al. 2019). Seabirds are commonly affected by several diseases, whose incidence and outbreaks may be facilitated by overcrowding in breeding colonies that favours the transmission of directly and indirectly (vector-) transmitted infectious diseases. Those common diseases include those caused by: (a) bacteria—gastrointestinal (*Salmonella* sp., *Escherichia coli*, *Campylobacter* sp.), *Chlamydia* sp., *Pasteurella multocida* (the etiologic agent of avian cholera) and *Clostridium botulinum* (that causes avian botulism); (b) gastrointestinal helminths; (c) fungi (e.g., aspergillosis); (d) protozoans (e.g., avian malaria, coccidiosis); and (e) virus, such as avian influenza, herpesvirus and arthropod-borne viral infections (e.g., West Nile virus, avian pox), reviewed in (Friend 1999,

Woods et al. 2009, Miller and Fowler 2015, Khan et al. 2019). Friend (1999) reports that most of the above mentioned diseases are frequently found in gulls and terns, with a few exceptions, such as avian pox (frequently detected in more pelagic species, e.g., Common Murres) and coccidiosis (often detected in gulls but also loons, cormorants, puffins and shearwaters). Seabirds inhabiting geographically distant polar regions are likely to be less affected by disease, given the harsh environmental conditions for some parasites and the lower biodiversity of those areas, which preclude the completion of the life cycle of some parasites that cannot establish there. However, long-distance dispersal of pathogens by other bird species visiting those areas might be one source for spill over that cannot be neglected (de Souza Petersen et al. 2017). If an exotic parasite is able to establish itself in these regions it may have highly detrimental consequences for these naïve new host populations (e.g., avian malaria infecting highly susceptible penguin populations associated with high mortality; Friend 1999, Khan et al. 2019). For example, since the 1980s, chicks of the endangered Amsterdam Albatross (*Diomedea amsterdamensis*) breeding on the remote Amsterdam Island experience recurrent and dramatic mortality episodes attributed to avian cholera, that only a recent vaccination program appeared to be able to eradicate or reduce its detrimental effects (Bourret et al. 2018).

Some physiological biomarkers can act as general indicators of disease presence and infection, which can be considered a direct consequence of such pressure, or that are not simultaneously affected by other environmental stress factors (e.g., food availability). However, many infections affecting wildlife are chronic and show no clinical significance or major impacts on population trends. Some physiological biomarkers usually used to assess disease are those related to immune function, including immunocompetence assessments, bacterial-killing capacity, leucocyte counts, inflammatory proteins and Erythrocyte Sedimentation Rate (ESR) (Norris and Evans 2000, Johnstone et al. 2015). ESR is an unspecific biomarker of inflammation and infection, as it reflects increased immunoglobulins and acute phase proteins in circulation in blood, which increase the sedimentation velocity of red blood cells in a sample placed vertically in a heparinised capillary tube (Johnstone et al. 2015). Changes in leucocyte profiles are often used as disease biomarkers in clinical practice because: (a) the number of white blood cells increases (leucocytosis) with inflammatory processes caused by bacterial, fungal or protozoan infections, (b) bacterial infections are usually characterised by heterophilia and, occasionally, a lymphocytic response, (c) fungal infections also cause marked heterophilia, frequently lymphopenia and, when chronic, monocytosis, and (d) protozoan infections, such as avian malaria, result in marked lymphocytosis (Stoskopf and Kennedy-Stoskopf 1986, Campbell 1995, Woods et al. 2009).

Seabirds are threatened by habitat change and destruction and this may further affect disease incidence and also severity, by increasing the impact of ubiquitous chronic parasites. Habitat degradation can: (a) impose nutritional challenges (discussed above) which may compromise immunocompetence due to lower energy allocated to immune function, making individuals more susceptible to disease, (b) lead to higher exposure to toxic chemicals, including those that act as immunosuppressors, (c) lead to losses of suitable habitat for breeding causing colony overcrowding and promoting the spread of infectious diseases and ectoparasite transmission, among others.

Overcrowding may also be responsible for increased aggressive behaviours, resulting in injury and inflammation which may lead to a high general stress level confirmed by increased H:L ratios, as observed in Adélie Penguins in crowded colonies (Vleck et al. 2000). The ectoparasite infestation of Black-legged Kittiwakes is a good example of how overcrowding in breeding colonies can contribute to parasitism and associated debilitating effects. This gull species usually breeds in dense colonies often infested with hard ticks (*Ixodes uriae*). Colonies showing a decreasing population trend had a higher prevalence and density of ticks, and this was hypothesised to be a consequence of lower recruitment in highly infested colonies (Boulinier and Danchin 1996). Hematophagous parasites such as ticks are known to detrimentally affect their hosts' health due to blood loss, which may lead to anaemia (that can be diagnosed by analysing erythrocyte indices—red blood cell's number and size, haematocrit and haemoglobin levels (Wanless et al. 1997, Fair et al. 2007, Johnstone et al. 2015), leading to nest desertion, chick mortality and even adult mortality. For example, a King Penguin (*Aptenodytes patagonicus*) was found to have lost an estimated volume of 350 mL of blood due to an infestation of over 4000 ticks during an incubation shift (Gauthier-Clerc et al. 1998). Anaemia may also be a consequence of blood parasites such as avian malaria haemosporidians which infect and destroy red blood cells (Campbell 1994, Johnstone et al. 2015, Minias 2015, Vanstreels et al. 2019), and were pointed as the cause of high mortality in penguins (Friend and Franson 1999, Grilo et al. 2016).

In some studies, the number of parasites has been linked to the levels of contaminants in tissues. For instance, the number of gastrointestinal nematodes positively correlated with PCB concentrations in the livers of Glaucous Gulls (*Larus hyperboreus*) (Sagerup et al. 2000), and with tissue mercury concentrations in Common Eiders (*Somateria mollissima*) (Wayland et al. 2001). However, the link between contaminant levels and immunosuppression was not unequivocally shown (Wayland et al. 2002), as suggested by the studies of Grasman et al. (1996), Grasman et al. (2000a), and Grasman et al. (2000b). These studies, using Herring Gulls (*Larus argentatus*) and Caspian Terns (*Sterna caspia*) as study model species, showed that chick cell-mediated immune response, measured with the phytohemagglutinin skin test (Lochmiller et al. 1993), was negatively correlated with PCB levels in eggs of the same clutch, and contamination-induced immunosuppression was probably the cause of increased prevalence of *Pasteurella multocida* eye infections in fish-eating birds from the Great Lakes. Similarly, Bustnes et al. (2004) found significant negative correlations between organochlorine levels and humoral immune response in female Glaucous Gulls in the Arctic. A case study that evaluated disease in a conservation context by addressing causes and consequences of a disease outbreak (at the organismal and population levels) is the study of an emergent herpesvirus disease that caused high nestling mortality in one of the most important south American populations of Magnificent Frigatebirds (*Fregata magnificens*) breeding in French Guiana (de Thoisy et al. 2009). In this species, abnormalities, i.e., nodular proliferative skin lesions, occurring in nestlings suffering from this disease (Fig. A3.3) were associated with haptoglobin levels (an inflammation biomarker) and lipid peroxidation (a biomarker of oxidative damage), and those were linked with short-term survival probability of nestlings (Sebastiano et al. 2016, 2017a,b). Increased levels of plasma haptoglobin concentration are often a sign of the onset of a non-specific immune response, but this molecule has also a role as an antioxidant (Matson et al. 2012). The potential stress factors that contributed to the emergence of this viral disease in frigatebird colonies might include high exposure to mercury contamination from naturally high soil levels and previous gold mining in the area (Sebastiano et al. 2016), and/or low food availability from the collapse of local fisheries (Sebastiano et al. 2019).

Finally, beyond the identification of physiological biomarkers that can be used to assess the impact of pathogens on seabirds, it is important to assess whether these biomarkers might also be used to assess the resilience of seabirds from the disease or the effects of vaccination programs.

Figure A3.3. Chick of Magnificent Frigatebird (*Fregata magnificens*) showing clinical signs of a herpesvirus disease. Photo credit: David Costantini.

Climate change and heat stress

Climate change can affect seabirds indirectly, due to changes in oceanic processes and spatio-temporal availability of prey (Schreiber and Burger 2001, Grémillet and Boulinier 2009), but also due to changes in habitat quality and disease incidence (Dutta 2017). However, global warming and the increased frequency of heat waves, may also affect seabirds directly. Those species breeding at high latitudes are expected to be exposed to the most severe temperature increases, while low-latitude species will be challenged because they already live close to their thermal limits, and additionally may suffer from water shortages (Wingfield et al. 2017, Cook et al. 2020). Seabirds are particularly exposed to high incident solar radiation, while foraging at sea and, some species, at their breeding sites, such as those that use open or exposed areas for breeding, often characterized by low lying vegetation, in cliffs, on the ground, on sea ice or in small shrubs/trees. Thermoregulatory constraints have been documented on distribution, habitat selection, reproductive success and survival in seabirds (reviewed in Oswald and Arnold 2012).

To cope with elevated temperatures, seabirds may adopt cooling behaviours by fluffing their feathers or drooping their wings to increase air circulation around them, pant (gular flutter) to increase evaporative heat loss (which is potentially measurable through bio-loggers, see above), and seek shade amongst other behaviours (Lustick 1984, Oswald and Arnold 2012, Cook et al. 2020). However, heat loss via evaporative cooling may be especially costly in terms of water regulation for seabirds (Hochscheid et al. 2002), because they are osmotically challenged from drinking salt water (despite having specialised salt glands), and may have limited access to fresh water (Oswald and Arnold 2012). Dehydration indicators can be useful physiological biomarkers to evaluate heat stress. Those include an increase in plasma protein levels, haematocrit, sodium and chloride ions and blood nitrogen urea, which are often assessed in veterinary clinical panels (Coles 1997, Shmueli et al. 2000, Thrall et al. 2012). Most studies on avian heat stress have been performed in broilers. Those showed increased H:L ratio, higher basophil levels and oxidative damage to cells (Altan et al. 2003). Many of these changes are mediated by an increased CORT release (Xu et al. 2018), which may have further implications in terms of behaviour, including abandonment of the breeding attempt, as shown in fasting Adélie Penguins (Spée et al. 2011).

Heat Shock Proteins (HSPs) are involved in the facilitation of repair of denatured proteins and tagging irreparable proteins for degradation. Some are constitutively expressed and present at normal temperatures with important roles in normal cell function, and others are synthesized in response to stressors including heat, through the activation of the HPA axis, enhancing thermal tolerance (Herring and Gawlik 2007, Johnstone et al. 2012). In marine ecosystems, HSPs have been mostly used to indicate heat stress in invertebrates (Dahlhoff 2004). HSPs have been assessed in different species and populations of penguins in the Antarctic peninsula, and their pattern of variation was suggested to be a consequence of temperature variation, human impact, immunity and UV radiation (Barbosa et al. 2007). Therefore, consequences of coping with heat stress constitute a conservation physiology field of research that warrants further attention using seabirds as model study species.

Final notes and recommendations

We overviewed several physiological biomarkers that can be evaluated in seabirds in a conservation context for a better mechanistic understanding of the overall health state and demographic trends of a population, using individual sampling as a strategy. Most of these biomarkers are evaluated in blood samples because this is a medium for the circulation of several metabolites, hormones, enzymes and cells that provide valuable information on an organism's physiological state in short- to a long-term timeframe. Recently, ecologists started using point of care devices that enable the obtention of several biochemical metrics in a matter of minutes from sampling (Stoot et al. 2014, Morales et al. 2019), but the use of these devices needs careful validation for the study species (Stoot et al. 2014). These can be especially useful when working at remote locations where laboratory facilities are not available, and where preservation and later transport of biological samples (generally at freezing temperatures) can be challenging. Collecting blood samples, although implying capturing and holding the birds by trained and accredited personnel, is relatively little invasive and a large sample size can be obtained in colonial seabirds with high breeding site fidelity (that allows for longitudinal studies). The drawback is that, for most seabird species, samples can only be obtained during the breeding season, although collection of feathers and other tissues, such as claws, may allow to measure physiological biomarkers that reflect the physiological state of birds during periods outside the breeding season. Feathers and faeces constitute other biological samples that can be obtained in a least invasive manner, and can be used for physiological evaluations of hormones and their metabolites, levels of contaminants and stable isotopes (that indicate foraging area and trophic niche, Ceia et al. of this book).

Suites of physiological biomarkers are routinely used in human and veterinary clinical practice and provide a more comprehensive insight into the health and stress status of an individual (Thrall et al. 2012, Madliger et al. 2018). Although causes for variation in these biomarkers can be identified by analogy with knowledge derived from veterinary practice (or even human health practice), there is a general lack of information on reference intervals for most seabird species in natural settings, and how these physiological biomarkers vary with age, sex and other intrinsic factors. When investigating wild populations, detecting subtle variation caused by stressors, with yet no overt signs of impaired health or morbidity, must rely on the comparison of mean values in contrasting circumstances. The sample size required in order to take robust conclusions, depends on the inter-individual variation and how strongly the stressors affect a given trait (Mallory et al. 2010). As pointed out by Mallory et al. (2010), common indices of seabird health remain largely unstudied or unreported, and few research programs have made simultaneous observations of multiple indices of health in this group. Therefore, there is a need for a conservation-physiology database with baseline information on variation of biomarkers in healthy and stressed seabirds and other animals, and on how this variation translates into different fitness outcomes.

The most studied seabird groups in terms of ecophysiology and biomonitoring include Sphenisciformes (penguins), Procellariiformes (petrels and albatrosses), Pelecaniformes (gannets and cormorants), and Charadriiformes (gulls and auks), due to their utility as sentinel taxa of wide spatial scale processes and wide distributional range (Sorenson et al. 2017). Many of these seabirds are iconic (e.g., penguins and albatrosses) and will be more effective at raising public awareness towards conservation. Studies are also biased towards species with accessible breeding colonies and nesting sites, such as coastal species, that present lower logistic limitations (especially when dealing with sensitive samples, that usually need at least a minimal laboratory processing). Exceptional research efforts have been conducted in areas in need of closer surveillance, such as those known to be threatened by anthropogenic pressures or because they are the study systems as close as possible to a pristine habitat (e.g., the Antarctic). On the other hand, there is a general lack of studies in tropical compared to temperate and polar ecosystems, because they are usually distant to research capacitated human and logistic resources, even though seabirds species occurring in these regions have usually poor or unknown conservation status, and are highly prone to suffer from climate change (Oro 2014) and nutritional stress (Behrenfeld et al. 2006).

A conservation physiology approach should, ideally, entail collaboration not only among experts in different physiological biomarkers, but also among different research areas including parasitology, veterinary medicine, toxicology, genetics, behavioural ecology and evolutionary biology (Mallory et al. 2010), so that information can be better integrated to understand how seabirds cope with environmental changes. This will allow to disentangle more easily the effects of different pressures affecting marine ecosystems, identify the causes of health decline or stress in seabird populations, help modelling seabird population statuses and dynamics, and plan and implement conservation measures with concrete recommendations. Recently developed and increasingly implemented genomic, transcriptomic and proteomic tools allowing for the screening and quantification of thousands of genes and proteins related with an array of physiological and biochemical responses, including those reflecting immunity, condition, metabolic and nutritional state, will be very useful in ecological and conservation studies to evaluate functional differences between populations, local adaptation, disease incidence and toxicological processes, among others (Madliger et al. 2016). Quantification of physiological biomarkers needs to be integrated in long-term studies to understand their relationships with seabird fitness, and long-lived seabirds with high fidelity to breeding sites present an excellent opportunity for these endeavours.

Acknowledgments

This work benefitted from funding by the strategic program of MARE, financed by FCT (UID/MAR/04292/2020), through national funds. We would like to thank Michael Beaulieu for his insightful and constructive comments that helped us to improve this chapter.

References cited

Altan, O., A. Pabuçcuoğlu, A. Altan, S. Konyalioğlu and H. Bayraktar. 2003. Effect of heat stress on oxidative stress, lipid peroxidation and some stress parameters in broilers. Br. Poult. Sci. 44: 545–50. doi: 10.1080/0007166031000161 8334.

Angelier, F. and O. Chastel. 2009. Stress, prolactin and parental investment in birds: A review. Gen. Comp. Endocrinol. 163: 142–8. doi: 10.1016/j.ygcen.2009.03.028.

Angelier, F., D. Costantini, P. Blévin and O. Chastel. 2018. Do glucocorticoids mediate the link between environmental conditions and telomere dynamics in wild vertebrates? A review. Gen. Comp. Endocrinol. 256: 99–111. doi: 10.1016/j.ygcen.2017.07.007.

Araújo, P.M., I. Viegas, A.D. Rocha, A. Villegas, J.G. Jones, L. Mendonça et al. 2019. Understanding how birds rebuild fat stores during migration: Insights from an experimental study. Sci. Rep. 9: 10065. doi: 10.1038/s41598-019-46487-z.

Arnold, J.M., R. Ordonez, D.A. Copeland, R. Nathan, J.M. Scornavacchi, D.J. Tyerman et al. 2011. Simple and inexpensive devices to measure heart rates of incubating birds. J. Field Ornithol. 82: 288–296. doi:10.1111/j.1557-9263.2011.00332.x.

Banerjee, B.D., V. Seth and R.S. Ahmed. 2001. Pesticide-induced oxidative stress: perspectives and trends. Rev. Environ. Health. 16: 1–40. doi: 10.1515/reveh.2001.16.1.1.

Barbosa, A., S. Merino, J. Benzal, J. Martínez and S. García-Fraile. 2007. Population variability in heat shock proteins among three Antarctic penguin species. Polar Biol. 30: 1239–1244. doi: 10.1007/s00300-007-0284-0.

Barbosa, A., E. de Mas, J. Benzal, J. Ines Diaz, M. Motas, S. Jerez et al. 2013. Pollution and physiological variability in gentoo penguins at two rookeries with different levels of human visitation. Antarct. Sci. 25: 329–338. doi: 10.1017/S0954102012000739.

Barrett, R.T., K. Camphuysen, T. Anker-Nilssen, J.W. Chardine, R.W. Furness, S. Garthe et al. 2007. Diet studies of seabirds: a review and recommendations. ICES J. Mar. Sci. 64: 1675–1691. doi: 10.1093/icesjms/fsm152.

Bauch, C., S. Kreutzer and P.H. Becker. 2010. Breeding experience affects condition: blood metabolite levels over the course of incubation in a seabird. J. Comp. Physiol. B 6: 835–45. doi: 10.1007/s00360-010-0453-2.

Bauch, C., P.H. Becker and S. Verhulst. 2013. Telomere length reflects phenotypic quality and costs of reproduction in a long-lived seabird. Proc. Biol. Sci. 280: 20122540–20122540. doi: 10.1098/rspb.2012.2540.

Beaulieu, M., A.-M. Thierry, D. González-Acuña and M.J. Polito. 2013. Integrating oxidative ecology into conservation physiology. Conserv. Physiol. 1: cot004. doi: 10.1093/conphys/cot004.

Beaulieu, M. and D. Costantini. 2014. Biomarkers of oxidative status: missing tools in conservation physiology. Conserv. Physiol. 2: cou014. doi: 10.1093/conphys/cou014.

Beaulieu, M., D. González-Acuña, A.-M. Thierry and M.J. Polito. 2015. Relationships between isotopic values and oxidative status: insights from populations of gentoo penguins. Oecologia. 177: 1211–1220. doi: 10.1007/s00442-015-3267-9.

Behrenfeld, M.J., R.T. O'Malley, D.A. Siegel, C.R. McClain, J.L. Sarmiento, G.C. Feldman et al. 2006. Climate-driven trends in contemporary ocean productivity. Nature 444: 752–755. doi: 10.1038/nature05317.

Bichet, C., S. Bouwhuis, C. Bauch, S. Verhulst, P.H. Becker and O. Vedder. 2020. Telomere length is repeatable, shortens with age and reproductive success, and predicts remaining lifespan in a long-lived seabird. Mol. Ecol. 29: 429–441. doi: 10.1111/mec.15331.

Blévin, P., S. Tartu, H.I. Ellis, O. Chastel, P. Bustamante, C. Parenteau et al. 2017. Contaminants and energy expenditure in an Arctic seabird: Organochlorine pesticides and perfluoroalkyl substances are associated with metabolic rate in a contrasted manner. Env. Res. 157: 118–126. doi: 10.1016/j.envres.2017.05.022.

Bonisoli-Alquati, A. 2014. Avian genetic ecotoxicology: DNA of the canary in a coalmine. Cur. Zool. 60: 285–298. doi: 10.1093/czoolo/60.2.285.

Boulinier, T. and E. Danchin. 1996. Population trends in Kittiwake *Rissa tridactyla* colonies in relation to tick infestation. Ibis 138: 326–334. doi: 10.1111/j.1474-919X.1996.tb04345.x.

Bourret, V., A. Gamble, J. Tornos, A. Jaeger, K. Delord, C. Barbraud et al. 2018. Vaccination protects endangered albatross chicks against avian cholera. Conserv. Lett. 11: e12443. doi:10.1111/conl.12443.

Breuner, C.W., B. Delehanty and R. Boonstra. 2013. Evaluating stress in natural populations of vertebrates: total CORT is not good enough. Funct. Ecol. 27: 24–36. doi: 10.1111/1365-2435.12016.

Brewer, J.H., K.M. O'Reilly and C.L. Buck. 2008. Effects of investigator disturbance on corticosterone concentrations of Black-legged Kittiwake chicks. J. Field Ornithol. 79: 391–398. doi: 10.1111/j.1557-9263.2008.00187.x.

Briggs, K.T., S.H. Yoshida and M.E. Gershwin. 1996. The influence of petrochemicals and stress on the immune system of seabirds. Regul. Toxicol. Pharmacol. 23: 145–155. doi:10.1006/rtph.1996.0036.

Burger, C., A. Schubert, S. Heinänen, M. Dorsch, B. Kleinschmidt, R. Žydelis et al. 2019. A novel approach for assessing effects of ship traffic on distributions and movements of seabirds. J. Environ. Manag. 251: 109511. doi: 10.1016/j.jenvman.2019.109511.

Burger, J. and M. Gochfeld. 2004. Marine birds as sentinels of environmental pollution. EcoHealth 1: 263–274. doi: 10.1007/s10393-004-0096-4.

Bustnes, J.O., V. Bakken, K.E. Erikstad, F. Mehlum and J.U. Skaare. 2001. Patterns of incubation and nest-site attentiveness in relation to organochlorine (PCB) contamination in glaucous gulls. J. Appl. Ecol. 38: 791–801. doi: 10.1046/j.1365-2664.2001.00633.x.

Bustnes, J.O., S.A. Hanssen, I. Folstad, K.E. Erikstad, D. Hasselquist and J.U. Skaare. 2004. Immune function and organochlorine pollutants in arctic breeding glaucous gulls. Arch. Environm. Contam. Toxicol. 47: 530–541. doi: 10.1007/s00244-003-3203-6.

Cairns, D.K. 1989. The regulation of seabird colony size: A hinterland model. Am. Nat. 134: 141–146.

Campbell, T.W. 1994. Hematology. pp. 176–191. *In*: Ritchie, B.W., G.J. Harrison and L.R. Harrison [eds.]. Avian Medicine: principles and applications. Wingers Publishing, Florida, USA.

Campbell, T.W. 1995. Avian hematology and cytology. Iowa State University Press, Ames, USA.

Carey, M. 2009. The effects of investigator disturbance on procellariiform seabirds: A review. N. Z. J. Zool. 36: 367–377. doi: 10.1080/03014220909510161.

Cilulko, J., P. Janiszewski, M. Bogdaszewski and E. Szczygielska. 2013. Infrared thermal imaging in studies of wild animals. Eur. J. Wildl. Res. 59: 17–23. https://doi.org/10.1007/s10344-012-0688-1.

Coles, B.H. 1997. Essentials of Avian Medicine and Surgery. 3rd edition. Blackwell Publishing, Oxford, U.K.

Colombo, S.M., A. Wacker, C.C. Parrish, M.J. Kainz and M.T. Arts. 2017. A fundamental dichotomy in long-chain polyunsaturated fatty acid abundance between and within marine and terrestrial ecosystems. Environm. Rev. 25: 163–174. doi: 10.1139/er-2016-0062.

Colominas-Ciuró, R., M. Bertellotti, V.L. D'Amico, E. Carabajal, J. Benzal, V. Vidal et al. 2021. Diet, antioxidants and oxidative status in pygoscelid penguins. Mar. Ecol. Prog. Ser. 665: 201–216. https://doi.org/10.3354/meps13651.

Connan, M., B.T. Bonnevie, C. Hagen, C.D. van der Lingen and C. McQuaid. 2017. Diet specialization in a colonial seabird studied using three complementary dietary techniques: effects of intrinsic and extrinsic factors. Mar. Biol. 164: 171. doi: 10.1007/s00227-017-3201-2.

Cook, T.R., R. Martin, J. Roberts, H. Häkkinen, P. Botha, C. Meyer et al. 2020. Parenting in a warming world: Thermoregulatory responses to heat stress in an endangered seabird. Conserv. Physiol. 8: coz109. doi: 10.1093/conphys/coz109.

Cooke, S.J., L. Sack, C.E. Franklin, A.P. Farrell, J. Beardall, M. Wikelski et al. 2013. What is conservation physiology? Perspectives on an increasingly integrated and essential science. Conserv. Physiol. 1: cot001. doi: 10.1093/conphys/cot001.

Costantini, D. 2008. Oxidative stress in ecology and evolution: lessons from avian studies. Ecol. Lett. 11: 1238–1251. doi: 10.1111/j.1461-0248.2008.01246.x.

Costantini, D., A. Meillere, A. Carravieri, V. Lecomte, G. Sorci, B. Faivre et al. 2014. Oxidative stress in relation to reproduction, contaminants, gender and age in a long-lived seabird.Oecologia 175: 1107–16. doi: 10.1007/s00442-014-2975-x.

Costantini, D. and G. Dell'Omo. 2015.Oxidative stress predicts long-term resight probability and reproductive success in Scopoli's shearwater (*Calonectris diomedea*). Conserv. Physiol. 3: cov024. doi: 10.1093/conphys/cov024.

Costantini, D., A. Goutte, C. Barbraud, B. Faivre, G. Sorci, H. Weimerskirch et al. 2015. Demographic responses to oxidative stress and inflammation in the wandering albatross (*Diomedea exulans*). PLoS ONE 10: e0133967. doi: 10.1371/journal.pone.0133967.

Costantini, D., M. Sebastiano, M.S. Müller, I. Eulaers, P. Ambus, G. Malarvannan et al. 2017. Individual variation of persistent organic pollutants in relation to stable isotope ratios, sex, reproductive phase and oxidative status in Scopoli's shearwaters (*Calonectris diomedea*) from the Southern Mediterranean. Sci. Total Environ. 598: 179–187. doi: 10.1016/j.scitotenv.2017.04.014.

Costantini, D. 2019. Understanding diversity in oxidative status and oxidative stress: the opportunities and challenges ahead. J. Exp. Biol. 222: jeb194688. doi: 10.1242/jeb.194688.

Costantini, D., P. Blevin, D. Herzke, B. Moe, G.W. Gabrielsen, J.O. Bustnes et al. 2019. Higher plasma oxidative damage and lower plasma antioxidant defences in an Arctic seabird exposed to longer perfluoroalkyl acids. Environ. Res. 168: 278–285. doi: 10.1016/j.envres.2018.10.003.

Culik, B., D. Adelung and A.J. Woakes. 1990. The effect of disturbance on the heart rate and behaviour of Adélie Penguins (*Pygoscelis adeliae*) during the breeding season. pp. 177–182. *In*: Hempel, G. and K.R. Kerry [eds.]. Antarctic Ecosystems. Springer, Berlin and Heidelberg, Germany.doi: 10.1007/978-3-642-84074-6_18.

D'Amico, V.L., N. Coria, M.G. Palacios, A. Barbosa and M. Bertellotti. 2016. Physiological differences between two overlapped breeding Antarctic penguins in a global change perspective. Polar Biol. 39: 57–64. doi: 10.1007/s00300-014-1604-9.

Dahlhoff, E. 2004. Biochemical indicators of stress and metabolism: Applications for marine ecological studies. Annual Rev. Physiol. 66: 183–207. doi: 10.1146/annurev.physiol.66.032102.114509.

Daszak, P. 2000. Emerging infectious diseases of wildlife—Threats to biodiversity and human health. Science 287: 443–449. doi: 10.1126/science.287.5452.443.

Davis, A.K., D.L. Maney and J.C. Maerz. 2008. The use of leukocyte profiles to measure stress in vertebrates: A review for ecologists. Funct. Ecol. 22: 760–772. doi: 10.1111/j.1365-2435.2008.01467.x.

De Mas, E., J. Benzal, S. Merino, F. Valera, M.J. Palacios, J.J. Cuervo et al. 2015. Erythrocytic abnormalities in three Antarctic penguin species along the Antarctic Peninsula: biomonitoring of genomic damage. Polar Biol. 38: 1067–1074. doi: 10.1007/s00300-015-1667-2.

de Souza P., E. Jansen de Araujo, L. Krüger, M.M. Seixas, T. Ometto, L.M. Thomazelli et al. 2017. First detection of avian influenza virus (H4N7) in Giant Petrel monitored by geolocators in the Antarctic region. Mar. Biol. 164: 62. doi: 10.1007/s00227-017-3086-0.

de Thoisy, B., A. Lavergne, J. Semelin, J.F. Pouliquen, F. Blanchard, E. Hansen et al. 2009. Outbreaks of disease possibly due to a natural avian herpesvirus infection in a colony of young Magnificent Frigatebirds (*Fregata magnificens*) in French Guiana. J. Wildl. Dis. 45: 802–807. doi: 10.7589/0090-3558-45.3.802.

Dias, M.P., R. Martin, E.J. Pearmain, I.J. Burfield, C. Small, R.A. Phillips et al. 2019. Threats to seabirds: A global assessment. Biol. Conserv. 237: 525–537. doi: 10.1016/j.biocon.2019.06.033.

Dobson, A. and J. Foufopoulos. 2001. Emerging infectious pathogens of wildlife. Philos. Trans. R. Soc. Lond. B. 356: 1001–1012. doi: 10.1098/rstb.2001.0900.

Duarte, J.A., F. Carvalho, M. Pearson, J.D. Horton, J.D. Browning, J.G. Jones et al. 2014. A high-fat diet suppresses *de novo* lipogenesis and desaturation but not elongation and triglyceride synthesis in mice. J. Lipid Res. 55: 2541–53. doi: 10.1194/jlr.M052308.

Dutta, H. 2017. Insights into the impacts of four current environmental problems on flying birds. Energy Ecol. Environ. 2: 329–349. doi: 10.1007/s40974-017-0075-6.

Einor, D., A. Bonisoli-Alquati, D. Costantini, T. Mousseau and A.P. Møller. 2016. Ionizing radiation, antioxidant response and oxidative damage: A meta-analysis. Sci. Total Envirn. 548-549: 463–471. doi: 10.1016/j.scitotenv.2016.01.027.

Ellenberg, U., A.N. Setiawan, A. Cree, D.M. Houston and P.J. Seddon. 2007. Elevated hormonal stress response and reduced reproductive output in Yellow-eyed penguins exposed to unregulated tourism. Gen. Comp. Endocrinol. 152: 54–63. doi: 10.1016/j.ygcen.2007.02.022.

Ellenberg, U., T. Mattern and P.J. Seddon. 2013. Heart rate responses provide an objective evaluation of human disturbance stimuli in breeding birds. Conserv. Physiol. 1: cot013. doi: 10.1093/conphys/cot013.

Erbe, C., S.A. Marley, R.P. Schoeman, J.N. Smith, L.E. Trigg and C.B. Embling. 2019. The effects of ship noise on marine mammals—A review. Front. Mar. Sci. 6: 606. https://doi.org/10.3389/fmars.2019.00606.

Fair, J.M., S. Whitaker and B. Pearson. 2007. Sources of variation in haematocrit in birds. Ibis 149: 535–552. doi: 10.1111/j.1474-919X.2007.00680.x.

Fairhurst, G.D., A.L. Bond, K.A. Hobson and R.A. Ronconi. 2015. Feather-based measures of stable isotopes and corticosterone reveal a relationship between trophic position and physiology in a pelagic seabird over a 153-year period. Ibis 157: 273–283. https://doi.org/10.1111/ibi.12232.

Foote, C.G., F. Daunt, J. González-Solís, L. Nasir, R.A. Phillips and P. Monaghan. 2011. Individual state and survival prospects: age, sex, and telomere length in a long-lived seabird. Behav. Ecol. 22: 156–161. doi: 10.1093/beheco/arq178.

Fossi, C., C. Leonzio and S. Focardi. 1989. Seasonal variation of mixed-function oxidase activity in a population of Yellow-legged Herring gull: Relationship to sexual cycle and pollutants. Mar. Environ. Res. 28: 35–39. doi: 10.1016/0141-1136(89)90180-3.

Fossi, M.C., S. Casini and L. Marsili. 1999. Nondestructive biomarkers of exposure to endocrine disrupting chemicals in endangered species of wildlife. Chemosphere 39: 1273–1285. doi:10.1016/S0045-6535(99)00195-2.

Fowler, G.S. 1999. Behavioral and hormonal responses of Magellanic penguins (*Spheniscus magellanicus*) to tourism and nest site visitation. Biol. Conserv. 90: 143–149. doi:10.1016/S0006-3207(99)00026-9.

Fox, G., R. Norstrom, D. Wigfield and S. Kennedy. 1988. Porphyria in herring gulls: A biochemical response to chemical contamination of Great Lakes food chains. Environm. Toxicol. Chem. 7: 831–839. doi: 10.1002/etc.5620071005.

Franson, J., T. Hollmen, R. Poppenga, M. Hario, M. Kilpi and M. Smith. 2000. Selected trace elements and organochlorines: Some findings in blood and eggs of nesting common eiders (*Somateria mollissima*) from Finland. Environm. Toxicol. Chem. 19: 1340–1347. doi: 10.1002/etc.5620190517.

Friend, M. and J.C. Franson. 1999. Field manual of wildlife diseases: General field procedures and diseases of birds. USGS Biological Resources Division National Wildlife Health Center, Madison, USA.

García-Tarrasón, M., C. Sanpera, L. Jover and D. Costantini. 2014. Levels of antioxidants in breeding female Audouin's gulls and their deposition in eggs across different environments. J. Exp. Mar. Biol. Ecol. 453: 116–122.doi: 10.1016/j.jembe.2014.01.012.

Gasparini, J., A. Roulin, V.A. Gill, S.A. Hatch and T. Boulinier. 2006. In kittiwakes food availability partially explains the seasonal decline in humoral immunocompetence. Funct. Ecol. 20: 457–463.https://doi.org/10.1111/j.1365-2435.2006.01130.x.

Gauthier-Clerc, M., Y. Clerquin and Y. Handrich. 1998. Hyperinfestation by ticks *Ixodes uriae*: A possible cause of death in adult King Penguins, a Long-Lived Seabird. Col. Waterbirds 21: 229–233. doi: 10.2307/1521910.

Giese, M., R. Handsworth and R. Stephenson. 1999. Measuring resting heart rates in penguins using an artificial egg. J. Field Ornithol. 70: 49–54. http://www.jstor.org/stable/4514381.

Grasman, K.A., G.A. Fox, P.F. Scanlon and J.P. Ludwig. 1996. Organochlorine-associated immunosuppression in prefledgling Caspian terns and herring gulls from the Great Lakes: An ecoepidemiological study. Environm. Health Persp. 104: 829–842. doi: 10.1289/ehp.96104s4829.

Grasman, K.A., M. Armstrong, D.L. Hammersley, P.F. Scanlon and G.A. Fox. 2000a. Geographic variation in blood plasma protein concentrations of young herring gulls (*Larus argentatus*) and Caspian terns (*Sterna caspia*) from the Great Lakes and Lake Winnipeg. Comp. Biochem. Physiol. C. 125: 365–375. doi:10.1016/S0742-8413(99)00118-8.

Grasman, K.A., P.F. Scanlon and G.A. Fox. 2000b. Geographic variation in hematological variables in adult and prefledgling herring gulls (*Larus argentatus*) and possible associations with organochlorine exposure. Arch. Environm. Cont. Toxicol. 38: 244–253. doi: 10.1007/s002449910032.

Grasman, K.A. 2002. Assessing immunological function in toxicological studies of avian wildlife. Integr. Comp. Biol. 42: 34–42. doi: 10.1093/icb/42.1.34.

Grémillet, D. and T. Boulinier. 2009. Spatial ecology and conservation of seabirds facing global climate change: A review. Mar. Ecol. Prog. Ser. 391: 121–137. doi: 10.3354/meps08212.

Grilli, M., M. Pari and A. Ibañez. 2018. Poor body conditions during the breeding period in a seabird population with low breeding success. Mar. Biol. 165: 142. doi: 10.1007/s00227-018-3401-4.

Grilo, M.L., R.E.T. Vanstreels, R. Wallace, D. García-Párraga, É.M. Braga, J. Chitty et al. 2016. Malaria in penguins—current perceptions. Avian Pathol. 45: 393–407. doi: 10.1080/03079457.2016.1149145.

Grimaldi, W., J. Jabour and E.J. Woehler. 2010. Considerations for minimising the spread of infectious disease in Antarctic seabirds and seals. Polar Rec. 47: 56–66. doi: 10.1017/S0032247410000100.

Guglielmo, C.G. 2010. Move that fatty acid: Fuel selection and transport in migratory birds and bats. Integr. Comp. Biol. 50: 336–345. doi: 10.1093/icb/icq097.

Halliwell, B. and J.M.C. Gutteridge. 2015. Free Radicals in Biology and Medicine. 5 ed. Oxford University Press, Oxford, UK.

Hansen, K.A., A. Hernandez, T.A. Mooney, K. Sørensen, M.H. Rasmussen and M. Wahlberg. 2020. The common murre (*Uria aalge*), an auk seabird, reacts to underwater sound. J. Acoust. Soc. Am. 147: 4069. doi: 10.1121/10.0001400.

Hansen, P.-D. 2003. Biomarkers. pp. 203–220. *In*: Markert, B., A.M. Breure and H.G. Zechmeister [eds.]. Bioindicators and biomonitors—principles, concepts and applications. Elsevier, Amsterdam, Netherlands.

Hatch, S.A., G.V. Byrd, D.B. Irons and G.L. Hunt. 1993. Status and ecology of kittiwakes (*Rissa tridactyla* and *R. brevirostris*) in the North Pacific. pp. 140–153. *In*: Briggs, K.T., K. Vermeer, K.H. Morgan and D. Siegel-Causey [eds]. The Status, Ecology, and Conservation of Marine Birds of the North Pacific. Canadian Wildlife Service, Ottawa, Canada.

Hebert, C.E., D.V. Chip Weseloh, L.T. Gauthier, M.T. Arts and R.J. Letcher. 2009. Biochemical tracers reveal intra-specific differences in the food webs utilized by individual seabirds. Oecol. 160: 15–23. doi: 10.1007/s00442-009-1285-1.

Hennessy, A.A., D.A. Kenny, C.J. Byrne, S. Childs, R.P. Ross, R. Devery et al. 2021. Fatty acid concentration of plasma, muscle, adipose and liver from beef heifers fed an encapsulated n-3 polyunsaturated fatty acid supplement. Animal 15: 100039. doi:10.1016/j.animal.2020.100039.

Hennig, B., R. Slim, M. Toborek and L.W. Robertson. 1999. Linoleic acid amplifies polychlorinated biphenyl-mediated dysfunction of endothelial cells. J. Biochem. Mol. Toxicol. 13: 83–91. doi: 10.1002/(sici)1099-0461(1999)13:2<83::aid-jbt4>3.0.co;2-7.

Hennig, B., P. Meerarani, R. Slim, M. Toborek, A. Daugherty, A.E. Silverstone et al. 2002. Proinflammatory properties of coplanar PCBs: *in vitro* and *in vivo* evidence. Toxicol. Appl. Pharmacol. 181: 174–83. doi: 10.1006/taap.2002.9408.

Herborn, K.A., B.J. Heidinger, W. Boner, J.C. Noguera, A. Adam, F. Daunt and P. Monaghan. 2014. Stress exposure in early post-natal life reduces telomere length: an experimental demonstration in a long-lived seabird. Proc. Biol. Sci. B. 281: 20133151-20133151. doi: 10.1098/rspb.2013.3151.

Herborn, K.A., J. Graves, P. Jerem, N. Evans, R. Nager, D. McCafferty et al. 2015. Skin temperature reveals the intensity of acute stress. Physiol. Behav. 152: 225–230. doi: 10.1016/j.physbeh.2015.09.032.

Herring, G. and D.E. Gawlik. 2007. The role of stress proteins in the study of allostatic overload in birds: Use and applicability to current studies in avian ecology. Sci. World J. 7: 1596–602. doi: 10.1100/tsw.2007.242.

Hochscheid, S., D. Grémillet, S. Wanless and M.A. du Plessis. 2002. Black and white under the South African sun: are juvenile Cape gannets heat stressed? J. Therm. Biol. 27: 325–332. doi: 10.1016/S0306-4565(01)00097-3.

Holmes, W.N. and J.G. Phillips. 1976. The adrenal cortex of birds. pp. 293–420. *In*: Chester-Jones, I. and I.W. Henderson [Eds.]. General, Comparative, and Clinical Endocrinology of the Adrenal Cortex. Academic Press, London, UK.

Isaksson, C. 2010. Pollution and its impact on wild animals: A meta-analysis on oxidative stress. EcoHealth 7: 342–350. doi: 10.1007/s10393-010-0345-7.

Iverson, S.J., C. Field, W.D. Bowen and W. Blanchard. 2004. Quantitative fatty acid signature analysis: A new method of estimating predator diets. Ecol. Monogr. 74: 211–235. doi:10.1890/02-4105.

Iverson, S.J. 2009. Tracing aquatic food webs using fatty acids: From qualitative indicators to quantitative determination. pp. 281–308. *In*: Kainz, M., M.T. Brett and M.T. Arts [eds.]. Lipids in Aquatic Ecosystems. Springer, New York, USA.

Jara-Carrasco, S., R. Barra, W. Espejo, J.E. Celis, D. González-Acuña, G. Chiang et al. 2016. Persistent organic pollutants and porphyrin levels in excreta of penguin colonies from the Antarctic Peninsula area. Polar Rec. 53: 79–87. doi: 10.1017/S0032247416000607.

Jenssen, B.M. 2006. Endocrine-disrupting chemicals and climate change: A worst-case combination for arctic marine mammals and seabirds? Environ. Health Perspect. 114 Suppl 1: 76–80. doi: 10.1289/ehp.8057.

Jerem, P., S. Jenni-Eiermann, D. McKeegan, D.J. McCafferty and R.G. Nager. 2019. Eye region surface temperature dynamics during acute stress relate to baseline glucocorticoids independently of environmental conditions. Physiol. Behav. 210: 112627. https://doi.org/10.1016/j.physbeh.2019.112627.

Johnstone, C., R. Reina and A. Lill. 2012. Interpreting indices of physiological stress in free-living vertebrates. J. Comp. Physiol. B 182: 861–879. doi: 10.1007/s00360-012-0656-9.

Johnstone, C.P., A. Lill and R.D. Reina. 2015. Use of erythrocyte indicators of health and condition in vertebrate ecophysiology: a review and appraisal. Biol. Rev. 92: 150–158. doi: 10.1111/brv.12219.

Jouventin, P. and H. Weimerskirch. 1990. Long-term changes in seabird and seal populations in the southern ocean. pp. 208–213. *In*: Kerry, K.R. and G. Hempel [eds.]. Antarctic Ecosystems. Springer, Berlin, Germany. doi: 10.1007/978-3-642-84074-6_22.

Keogan, K., F. Daunt, S. Wanless, R.A. Phillips, C.A. Walling, P. Agnew et al. 2018. Global phenological insensitivity to shifting ocean temperatures among seabirds. Nat. Clim. Change 8: 313–318. doi: 10.1038/s41558-018-0115-z.

Khan, J., J. Provencher, M. Forbes, M. Mallory, C. Lebarbenchon and K. McCoy. 2019. Parasites of seabirds: A survey of effects and ecological implications. pp. 1–50. *In*: Sheppard, C. [ed.]. Adv. Mar. Biol. 82: 1–50. doi: 1016/bs.amb.2019.02.001.

King, L.E., S.R. de Solla, J.R. Marentette, R.A. Lavoie, T.K. Kyser, L.M. Campbell et al. 2017. Fatty acids, stable isotopes, and regurgitate reveal diet differences between Lake Ontario and Lake Erie double-crested cormorants (*Phalacrocorax auritus*). J. Great Lakes Res. 43: 132–140. doi:1016/j.jglr.2017.03.004.

Kitaysky, A.S., J.C. Wingfield and J.F. Piatt. 1999. Dynamics of food availability, body condition and physiological stress response in breeding Black-legged Kittiwakes. Funct. Ecol. 13: 577–584. doi: 10.1046/j.1365-2435.1999.00352.x.

Kitaysky, A.S., J.F. Piatt and J.C. Wingfield. 2007. Stress hormones link food availability and population processes in seabirds. Mar. Ecol. Progr. Ser. 352: 245–258. doi: 10.3354/meps07074.

Kitaysky, A.S., J.F. Piatt, S.A. Hatch, E.V. Kitaiskaia, Z.M. Benowitz-Fredericks, M.T. Shultz et al. 2010. Food availability and population processes: Severity of nutritional stress during reproduction predicts survival of long-lived seabirds. Funct. Ecol. 24: 625–637. doi: 10.1111/j.1365-2435.2009.01679.x.

Kleinschmidt, B., C. Burger, M. Dorsch, G. Nehls, S. Heinänen, J. Morkūnas et al. 2019. The diet of red-throated divers (*Gavia stellata*) overwintering in the German Bight (North Sea) analysed using molecular diagnostics. Mar. Biol. 166: 77. doi: 10.1007/s00227-019-3523-3.

Koivula, M.J. and T. Eeva. 2010. Metal-related oxidative stress in birds. Environ. Pollut. 158: 2359–2370. doi: 10.1016/j.envpol.2010.03.013.

Kulaszewicz, I., K. Wojczulanis-Jakubas and D. Jakubas. 2017. Trade-offs between reproduction and self-maintenance (immune function and body mass) in a small seabird, the little auk. J. Avian Biol. 48: 371–379. doi: 10.1111/jav.01000.

Kursa, M. and V. Bezrukov. 2007. Health status in an antarctic top predator: Micronuclei frequency and white blood cell differentials in the South Polar Skua (*Catharacta maccormicki*). Polarforschung 77: 1–5. doi: 10.2312/polarforschung.77.1.1.

Lochmiller, R.L., M.R. Vestey and J.C. Boren. 1993. Relationship between protein nutritional status and immunocompetence in northern bobwhite chicks. Auk. 110: 503–510. doi: 10.2307/4088414.

Lustick, S. 1984. Thermoregulation in adult seabirds. pp. 183–201. *In*: Rahn, H. and G.C. Whittow [eds.]. Seabird Energetics. Springer, Boston, MA, USA. doi: 10.1007/978-1-4684-4859-7_9.

Maceda-Veiga, A., J. Figuerola, A.M. Silvestre, G. Viscor, N. Ferrari and M. Pacheco. 2015. Inside the redbox: Applications of haematology in wildlife monitoring and ecosystem health assessment. Sci. Total Environ. 514: 322–332. doi: 10.1016/j.scitotenv.2015.02.004.

Madliger, C.L., S.J. Cooke, E.J. Crespi, J.L. Funk, K.R. Hultine, K.E. Hunt et al. 2016. Success stories and emerging themes in conservation physiology. Conserv. Physiol. 4: cov057. doi: 10.1093/conphys/cov057.

Madliger, C., O. Love, K. Hultine and S. Cooke. 2018. The conservation physiology toolbox: Status and opportunities. Conserv. Physiol. 6: coy092. doi: 10.1093/conphys/coy029.

Maillet, D. and J.-M. Weber. 2006. Performance-enhancing role of dietary fatty acids in a long-distance migrant shorebird: The semipalmated sandpiper. J. Exp. Biol. 209: 2686–95. doi: 10.1242/jeb.02299.

Mallory, M.L., S.A. Robinson, C.E. Hebert and M.R. Forbes. 2010. Seabirds as indicators of aquatic ecosystem conditions: A case for gathering multiple proxies of seabird health. Mar. Pollut. Bull. 60: 7–12. doi: 10.1016/j.marpolbul.2009.08.024.

Matson, K.D., N.P. Horrocks, M.A. Versteegh and B.I. Tieleman. 2012. Baseline haptoglobin concentrations are repeatable and predictive of certain aspects of a subsequent experimentally-induced inflammatory response. Comp. Biochem. Physiol. A. 162: 7–15. doi:10.1016/j.cbpa.2012.01.010.

Maxwell, M.H. 1993. Avian blood leukocyte responses to stress. World's Poul. Sci. J. 49: 34–43. doi:10.1079/WPS19930004.

McCafferty, D.J. 2013. Applications of thermal imaging in avian science. Ibis. 155: 4–15. doi: 10.1111/ibi.12010.

McWilliams, S.R., C. Guglielmo, B. Pierce and M. Klaassen. 2004. Flying, fasting, and feeding in birds during migration: a nutritional and physiological ecology perspective. J. Avian Biol. 35: 377–393. doi: 10.1111/j.0908-8857.2004.03378.x.

Medeiros, R., J.A. Ramos, V.H. Paiva, A. Almeida, P. Pedro and S. Antunes. 2007. Signage reduces the impact of human disturbance on little tern nesting success in Portugal. Biol. Conserv. 135: 99–106. doi: 10.1016/j.biocon.2006.10.001.

Miller, E. and M. Fowler. 2015. Fowler's Zoo and Wild Animal Medicine. Vol. 8: Sauders, Philadelphia, PA, USA.

Minias, P. 2015. The use of haemoglobin concentrations to assess physiological condition in birds: A review. Conserv. Physiol. 3: cov007. doi: 10.1093/conphys/cov007.

Morales, A., B. Frei, C. Leung, R. Titman, S. Whelan, Z.M. Benowitz-Fredericks et al. 2019. Point-of-care blood analyzers measure the nutritional state of eighteen free-living bird species. Comp. Biochem. Physiol. A. 240: 110594. doi:10.1016/j.cbpa.2019.110594.

Mörner, T., D.L. Obendorf, M. Artois and M.H. Woodford. 2002. Surveillance and monitoring of wildlife diseases. Rev. Sci. Tech. 21: 67–76. doi: 10.20506/rst.21.1.1321.

Müller, W.-U., M. Nüsse, B.M. Miller, A. Slavotinek, S. Viaggi and C. Streffer. 1996. Micronuclei: A biological indicator of radiation damage. Mutat. Res. Rev. Genet. 366: 163–169. doi: 10.1016/S0165-1110(96)90037-8.

Newman, S.H., A. Chmura, K. Converse, A.M. Kilpatrick, N. Patel, E. Lammers et al. 2007. Aquatic bird disease and mortality as an indicator of changing ecosystem health. Mar. Ecol. Progr. Ser. 352: 299–309. doi: 10.3354/meps07076.

Nimon, A.J., R.C. Schroter and B. Stonehouse. 1995. Heart rate of disturbed penguins. Nature 374: 415–415. doi: 10.1038/374415a0.

Nimon, A.J., R.C. Schroter and R.K. Oxenham. 1996. Artificial eggs: measuring heart rate and effects of disturbance in nesting penguins. Physiol. Behav. 60: 1019–22. doi: 10.1016/0031-9384(96)00079-0.

Norris, K. and M.R. Evans. 2000. Ecological immunology: life history trade-offs and immune defence in birds. Behav. Ecol. 11: 19–26. doi: 10.1093/beheco/11.1.19.

Oost, R., C. Porte and N. Van den Brink. 2005. Biomarkers in environmental assessment. pp. 87–152. *In*: den Besten, P.J. and M. Munawar [eds.]. Ecotoxicological testing of marine and freshwater ecosystems; emerging techniques, trends, and strategies.CRC Press, Boca Raton, FL, USA.

Orians, G.H. and N.E. Pearson. 1979. On the theory of central place foraging. pp. 155–177. *In*: Stairs, G.R., J. Horn, and R.D. Mitchell [eds.]. Analysis of Ecological Systems. Ohio State Press, Columbus.Ohio, USA.

Oro, D. 2014. Seabirds and climate: Knowledge, pitfalls, and opportunities. Front. Ecol. Evol. 2: 79. doi: 10.3389/fevo.2014.00079.

Oropesa, A.-L., M. Pérez-López, D. Hernández, J.-P. García, L.-E. Fidalgo, A. López-Beceiro et al. 2007. Acetylcholinesterase activity in seabirds affected by the Prestige oil spill on the Galician coast (NW Spain). Sci. Total Environ. 372: 532–538. doi:10.1016/j.scitotenv.2006.09.006.

O'Sullivan, R.J. and J. Karlseder. 2010. Telomeres: Protecting chromosomes against genome instability. Nat. Rev. Mol. Cell. Biol. 11: 171–181. doi: 10.1038/nrm2848.

Oswald, S.A. and J.M. Arnold. 2012. Direct impacts of climatic warming on heat stress in endothermic species: Seabirds as bioindicators of changing thermoregulatory constraints. Integr. Zool. 7: 121–136. doi: 10.1111/j.1749-4877.2012.00287.x.

Oudi, A., M. Ali Chokri, A. Hammouda, R. Chaabane, R. Badraoui, A. Besnard et al. 2019. Physiological impacts of pollution exposure in seabird's progeny nesting in a Mediterranean contaminated area. Mar. Pollut. Bull. 142: 196–205. doi: 10.1016/j.marpolbul.2019.02.056.

Palacios, M.G., V.L. D'Amico and M. Bertellotti. 2018. Ecotourism effects on health and immunity of Magellanic penguins at two reproductive colonies with disparate touristic regimes and population trends. Conserv. Physiol. 6: coy060. doi: 10.1093/conphys/coy060.

Palsdottir, H. and C. Hunte. 2004. Lipids in membrane protein structures. Biochim. Bioph. Acta Biomembranes 1666: 2–18. doi: 10.1016/j.bbamem.2004.06.012.

Parkinson, A. 1996. Biotransformation of xenobiotics. pp. 161–304. *In*: Klassen, C.D. [ed.]. Casarett and Doull's toxicology: the basic science of poisons. McGraw-Hill, New York, NY, USA.

Parrish, C.C. 2009. Essential fatty acids in aquatic food webs. pp. 309–326. *In*: Kainz, M., M.T. Brett and M.T. Arts [eds.]. Lipids in Aquatic Ecosystems. Springer, New York, NY, USA. doi: 10.1007/978-0-387-89366-2_13.

Poisbleau, M., L. Demongin, F. Angelier, S. Dano, A. Lacroix and P. Quillfeldt. 2009. What ecological factors can affect albumen corticosterone levels in the clutches of seabirds? Timing of breeding, disturbance and laying order in rockhopper penguins (*Eudyptes chrysocome chrysocome*). Gen. Comp. Endocrinol. 162: 139–145. doi:10.1016/j.ygcen.2009.03.022.

Price, E.R. and C.G. Guglielmo. 2009. The effect of muscle phospholipid fatty acid composition on exercise performance: a direct test in the migratory white-throated sparrow (*Zonotrichia albicollis*). Am. J. Physiol. Regul. Integr. Comp. Physiol. 297: R775–R782. doi: 10.1152/ajpregu.00150.2009.

Quillfeldt, P. and E. Möstl. 2003. Resource allocation in Wilson's storm-petrels *Oceanites oceanicus* determined by measurement of glucocorticoid excretion. Acta Ethol. 5: 115–122. doi: 10.1007/s10211-003-0074-9.

Quillfeldt, P., J.F. Masello and E. Möstl. 2004. Blood chemistry in relation to nutrition and ectoparasite load in Wilson's storm-petrels *Oceanites oceanicus*. Polar Biol. 27: 168–176. doi: 10.1007/s00300-003-0572-2.

Quirós, L., X. Ruiz, C. Sanpera, L. Jover and B. Piña. 2008. Analysis of micronucleated erythrocytes in heron nestlings from reference and impacted sites in the Ebro basin (N.E. Spain). Environm. Pollut. 155: 81–87. doi: 10.1016/j.envpol.2007.10.030.

Rattner, B.A., V.P. Eroschenko, G.A. Fox, D.M. Fry and J. Gorsline. 1984. Avian endocrine responses to environmental pollutants. J. Exp. Zool. 232: 683–689. doi: 10.1002/jez.1402320337.

Rocha, A.D., P.M. Araújo, F.R. Martinho, J.A. Ramos and J.A. Masero. 2016. A non-lethal biopsy technique for sampling subcutaneous adipose tissue of small and medium-sized birds. J. Field Ornithol. 87: 213–221. doi:10.1111/jofo.12145.

Romero, L.M. 2004. Physiological stress in ecology: Lessons from biomedical research. Trends Ecol. Evol. 19: 249–255. doi: 10.1016/j.tree.2004.03.008.

Sagerup, K., E.O. Henriksen, A. Skorping, J.U. Skaare and G.W. Gabrielsen. 2000. Intensity of parasitic nematodes increases with organochlorine levels in the glaucous gull. J. Appl. Ecol. 37: 532–539. doi: 10.1046/j.1365-2664.2000.00521.x.

Schlezinger, J.J., R.D. White and J.J. Stegeman. 1999. Oxidative inactivation of cytochrome P-450 1A (CYP1A) stimulated by 3,3',4,4'-tetrachlorobiphenyl: production of reactive oxygen by vertebrate CYP1As. Mol. Pharmacol. 56: 588–597. doi: 10.1124/mol.56.3.588.

Schreiber, E.A. and J. Burger. 2001. Biology of Marine Birds.CRC Press, Boca Raton, FL, USA.

Sebastiano, M., P. Bustamante, D. Costantini, I. Eulaers, G. Malarvannan, P. Mendez-Fernandez et al. 2016. High levels of mercury and low levels of persistent organic pollutants in a tropical seabird in French Guiana, the Magnificent frigatebird, *Fregata magnificens*. Environ. Pollut. 214: 384–393. doi: 10.1016/j.envpol.2016.03.070.

Sebastiano, M., M. Eens, H. Abd Elgawad, B. de Thoisy, V. Lacoste, K. Pineau et al. 2017a. Oxidative stress biomarkers are associated with visible clinical signs of a disease in frigatebird nestlings. Sci. Rep. 7: 1599. doi: 10.1038/s41598-017-01417-9.

Sebastiano, M., M. Eens, F. Angelier, K. Pineau, O. Chastel and D. Costantini. 2017b. Corticosterone, inflammation, immune status and telomere length in frigatebird nestlings facing a severe herpesvirus infection. Conserv. Physiol. 5: cow073. doi: 10.1093/conphys/cow073.

Sebastiano, M., M. Eens, K. Pineau, O. Chastel and D. Costantini. 2019. Food supplementation protects Magnificent frigatebird chicks against a fatal viral disease. Conserv. Lett. 12: e12630. doi:10.1111/conl.12630.

Sebastiano, M., F. Angelier, P. Blévin, C. Ribout, K. Sagerup, S. Descamps et al. 2020. Exposure to PFAS is associated with telomere length dynamics and demographic responses of an arctic top predator. Environ. Sci. Technol. 54: 10217–10226. doi: 10.1021/acs.est.0c03099.

Sebastiano, M., W. Jouanneau, P. Blévin, F. Angelier, C. Parenteau, J. Gernigon et al. 2021. High levels of fluoroalkyl substances and potential disruption of thyroid hormones in three gull species from South Western France. Sci. Total Environ. 765: 144611. doi:10.1016/j.scitotenv.2020.144611.

Seebacher, F. and C. Franklin. 2012. Determining environmental causes of biological effects: The need for a mechanistic physiological dimension in conservation biology. Philos. Trans. R. Soc. Lond. B. Biol. Sci. 367: 1607–1614. doi: 10.1098/rstb.2012.0036.

Shmueli, M., I. Izhaki, O. Zinder and Z. Arad. 2000. The physiological state of captive and migrating Great White Pelicans (*Pelecanus onocrotalus*) revealed by their blood chemistry. Comp. Biochem. Physiol. A 125: 25–32. doi:10.1016/S1095-6433(99)00162-2.

Siegel, H.S. 1980. Physiological stress in birds. Bioscience 30: 529–534. doi: 10.2307/1307973.

Smith, K.F., K. Acevedo-Whitehouse, K. and A.B. Pedersen. 2009. The role of infectious diseases in biological conservation. Anim. Conserv. 12: 1–12. doi: 10.1111/j.1469-1795.2008.00228.x.

Soldatini, C., Y.V. Albores-Barajas, M. Tagliavia, B. Massa, L. Fusani and V. Canoine. 2015. Effects of human disturbance on cave-nesting seabirds: The case of the storm petrel. Conserv. Physiol. 3: cov041. doi: 10.1093/conphys/cov041.

Soldatini, C., M. Sebastiano, Y.V. Albores-Barajas, H. Abdelgawad, P. Bustamante and D. Costantini. 2020. Mercury exposure in relation to foraging ecology and its impact on the oxidative status of an endangered seabird. Sci. Total Environ. 724: 138131. doi:10.1016/j.scitotenv.2020.138131.

Sorenson, G.H., C.J. Dey, C.L. Madliger and O.P. Love. 2017. Effectiveness of baseline corticosterone as a monitoring tool for fitness: a meta-analysis in seabirds. Oecologia 183: 353–365. doi: 10.1007/s00442-016-3774-3.

Spée, M., L. Marchal, D. Lazin, Y. Le Maho, O. Chastel, M. Beaulieu et al. 2011. Exogenous corticosterone and nest abandonment: A study in a long-lived bird, the Adélie penguin. Horm. Behav. 60: 362–70. doi: 10.1016/j.yhbeh.2011.07.003.

Stoot, L.J., N.A. Cairns, F. Cull, J.J. Taylor, J.D. Jeffrey, F. Morin et al. 2014. Use of portable blood physiology point-of-care devices for basic and applied research on vertebrates: A review. Conserv. Physiol. 2: cou011. doi: 10.1093/conphys/cou011.

Storey, A.E., M.G. Ryan, M.G. Fitzsimmons, A.-L. Kouwenberg, L.S. Takahashi, G.J. Robertson et al. 2017. Balancing personal maintenance with parental investment in a chick-rearing seabird: Physiological indicators change with foraging conditions. Conserv. Physiol. 5: cox055.doi: 10.1093/conphys/cox055.

Stoskopf, M.K. and S. Kennedy-Stoskopf. 1986. Zoo and wild animal medicine. W B Saunders, Philadelphia, PA, USA.

Stout, J.F. and E.R. Schwab. 1980. Telemetry of heart rate as a measure of the effectiveness of dispersal inducing stimuli in seagulls. pp. 603–610. *In*: Amlaner, C.J. and D.W. Macdonald [eds.]. A Handbook on Biotelemetry and Radio Tracking. Pergamon Press, Oxford, UK. doi: 10.1016/B978-0-08-024928-5.50078-6.

Tartu, S., F. Angelier, J.C. Wingfield, P. Bustamante, P. Labadie, H. Budzinski et al. 2015. Corticosterone, prolactin and egg neglect behavior in relation to mercury and legacy POPs in a long-lived Antarctic bird. Sci. Total Environm. 505: 180–188. doi: 10.1016/j.scitotenv.2014.10.008.

Thrall, M.A., G. Weiser, R. Allison and T.W. Campbell. 2012.Veterinary hematology and clinical chemistry. John Wiley & Sons, Inc., Ames, Iowa, USA.

Totzke, U., M. Fenske, O. Hüppop, H. Raabe and N. Schach. 1999. The influence of fasting on blood and plasma composition of Herring Gulls (*Larus argentatus*). Physiol. Biochem. Zool. 72: 426–437. doi: 10.1086/316675.

Troisi, G., L. Borjesson, S. Bexton and I. Robinson. 2007. Biomarkers of polycyclic aromatic hydrocarbon (PAH)-associated hemolytic anemia in oiled wildlife. Environm. Res. 105: 324–329. doi: 10.1016/j.envres.2007.06.007.

Vanstreels, R.E.T., D.d.-A. Dutra, F.C. Ferreira-Junior, R. Hurtado, L. Egert, L.F.S.P. Mayorga et al. 2019. Epidemiology, hematology, and unusual morphological characteristics of *Plasmodium* during an avian malaria outbreak in penguins in Brazil. Parasitol. Res. 118: 3497–3508. doi: 10.1007/s00436-019-06459-8.

Veghte, J.H. and C.F. Herreid. 1965. Radiometric determination of feather insulation and metabolism of arctic birds. Physiol. Zool. 38: 267–275.

Verreault, J., J.U. Skaare, B.M. Jenssen and G.W. Gabrielsen. 2004. Effects of organochlorine contaminants on thyroid hormone levels in Arctic breeding glaucous gulls, *Larus hyperboreus*. Environm. Health Persp. 112: 532–537. doi: 10.1289/ehp.6756.

Verreault, J., N. Verboven, G.W. Gabrielsen, R.J. Letcher and O. Chastel. 2008. Changes in prolactin in a highly organohalogen contaminated Arctic top predator seabird, the glaucous gull. Gen. Comp. Endocrinol. 156: 569–76. doi: 10.1016/j.ygcen.2008.02.013.

Viblanc, V.A., A.D. Smith, B. Gineste and R. Groscolas. 2012. Coping with continuous human disturbance in the wild: insights from penguin heart rate response to various stressors. BMC Ecol. 12: 10. doi: 10.1186/1472-6785-12-10.

Viegas, I., P.M. Araújo, A.D. Rocha, A. Villegas, J.G. Jones, J.A. Ramos et al. 2017. Metabolic plasticity for subcutaneous fat accumulation in a long-distance migratory bird traced by ^2H_2O. J. Exp. Biol. 220: 1072–1078. doi: 10.1242/jeb.150490.

Vleck, C.M., N. Vertalino, D. Vleck and T.L. Bucher. 2000. Stress, corticosterone, and heterophil to lymphocyte ratios in free-living Adélie Penguins. Condor 102: 392–400. doi: 10.1093/condor/102.2.392.

Walcott, S.M., A.L. Kirkham and J.M. Burns. 2020. Thermoregulatory costs in molting Antarctic Weddell seals: impacts of physiological and environmental conditions. Conserv. Physiol. 8: coaa022. doi: 10.1093/conphys/coaa022.

Walker, B.G., P.D. Boersma and J.C. Wingfield. 2006. Habituation of adult magellanic penguins to human visitation as expressed through behavior and corticosterone secretion. Conserv. Biol. 20: 146–154. doi: 10.1111/j.1523-1739.2005.00271.x.

Walker, C.H. 1990. Persistent pollutants in fish-eating sea birds—bioaccumulation, metabolism and effects. Aquatic Toxicol. 17: 293–324. doi: 10.1016/0166-445X(90)90014-G.

Walton, P., C.M.R. Turner, G. Austin, M.D. Burns and P. Monaghan. 1997. Sub-lethal effects of an oil pollution incident on breeding kittiwakes *Rissa tridactyla*. Mar. Ecol. Progr. Ser. 155: 261–268. doi: 10.3354/meps155261.

Wanless, S., T.R. Barton and M.P. Harris. 1997. Blood hematocrit measurements of 4 species of North Atlantic seabirds in relation to levels of infestation by the tick *Ixodes uriae*. Col. Waterbirds 20: 540–544.

Watson, H., M. Bolton and P. Monaghan. 2014. Out of sight but not out of harm's way: Human disturbance reduces reproductive success of a cavity-nesting seabird. Biol. Conserv. 174: 127–133. doi: 10.1016/j.biocon.2014.03.020.

Watson, H., E. Videvall, M.N. Andersson and C. Isaksson. 2017. Transcriptome analysis of a wild bird reveals physiological responses to the urban environment. Sci. Rep. 7: 44180. doi: 10.1038/srep44180.

Wayland, M., H.G. Gilchrist, D.L. Dickson, T. Bollinger, C. James, R.A. Carreno et al. 2001. Trace elements in king eiders and common eiders in the canadian arctic. Arch. Environm. Contam.Toxicol. 41: 491–500. doi: 10.1007/s002440010276.

Wayland, M., H.G. Gilchrist, T. Marchant, J. Keating and J.E. Smits. 2002. Immune function, stress response, and body condition in arctic-breeding common eiders in relation to cadmium, mercury, and selenium concentrations. Environm. Res. 90: 47–60. doi: 10.1006/enrs.2002.4384.

Weimerskirch, H., A. Prudor and Q. Schull. 2018. Flights of drones over sub-Antarctic seabirds show species- and status-specific behavioural and physiological responses. Polar Biol. 41: 259–266.doi: 10.1007/s00300-017-2187-z.

Will, A.P., Y. Suzuki, K.H. Elliot, S.A. Hatch, Y. Watanuki and A.S. Kitaysky. 2014. Feather corticosterone reveals developmental stress in seabirds. J. Exp. Biol. 217: 2371–2376. doi: 10.1242/jeb.098533.

Williams, CT., S.J. Iverson and C.L. Buck. 2009. The effects of diet and caloric restriction on adipose tissue fatty acid signatures of tufted puffin (*Fratercula cirrhata*) nestlings. J. Comp. Physiol. B. 179: 711–20. doi: 10.1007/s00360-009-0354-4.

Williams, C.T. and C.L. Buck. 2010. Using fatty acids as dietary tracers in seabird trophic ecology: Theory, application and limitations. J. Ornithol. 151: 531–543. doi: 10.1007/s10336-010-0513-0.

Wingfield, J.C. and A.S. Kitaysky. 2002. Endocrine responses to unpredictable environmental events: Stress or anti-stress hormones? Integr. Comp. Biol. 42: 600–609. doi: 10.1093/icb/42.3.600.

Wingfield, J.C., J.H. Perez, J.S. Krause, K.R. Word, P.L. Gonzalez-Gomez, S. Lisovski et al. 2017. How birds cope physiologically and behaviourally with extreme climatic events. Philos. Trans. R. Soc. Lond. B. Biol. Sci. 372: 20160140. doi: 10.1098/rstb.2016.0140.

Wojczulanis-Jakubas, K., D. Jakubas, O. Chastel and I. Kulaszewicz. 2015. A big storm in a small body: Seasonal changes in body mass, hormone concentrations and leukocyte profile in the little auk (*Alle alle*). Polar Biol. 38: 1203–1212. doi: 10.1007/s00300-015-1687-y.

Wolins, N.E., D.L. Brasaemle and P.E. Bickel. 2006. A proposed model of fat packaging by exchangeable lipid droplet proteins. FEBS Lett. 580: 5484–5491. doi:10.1016/j.febslet.2006.08.040.

Woods, J.S. 1995. Porphyrin metabolism as indicator of metal exposure and toxicity. pp. 19–52. *In*: Goyer, R.A. and M.G. Cherian [eds.]. Toxicology of Metals: Biochemical Aspects. Springer, Berlin and Heidelberg, Germany.

Woods, R., H.I. Jones, J. Watts, G.D. Miller and G.R. Shellam. 2009. pp. 35–55. *In*: Kerry, K.R. and M. Riddle [eds.]. Health of Antarctic Wildlife. Springer, Berlin and Heidelberg, Germany.

World Health Organization and International Programme on Chemical Safety. 1993. Biomarkers and risk assessment: Concepts and principles/published under the joint sponsorship of the United Nations environment Programme, the International Labour Organisation, and the World Health Organization. World Health Organization. https://apps.who.int/iris/handle/10665/39037.

Xu, Y., X. Lai, Z. Li, X. Zhang and Q. Luo. 2018. Effect of chronic heat stress on some physiological and immunological parameters in different breed of broilers. Poultry Sci. 97: 4073–4082. doi: 10.3382/ps/pey256.

Yorio, P. and P.D. Boersma. 1992. The effects of human disturbance on Magellanic Penguin *Spheniscus magellanicus* behaviour and breeding success. Bird Conserv. Int. 2: 161–173. doi: 10.1017/S0959270900002410.

Yorio, P., E. Frere, P. Gandini and A. Schiavini. 2001. Tourism and recreation at seabird breeding sites in Patagonia, Argentina: Current concerns and future prospects. Bird Conserv. Int. 11: 231–245. doi: 10.1017/S0959270901000314.

Young, M.J., L. Dutoit, F. Robertson, Y. van Heezik, P.J. Seddon and B.C. Robertson. 2020. Species in the faeces: DNA metabarcoding as a method to determine the diet of the endangered yellow-eyed penguin. Wildl. Res. 47: 509–522, 14. doi: 10.1071/WR19246.

Young, R.C., A.S. Kitaysky, M.F. Haussmann, S. Descamps, R.A. Orben, K.H. Elliott et al. 2013. Age, Sex, and Telomere Dynamics in a Long-Lived Seabird with Male-Biased Parental Care. PLOS ONE 8: e74931. 10.1371/journal.pone.0074931.

Zúñiga-González, G., O. Torres-Bugarín, A. Zamora-Perez, B.C. Gómez-Meda, M.L. Ramos Ibarra, S. Martínez-González et al. 2001. Differences in the number of micronucleated erythrocytes among young and adult animals including humans: Spontaneous micronuclei in 43 species. Mutat. Res. Genet. Toxicol. Environ. *Mutagen.* 494: 161–167. doi:10.1016/S1383-5718(01)00180-2.

CHAPTER 4

Tracking Seabirds for Conservation and Marine Spatial Planning

Jorge M. Pereira,[1,*] *Vitor H. Paiva*,[1] *Lucas Krüger*[2] and *Stephen C. Votier*[3]

Introduction

The global oceans are in crisis. The rapidity of environmental change requires novel approaches to conservation and management (Palumbi et al. 2009, Steffen et al. 2015, Lenton et al. 2019). Marine megafauna has in particular been severely impacted during the Anthropocene but are also important ecosystem sentinels (Hazen et al. 2019).

Seabirds have been the focus of scientific research for several decades and therefore particularly well-studied in comparison with other marine animals. Seabirds travel widely and feed across many trophic levels but are also relatively prominent and accessible when they return to land to breed, making them effective sentinels (Parsons et al. 2008). Seabird monitoring enables information gathering across a range of spatio-temporal scales. For instance, diet and demographic monitoring have been used as indicators of ecosystem health and to prioritise conservation goals (Furness 1997, Piatt and Sydeman 2007). Moreover, they may act as important umbrella species facilitating protection of habitats and less charismatic taxa (Lescroël et al. 2016). In this way, seabird tracking has emerged as a powerful complement with the ability to monitor biodiversity and reveal key information about the potential winners and losers of global change (Brisson-Curadeau et al. 2017, Hays et al. 2019).

This chapter provides an overview of the recent advances in tracking devices used to study seabird distribution. Main applications of tracking for resolving seabird conservation and marine spatial planning issues are also addressed. Lastly, we discuss the future of seabird tracking and the challenges for researchers and policy makers to deal with the increasing volume and complexity of movement data in coming years.

[1] University of Coimbra, MARE - Marine and Environmental Sciences Centre, Department of Life Sciences, 3000-456 Coimbra, Portugal.
Email: vitorpaiva@ci.uc.pt
[2] Instituto Antartico Chileno, Plaza Muñoz Gamero, 1055 Punta Arenas, Chile.
Email: lkruger@inach.cl
[3] Heriot-Watt University, The Lyell Centre, EH14 4AP Edinburgh, UK.
Email: S.Votier@hw.ac.uk
* Corresponding author: jpereira.ecotop.uc@gmail.com

Advances in tracking devices to study seabird distribution

For much of the 20th Century, seabird research focused on demography, behaviour and physiology at their breeding colonies (Piatt and Sydeman 2007), as well as opportunistic research from boats (Hudson and Furness 1989). In the early 1970's however, the first seabirds were fitted with radio transmitters to study colony attendance patterns and short-term foraging movements (Southern 1970). These devices provided novel insights and were relatively inexpensive, but were also limited in terms of detection range (approximately 15–20 km) and precision (Burguer and Shaffer 2008, Ponchon et al. 2013, López-López 2016).

One of the most significant advances in seabird ecology was the development and deployment of Platform Terminal Transmitters (PTTs) in the early 1990's. This led to detailed reconstructions of albatross and penguin at-sea behaviour in the Southern Ocean (Jouventin and Weimerskirch 1990, Ancel et al. 1992, Weimerskirch et al. 1993). PTTs (Fig. A4.1) transmit radio signals to the Argos Satellite System and variations in doppler shift created by the moving satellite allows location (to within ~ 500 m) to be estimated in near real time without the need to recover devices. It was also possible to incorporate solar-panels to increase battery capacity for long-term tracking (Wilson et al. 2002), and to study hitherto cryptic behaviours such as dispersal and immature movement (Weimerskirch et al. 2006, Votier et al. 2011, Péron and Grémillet 2013).

Following the success of PTTs, archival Global Positioning System (GPS) tags were another significant milestone in the study of seabird movement ecology (Fig. A4.2A). These devices record

Figure A4.1. An adult Wandering albatross (*Diomedea exulans*) from Bird Island in South Georgia equipped with a Platform Terminal Transmitter (PTT) on its back (photo courtesy of José C. Xavier).

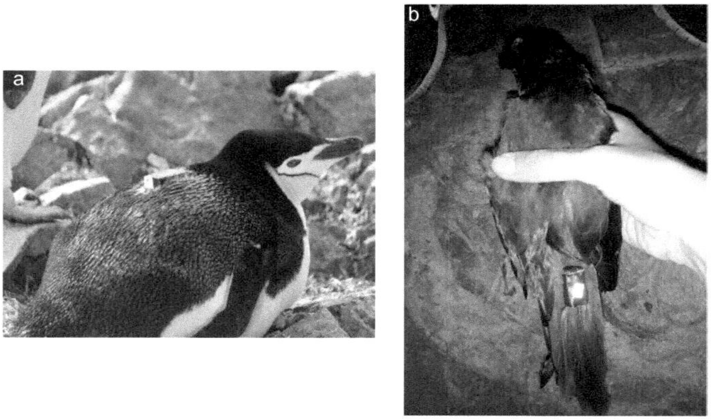

Figure A4.2. **(A)** An adult Chinstrap penguin (*Pygoscelis antarcticus*) from Harmony Point, Nelson Island, in the Maritime Antarctic Peninsula, equipped with an archival Global Positioning System (GPS) tag on its back (photo courtesy of Lucas Krüger). **(B)** An adult Bulwer's petrel (*Bulweria bulwerii*) from Raso Islet in the Cabo Verde archipelago, equipped with a 2 g archival GPS tag (photo courtesy of Biosfera I). Archival GPS tags are smaller, lighter and more affordable when compared to PTTs, enabling to track a greater number of seabirds, including many poorly studied smaller species. Nevertheless, due to battery constraints these tags are mainly used to track short-term movements of individual seabirds during the breeding season.

precision locations (10 m maximum error) (Bridge et al. 2011), potentially at high frequencies (10 Hz and above) (Gibb et al. 2017). In the past 20 years or so, not only did these devices decrease in size but they also became significantly cheaper, driven by mass-produced components and demand for recreational use (e.g., for hiking and pet tracking). In addition to these inexpensive tags repurposed by researchers, several companies manufactured devices specifically designed for tracking birds, including smaller and lighter tags weighing as little as 1 g (Fig. A4.2B). This led to the mass deployment of GPS tags across a wide range of locations and approximately 216 seabird species from 17 families over the last four decades (Bernard et al. 2021). However, there is very little information on the at-sea distribution of smaller species such as gadfly-petrels, storm-petrels, diving-petrels and prions.

While most GPS tags are archival with a primary-cell battery, some also transmit and have photovoltaic cells greatly increasing their utility. GPS-PTTs relay positions via satellite, while others upload information via the Global System for Mobile Communications (GPS-GSM) or relay information vary a base-station (Bouten et al. 2013). Solar-powered GPS-GSM tags (Fig. A4.3) have the potential to remain active during long periods of time, providing fine-scale longitudinal movement information throughout the annual cycle in near-real time (Navarro et al. 2016). However, GPS-GSM tags tend to be expensive (e.g., > 500 €), by an order of magnitude more than archival GPS loggers (e.g., < 50 €). Overall, archival GPS loggers are more suited for use in short-term studies during the breeding season, when seabirds return repeatedly to the nest site while central-place foraging (Sansom et al. 2018). As a consequence, fine-scale information on at-sea distribution of non-breeding seabirds is generally poorly represented or virtually absent, especially during immaturity (Péron and Grémillet 2013).

Geolocation has been instrumental in the study of long-range seabird movements, revealing, for example, remarkable ocean-wandering migration feats (Croxall 2005, Shaffer et al. 2006, Egevang et al. 2010). Geolocators (Fig. A4.4) record ambient light intensity multiple times daily which can be used, together with an internal clock, to calculate day length and midday/midnight and in turn, latitude and longitude (Wilson et al. 1992, Phillips et al. 2005). This generates twice daily location estimates with average error of 185 ± 115 km, although the method works only poorly near to the poles and during the equinoxes (Phillips et al. 2004). In addition, most models are archival necessitating device recovery. Despite these limitations, geolocators have a number of advantages. Because they can store up to 60 months of data, means it is possible to investigate the degree of inter-annual migratory site fidelity (Phillips et al. 2005, Grecian et al. 2019). Moreover, devices with temperature/saltwater sensors provide the opportunity not only to improve location precision by integrating remotely sensed environmental data (Merkel et al. 2016), but also to link distribution with activity behaviour (Mackley et al. 2010). In comparison to satellite transmitters, geolocators

Figure A4.3. An adult Southern giant petrel (*Macronectes giganteus*) from Harmony Point, Nelson Island, in the Maritime Antarctic Peninsula, equipped with a solar-powered GPS transmitter on its back (photo courtesy of Denyelle Corá). Despite the high costs of acquisition, solar-powered GPS transmitters have the potential to provide fine-scale longitudinal movement information of individual birds throughout the annual cycle in near-real time.

Figure A4.4. An adult Macaronesian shearwater (*Puffinus baroli*) from Selvagem Grande, Madeiran archipelago, in the subtropical North Atlantic Ocean, equipped with a geolocator (GLS) on its leg (photo courtesy of Luís Ferreira). GLS tags are smaller, lighter and more affordable than the majority of tags and are capable of storing multiple years of data. As a consequence, these tags have been instrumental in the study of long-range movements of individual seabirds, revealing remarkable ocean-wandering migrations.

also tend to be smaller, lighter (the smallest are 0.3 g) and more affordable (e.g., > 250 €), although the cost and size of tags is changing rapidly.

Overall, device choice is a major consideration in study design involving trade-offs between required spatio-temporal accuracy, deployment duration, size of the study species, likelihood of tag recovery, potential device effects and cost of acquisition. Tracking devices are now very popular in seabird ecological studies, and hundreds of papers reporting data collecting with the different devices are published annually (Bernard et al. 2021). The technical details and applications of tracking devices were reviewed in detail by (Burguer and Shaffer 2008, Bouten et al. 2013, López-López 2016). Miniaturisation has also allowed the deployment of multiple devices to record detailed information on physiology (e.g., heart-rate sensors, stomach-temperature probes, piezoelectric-film probes), local environment (e.g., immersion loggers, compass-temperature loggers, video cameras) or behaviour other than movement (e.g., accelerometers, magnetometers, time-depth recorders, beak sensors, video cameras) (Ropert-Coudert and Wilson 2005, Yoda 2019). When combined with positional loggers these can generate valuable new biological insight, as well as opening the potential that seabirds can act as live environmental monitors (Yonehara et al. 2016). However, at present, such multi-logger deployments are only appropriate for larger seabird species.

Seabird tracking studies are based on the assumption that birds are not influenced by device attachment to such a degree that this fundamentally alters their behaviour. Earlier studies focused on tag weight as a proportion of body mass such as the 3% body mass guidance (Phillips et al. 2003). However, more recently most studies report no effects of the devices on seabird behaviour, but exact data is often not detailed. Previous meta-analysis have documented the impacts of device deployments on breeding success, physiology, foraging behaviour, survival and fitness in a variety of bird species (Barron et al. 2010, Bodey et al. 2018, Geen et al. 2019), even when the devices were below 3% of birds' body mass (Sun et al. 2020). Hence, careful consideration in the study design is vital to evaluate the impacts of device deployments on seabirds (Chivers et al. 2016), including not only the effects on breeding performance by directly comparing groups of untagged and tagged birds (Ludynia et al. 2012), but also evaluate whether devices may lead to short- and long-term behavioural changes (Gillies et al. 2020).

Tracking seabirds for conservation and marine spatial planning

Seabird populations have declined steeply in recent decades and their status continues to deteriorate (Paleczny et al. 2015, Grémillet et al. 2018). The three most grave threats to seabirds are alien species, bycatch and climate change, although there are also pressures from overfishing, pollution and energy production (Dias et al. 2019a). Strategies and tools are therefore required to identify, predict and mitigate the conflicts that arise between anthropogenic activities and seabirds. Seabird tracking has emerged as a powerful tool to aid conservation and inform marine spatial planning (Lennox et al. 2019, Hays et al. 2019). By identifying important seabird habitats (e.g., migratory routes, foraging grounds and breeding areas) tracking can highlight areas for protection and contribute to a more sustainable exploitation of marine resources (Lascelles et al. 2012, Ronconi et al. 2012). In this context, the use of seabirds as indicators of marine spatial planning is gathering momentum to put this group of animals as central users of the global oceans and indicators of ocean health. It is now time to look forward to the key research directions that these new efforts are taking. Below, we consider some of the main applications of tracking to understand how or where seabirds may be exposed to anthropogenic threats and towards the prioritisation of conservation goals.

Environmental and climate change

To elucidate the ecological consequences of climate change, a detailed understanding of where seabirds go and the drivers of movement is required to predict future consequences. Monitoring the distribution and activity patterns of seabirds chimes with calls for marine spatial planning to become more fluid to reflect the dynamic nature of marine environments (Maxwell et al. 2015). Yet, practical examples where tracking led to clearly identifiable real-world changes in conservation or marine spatial planning efforts are scarce (Hays et al. 2019). One of the most frequent strategies used to understand seabirds' response to global change is to combine tracking with environmental monitoring such as from satellite imagery. Such an approach has the potential to study seabird-environmental interactions remotely and assess inter-annual migratory/foraging fidelity (Wakefield et al. 2015, Grecian et al. 2019) or shifts in distribution ranges due to detrimental environmental conditions induced by climate change (Poloczanska et al. 2016). Therefore, understanding shifts in seabird habitat preferences through long-term series with changing environmental and oceanographic conditions is key for implementing dynamic ecosystem-level management (McGowan et al. 2013).

The physical dynamics of the marine environment, influenced by wind patterns and currents, determine the occurrence of fronts where mobile pelagic prey concentrate and are the target of seabirds (Wakefield et al. 2009, Scales et al. 2014). Changes in the strength and persistence of fronts may be linked with a reduced foraging and breeding success, because prey disperse into a wider area (Ropert-Coudert et al. 2009). It is unclear how climate change will alter ocean fronts, but seabird tracking is likely to prove important for understanding any consequences of change.

Other environmental changes affecting some seabird communities is the increased frequency, severity and duration of extreme climatic events such as marine heatwaves, increased storminess and effects of changing wind. For example, GPS tracking breeding Black-legged Kittiwakes (*Rissa tridactyla*) on Middleton Island, Alaska, revealed increased foraging effort during and immediately after a heatwave event, with possible population-level consequences (Osborne et al. 2020). Moreover, GPS tracking of Great Frigatebirds (*Fregata minor*) and Red-footed Boobies (*Sula sula*) breeding on Europa Island, located in the Mozambique Channel, revealed that both species adjusted their behaviour to avoid the centre of tropical cyclones (Weimerskirch and Prudor 2019). Although extreme climatic events are poorly studied, long-term tracking data may reveal the mechanisms underpinning the effects of those events, particularly when combined with data on breeding parameters (Jenouvrier et al. 2015, Ropert-Coudert et al. 2018, Pereira et al. 2020). There is a greater need to integrate monitoring of demography with at-sea behaviour from biologging (Carneiro et al. 2020).

Interaction with fisheries and marine traffic

Tracking has provided crucial information on how fisheries influence seabirds (Le Bot et al. 2018). For instance, by comparing the distribution of tracked Procellariiformes with spatio-temporal patterns of fishing effort in the Southern Ocean provided important evidence for the catastrophic impact of bycatch in seabird populations (Phillips et al. 2006). Moreover, precision tracking of individual seabirds and vessels revealed the extent to which birds are attracted to vessels to feed on discards (Votier et al. 2010, 2013, Patrick et al. 2015), or not in some cases (Granadeiro et al. 2011, Cianchetti-Benedetti et al. 2018, Clark et al. 2020). This has also enabled us to quantify the ecological footprint of individual boats (Bodey et al. 2014), as well as the impact of fishing activity on broad-scale search patterns (Bartumeus et al. 2010). Tracking is also important for assessing how competition with fisheries impacts seabird movements. For example, experimental fisheries closures close to African penguin (*Spheniscus demersus*) colonies created fluctuations in foraging fish availability, and revealed increased foraging effort of tracked adults as well as decreased offspring condition as forage fish abundance declined (Pichegru et al. 2010, 2012, Campbell et al. 2019). However, long-term tracking studies together with population dynamic studies are necessary to evaluate population consequences of seabird-fisheries competition (Clay et al. 2019).

Vessel location data can be obtained either through the use of Vessel Monitoring System (VMS) or Automatic Identification System (AIS) programs, although this approach is limited by data resolution and accessibility due to poor coverage in the high seas and because of Illegal Unreported and Unregulated (IUU) fishing (Corbeau et al. 2019). However, global initiatives such as Global Fishing Watch (Kroodsma et al. 2018) have improved transparency of fishing activity data by making it publicly accessible, enabling a more complete understanding of seabird-fishery interactions (Pereira et al. 2021, Fischer et al. 2021, Orben et al. 2021). Tags designed to detect the presence of vessels through radio emissions produced by marine radar offer information on vessels not using VMS or AIS, and hence have the potential to detect IUU fishing (Weimerskirch et al. 2018, 2020, Grémillet et al. 2019). Simultaneously, accelerometers, geolocators and bird-borne cameras not only provide information on vessel interactions, but also details on the behaviour, nature of such associations such as prey capture and interactions with fishing gear (Votier et al. 2013, Krüger et al. 2017b, Cianchetti-Benedetti et al. 2018). Moreover, seabird tracking combined with diet sampling and stable isotopes also provide insight into the use of fishery discards (Votier et al. 2010, Matos et al. 2018, Giménez et al. 2021).

Marine traffic may also impact some seabirds via disturbance, and tracking could prove important understanding on this poorly studied stressor (Schwemmer et al. 2011). Yet, few studies have attempted to quantify the collision risk with marine traffic which may be illuminated at night (Merkel and Johansen 2011). Of the few studies in this area, precision tracking of individual seabirds and marine traffic provide fine-scale information on the bird-ships encounters and quantify the disturbance of marine traffic on seabirds (Burger et al. 2019, Lieske et al. 2020).

Offshore energy production

Seabird tracking data is important for quantifying potential impacts of offshore energy developments by estimating the collision risk (Soanes et al. 2013b) and costs of displacement (Dierschke et al. 2016). Research evaluating collision risk with marine renewable energy installations (MREIs), particularly offshore wind farms (Vanermen et al. 2020), estimate spatial overlap with tracked birds (Thaxter et al. 2015, Garthe et al. 2017, Peschko et al. 2020). This approach has benefits from Eulerian approaches (i.e., observations of birds from a fixed position) because it provides information about connectivity, age/sex and as it is not constrained by weather or time of day. Tagging can also provide information on flight heights, which is important for accurate collision risk modelling (Ross-Smith et al. 2016). GPS tags provide flight heights, but have low precision unless sampling at high frequencies (e.g., > 1 Hz) (Peschko et al. 2021). Barometric pressure loggers provide more reliable altitude estimates that can generate very different collision risk estimates compared with flight-height observations (Cleasby et al. 2015). Biologging is especially valuable for assessing displacement impacts in terms of movement and energetics (Heylen and Nachtsheim 2018, Williams et al. 2020). GPS tags and time-depth recorders are also valuable to assess impacts of other poorly studied MREIs such as tidal stream and wave energy installations (Grecian et al. 2010, Waggitt and Scott 2014). Finally, biologging may also be important for understanding any potential positive impacts of MREIs such as by creating roost sites, fish aggregating devices and protected areas at-sea (Inger et al. 2009).

Oil, gas and deep-sea mining platforms may also pose a threat to some seabirds. Yet, few studies have attempted to quantify seabird collision rates at these structures which may be illuminated at night (Ronconi et al. 2015, Miller et al. 2018). Of the limited research in this area, overlap between platforms and birds tracked with geolocation loggers provide only course-scale information (Hedd et al. 2011, Tranquilla et al. 2013, Fort et al. 2013) and cannot measure bird-platform interactions directly and at refined spatial scale (Ronconi et al. 2015, Bolton 2020).

Marine pollution

Pollution is a threat to some seabird populations, both on land (e.g., light pollution) and at-sea (e.g., oil, chemical pollutants and plastic), and tracking provides insight into these potential threats. For example, artificial light pollution may influence seabird mortality at night, especially for fledglings (Ainley et al. 2001, Rodríguez and Rodríguez 2009). Much of the information on birds killed or grounded annually are collected from rescue campaigns (Rodríguez et al. 2012), but key information such as the origin of birds attracted to lights, the drivers of attraction, and to what distances or light intensity thresholds seabirds are affected can only be obtained with tracking data (Rodríguez et al. 2015).

Movement information can be used to quantify the risk exposure to hydrocarbon pollution, such as oil spills, the distribution of plastics/microplastics or heavy metals. For example, tracking Black-vented Shearwaters (*Puffinus opisthomelas*) and Cory's Shearwaters (*Calonectris borealis*), combined with stable isotope analysis, pinpointed the areas of higher mercury contamination in the foraging ranges of both species during breeding (Soldatini et al. 2020) and non-breeding periods (Gatt et al. 2020), respectively. Moreover, tracking can also provide information on the risk exposure to other pollutants such as oil (Montevecchi et al. 2012) and chemical additives used to produce plastic (Miller et al. 2020). Tracking studies together with contaminant studies are necessary to assess the risk exposure of seabirds to marine pollution (Mallory et al. 2010).

Marine protected areas

Seabird tracking is proving instrumental for marine spatial planning and for identifying protected areas at-sea (Le Corre et al. 2012, Krüger et al. 2017a, Hays et al. 2019). This has become especially effective as there is a move towards multi-species, multi-colony projects across different institutions (Grecian et al. 2016, Fayet et al. 2017, Hindell et al. 2020). This is exemplified by BirdLife International's seabird tracking database, created in 2003, it is a major repository for seabird tracking data holding > 14 million locations from > 30,000 individual tracks of > 130 species breeding at approximately 362 colonies. The increasing volume of tracking data compiled in the seabird tracking database led to a major need to create standard methodological frameworks to analyse tracking data for multiple seabird species and colonies. To handle the dimension and complexity of the data compiled in the seabird tracking database, BirdLife International developed a generic protocol to identify important areas for seabirds at regional and global scales (Lascelles et al. 2016). Several recent studies have used such methodologies to identify areas with higher conservation importance around important seabird colonies (Dias et al. 2018, 2019b, Steinfurth et al. 2020) and in pelagic waters (Dias et al. 2017, Heerah et al. 2019). Such collaborative projects that breakdown international barriers are crucial if we are to create protected areas for seabirds at-sea (Handley et al. 2021), especially on high seas (Beal et al. 2021).

Challenges and future of seabird tracking

It is clear that tracking seabird movements have proved invaluable for ecological investigation and for resolving marine conservation issues (Heylen and Nachtsheim 2018). Looking forward, it seems likely that developments in hardware and analysis techniques are expected to lead to further discoveries (Hays et al. 2016, Sequeira et al. 2019a). The tendency of devices to become smaller and lighter will allow tracking of even smaller species and development of new sensors will provide more information on behaviour and physiology, as well as environmental conditions (Wilmers et al. 2015, Williams et al. 2020). However, one of the major constraints is battery optimization. Many seabird tracking devices contain lithium batteries, which are ideal for short-term studies but have a relatively short lifespan (usually three or four years for larger batteries), because of a limited number of charging-discharging cycles and vulnerabilities to moisture and salt ingress. Moreover, small batteries may become rapidly depleted if they are unable to locate a satellite, which is particularly problematic for burrow-nesting species, especially during long incubation shifts. GPS-GSM transmitters and other tracking devices equipped with photovoltaic solar-panels are expected to overcome some of the issues related with battery capacity. Soon, scientists will also be able to track birds from space using the International Cooperation for Animal Research Using Space (ICARUS) initiative. The ICARUS solar-powered transmitters are < 5 g enabling to track a greater number of seabirds, including many poorly-studied smaller species (Wikelski et al. 2007). It is particularly challenging to gather data during immaturity, before first reproduction. Tracking a single animal with a high spatio-temporal resolution to uncover in great detail how those behaviours happen remains challenging for most group of species. So far, studies on movements and at-sea behaviour of early-life stages are more common for medium and large sized seabirds, such as sulids, albatrosses, penguins, petrels and shearwaters (Votier et al. 2017, de Grissac et al. 2017, Orgeret et al. 2019, Louzao et al. 2021), but for smaller species, there is virtually no data (Bicknell et al. 2014).

The widespread use of tracking devices has also raised the questions of how many and for how long individuals should be tracked to represent the overall behaviour of the population (Sequeira et al. 2019b). Previous studies revealed the importance of sample sizes to make reliable predictions of a population's at-sea distributions (Soanes et al. 2013a, Lascelles et al. 2016). Small sample sizes may not represent the overall colony-level distributions, but rather depict just a small proportion of the population. Larger sample sizes will provide greater confidence in species- or population-level inferences.

Simultaneously, new analytical techniques and visualisation tools will add interpretation of movement data (Fan et al. 2014, Joo et al. 2020). Animal movements often contain noisy, non-linear, spatially and temporally correlated artifacts, requiring a high degree of analytical proficiency to infer behaviours at appropriate spatio-temporal scales (Patterson et al. 2017). Hence, scientists have developed several techniques to infer behavioural classifications from movement data. Traditional movement description approaches include First-Passage Time (FPT), speed/sinuosity thresholds, fractal analysis, residence time and positional entropy (Bennison et al. 2018). More recently, sophisticated techniques such as machine learning and state-space modelling are gaining momentum as powerful tools to infer animal behaviour from movement data (Patterson et al. 2008, Wang 2019, Wijeyakulasuriya et al. 2020). However, these techniques usually need substantial computational capacity, high expertise from the scientist and are also subject to a certain degree of subjectivity. To simplify the behavioural classification of animal movement data, some more user friendly procedures have been recently created (Garriga et al. 2016) and applied to a wide range of seabird species (de Grissac et al. 2017, Jones et al. 2018, De Pascalis et al. 2020).

Answering more complex questions may also require a further level of refinement and instrumentation. Seabird scientists are now able to incorporate several sensors into the tracking devices to measure, for instance tri-dimensional motion, pressure, wet/dry activity and direct quantification of behaviours with the use of miniaturised cameras. Multiple sensor-based tracking can provide detailed information of where, when and how birds use the surrounding habitat (Patterson et al. 2019), however with the high cost of effort over processing and analysing multiple data (Williams et al. 2020). In addition, the development of sophisticated techniques such as artificial intelligence and machine learning are emerging as cutting-edge and powerful tools to process large and complex datasets. The utilisation of the "training-and-testing" procedure has proven to be an efficient method for classifying and quantifying behavioural modes over large datasets composed of geographical fixes, accelerometer data and video records with high precision and accuracy (Browning et al. 2018, Yoda 2019, Korpela et al. 2020). However, these methods usually require the use of programming languages and/or processing software whose interfaces are not user friendly for those starting to learn it. With a shallow "non-machine" learning curve, scientists need to encourage a culture of data sharing and interdisciplinary collaborative work, as a lot of time and effort investment is necessary for scientists to actually become skilled on those methods (López-López 2016).

Seabirds may be effective as ocean sentinels (Hazen et al. 2019), yet there are still many limitations for them as effective monitoring tools. For example, marine mammals have been used as oceanographic sampling platforms—their large size permitting scientists to equip tags such as Conductivity-Temperature-Depth (CTD) sensors measuring a variety of oceanographic variables in remote areas (Treasure et al. 2017, Nakanowatari et al. 2017, Teilmann et al. 2020). The use of seabirds in this type of remote sensing of the marine environment is still in its infancy, but with huge potential (Harcourt et al. 2019). The use of cameras coupled to GPS devices can give scientists access to environmental conditions on foraging areas, about the fish swarms being targeted by seabirds, interactions with other species and with human activities (Votier et al. 2013, Tremblay et al. 2014, Mattern et al. 2018). In addition, it allows us to identify previously unknown foraging behaviours, like underwater vocalisations (Thiebault et al. 2019) or the consumption of difficult to identify prey, like jellyfish (Thiebot et al. 2017).

Summary and conclusions

Tracking has fundamentally altered our understanding of the at-sea behaviour of seabirds, providing novel insights for a previously cryptic part of their lives. This has been transformative in terms of conservation and marine spatial planning as it provides evidence that can be used to quantify and, in some cases, mitigate direct threats to seabird populations. It also provides us with the chance to horizon scan and cements seabirds as marine sentinels and flagships for change.

Our chapter provides an overview of a very large body of research, and it seems likely that this field will grow in future as tags are becoming cheaper and analytical tools are shared more widely. Central to tracking being an effective tool for applied ecology is a detailed understanding of device effects. Collaboration is also crucial-seabirds are global travellers, and it is unrealistic that small research groups will have the resources to be able to operate at the necessary scales to be able to quantify this. While seabird population declines are a cause for deep concern, the response of the seabird research community to tracking and its application to conservation is most encouraging. Long may we continue to break down borders.

References cited

Ainley, D.G., R. Podolsky, L. Deforest and G. Spencer. 2001. The status and population trends of the Newell's shearwater on Kaua'i: Insights from Modeling. Stud. AvianBiol. 22: 108–123. doi: https://sora.unm.edu/node/139607.

Ancel, A., G.L. Kooyman, P.J. Ponganis, J.-P. Gendner, J. Lignon, X. Mestre et al. 1992. Foraging behaviour of emperor penguins as a resource detector in winter and summer. Nature 360: 336–339. doi: 10.1038/360336a0.

Barron, D.G., J.D. Brawn and P.J. Weatherhead. 2010. Meta-analysis of transmitter effects on avian behaviour and ecology. Methods Ecol. Evol. 1: 180–187. doi: 10.1111/j.2041-210X.2010.00013.x.

Bartumeus, F., L. Giuggioli, M. Louzao, V. Bretagnolle, D. Oro and S.A. Levin. 2010. Fishery discards impact on seabird movement patterns at regional scales. Curr. Biol. 20: 215–222. doi: 10.1016/j.cub.2009.11.073.

Beal, M., M.P. Dias, R.A. Phillips, S. Oppel, C. Hazin, E.J. Pearmaim et al. 2021. Global political responsibility for the conservation of albatrosses and large petrels. Sci. Adv. 7: eabd7225. doi: 10.1126/sciadv.abd7225.

Bennison, A., S. Bearhop, T.W. Bodey, S.C. Votier, W.J. Grecian, E.D. Wakefield et al. 2018. Search and foraging behaviors from movement data: A comparison of methods. Ecol. Evol. 8: 13–24. doi: 10.1002/ece3.3593.

Bernard, A., A.S.L. Rodrigues, V. Cazalis and D. Grémillet. 2021. Toward a global strategy for seabird tracking. Conserv. Lett. e12804. doi: 10.1111/conl.12804.

Bicknell, A.W.J., M.E. Knight, D.T. Bilton, M. Campbell, J.B. Reid, J. Newton et al. 2014. Intercolony movement of pre-breeding seabirds over oceanic scales: implications of cryptic age-classes for conservation and metapopulation dynamics. Divers. Distrib. 20: 160–168. doi: 10.1111/ddi.12137.

Bodey, T.W., M.J. Jessopp, S.C. Votier, H.D. Gerritsen, I.R. Cleasby, K.C. Hamer et al. 2014. Seabird movement reveals the ecological footprint of fishing vessels. Curr. Biol. 24: R514–R515. doi: 10.1016/j.cub.2014.04.041.

Bodey, T.W., I.R. Cleasby, F. Bell, N. Parr, A. Schultz, S.C. Votier et al. 2018. A phylogenetically controlled meta-analysis of biologging device effects on birds: Deleterious effects and a call for more standardized reporting of study data. Methods Ecol. Evol. 9: 946–955. doi: 10.1111/2041-210X.12934.

Bolton, M. 2020. GPS tracking reveals highly consistent use of restricted foraging areas by European Storm-petrels *Hydrobates pelagicus* breeding at the largest UK colony: Implications for conservation management. BirdConserv. Int. 31: 35–52. DOI: 10.1017/S0959270920000374.

Bouten, W., E.W. Baaij, J. Shamoun-Baranes and K.C.J. Camphuysen. 2013. A flexible GPS tracking system for studying bird behaviour at multiple scales. J. Ornithol. 154: 571–580. doi: 10.1007/s10336-012-0908-1.

Bridge, E.S., K. Thorup, M.S. Bowlin, P.B. Chilson, R.H. Diehl, R.W. Fléron et al. 2011. Technology on the move: Recent and forthcoming innovations for tracking migratory birds. Bioscience 61: 689–698. doi: 10.1525/bio.2011.61.9.7.

Brisson-Curadeau, E., A. Patterson, S. Whelan, T. Lazarus and K.H. Elliott. 2017. Tracking cairns: Biologging improves the use of seabirds as sentinels of the sea. Front. Mar. Sci. 4: 373. doi: 10.3389/fmars.2017.00357.

Browning, E., M. Bolton, E. Owen, S. Shoji, T. Guilford and R. Freeman. 2018. Predicting animal behaviour using deep learning: GPS data alone accurately predict diving in seabirds. Methods Ecol. Evol. 9: 681–692. doi: 10.1111/2041-210X.12926.

Burger, C., A. Schubert, S. Heinänen, M. Dorsch, B. Kleinschmidt, R. Žydelis et al. 2019. A novel approach for assessing effects of ship traffic on distributions and movements of seabirds. J. Environ. Manage. 251: 109511. doi: 10.1016/j.jenvman.2019.109511.

Burguer, A.E. and S.A. Shaffer. 2008. Application of tracking and data-logging technology in research and conservation of seabirds. Auk 125: 253–264. doi: 10.1525/auk.2008.1408.

Campbell, K.J., A. Steinfurth, L.G. Underhill, J.C. Coetzee, B.M. Dyer, K. Ludynia et al. 2019. Local forage fish abundance influences foraging effort and offspring condition in an endangered marine predator. J. Appl. Ecol. 56: 1751–1760. doi: 10.1111/1365-2664.13409.

Carneiro, A.P.B., E.J. Pearmain, S. Oppel, T.A. Clay, R.A. Phillips, A. Bonnet-Lebrun et al. 2020. A framework for mapping the distribution of seabirds by integrating tracking, demography and phenology. J. Appl. Ecol. 57: 514–525. doi: 10.1111/1365-2664.13568.

Chivers, L.S., S.A. Hatch and K.H. Elliott. 2016. Accelerometry reveals an impact of short-term tagging on seabird activity budgets. Condor 118: 159–168. doi: 10.1650/CONDOR-15-66.1.

Cianchetti-Benedetti, M., G. Dell'Omo, T. Russo, C. Catoni and P. Quillfeldt. 2018. Interactions between commercial fishing vessels and a pelagic seabird in the southern Mediterranean Sea. BMC Ecol. 18: 54. doi: 10.1186/s12898-018-0212-x.

Clark, B.L., F. Vigfúsdóttir, M.J. Jessopp, J.M. Burgos, T.W. Bodey and S.C. Votier. 2020. Gannets are not attracted to fishing vessels in Iceland-potential influence of a discard ban and food availability. ICESJ. Mar. Sci. 77: 692–700. doi: 10.1093/icesjms/fsz233.

Clay, T.A., C. Small, G.N. Tuck, D. Pardo, A.P.B. Carneiro, A.G. Wood et al. 2019. A comprehensive large-scale assessment of fisheries bycatch risk to threatened seabird populations. J. Appl. Ecol. 56: 1882–1893. doi: 10.1111/1365-2664.13407.

Cleasby, I.R., E.D. Wakefield, S. Bearhop, T.W. Bodey, S.C. Votier and K.C. Hamer. 2015. Three-dimensional tracking of a wide-ranging marine predator: flight heights and vulnerability to offshore wind farms. J. Appl. Ecol. 52: 1474–1482. doi: 10.1111/1365-2664.12529.

Corbeau, A., J. Collet, M. Fontenille and H. Weimerskirch. 2019. How do seabirds modify their search behaviour when encountering fishing boats? PLoS One 14: e0222615. doi: 10.1371/journal.pone.0222615.

Croxall, J.P. 2005. Global circumnavigations: Tracking year-round ranges of non-breeding albatrosses. Science 307: 249–250. doi: 10.1126/science.1106042.

de Grissac, S, F. Bartumeus, S.L. Cox and H. Weimerskirch. 2017. Early-life foraging: Behavioral responses of newly fledged albatrosses to environmental conditions. Ecol. Evol. 7: 6766–6778. doi: 10.1002/ece3.3210.

De Pascalis, F., S. Imperio, A. Benvenuti, C. Catoni, D. Rubolini and J.C. Cecere. 2020. Sex-specific foraging behaviour is affected by wind conditions in a sexually size dimorphic seabird. Anim. Behav. 166: 207–218. doi: 10.1016/j.anbehav.2020.05.014.

Dias, M.P., S. Oppel, A.L. Bond, A.P.B. Carneiro, R.J. Cuthbert, J. González-Solís et al. 2017. Using globally threatened pelagic birds to identify priority sites for marine conservation in the South Atlantic Ocean. Biol. Conserv. 211: 76–84. doi: 10.1016/j.biocon.2017.05.009.

Dias, M.P., A.P.B. Carneiro, V. Warwick-Evans, C. Harris, K. Lorenz, B.G. Lascelles et al. 2018. Identification of marine Important Bird and Biodiversity Areas for penguins around the South Shetland Islands and South Orkney Islands. Ecol. Evol. 8: 10520–10529. doi: 10.1002/ece3.4519.

Dias, M.P., R. Martin, E.J. Pearmain, I.J. Burfield, C. Small, R.A. Phillips et al. 2019a. Threats to seabirds: A global assessment. Biol. Conserv. 237: 525–537. doi: 10.1016/j.biocon.2019.06.033.

Dias, M.P., V. Warwick-Evans, A.P.B. Carneiro, C. Harris, B.G. Lascelles, H.L. Clewlow et al. 2019b. Using habitat models to identify marine important bird and biodiversity areas for Chinstrap Penguins *Pygoscelis antarcticus* in the South Orkney Islands. Polar Biol. 42: 17–25. doi: 10.1007/s00300-018-2404-4.

Dierschke, V., R.W. Furness and S. Garthe. 2016. Seabirds and offshore wind farms in European waters: Avoidance and attraction. Biol. Conserv. 202: 59–68. doi: 10.1016/j.biocon.2016.08.016.

Egevang, C., I.J. Stenhouse, R.A. Phillips, A. Petersen, J.W. Fox and J.R.D. Silk. 2010. Tracking of Arctic terns Sterna paradisaea reveals longest animal migration. Proc. Natl. Acad. Sci. 107: 2078–2081. doi: 10.1073/pnas.0909493107.

Fan, J., F. Han and H. Liu. 2014. Challenges of big data analysis. Natl. Sci. Rev. 1: 293–314. doi: 10.1093/nsr/nwt032.

Fayet, A.L., R. Freeman, T. Anker-Nilssen, A. Diamond, K.E. Erikstad, D. Fifield et al. 2017. Ocean-wide Drivers of migration strategies and their influence on population breeding performance in a declining seabird. Curr. Biol. 27: 3871–3878. doi: 10.1016/j.cub.2017.11.009.

Fischer, J., I. Debski, D. Spitz, G. Taylor and H. Wittmer. 2021. Year-round offshore distribution, behaviour, and overlap with commercial fisheries of a Critically Endangered small petrel. Mar. Ecol. Prog. Ser. 660: 171–187. doi: 10.3354/meps13586.

Fort, J., B. Moe, H. Strøm, D. Grémillet, J. Welcker, J. Schultner et al. 2013. Multicolony tracking reveals potential threats to little auks wintering in the North Atlantic from marine pollution and shrinking sea ice cover. Divers. Distrib. 19: 1322–1332. doi: 10.1111/ddi.12105.

Furness, R. 1997. Seabirds as monitors of the marine environment. ICES J. Mar. Sci. 54: 726–737. doi: 10.1006/jmsc.1997.0243.

Garriga, J., J.R.B. Palmer, A. Oltra and F. Bartumeus. 2016. Expectation-maximization binary clustering for behavioural annotation. PLoS One 11: e0151984. doi: 10.1371/journal.pone.0151984.

Garthe, S., N. Markones and A.M. Corman. 2017. Possible impacts of offshore wind farms on seabirds: A pilot study in Northern Gannets in the southern North Sea. J. Ornithol. 158: 345–349. doi: 10.1007/s10336-016-1402-y.

Gatt, M.C., B. Reis, J.P. Granadeiro, E. Pereira and P. Catry. 2020. Generalist seabirds as biomonitors of ocean mercury: The importance of accurate trophic position assignment. Sci. Total Environ. 740: 140159. doi: 10.1016/j.scitotenv.2020.140159.

Geen, G.R., R.A. Robinson and S.R. Baillie. 2019. Effects of tracking devices on individual birds—A review of the evidence. J. Avian Biol. 50: e01823. doi: 10.1111/jav.01823.

Gibb, R., A. Shoji, A.L. Fayet, C.M. Perrins, T. Guilford and R. Freeman. 2017. Remotely sensed wind speed predicts soaring behaviour in a wide-ranging pelagic seabird. J. R. Soc. Interface 14: 20170262. doi: 10.1098/rsif.2017.0262.

Gillies, N., A.L. Fayet, O. Padget, M. Syposz, J. Wynn, S. Bondet al. 2020. Short-term behavioural impact contrasts with long-term fitness consequences of biologging in a long-lived seabird. Sci. Rep. 10: 15056. doi: 10.1038/s41598-020-72199-w.

Giménez, J., G.E. Arneill, A. Bennison, E. Pirotta, H.D. Gerritsen, T.W. Bodey et al. 2021. Sexual mismatch between vessel-associated foraging and discard consumption in a marine top predator. Front. Mar. Sci. 8: 636468. doi: 10.3389/fmars.2021.636468.

Granadeiro, J.P., R.A. Phillip, P. Brickle and P. Catry. 2011. Albatrosses following fishing vessels: How badly hooked are they on an easy meal? PLoS One 6: e17467. doi: 10.1371/journal.pone.0017467.

Grecian, W.J., R. Inger, M.J. Attrill, S. Bearhop, B.J. Godely, M.J. Witt et al. 2010. Potential impacts of wave-powered marine renewable energy installations on marine birds. Ibis 152: 683–697. doi: 10.1111/j.1474-919X.2010.01048.x.

Grecian, W.J., M.J. Witt, M.J. Attrill, S. Bearhop, P.H. Becker, C. Egevang et al. 2016. Seabird diversity hotspot linked to ocean productivity in the canary current large marine ecosystem. Biol. Lett. 12: 20160024. doi: 10.1098/rsbl.2016.0024.

Grecian, WJ., H.J. Williams, S.C. Votier, S. Bearhop, I.R. Cleasby, D. Grémillet et al. 2019. Individual spatial consistency and dietary flexibility in the migratory behavior of Northern Gannets Wintering in the Northeast Atlantic. Front. Ecol. Evol. 7: 214. doi: 10.3389/fevo.2019.00214.

Grémillet, D., A. Ponchon, M. Paleczny, M.-L.D. Palomares, V. Karpouzi and D. Pauly. 2018. Persisting worldwide seabird-fishery competition despite seabird community decline. Curr. Biol. 28: 4009–4013. doi: 10.1016/j.cub.2018.10.051.

Grémillet, D., J. Collet, H. Weimerskirch, N. Courbin, P.G. Ryan and L. Pichegru. 2019. Radar detectors carried by Cape gannets reveal surprisingly few fishing vessel encounters. PLoS One 14: e0210328. doi: 10.1371/journal.pone.0210328.

Handley, J., M.-M. Rouyer, E.J. Pearmain, V. Warwick-Evans, K. Teschke, J.T. Hinke et al. 2021. Marine important bird and biodiversity areas for Penguins in Antarctica, targets for conservation action. Front. Mar. Sci. 7: 602972. doi: 10.3389/fmars.2020.602972.

Harcourt, R., A.M.M. Sequeira, X. Zhang, F. Roquet, K. Komatsu, M. Heupel et al. 2019. Animal-borne telemetry: An integral component of the ocean observing toolkit. Front. Mar. Sci. 6: 326. doi: 10.3389/fmars.2019.00326.

Hays, G.C., L.C. Ferreira, A.M.M. Sequeira, M.G. Meekan, C.M. Duarte, H. Bailey et al. 2016. Key questions in marine megafauna movement ecology. Trends Ecol. Evol. 31: 463–475. doi: 10.1016/j.tree.2016.02.015.

Hays, G.C., H. Bailey, S.J. Bograd, W.D. Bowen, C. Campagna, R.H. Carmichael et al. 2019. Translating marine animal tracking data into conservation policy and management. Trends Ecol. Evol. 34: 459–473. doi: 10.1016/j.tree.2019.01.009.

Hazen, E.L., B. Abrahms, S. Brodie, G. Carroll, M.G. Jacox, M.S. Savoca et al. 2019. Marine top predators as climate and ecosystem sentinels. Front. Ecol. Environ. 17: 565–574. doi: 10.1002/fee.2125.

Hedd, A., W.A. Montevecchi, L.M. Tranquilla, C.M. Burke, D.A. Fifield, G.J. Robertson et al. 2011. Reducing uncertainty on the Grand Bank: Tracking and vessel surveys indicate mortality risks for common murres in the North-West Atlantic. Anim. Conserv. 14: 630–641. doi: 10.1111/j.1469-1795.2011.00479.x.

Heerah, K., M.P. Dias, K. Delord, S. Oppel, C. Barbraud, H.Weimerskirch et al. 2019. Important areas and conservation sites for a community of globally threatened marine predators of the Southern Indian Ocean. Biol. Conserv. 234: 192–201. doi: 10.1016/j.biocon.2019.03.037.

Heylen, B.C. and D.A. Nachtsheim. 2018. Bio-telemetry as an essential tool in movement ecology and marine conservation. pp. 83–107. *In*: Jungblut, S., V. Liebich and M. Bode [eds.]. YOUMARES 8 - Oceans Across Boundaries: Learning from each other. Springer International Publishing, Kiel, Germany. doi: https://doi.org/10.1007/978-3-319-93284-2_7.

Hindell, M.A., R.R. Reisinger, Y. Ropert-Coudert, L.A. Hückstädt, P.N. Trathan, H. Bornemann et al. 2020. Tracking of marine predators to protect Southern Ocean ecosystems. Nature 580: 87–92. doi: 10.1038/s41586-020-2126-y.

Hudson, A.V. and R.W. Furness. 1989. The behaviour of seabirds foraging at fishing boats around Shetland. Ibis 131: 225–237. doi: 10.1111/j.1474-919X.1989.tb02765.x.

Inger, R., M.J. Attrill, S. Bearhop, A.C. Broderick, W.J. Grecian, D.J. Hodgson et al. 2009. Marine renewable energy: Potential benefits to biodiversity? An urgent call for research. J. Appl. Ecol. 46: 1145–1153. doi: 10.1111/j.1365-2664.2009.01697.x.

Jenouvrier, S., C. Péron and H. Weimerskirch. 2015. Extreme climate events and individual heterogeneity shape life-history traits and population dynamics. Ecol. Monogr. 85: 605–623. doi: 10.1890/14-1834.1.sm.

Jones, T.B., S.C. Patrick, J.P.Y. Arnould, M.A. Rodríguez-Malagón, M.R. Wells and J.A. Green. 2018. Evidence of sociality in the timing and location of foraging in a colonial seabird. Biol. Lett. 14: 20180214. doi: 10.1098/rsbl.2018.0214.

Joo, R., M.E. Boone, T.A. Clay, S.C. Patrick, S. Clusella-Trullas and M. Basille. 2020. Navigating through the R packages for movement. J. Anim. Ecol. 89: 248–267. doi: 10.1111/1365-2656.13116.

Jouventin, P. and H. Weimerskirch. 1990. Satellite tracking of Wandering albatrosses. Nature 343: 746–748. doi: 10.1038/343746a0.

Korpela, J., H. Suzuki, S. Matsumoto, Y. Mizutani, M. Samejima, T. Maekawa et al. 2020. Machine learning enables improved runtime and precision for bio-loggers on seabirds. Commun. Biol. 3: 633. doi: 10.1038/s42003-020-01356-8.

Kroodsma, D.A., J. Mayorga, T. Hochberg, N.A. Miller, K. Boerder, F. Ferretti et al. 2018. Tracking the global footprint of fisheries. Science 359: 904–908. doi: 10.1126/science.aao5646.

Krüger, L., J.A. Ramos, J.C. Xavier, D. Grémillet, J. González-Solís, Y. Kolbeinsson et al. 2017a. Identification of candidate pelagic marine protected areas through a seabird seasonal-, multispecific- and extinction risk-based approach. Anim. Conserv. 20: 409–424. doi: 10.1111/acv.12339.

Krüger, L., V.H. Paiva, M.V. Petry and J.A. Ramos. 2017b. Strange lights in the night: using abnormal peaks of light in geolocator data to infer interaction of seabirds with nocturnal fishing vessels. Polar Biol. 40: 221–226. doi: 10.1007/s00300-016-1933-y.

Lascelles, B.G., G.M. Langham, R.A. Ronconi and J.B. Reid. 2012. From hotspots to site protection: Identifying marine protected areas for seabirds around the globe. Biol. Conserv. 156: 5–14. doi: 10.1016/j.biocon.2011.12.008.

Lascelles, B.G., P.R. Taylor, M.G.R. Miller, M.P. Dias, S. Oppel, L. Torres et al. 2016. Applying global criteria to tracking data to define important areas for marine conservation. Divers. Distrib. 22: 422–431. doi: 10.1111/ddi.12411.

Le Bot, T., A. Lescroël and D. Grémillet. 2018. A toolkit to study seabird–fishery interactions. ICES J. Mar. Sci. 75: 1513–1525. doi: 10.1093/icesjms/fsy038.

Le Corre, M., A. Jaeger, P. Pinet, M.A. Kappes, H. Weimerskirch, T. Catry et al. 2012. Tracking seabirds to identify potential Marine Protected Areas in the tropical western Indian Ocean. Biol. Conserv. 156: 83–93. doi: 10.1016/j.biocon.2011.11.015.

Lennox, R.J., C. Engler-Palma, K. Kowarski, A. Filous, R. Whitlock, S.J. Cooke et al. 2019. Optimizing marine spatial plans with animal tracking data. Can. J. Fish. Aquat. Sci. 76: 497–509. doi: 10.1139/cjfas-2017-0495.

Lenton, T.M., J. Rockström, O. Gaffney, S. Rahmstorf, K. Richardson, W. Steffen et al. 2019. Climate tipping points—too risky to bet against. Nature 575: 592–595. doi: 10.1038/d41586-019-03595-0.

Lescroël, A., R. Mathevet, C. Péron, M. Authier, P. Provost, A. Takahashi et al. 2016. Seeing the ocean through the eyes of seabirds: A new path for marine conservation? Mar. Policy 68: 212–220. doi: 10.1016/j.marpol.2016.02.015.

Lieske, D.J., L.M. Tranquilla, R.A. Ronconi and S. Abbott. 2020. "Seas of risk": Assessing the threats to colonial-nesting seabirds in Eastern Canada. Mar. Policy 115: 103863. doi: 10.1016/j.marpol.2020.103863.

López-López, P. 2016. Individual-based tracking systems in ornithology: Welcome to the era of big data. Ardeola 63: 103–136. doi: 10.13157/arla.63.1.2016.rp5.

Louzao, M., K. Delord, D. García, I. Afán, J.M. Arcos and H. Weimerskirch. 2021. First days at sea: Depicting migration patterns of juvenile seabirds in highly impacted seascapes. PeerJ. 9: e11054. doi: 10.7717/peerj.11054.

Ludynia, K., N. Dehnhard, M. Poisbleau, L. Demongin, J.F. Masello and P. Quillfeldt. 2012. Evaluating the impact of handling and logger attachment on foraging parameters and physiology in southern Rockhopper Penguins. PLoS One 7: e50429. doi: 10.1371/journal.pone.0050429.

Mackley, E., R.A. Phillips, J.R.D. Silk, E.D. Wakefield, V. Afanasyev, J. Fox et al. 2010. Free as a bird? Activity patterns of albatrosses during the nonbreeding period. Mar. Ecol. Prog. Ser. 406: 291–303. doi: 10.3354/meps08532.

Mallory, M.L., S.A. Robinson, C.E. Hebert and M.R. Forbes. 2010. Seabirds as indicators of aquatic ecosystem conditions: A case for gathering multiple proxies of seabird health. Mar. Pollut. Bull. 60: 7–12. doi: 10.1016/j.marpolbul.2009.08.024.

Matos, D.M., J.A. Ramos, J.G. Calado, F.R. Ceia, J. Hey and V.H. Paiva. 2018. How fishing intensity affects the spatial and trophic ecology of two gull species breeding in sympatry. ICES J. Mar. Sci. 75: 1949–1964. doi: 10.1093/icesjms/fsy096.

Mattern, T., M.D. McPherson, U. Ellenberg, Y. van Heezik and P.J. Seddon. 2018. High definition video loggers provide new insights into behaviour, physiology, and the oceanic habitat of a marine predator, the yellow-eyed penguin. PeerJ. 6: e5459. doi: 10.7717/peerj.5459.

Maxwell, S.M., E.L. Hazen, R.L. Lewison, D.C. Dunn, H. Bailey, S.J. Bograd et al. 2015. Dynamic ocean management: Defining and conceptualizing real-time management of the ocean. Mar. Policy 58: 42–50. doi: 10.1016/j.marpol.2015.03.014.

McGowan, J., E. Hines, M. Elliott, J. Howar, A. Dransfield, N. Nur et al. 2013. Using seabird habitat modeling to inform marine spatial planning in central California's national marine sanctuaries. PLoS One 8: e71406. doi: 10.1371/journal.pone.0071406.

Merkel, B., R.A. Phillips, S. Descamps, N.G. Yoccoz, B. Moe and H. Strøm. 2016. A probabilistic algorithm to process geolocation data. Mov. Ecol. 4: 26. doi: 10.1186/s40462-016-0091-8.

Merkel, F.R. and K.L. Johansen. 2011. Light-induced bird strikes on vessels in Southwest Greenland. Mar. Pollut. Bull. 62: 2330–2336. doi: 10.1016/j.marpolbul.2011.08.040.

Miller, A., J.E. Elliott, L.K. Wilson, K.W. Elliott, K.G. Drouillar, J. Verreault et al. 2020. Influence of overwinter distribution on exposure to persistent organic pollutants (POPs) in seabirds, ancient murrelets (*Synthliboramphus antiquus*), breeding on the Pacific coast of Canada. Environ. Pollut. 259: 113842. doi: 10.1016/j.envpol.2019.113842.

Miller, K.A., K.F. Thompson, P. Johnston and D. Santillo. 2018. An overview of seabed mining including the current state of development, environmental impacts, and knowledge gaps. Front. Mar. Sci. 4: 418. doi: 10.3389/fmars.2017.00418.

Montevecchi, W.A., A. Hedd, L.M. Tranquilla, D.A. Fifield, C.M. Burke, P.M. Regular et al. 2012. Tracking seabirds to identify ecologically important and high risk marine areas in the western North Atlantic. Biol. Conserv. 156: 62–71. doi: 10.1016/j.biocon.2011.12.001.

Nakanowatari, T., K.I. Ohshima, V. Mensah, Y. Mitani, K. Hattori, M. Kobayashi et al. 2017. Hydrographic observations by instrumented marine mammals in the Sea of Okhotsk. Polar Sci. 13: 56–65. doi: 10.1016/j.polar.2017.06.001.

Navarro, J., D. Grémillet. I. Afán, F. Ramírez, W. Bouten and M.G. Forero. 2016. Feathered detectives: Real-time GPS tracking of scavenging gulls pinpoints illegal waste dumping. PLoS One 11: e0159974. doi: 10.1371/journal.pone.0159974.

Orben, R.A., J. Adams, M. Hester, S.A. Shaffer, R.M. Suryan, T. Deguchi et al. 2021. Across borders: External factors and prior behaviour influence North Pacific albatross associations with fishing vessels. J. Appl. Ecol. doi: 10.1111/1365-2664.13849.

Orgeret, F., C. Péron, M.R. Enstipp, K. Delord, H. Weimerskirch and C.A. Bost. 2019. Exploration during early life: distribution, habitat and orientation preferences in juvenile king penguins. Mov. Ecol. 7: 29. doi: 10.1186/s40462-019-0175-3.

Osborne, O., P. O'Hara, S. Whelan, P. Zandbergen, S. Hatch and K. Elliott. 2020. Breeding seabirds increase foraging range in response to an extreme marine heatwave. Mar. Ecol. Prog. Ser. 646: 161–173. doi: 10.3354/meps13392.

Paleczny, M., E. Hammill, V. Karpouzi and D. Pauly. 2015. Population trend of the world's monitored seabirds, 1950–2010. PLoS One 10: e0129342. doi: 10.1371/journal.pone.0129342.

Palumbi, S.R., P.A. Sandifer, J.D. Allan, M.W. Beck, D.G. Fautin, M.J. Fogarty et al. 2009. Managing for ocean biodiversity to sustain marine ecosystem services. Front. Ecol. Environ. 7: 204–211. doi: 10.1890/070135.

Parsons, M., I. Mitchell, A. Butler, N. Ratcliffe, M. Frederiksen, S. Foster et al. 2008. Seabirds as indicators of the marine environment. ICES J. Mar. Sci. 65: 1520–1526. doi: 10.1093/icesjms/fsn155.

Patrick, S.C., S. Bearhop, T.W. Bodey, W.J. Grecian, K.C. Hamer, J. Lee et al. 2015. Individual seabirds show consistent foraging strategies in response to predictable fisheries discards. J. Avian Biol. 46: 431–440. doi: 10.1111/jav.00660.

Patterson, A., H.G. Gilchrist, L. Chivers, S. Hatch and K. Elliott. 2019. A comparison of techniques for classifying behavior from accelerometers for two species of seabird. Ecol. Evol. 9: 3030–3045. doi: 10.1002/ece3.4740.

Patterson, T.A., L. Thomas, C. Wilcox, O. Ovaskainen and J. Matthiopoulos. 2008. State-space models of individual animal movement. Trends Ecol. Evol. 23: 87–94. doi: 10.1016/j.tree.2007.10.009.

Patterson, T.A., A. Parton, R. Langrock, P.G. Blackwell, L. Thomas and R. King. 2017. Statistical modelling of individual animal movement: An overview of key methods and a discussion of practical challenges. AStA. Adv. Stat. Anal. 101: 399–438. doi: 10.1007/s10182-017-0302-7.

Pereira, J.M., J.A. Ramos, A.M. Marques, F.R. Ceia, L. Krüger, S.C. Votier et al. 2021. Low spatial overlap between foraging shearwaters during the breeding season and industrial fisheries off the west coast of Portugal. Mar. Ecol. Prog. Ser. 657: 209–221. doi: 10.3354/meps13549.

Pereira, J.M., V.H. Paiva, F.R. Ceia and J.A. Ramos. 2020. Facing extremes: Cory's shearwaters adjust their foraging behaviour differently in response to contrasting phases of North Atlantic Oscillation. Reg. Environ. Chang. 20: 77. doi: 10.1007/s10113-020-01662-1.

Péron, C. and D. Grémillet. 2013. Tracking through life stages: Adult, immature and juvenile autumn migration in a long-lived seabird. PLoS One 8: e72713. doi: 10.1371/journal.pone.0072713.

Peschko, V., M. Mercker and S. Garthe. 2020. Telemetry reveals strong effects of offshore wind farms on behaviour and habitat use of common guillemots (*Uria aalge*) during the breeding season. Mar. Biol. 167: 118. doi: 10.1007/s00227-020-03735-5.

Peschko, V., B. Mendel, M. Mercker, J. Dierschke and S. Garthe. 2021. Northern gannets (*Morus bassanus*) are strongly affected by operating offshore wind farms during the breeding season. J. Environ. Manage. 279: 111509. doi: 10.1016/j.jenvman.2020.111509.

Phillips, R.A., J.C. Xavier and J.P. Croxall. 2003. Effects of satellite transmitters on albatrosses and petrels. Auk 120: 1082–1090. doi: 10.1093/auk/120.4.1082.

Phillips, R.A., J.R.D. Silk, J.P. Croxall, V. Afanasyev and D. Briggs. 2004. Accuracy of geolocation estimates for flying seabirds. Mar. Ecol. Prog. Ser. 266: 265–272. doi: 10.3354/meps266265.

Phillips, R.A., J.R.D. Silk, J.P. Croxall, V. Afanasyev and V.J. Bennett. 2005. Summer distribution and migration of nonbreeding albatrosses: Individual consistencies and implications for conservation. Ecology 86: 2386–2396. doi: 10.1890/04-1885.

Phillips, R.A., J.R.D. Silk, J.P. Croxall and V. Afanasyev. 2006. Year-round distribution of white-chinned petrels from South Georgia: Relationships with oceanography and fisheries. Biol. Conserv. 129: 336–347. doi: 10.1016/j.biocon.2005.10.046.

Piatt, I. and W. Sydeman. 2007. Seabirds as indicators of marine ecosystems. Mar. Ecol. Prog. Ser. 352: 199–204. doi: 10.3354/meps07070.

Pichegru, L., D. Grémillet, R.J.M. Crawford and P.G. Ryan. 2010. Marine no-take zone rapidly benefits endangered penguin. Biol. Lett. 6: 498–501. doi: 10.1098/rsbl.2009.0913.

Pichegru, L, P.G. Ryan, R. van Eeden, T. Reid, F. Grémillet and R. Wanless. 2012. Industrial fishing, no-take zones and endangered penguins. Biol. Conserv. 156: 117–125. doi: 10.1016/j.biocon.2011.12.013.

Poloczanska, E.S., M.T. Burrows, C.J. Brown, J.G. Molinos, B.S. Halpern, O. Hoegh-Guldberg et al. 2016. Responses of marine organisms to climate change across oceans. Front. Mar. Sci. 3: 62. doi: 10.3389/fmars.2016.00062.

Ponchon, A., D. Grémillet, B. Doligez, T. Chambert, T. Tveraa, J. González-Solís et al. 2013. Tracking prospecting movements involved in breeding habitat selection: Insights, pitfalls and perspectives. Methods Ecol. Evol. 4: 143–150. doi: 10.1111/j.2041-210x.2012.00259.

Rodríguez, A. and B. Rodríguez. 2009. Attraction of petrels to artificial lights in the Canary Islands: Effects of the moon phase and age class. Ibis. 151: 299–310. doi: 10.1111/j.1474-919X.2009.00925.x.

Rodríguez, A., B. Rodríguez and M.P. Lucas. 2012. Trends in numbers of petrels attracted to artificial lights suggest population declines in Tenerife, Canary Islands. Ibis 154: 167–172. doi: 10.1111/j.1474-919X.2011.01175.x.

Rodríguez, A., B. Rodríguez and J.J. Negro. 2015. GPS tracking for mapping seabird mortality induced by light pollution. Sci. Rep. 5: 10670. doi: 10.1038/srep10670.

Ronconi, R.A., B.G. Lascelles, G.M. Langham, J.B. Reid and D. Oro. 2012. The role of seabirds in marine protected Area identification, delineation, and monitoring: Introduction and synthesis. Biol. Conserv. 156: 1–4. doi: 10.1016/j.biocon.2012.02.016.

Ronconi, R.A., K.A. Allard and P.D. Taylor. 2015. Bird interactions with offshore oil and gas platforms: Review of impacts and monitoring techniques. J. Environ. Manage. 147: 34–45. doi: 10.1016/j.jenvman.2014.07.031.

Ropert-Coudert, Y. and R.P. Wilson. 2005. Trends and perspectives in animal-attached remote sensing. Front. Ecol. Environ. 3: 437–444. doi: 10.1890/1540-9295(2005)003[0437:TAPIAR]2.0.CO;2.

Ropert-Coudert, Y., A. Kato and A. Chiaradia. 2009. Impact of small-scale environmental perturbations on local marine food resources: a case study of a predator, the little penguin. Proc. R. Soc. B Biol. Sci. 276: 4105–4109. DOI: 10.1098/rspb.2009.1399.

Ropert-Coudert, Y., A. Kato, K. Shiomi, C. Barbraud, F. Angelier, K. Delord et al. 2018. Two recent massive breeding failures in an Adélie Penguin colony call for the creation of a marine protected area in D'Urville Sea/Mertz. Front. Mar. Sci. 5: 264. doi: 10.3389/fmars.2018.00264.

Ross-Smith, V.H., C.B. Thaxter, E.A. Masden, J. Shamoun-Baranes, N.H.K. Burton, L.J. Wright et al. 2016. Modelling flight heights of lesser black-backed gulls and great skuas from GPS: A Bayesian approach. J. Appl. Ecol. 53: 1676–1685. doi: 10.1111/1365-2664.12760.

Sansom, A., L.J. Wilson, R.W.G. Caldow and M. Bolton. 2018. Comparing marine distribution maps for seabirds during the breeding season derived from different survey and analysis methods. PLoS One 13: e0201797. doi: 10.1371/journal.pone.0201797.

Scales, K.L., P.I. Miller, L.A. Hawkes, S.N. Ingram, D.W. Sims and S.C. Votier. 2014. On the front line: Frontal zones as priority at-sea conservation areas for mobile marine vertebrates. J. Appl. Ecol. 51: 1575–1583. doi: 10.1111/1365-2664.12330.

Schwemmer, P., B. Mendel, N. Sonntag, V. Dierschke and S. Garthe. 2011. Effects of ship traffic on seabirds in offshore waters: implications for marine conservation and spatial planning. Ecol. Appl. 21: 1851–1860. doi: 10.1890/10-0615.1.

Sequeira, A.M.M., G.C. Hays, D.W. Sims, V.M. Eguíluz, J.P. Rodríguez, M.R. Heupel et al. 2019a. Overhauling ocean spatial planning to improve marine megafauna conservation. Front. Mar. Sci. 6: 639. doi: 10.3389/fmars.2019.00639.

Sequeira A.M.M., M.R. Heupel, M.-A. Lea, V.M. Eguíluz, C.M. Duarte, M.G. Meekan et al. 2019b. The importance of sample size in marine megafauna tagging studies. Ecol. Appl. 29: e01947doi: 10.1002/eap.1947.

Shaffer, S.A., Y. Tremblay, H. Weimerskirch, D. Scott, D.R. Thompson, P.M. Sagar et al. 2006. Migratory shearwaters integrate oceanic resources across the Pacific Ocean in an endless summer. Proc. Natl. Acad. Sci. 103: 12799–12802. doi: 10.1073/pnas.0603715103.

Soanes, L.M., J.P.Y. Arnould, S.G. Dodd, M.D. Sumner and J.A. Green. 2013a. How many seabirds do we need to track to define home-range area? J. Appl. Ecol. 50: 671–679. doi: 10.1111/1365-2664.12069.

Soanes, L.M., P.W. Atkinson, R.D. Gauvain and J.A. Green. 2013b. Individual consistency in the foraging behaviour of Northern Gannets: Implications for interactions with offshore renewable energy developments. Mar. Policy 38: 507–514. doi: 10.1016/j.marpol.2012.08.006.

Soldatini, C., M. Sebastiano, Y.V. Albores-Barajas, H. Abdelgawad, P. Bustamante and D. Costantini. 2020. Mercury exposure in relation to foraging ecology and its impact on the oxidative status of an endangered seabird. Sci. Total. Environ. 724: 138131. doi: 10.1016/j.scitotenv.2020.138131.

Southern, W. E. 1970. En route behavior of homing herring gulls as determined by radio-tracking. Wilson Bull. 82: 189–200. doi: https://www.jstor.org/stable/4159952?seq=1.

Steffen, W., K. Richardson, J. Rockstrom, S.E. Cornell, I. Fetzer, E.M. Bennett et al. 2015. Planetary boundaries: Guiding human development on a changing planet. Science 347: 1259855–1259855. doi: 10.1126/science.1259855.

Steinfurth, A., S. Oppel, M.P. Dias, T. Starnes, E. Pearmain, B. Dilley et al. 2020. Important marine areas for the conservation of northern rockhopper penguins within the Tristan da Cunha Exclusive Economic Zone. Endanger. Species Res. 43: 409–420. doi: 10.3354/esr01076.

Sun, A., S. Whelan, S. Hatch and K. Elliott. 2020. Tags below three percent of body mass increase nest abandonment by rhinoceros auklets, but handling impacts decline as breeding progresses. Mar. Ecol. Prog. Ser. 643: 173–181. doi: 10.3354/meps13341.

Teilmann, J., M.D. Agersted and M.P. Heide-Jørgensen. 2020. A comparison of CTD satellite-linked tags for large cetaceans—Bowhead whales as real-time autonomous sampling platforms. Deep Sea Res. Part I Oceanogr. Res. Pap. 157: 103213. doi: 10.1016/j.dsr.2020.103213.

Thaxter, C.B., V.H. Ross-Smith, W. Bouten, N.A. Clark, G.J. Conway, M.M. Rehfisch et al. 2015. Seabird-wind farm interactions during the breeding season vary within and between years: A case study of lesser black-backed gull *Larus fuscus* in the UK. Biol. Conserv. 186: 347–358. doi: 10.1016/j.biocon.2015.03.027.

Thiebault, A., I. Charrier, T. Aubin, D.B. Green and P.A. Pistorius. 2019. First evidence of underwater vocalisations in hunting penguins. PeerJ. 7: e8240. doi: 10.7717/peerj.8240.

Thiebot, J.-B., J.P.Y. Arnould, A. Gómez-Laich, K. Ito, A. Kato, T. Mattern et al. 2017. Jellyfish and other gelata as food for four penguin species—insights from predator-borne videos. Front. Ecol. Environ. 15: 437–441. doi: 10.1002/fee.1529.

Tranquilla, L.M., W.A. Montevecchi, A. Hedd, D. Fifield, C. Burke, P. Smith et al. 2013. Multiple-colony winter habitat use by murres *Uria* spp. in the Northwest Atlantic Ocean: Implications for marine risk assessment. Mar. Ecol. Prog. Ser. 472: 287–303. doi: 10.3354/meps10053.

Treasure, A., F. Roquet, I. Ansorge, M. Bester, L. Boehme, H. Bornemann et al. 2017 Marine mammals exploring the oceans pole to pole: A review of the MEOP consortium. Oceanography 30: 132–138. doi: 10.5670/oceanog.2017.234.

Tremblay, Y., A. Thiebault, R. Mullers and P.A. Pistorius. 2014. Bird-borne video-cameras show that seabird movement patterns relate to previously unrevealed proximate environment, Not Prey. PLoS One 9: e88424. doi: 10.1371/journal.pone.0088424.

Vanermen, N., W. Courtens, R. Daelemans, L. Lens, W. Müller, M. Van de walle et al. 2020. Attracted to the outside: A meso-scale response pattern of lesser black-backed gulls at an offshore wind farm revealed by GPS telemetry. ICES J. Mar. Sci. 77: 701–710. doi: 10.1093/icesjms/fsz199.

Votier, S.C., S. Bearhop, M.J. Witt, R. Inger, D. Thompson and J. Newton. 2010. Individual responses of seabirds to commercial fisheries revealed using GPS tracking, stable isotopes and vessel monitoring systems. J. Appl. Ecol. 47: 487–497. doi: 10.1111/j.1365-2664.2010.01790.x.

Votier, S.C., W.J. Grecian, S.C. Patrick and J. Newton. 2011. Inter-colony movements, at-sea behaviour and foraging in an immature seabird: results from GPS-PPT tracking, radio-tracking and stable isotope analysis. Mar. Biol. 158: 355–362. doi: 10.1007/s00227-010-1563-9.

Votier, S.C., A. Bicknell, S.L. Cox, K.L. Scales and S.C. Patrick. 2013. A Bird's eye view of discard reforms: Bird-borne cameras reveal seabird/fishery interactions. PLoS One 8: e57376. doi: 10.1371/journal.pone.0057376.

Votier, S.C., A.L. Fayet, S. Bearhop, T.W. Bodey, B.L. Clark, J. Grecian et al. 2017. Effects of age and reproductive status on individual foraging site fidelity in a long-lived marine predator. Proc. R. Soc. B Biol. Sci. 284: 20171068. doi: 10.1098/rspb.2017.1068.

Waggitt, J. and B. Scott. 2014. Using a spatial overlap approach to estimate the risk of collisions between deep diving seabirds and tidal stream turbines: A review of potential methods and approaches. Mar. Policy 44: 90–97. doi: 10.1016/j.marpol.2013.07.007.

Wakefield, E.D., R.A. Phillips and J. Matthiopoulos. 2009. Quantifying habitat use and preferences of pelagic seabirds using individual movement data: A review. Mar. Ecol. Prog. Ser. 391: 165–182. doi: 10.3354/meps08203.

Wakefield, ED., I.R. Cleasby, S. Bearhop, T.W. Bodey, R.D. Davies, P.I. Miller et al. 2015. Long-term individual foraging site fidelity-why some gannets don't change their spots. Ecology 96: 3058–3074. doi: 10.1890/14-1300.1.

Wang, G. 2019. Machine learning for inferring animal behavior from location and movement data. Ecol. Inform. 49: 69–76. doi: 10.1016/j.ecoinf.2018.12.002.

Weimerskirch, H. and A. Prudor. 2019. Cyclone avoidance behaviour by foraging seabirds. Sci. Rep. 9: 5400. doi: 10.1038/s41598-019-41481-x.

Weimerskirch, H., M. Salamolard, F. Sarrazin and P. Jouventin. 1993. Foraging strategy of Wandering Albatrosses through the breeding season: A study using satellite telemetry. Auk. 110: 325–342. doi: 10.2307/4088559.

Weimerskirch, H., S. Akesson and D. Pinaud. 2006. Postnatal dispersal of wandering albatrosses Diomedea exulans: Implications for the conservation of the species. J. Avian Biol. 37: 23–28. doi: 10.1111/j.0908-8857.2006.03675.x.

Weimerskirch, H., D.P. Filippi, J. Collet, S.M. Waugh and S.C. Patrick. 2018. Use of radar detectors to track attendance of albatrosses at fishing vessels. Conserv. Biol. 32: 240–245. doi: 10.1111/cobi.12965.

Weimerskirch. H., J. Collet, A. Corbeau, A. Pajot, F. Hoarau, C. Marteau et al. 2020. Ocean sentinel albatrosses locate illegal vessels and provide the first estimate of the extent of nondeclared fishing. Proc. Natl. Acad. Sci. 117: 3006–3014. doi: 10.1073/pnas.1915499117.

Wijeyakulasuriya, D.A., E.W. Eisenhauer, B.A. Shaby and E.M. Hanks. 2020. Machine learning for modeling animal movement. PLoS One 15: e0235750. doi: 10.1371/journal.pone.0235750.

Wikelski, M., E.W. Kays, N.J. Kasdin, K. Thorup, J.A. Smith and G.W. Swenson. 2007. Going wild: What a global small-animal tracking system could do for experimental biologists. J. Exp. Biol. 210: 181–186. doi: 10.1242/jeb.02629.

Williams, H.J., L.A. Taylor, S. Benhamou, A.I. Bijleveld, T.A. Clay and S. de Grissac et al. 2020. Optimizing the use of biologgers for movement ecology research. J. Anim. Ecol. 89: 186–206. doi: 10.1111/1365-2656.13094.

Wilmers, C.C., B. Nickel, C.M. Bryce, J.A. Smith, R.E. Wheat and V. Yovovich. 2015. The golden age of bio-logging: How animal-borne sensors are advancing the frontiers of ecology. Ecology 96: 1741–1753. DOI: 10.1890/14-1401.1.

Wilson, R.P., J.J. Ducamp, W.G. Rees, B.M. Culik and K. Niekamp. 1992. Estimation of location: Global coverage using light intensity. pp. 131–134 *In*: Priede, I.G. and S.M. Swift [eds.]. Wildlife Telemetry: Remote Monitoring and Tracking of Animals. Ellis Horwood Ltd, Chichester, United Kingdom.doi: https://doi.org/10.1007/BF00044136.

Wilson, R.P., D. Grémillet, J. Syder, M. Kierspel, S. Garthe, H. Weimerskirch et al. 2002. Remote-sensing systems and seabirds: Their use, abuse and potential for measuring marine environmental variables. Mar. Ecol. Prog. Ser. 228: 241–261. doi: 10.3354/meps228241.

Yoda, K. 2019. Advances in bio-logging techniques and their application to study navigation in wild seabirds. Adv. Robot. 33: 108–117. doi: 10.1080/01691864.2018.1553686.

Yonehara, Y., Y. Goto, K. Yoda, Y. Watanuki, L.C. Young, H. Weimerskirch et al. 2016. Flight paths of seabirds soaring over the ocean surface enable measurement of fine-scale wind speed and direction. Proc. Natl. Acad. Sci. 113: 9039–9044. doi: 10.1073/pnas.1523853113.

Part B
Seabirds Interacting with Human Activities and Infrastructures

Chapter 1

Seabird and Fisheries Interactions

Nuno Oliveira,[1,*] *Jaime A. Ramos,*[2] *Joana Gomes Calado*[2] and *José Manuel Arcos*[3]

Introduction

Commercial fisheries caused dramatic changes in the structure and functioning of marine ecosystems worldwide, particularly after the industrialization of fisheries in the XX century (Williams 1998, Pauly et al. 2005, McCauley et al. 2015). Overfishing is visibly most significant in coastal waters (Jackson 2001), but when over pelagic fish stocks (Swartz et al. 2010) and in deep waters (Pauly et al. 2003) is also of great concern. Traditionally, seabirds have benefited fisheries because fishermen use seabird aggregations to locate fish schools. Even modern industrial tuna purse-seine fisheries locate fishing grounds using bird radars to detect seabird distribution and behaviour (Hall and Roman 2013). On the other hand, seabirds may also have negative effects on fisheries when they steal bait, cause gear damages or reduce the amount of catches (Montevecchi 2002). However, the influence of fisheries on seabirds are prevalent, and these can be both direct or indirect and either positive or negative (Montevecchi 2002). Direct and positive effects refer to the supply of large and patchily predictable food resources as discards, offal and the facilitation in capturing prey during fishing operations. Direct and negative effects encompass the competition between seabirds and fish stocks, and fishery related seabird injury and mortality. Indirect effects may be positive when fisheries reduce seabird competitors or negative, when fisheries deplete fish stocks and affect seabird communities in multiple ways, including the capture of species such as tuna that drive pelagic fish species to the surface and thus act as facilitators to seabirds (Ashmole and Ashmole 1967, see review by Veit and Harrison 2017). The different aspects of seabird-fishery interactions have been addressed in many papers, including reviews on interactions between fisheries and seabirds (Tasker et al. 2000, Montevecchi 2002), and on best practices to study competition between seabirds and fisheries (Sydeman et al. 2017) and seabird-fishery direct interactions (Le Bot et al. 2018).

[1] SPEA – Portuguese Society for the Study of Birds, Marine Conservation Department, Av. Columbano Bordalo Pinheiro 87 3º andar, 1070-062 Lisboa, Portugal.

[2] University of Coimbra, MARE - Marine and Environmental Sciences Centre, Department of Life Sciences, 3004-517 Coimbra, Portugal.
Emails: jramos@uc.pt; joana.gomes.calado@gmail.com

[3] SEO/BirdLife, Marine Programme, C/Murcia 2-6, local 13, 08026 Barcelona, Spain and Dept. Biologia Evolutiva, Ecologia i Ciències Ambientals, Universitat de Barcelona, Av Diagonal 643, 08028 Barcelona, Spain.

* Corresponding author: nuno.oliveira@spea.pt

First and foremost, the effects of fisheries on seabirds may vary at the species, population, sex and individual levels. This means that it may be difficult to predict how individuals of a specific population will react to fisheries in a specific location. Observations on board fishing boats show that species abundance behind fishing vessels, prey capture method (plunge-diving, surface seizing, kleptoparasitism), the dominance rank and aggressiveness of individuals in the huge seabird aggregation vary strongly between species and individuals (Hudson and Furness 1989, Cherel et al. 1996). It is often reported that the most generalist and dominant species forage more around fishing vessels than specialist species (Navarro et al. 2010, Camphuysen et al. 2015), although this might be not always the case (Arcos and Oro 2002b, Collet et al. 2017). Fisheries have an obvious influence on seabird foraging ecology, mostly in terms of (1) diet, by providing new prey items not normally available to surface-feeders and shallow-divers, such as bathypelagic and demersal species through fishery discards (Mariano-Jelicich et al. 2014, Alonso et al. 2015, Calado et al. 2021a), and (2) foraging distribution, because seabirds may become more attracted to fishing boats than to the open sea due to facilitated foraging opportunities (Arcos and Oro 1996, Cama et al. 2013). Fisheries also have the potential to shape population structure, for example when a specific sex or age class is disproportionally linked with fishing vessels. This is the case for Wandering Albatrosses (*Diomedea exulans*) breeding at South Georgia, where the overlap with pelagic longline fisheries and resulting mortality is female biased (Jiménez et al. 2016, Gianuca et al. 2017). Fisheries may reduce the prey density below a threshold level that is needed for successful seabird foraging (Cury et al. 2011), but well-designed studies are needed to evaluate the role of fisheries versus other factors in explaining when prey density drops below such a threshold level (Sydeman et al. 2017).

The review of Le Bot et al. (2018) identified the four main aspects in studies of seabird-fishery direct interactions that are essential to aid conservation strategies: (1) evaluate overlap between fisheries and foraging seabirds, (2) assess movement and behaviour of seabirds that attend fisheries, (3) quantify to which extent seabird populations make use of fishery subsidies and (4) evaluate the impact of seabird-fishery interactions on seabird demographic parameters. All modern techniques to study seabird distribution and foraging ecology are important to address seabird fishery interactions, including diet assessment (Ceia et al. 2022) and movement ecology (Pereira et al. 2022). Recent advances such as devices combining a GPS with a ship radar/detector of the automatic ship identification system are very promising to record encounters of seabirds with fishing vessels (Weimerskirch et al. 2017). This chapter provides an overview of at-sea seabird-fishery interactions, focusing on the effects that two main fishery characteristics have on seabird populations, discards and seabird bycatch, and on how to reduce these effects using legislation and mitigation measures. This will provide a framework for the need to involve the fisheries sector in the understanding of seabird-fisheries interactions, and in raising awareness, capacity building and collaboration with fishermen for the implementation of solutions to mitigate the negative impacts of fisheries on seabird populations.

Effects of discards on seabirds

Fishery discards are defined as the portion of the catch which is not retained on board during commercial fishing operations and is returned to the sea, often dead or dying (Kelleher 2005). Discards do not include post-harvest waste, such as offal, but include slipped fish, i.e., fish caught in a net and released deliberately into the sea without being brought aboard (Kelleher 2005). Slipping is typically performed in purse-seine fisheries, where partial and total catches might be released (Borges et al. 2001, Stratoudakis and Marçalo 2002), and is a manoeuvre known to often cause physiological responses, physical damage and behavioural impairments to fish (Marçalo et al. 2010, Marçalo et al. 2013). Fishermen can discard both target and non-target (i.e., bycatch) species mainly due to economic and regulatory reasons (Table B1.1).

Table B1.1. Main motives for discarding (after Kelleher 2005, Fernandes et al. 2015, Gilman et al. 2017).

Categories	Motives
Economic	Little or no market for species/sizes/sex
	Highgrading (i.e., discarding lower-value catch)
	Damaged catch
	Species that can damage the rest of the catch
	Poisonous/spoiled
Regulatory	Exceeded quotas
	Minimum landing size
	Limits on catch composition (percentages by species)
	Bycatch limits
	Protected and unmarketable species/sexes (including seasonal bans) Specific seasonal bans
Other	Insufficient space to retain all catch

Seabirds are the most conspicuous scavengers feeding on fishery discards (Furness et al. 2007). In the 1990's, Garthe et al. (1996) estimated a total amount of 789000 t of fishery waste in the North Sea, potentially benefiting about 5.9 million seabird individuals, which consumed 310 000 t of fishery waste. In the North Sea seabird community populations of scavenging gulls, kittiwakes, skuas, fulmars and gannets have become strongly dependent on food sources associated with fishery waste. A similar pattern is apparent in other areas of the Southern hemisphere where species such as the Black-browed Albatross (*Thalassarche melanophris*) rely strongly on fishery discards in areas such as the Falkland Islands and Iles Kerguelen, presenting a higher breeding success in areas where fishery discards are present in their diet (McInnes et al. 2017). Likewise, extensive consumption of discards has been reported in temperate regions such as the Mediterranean (Oro and Ruiz 1997), by several opportunistic species such as gulls, but also specialised and threatened species such as the Critically Endangered Balearic Shearwater (*Puffinus mauretanicus*) (Oro and Ruiz 1997, Arcos and Oro 2002b). However, tropical seabirds appear to rely less on fishery discards, but more studies are needed in tropical areas. Blaber et al. (1995) studied the diet of 12 tropical seabird species in the Great Barrier Reef, Australia, and found that up to 20% of the diet of terns was made of discarded taxa, particularly during the open trawling season. On the other hand, the Masked Bobby (*Sula dactylatra*), Red-footed Booby (*Sula sula*) and Great Frigatebird (*Fregata minor*) did not consume fishery discards. In the Indian Ocean also foraging Great Frigatebirds do not overlap with fisheries (Weimerskirch et al. 2010).

Exploiting such abundant and easily accessible food will decrease the time and energy spent foraging (Emlen 1966), at least when natural prey is scarce. Fishery discards can be beneficial both during the (1) breeding season, allowing greater nest attendance and chick food provisioning, thereby increasing breeding success (Oro et al. 1995, Oro 1996), and (2) the non-breeding season, enabling a higher body condition and survival of individuals (Berón et al. 2013). Fishery discards have been shown to potentially promote seabird population growth (Tasker et al. 2000, Bicknell et al. 2013, Sherley et al. 2020). However, in areas where discards are of lower nutritional value, mainly composed of demersal fish species, they may constitute ecological traps for seabirds, and may not compensate the depletion of natural small pelagic fish prey (Le Bot et al. 2019). This negative effect of the switch from high-lipid to low-lipid prey was named the junk-food hypothesis (Alverson 1992); it was proposed for several seabird populations (Grémillet et al. 2008) and received support from controlled conditions in captivity (Romano et al. 2006). Overall, fishery discards have shaped seabird communities for many areas, to the benefit of generalist predators (Tasker et al. 2000, Votier et al. 2004), and the large increase of gull populations in the northern hemisphere is perhaps the best example of how fishery discards have changed seabird communities (Tasker et al. 2000).

Opportunistic species such as the Yellow-legged Gull (*Larus michahellis*) interact with different types of fisheries. For example, in the western Portuguese coast, where purse-seiners are very important in terms of landings, they were the most abundant species interacting with fishing boats, regardless of fishery catch quantities (Calado et al. 2021b), suggesting that gulls were foraging through facilitated foraging opportunities (Arcos and Oro 2002a, González-Solís 2003). In the Ebro Delta (Mediterranean Sea), where trawlers dominate the fisheries, yellow-legged gulls were also the most abundant species, presumably benefiting from the high discard availability provided by trawlers (Arcos et al. 2001). Overall, dietary analyses indicate that the most consumed fish species by yellow-legged gulls across four breeding colonies of the western Iberian Peninsula match the fish species that are most landed by local fisheries (Calado et al. 2021a), suggesting a strong association between gull diet and fishing activities. Additionally, pelagic fish landings seemed to affect gull stable isotope values and blood biochemical parameters, suggesting that foraging on discards have trophic and nutritional effects (Calado et al. 2021a).

Bycatch of seabirds by fisheries

Bycatch is defined as the incidental (unintentional) capture of large marine biota (usually applied to seabirds, mammals, turtles and large non-commercial predatory fish) in fishing gear. It is considered as one of the most severe and widespread threats to seabirds at sea, as it causes direct mortality and therefore affects adult survival, the most sensitive parameter of these long-lived animals (Dias et al. 2019). Seabird bycatch gained global awareness in the 1990s when it was associated to the decline of several albatross species from the Southern Oceans, mainly due to pelagic longlines (Croxall et al. 1990, Brothers 1991).

However, the problem is widespread across the world, and involves different types of gear (Table B1.2). Particular attention has been given to both pelagic and bottom longlines, as well as to static nets, but it also occurs in trawlers, purse-seiners, fishing traps and others, including recreational fisheries. The species involved may differ according to the region and type of gear; surface-feeding seabirds such as albatrosses and petrels are more affected by longlines, usually offshore, whereas for diving species such as cormorants, seaducks and penguins, static nets in shallow waters are of major

Table B1.2. The different risks of seabird bycatch in relation to fishing gear.

Fishing gear	Risk of bycatch
Bottom and pelagic longlines	Hook entanglement during setting and until longlines reach a depth out of range for diving seabirds; to a lesser extent during hauling.
Small-scale demersal longlines	Hook entanglement usually during setting and to a lesser extent during hauling; but in some fisheries setting the line in shallow waters may also occur largely during the period longlines are soaked in the water and at a diving range for foraging seabirds.
Trawls	Collisions with nets and cables, warp cables and paravanes; net entanglement during hauling; net entanglement during the entire operation in the case of pelagic trawls.
Drift gillnets	Net entanglement along the entire operation, but higher risk during the period gillnets are soaked in water.
Set gillnets	Net entanglement is often likely to occur during net setting when nets are set with fish remains from the previous net event; or along the entire operation (during hauling) in fishing grounds at a diving range for foraging seabirds.
Purse-seines	Seabirds are blocked by the concentrated fish school during the ending phase of hauling; are drowned when attempting to come to the surface underneath a net fold that has formed underwater; or get crushed through the hauling hydraulic winches (Triplex Power-Blocks) as some lean on the net being hauled.

concern. Globally, seabird deaths have been estimated in 160,000–320,000 per year for longlines (Anderson et al. 2011) and 400,000 for static nets (Zydelis et al. 2013), while no overall assessment is available for other gears. Therefore, well over half a million seabirds are estimated to die annually from bycatch, a figure that could be far larger if all types of gears were assessed at the global scale, and the multiple knowledge gaps were covered.

A first step to understand seabird bycatch is to collect reliable information of its occurrence, as to provide estimates of birds affected at the species level, as well as to understand which factors may influence their capture (details on fishing practices, temporal and spatial patterns, environmental conditions, etc.). This poses a major challenge when facing a commercial activity that has traditionally paid little attention to side-effects on the ecosystem, and that operates far away from the view of the general public or even from seabird researchers. However, increasing awareness on ecosystem impact of fisheries beyond the targeted fish stocks (Tegner and Dayton 1999, Estes et al. 2011) is increasingly pushing management bodies to address issues such as seabird bycatch.

For industrial fisheries, which often consist of relatively few, large vessels, it is common practice to have on board observer programmes, where independent observers collect information on the fishing activity, and this may include the occurrence of seabird bycatch. A pioneering programme addressing seabird bycatch was that of the Commission for the Conservation of Antarctic Marine Living Resources (CCMLR), in the Southern Oceans, but, several Regional Fisheries Management Organizations (RFMOs) have followed this example and start collecting information on bycatch. Far more difficult is to cover small-scale fisheries, particularly those of artisanal type consisting of hundreds or thousands of small vessels, for which an observer programme is out of scope (Salas et al. 2007). For these fisheries, questionnaires have helped to provide preliminary assessments of seabird bycatch, although information should be taken as just indicative, and to likely underestimate bycatch (as fishermen may be reluctant to provide sensitive information to strangers, e.g., Oliveira et al. 2015). An alternative approach that might provide more reliable information is to deliver self-reporting logbooks to fishermen, where bycatch is recorded whenever it occurs, along with information on the fishing activity on a daily basis. The use of cameras onboard is also being tested in several fisheries as an alternative to bringing observers onboard (Bartholomew et al. 2018).

To the problems described above, it is important to mention the difficulties inherent to the statistical nature of bycatch events. Although some fisheries capture seabirds on a regular basis, in many cases bycatch is a rather infrequent phenomenon, in which a few (occasional) events may involve tens or even hundreds of birds at a time, due to the gregarious foraging behaviour of many species (Laneri et al. 2010). This irregularity makes it difficult to detect these cases if there is no systematic recording programme in place, and to infer estimates of seabirds affected by a given fishery. On the other hand, even if the occasional bycatch events did only involve single or few birds at a time, and was not perceived by fishermen as a problem, for small seabird populations facing bycatch by small-scale fisheries (consisting of too many vessels) the impact might be enough as to cause severe declines. Moreover, these types of fisheries are often multi-gear (i.e., alternate different gears), and their fishing practices differ in too many subtleties, which may influence the probability of bycatch. For such fisheries it is particularly complex to find regular patterns of bycatch, and hence to provide any estimates of its occurrence and impact.

Another challenge is to assess the impact of bycatch on seabird populations, which may differ between gear types, regions and seabird species. While the large numbers of bycaught seabirds reported or estimated for some fleets and regions are of direct concern, the key issue here is to relate such mortality with demographic parameters and population changes, but few studies provide sound information on this issue. For instance, Genovart et al. (2018) related different adult survival rates of Scopoli's Shearwaters (*Calonectris diomedea*) from three western Mediterranean colonies with the risk of bycatch for each colony, as estimated from the overlap between the foraging grounds of the birds at each colony (as revealed by GPS tracking) and the longline fishing effort in the region. Regular et al. (2013) followed another approach, by assessing population changes in eastern Canada seabird communities after the closure of an important gillnet fishery in 1992. As predicted, diving species increased, while surface-feeding scavenging species declined (likely due to a reduction on the availability of discards).

Effects of the European Union discard ban policy on seabirds

The recent collapse of coastal fish stocks in several areas of the world raises serious concerns for humans and the environment (Pauly and Christensen 1995). Despite the technological advances and exploitation of new fishing resources, the catch per unit effort is decreasing, reflecting overexploitation of marine resources and fish depletion (Thurstan et al. 2010, Rousseau et al. 2019), with large impacts on other organisms such as plankton, predatory fish, cetaceans and seabirds (Smith et al. 2011). Despite these concerns, most fisheries throughout the world discard large amounts of unwanted fish, leading to stronger impacts on marine ecosystems. As part of the reform of the Common Fisheries Policy, and in order to attain sustainable fisheries and revitalize the fisheries economic sector (Leitão and Baptista 2017), the European Union (EU) is imposing an obligation to land the unwanted catch, known as the discard ban policy. We focus on the implications that this policy can have for seabirds, as an example of legislation with potential large scale beneficial effects for the marine environment (for further details and analysis of this policy see Uhlmann et al. 2019).

The discard ban policy prohibits the discard of species subject to catch limits (i.e., Total Allowable Catches, TACs), which includes most commercial species such as mackerels and anchovies that are very attractive for seabirds. On the other hand, species that are not subjected to a minimum landing size can still be discarded. Other exceptions include species with a high survival probability in relation to capture and discarding processes, and limited volumes of permissible discards under certain conditions. This so called *de minimis* exemption was included because the policy recognizes that achieving zero discards by some fisheries is impossible, and a minimum discarding allowance of up to 5% is permitted in relation to the different discard levels of the fisheries that operate in each region (Uhlmann et al. 2019). However, to apply for exemptions the stakeholders of the fisheries sector need to prepare and apply a discard management plan (Damalas 2015). Overall, the characteristics of the discard ban policy may lead to short-and-medium term regional differences in the effects on seabird communities.

Without the easy meal provided by discards, it is obvious that frequent scavengers around fishing vessels will be affected. The plastic and opportunistic behaviour of the scavenging seabirds suggests that they should search for other food resources and perhaps intensify the use of alternative habitats. Votier et al. (2004) showed that the Great Skua (*Stercorarius skua*), a generalist predator and scavenger, feeds more on discards when they are available, and increases the predation pressure on other bird species when discards are reduced, particularly if such a reduction coincides with a low availability of small pelagic fish. Several recent studies addressed the short-and-medium term impacts of the discard ban policy on opportunistic gull species. Generalist species such as the Yellow-legged Gull may increase the consumption of alternative prey species, and the use of alternative habitats such as terrestrial and urban environments (Bicknell et al. 2013). In conditions of lower fish availability, Yellow-legged Gulls are expected to increase the consumption of marine invertebrates, refuse and even fleshy fruits (Matos et al. 2018, Calado et al. 2021a). Intra- and inter-specific kleptoparasitism as well as predation on other seabird species are likely to increase, with a possible consequent impact on vulnerable species such as the Audouin's Gull (*Ichthyaetus audouinii*) (Arcos et al. 2001, Martínez-Abraín et al. 2003). In contrast, when faced with a reduction in the availability of discards, the more piscivorous specialist Audouin's Gull may increase marine foraging effort (i.e., foraging time and ranges) and changes its at-sea distribution (González-Solís et al. 1997, Bécares et al. 2015). Dietary changes may also occur, particularly a decrease in the consumption of demersal fish and an increase in the consumption of epipelagic fish and pelagic crabs (Matos et al. 2018).

The discard ban may also enhance the attraction of seabirds to fishing gear that provide alternative food sources, such as baited hooks from longlines, but with an increased risk of bycatch: Soriano-Redondo et al. (2016) showed that the probability of Scopoli's Shearwaters to interact with longliners decreases exponentially with the number of trawlers in the area. In order to evaluate the true impact of a change in legislation such as the discard ban policy long term studies are needed. This will be necessary to control other potentially long-time visible confounding factors such as environmental changes, low reproductive success and high survival rates of seabirds which imply obvious changes in demography. Species with very large populations will adjust to the change, and may decrease without any conservation concerns. However, small populations that are of conservation concern either locally or regionally may be more affected, and policy measures may be necessary to reduce the impact of the discard ban. Ultimately, the effects of the discard ban should be compensated in the medium-long term if the policy is applied in a way that accomplishes its main aim, i.e., reducing overfishing and hence allowing fish stocks to recover. In the long-term the discard ban may lead to a readjustment of seabird communities, in favour of specialist species.

Mitigation measures

Bycatch mortality has led to a sharp decline in many seabird species around the world. In order to solve such a negative impact on seabirds, several mitigation measures have been tested (Williams et al. 2012). The first measures were tested, as expected, in those areas where the problem of bycatch was first identified. The initial tests date back to the early 1990's in the Southern Ocean (Bull 2007) to decrease associated bycatch mortality of albatrosses, following the proven strong mortality associated with pelagic longline fisheries. Recently, with increased knowledge and awareness that bycatch is a serious threat to seabirds at a global level (Philips et al. 2016, Rodríguez et al. 2019), there were efforts to find effective solutions for each fishing gear.

In general, mitigation measures can be split into two large groups, operational measures and technical measures. The first are related to the operability of fishing gears or vessels, while the second imply a modification in the gear or using specific devices. The night setting is one of the most efficient operational measures (e.g., Sánchez and Belda 2003). Bearing in mind that the majority of seabirds feed mainly during the day, setting fishing nets or longlines during the night very effectively prevents seabird bycatch. Although, in the case of small longliners operating in shallow waters, it could make more sense to limit the entire operation to the night period (Cortés and González-Solis 2018). This is because birds tend to be caught during the period when the longline is setting in the water, rather than during setting or hauling, as it is being observed in south Ocean pelagic longlines. Another operational measure is related with depth at which fishing gears (namely gillnets and longlines) are set. Those gears set in deeper waters, near the bottom, are less likely to catch seabirds, than those set at shallow waters (Williams et al. 2012, Senko et al. 2013). Another easy way to reduce seabird attendance to the fishing vessels is to avoid discarding during or among fishing events. In general, seabird species are scavengers that are attracted to vessels when food is available (Pierre et al. 2014). While changing for less attractive bait (e.g., crabs, octopus or shrimps) may also reduce the catchability of seabirds, some fishermen find these types of baits less effective for certain target species. On the other hand, technical measures have explored one of the main sensory mechanisms used by birds—the vision (Martin and Crawford 2015). Vision is the first mechanism used by birds to detect the presence of a threat, be it in the shape of a predator or an object that the bird needs to deviate from during the flight. As such, the technical mitigation of seabird bycatch has focused mainly on measures that exploit this sensory field, either by means of direct deterring or by increasing the visibility of a certain barrier or threat. Scary-bird devices are examples of the former (Oliveira et al. 2020). By mimicking an avian predator, scary-bird devices have the function of alerting and deterring birds from a certain location, such as a fishing area or an active vessel. Scary-bird showed a great potential for deterring Yellow-Legged

Gulls and Lesser Black-backed Gulls (*Larus fuscus*) in purse seines in the Portuguese western coast. These devices have also shown great potential by apparently reducing Northern Gannet (*Morus bassanus*) and Cory's Shearwater (*Calonectris borealis*) bycatch in longlines and gillnets. Within the same study of Oliveira et al. (2020), another measure recently developed and tested was the high contrast panels. These panels are composed of strips of white and black cloth attached to the centre of gillnets. Their main purpose is to increase the probability of detection of gillnets by seabirds. However, its effectiveness is not clear. The first studies indicated a reduction in the bycatch rate, but more recent data point to a reduced effectiveness or even an inverse attraction effect, which in turn causes an increase in bycatch probability (Field et al. 2019). A slightly more technological and, therefore, more expensive measure, are LED light systems. These are composed of a LED coupled to a silicone support and arranged along the top main gillnet cable with a 10 m spacing. Being a very recent measure, clear results of the effect of this measure are not yet known (Cantlay et al. 2020, Bielli et al. 2020, Mangel et al. 2018).

In the case of industrial fisheries, several mitigation measures are widely accepted according to BirdLife International fact sheets (BirdLife International and ACAP 2014), namely for pelagic longlines and trawls. Such measures are meant to avoid interactions with seabirds (night setting and spatiotemporal closures of fishing activity), to reduce baited hook availability to seabirds (underwater setting funnel, weighted lines, thawed bait, line shooter, bait-casting machines and side-setting), to deter birds from reaching baited hooks (streamer liner, acoustic deterrents and water cannons), to make baited hooks less attractive/visible for seabirds (dumping of offal, artificial baits and blued dyed bait), to avoid cable strikes (deploy streamer lines and install snatch blocks), to reduce net entanglement (clean nets after every shot, avoid long times with the net on the water surface, by applying net binding and add weight to the net belly) and finally to improve offal management.

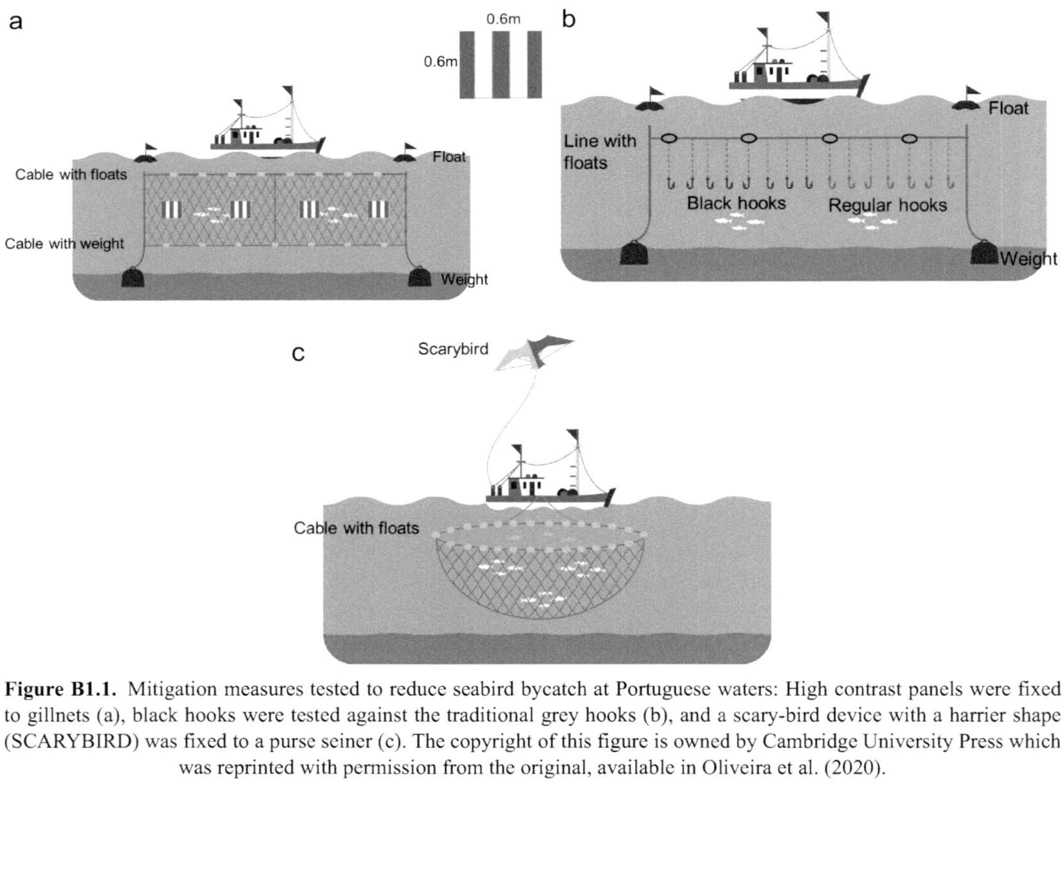

Figure B1.1. Mitigation measures tested to reduce seabird bycatch at Portuguese waters: High contrast panels were fixed to gillnets (a), black hooks were tested against the traditional grey hooks (b), and a scary-bird device with a harrier shape (SCARYBIRD) was fixed to a purse seiner (c). The copyright of this figure is owned by Cambridge University Press which was reprinted with permission from the original, available in Oliveira et al. (2020).

Longlines were the first type of gear to be tested for mitigation measures. However, there is no measure that can be applied effectively to the wide range of hook related fisheries around the world. If, on the one hand, hookpods have demonstrated great effectiveness in reducing the mortality of albatrosses caught on drift longlines in the Southern Ocean (Williams et al. 2012), they are no longer adaptable to demersal longlines. This last gear, used by the small artisanal fleets and some industrial vessels, although set to the bottom, operate in the water column at a depth easily reachable by diving seabirds. Because demersal longlines are set in shallow areas near the coastline, they are more susceptible to tidal variations and poor sea conditions, being regularly pushed against the rocky bottom. Such conditions make it hard to use hookpods or similar devices (Williams et al. 2012). Torilines are also structures that have shown very positive results in large pelagic longlines (Lokkeborg and Robertson 2002, da Rocha et al. 2021). However, they are less versatile structures that imply a rectilinear movement during setting or hauling, making them incompatible with the artisanal fleets that operate demersal longlines. Operating in shallow waters, small vessels set the longlines in a not straight line and often the fishing operation implies a sudden change in direction, mainly during hauling. Also, setting at greater depth, using a setting funnel, showed good results at the level of longlines operated by larger or smaller vessels, but the main constraints are related with bigger changes in the fishing operation or at the vessel structure. Finally, the reduction of excess material from purse seines, preventing birds to be drown when attempting to surface underneath a net fold that has formed underwater (Norriss et al. 2020), showed promising results to reduce seabird bycatch. There are still other technical measures that have been recently trialled or trials are ongoing, although the evidence of their functionality is not clear, such as, for example, dyed bait, odour deterrents, the lateral gear setting or laser and call-acoustic system devices used to deter birds from the vessel vicinity.

For a mitigation measure to be truly effective, it must be compatible with the target fishery, both in technical and economic terms (Cox et al. 2007). Modifications are expected to be more easily accepted by the fishing community, especially in terms of artisanal fleets, if they are simple, easy to operate, imply few changes to the fishing operation, be economically viable and have no negative effect on the landings (Avery et al. 2017). Thus, in addition to analysing the effectiveness of reducing bycatch, testing of mitigation measures should include an economic and acceptability analysis (Good et al. 2020). Only then would it be possible to obtain a solid assessment of the feasibility of a given mitigation measure.

Acknowledgments

This work benefited from the strategic program of MARE, financed by FCT (UIDB/04292/2020) through national funds. We thank Ana Marçalo for constructive comments on a draft version of this chapter.

References cited

Alverson, D.L. 1992. A review of commercial fisheries and the Steller sea lion (*Eumetopias jubatus*): the conflict arena. Rev. Aquat. Sci. 6: 203–256.

Alonso, H., A. Almeida, J.P. Granadeiro and P. Catry. 2015. Temporal and age-related dietary variations in a large population of yellow-legged gulls *Larus michahellis*: Implications for management and conservation. Eur. J. Wildl. Res. 61: 2010–011.doi: 10.1007/s10344-015-0958-9.

Anderson, O.R.J., C.J. Small, J.P. Croxall, E.K. Dunn, B.J. Sullivan, O. Yates et al. 2011. Global seabird bycatch in longline fisheries. Endanger. Species Res. 14: 91–106. doi: 10.3354/esr00347.

Arcos, J.M. and D. Oro. 1996. Changes in foraging range of Audouin's Gulls *Larus audouinii* in relation to a trawler moratorium in the western Mediterranean. Colon. Waterbirds 19: 128–131. doi: 10.2307/1521817.

Arcos, J.M., D. Oro and D. Sol. 2001. Competition between the yellow-legged gull *Larus cachinnans* and Audouin's gull *Larus audouinii* associated with commercial fishing vessels: The influence of season and fishing fleet. Mar. Biol. 139: 807–816. doi: 10.1007/s002270100651.

Arcos, J.M. and D. Oro. 2002a. Significance of nocturnal purse seine fisheries for seabirds: A case study off the Ebro Delta (NW Mediterranean). Mar. Biol. 141: 277–286. doi: 10.1007/s00227-002-0828-3.

Arcos, J.M. and D. Oro. 2002b. Significance of fisheries discards for a threatened Mediterranean seabird, the Balearic shearwater *Puffinus mauretanicus*. Mar. Ecol. Progr. Ser. 239: 209–220. doi: 10.3354/meps239209.

Ashmole, N.P. and M.J. Ashmole. 1967. Comparative feeding ecology of seabirds of a tropical oceanic island. Bull. Peabody Mus. Nat. Hist. 24: 1–131.

Avery, J.D., K. Aagaard, J.C. Burkhalter and O.J. Robinson. 2017. Seabird longline bycatch reduction devices increase target catch while reducing bycatch: A meta-analysis. J. Nat. Conserv. 38: 37–45. doi: 10.1016/j.jnc.2017.05.004.

Bartholomew, D.C., J.C. Mangel, J. Alfaro-Shigueto, S. Pingo, A. Jimenez and B.J. Godley. 2018. Remote electronic monitoring as a potential alternative to on-board observers in small-scale fisheries. Biol. Conserv. 219: 35–45.doi: 10.1016/j.biocon.2018.01.003.

Bécares, J., M. García-Tarrasón, D. Villero, S. Bateman, L. Jover, García-Matarranz et al. 2015. Modelling terrestrial and marine foraging habitats in breeding Audouin's gulls *Larus audouinii*: timing matters. PLoS One 10: e0120799. doi: 10.1371/journal.pone.0120799.

Belda, E.J. and A. Sánchez. 2001. Seabird mortality on longline fisheries in the western Mediterranean: Factors afecting bycatch and proposed mitigating measures. Biol. Conserv. 98: 357–363. doi: 10.1016/S0006-3207(00)00178-6.

Berón, M.P., J.P. Seco Pon, G.O. García, C.A. Paterlini, R. Mariano-Jelicich and M. Favero.2013. The diet of Olrog's Gull (*Larus atlanticus*) reveals an association with fisheries during the non-breeding season. Emu. 113: 69–76. doi: 10.1071/MU11100.

Bicknell, A.W.J., D. Oro, K.C.J. Camphuysen and S.C. Votier. 2013. Potential consequences of discard reform for seabird communities. J. Appl. Ecol. 50: 649–658. doi: 10.1111/1365-2664.12072.

Bielli, A., J. Alfaro-Shigueto, P.D. Doherty, B.J. Godley, C. Ortiz, A. Pasara et al. 2020. An illuminating idea to reduce bycatch in the Peruvian small-scale gillnet fishery. Biol. Conserv. 241: 108277. doi: 10.1016/j.biocon.2019.108277.

BirdLife International and the Agreement on the Conservation of Albatrosses and Petrels (ACAP). 2014. Bycatch Mitigation Introduction Sept 2014. https://www.birdlife.org/sites/default/files/attachments/ENG%20FS_Intro_SEPT14_w.pdf.

Blaber, S.J.M., D.A. Milton, G.C. Smith and M.J. Farmer. 1995. Trawl discards in the diets of tropical seabirds of the northern Great Barrier Reef, Australia. Mar. Ecol. Progr. Ser. 127: 1–13. doi: 10.3354/meps127001.

Borges, T.C., K. Erzini, L. Bentes, M.E. Costa, J.M.S. Gonçalves, P.G. Lino et al. 2001. By-catch and discarding practices in five Algarve (Southern Portugal) métiers. J. Appl. Ichthyol. 17: 104–114. doi: 10.1111/j.1439-0426.2001.00283.x.

Brothers, N.P. 1991. Albatross mortality and associated bait loss in the Japanese longline fishery in the Southern Ocean. Biol. Conserv. 55: 255–268. doi: 10.1016/0006-3207(91)90031-4.

Bull, L. 2007. A review of methodologies for mitigating incidental catch of seabirds in New Zealand fisheries. DOC Research and Development Series 263. Science and Technical Publishing, Department of Conservation, Wellington, New Zealand.

Calado, J.G., S.N. Verissimo, V.H. Paiva, R. Ramos, P.T. Vaz, D. Matos et al. 2021a. Influence of fisheries on the spatio-temporal feeding ecology of gulls along the western Iberian coast. Mar. Ecol. Progr. Ser. 661: 187–201. doi: 10.3354/meps1360.

Calado, J.G., J.A. Ramos, A. Almeida, N. Oliveira and V.H. Paiva. 2021b. Seabird-fishery interactions and bycatch at multiple gears in the Atlantic Iberian coast. Ocean Coast. Manag. 105306. doi: 10.1016/j.ocecoaman.2020.105306.

Cama, A., J. Bort, I. Christel, D.R. Vieites and X. Ferrer. 2013. Fishery management has a strong effect on the distribution of Audouin's gull. Mar. Ecol. Prog. Ser. 484: 279–286. doi: 10.3354/meps10312.

Camphuysen, K.C.J., J. Shamoun-Baranes, E.E. Van Loon and W. Bouten. 2015. Sexually distinct foraging strategies in an omnivorous seabird. Mar. Biol. 162: 1417–1428. doi: 10.1007/s00227-015-2678-9.

Cantlay, J.C., A.L. Bond, A.M. Wells-Berlin, R. Crawford, G.R. Martin, Y. Rouxel et al. 2020. Ineffectiveness of light emitting diodes as underwater deterrents for Long-tailed Ducks Clangula hyemalis. Global Ecol. Conserv. 23: e01102. doi: 10.1016/j.gecco.2020.e01102.

Ceia, F.R., J.C. Xavier, A.R. Carreiro, I. dos Santos and Y. Cherel. 2022. Conventional and modern approaches to study seabird trophic ecology and diet. pp. 19–35. *In*: Ramos, J.A. and L. Pereira [eds.]. Seabird Biodiversity and Human Activities. CRC Press, Boca Raton, FL.

Cherel, Y., H. Weimerskirch and G. Duhamel. 1996. Interactions between longline vessels and seabirds in Kerguelen waters and a method to reduce seabird mortality. Biol. Conserv. 75: 63–70. doi: 10.1016/0006-3207(95)00037-2.

Collet, J., S.C. Patrick and H. Weimerskirch. 2017. A comparative analysis of the behavioral response to fishing boats in two albatross species. Behav. Ecol. 28: 1337–1347. doi: 10.1093/beheco/arx097.

Cortés, V. and J. González-Solís. 2018. Seabird bycatch mitigation tests in artisanal demersal longliners of the western Mediterranean. PLoS ONE 13: e0196731. doi: 10.1371/journal.pone.0196731.

Cox, T.M., R.L. Lewison, R. Žydelis, L.B. Crowder, C. Safina and A.J. Read. 2007. Comparing effectiveness of experimental and implemented bycatch reduction measures: The ideal and the real. Conserv. Biol. 21: 1155–1164. doi: 10.1111/j.1523-1739.2007.00772.x.

Croxall, J.P., P. Rothery, S.P.C. Pickering and P.A. Prince. 1990. Reproductive performance, recruitment and survival of wandering albatrosses *Diomedea exulans* at Bird Island, South Georgia. J. Anim. Ecol. 59: 775–796. doi: 10.2307/4895.

Cury, P.M., I.L. Boyd, S. Bonhommeau, T. Anker-Nilssen, R.J.M. Crawford, R.W. Furness et al. 2011.Global seabird response to forage fish depletion-one-third for the birds. Science 334: 1703–1706. doi: 10.1126/science.1212928.

Da Rocha, N., S. Oppel, S. Prince, S. Matjila, T.M. Shaanika, C. Naomab et al. 2021. Reduction in seabird mortality in Namibian fisheries following the introduction of bycatch regulation. Biol. Conserv. 253: 108915. doi: 10.1016/j.biocon.2020.108915.

Damalas, D. 2015. Mission impossible: Discard management plans for the EU Mediterranean fisheries under the reformed Common Fisheries Policy. Fish. Res. 165: 96–99. doi: 10.1016/j.fishres.2015.01.006.

Dias, M.P., R. Martin, E.J. Pearmain, I.J. Burfield, C. Small, R.A. Phillips et al. 2019. Threats to seabirds: A global assessment. Biol. Conserv. 237: 525–537. doi: 10.1016/j.biocon.2019.06.033.

Emlen. J.M. 1966. The role of time and energy in food preference. Amer. Natur. 100: 611–617.

Estes, J.A., J. Terborgh, J.S. Brashares, M.E. Power, J. Berger, W.J. Bond et al. 2011. Trophic downgrading of planet earth. Science 333: 301–306. doi: 10.1126/science.1205106.

Fernandes, A.C., N. Pérez, N. Prista, J. Santos and M. Azevedo. 2015. Discards composition from Iberian trawl fleets. Mar. Policy 53: 33–44. doi: 10.1016/j.marpol.2014.10.012.

Field, R., R. Crawford, R. Enever, T. Linkowski, G. Martin, J. Morkūnas et al. 2019. High contrast panels and lights do not reduce bird bycatch in Baltic Sea gillnet fisheries. Global Ecology and Conservation 18. e00602. doi: 10.1016/j.gecco.2019.e00602.

Furness, R.W., A.E. Edwards and D. Oro. 2007. Influence of management practices and of scavenging seabirds on availability of fisheries discards to benthic scavengers. Mar. Ecol. Prog. Ser. 350: 235–244. doi: 10.3354/meps07191.

Garthe, S., C.J.K. Camphuysen and R.W. Furness. 1996. Amounts of discards by commercial fisheries and their significance as food for seabirds in the North Sea. Mar. Ecol. Prog. Ser. 136: 1–11. doi: 10.3354/meps136001.

Genovart, M., J. Bécares, J.M. Igual, A. Martínez-Abraín, R. Escandell, A. Sánchez et al. 2018. Differential adult survival at close seabird colonies: The importance of spatial foraging segregation and bycatch risk during the breeding season. Glob. Change Biol. 24: 1279–1290. doi: 10.1111/gcb.13997.

Gianuca, D., R.A. Phillips, S. Townley and S.C. Votier. 2017. Global Patterns of sex and age specific variation in seabird bycatch. Biol. Conserv. 205: 60–76. doi: 10.1016/j.biocon.2016.11.028.

Gilman, E., M. Weijerman, P. Suuronen and H.J. Link. 2017. Ecological data from observer programmes underpin ecosystem-based fisheries management. ICES J. Mar. Sci. 74: 1481–1495. doi: 10.1093/icesjms/fsx032.

González-Solís, J., D. Oro, L. Jover, X. Ruiz and V. Pedrocchi. 1997. Trophic niche width and overlap of two sympatric gulls in the southwestern mediterranean. Oecologia 112: 75–80. doi: 10.1007/s004420050285.

González-Solís, J. 2003. Impact of fisheries on activity, diet and predatory interactions between yellow-legged and Audouin's gulls breeding at the Chafarinas Islands. Sci. Mar. 67: 83–88. doi: 10.3989/scimar.2003.67s283.

Good, S.D., G.B. Baker, M. Gummery, S.C. Votier and R.A. Phillips. 2020. National Plans of Action (NPOAs) for reducing seabird bycatch: Developing best practice for assessing and managing fisheries impacts. Biol. Conserv. 247: 108592. doi: 10.1016/j.biocon.2020.108592.

Grémillet, D., L. Pichegru, G. Kuntz, A.G. Woakes, S. Wilkinson, R.J. Crawford et al. 2008. A junk-food hypothesis for gannets feeding on fishery waste. Proc. R. Soc. B Biol. Sci. 275: 1149–1156. doi: 10.1098/rspb.2007.1763.

Hall, M. and M. Roman. 2013. Bycatch and non-tuna catch in the tropical tuna purse seine fisheries of the world. - FAO Fisheries and Aquaculture Technical Paper No. 568.

Hudson, A.V. and R.W. Furness. 1989. The behaviour of seabirds foraging at fishing boats around Shetland. Ibis 131: 225–237. doi: 10.1111/j.1474-919X.1989.tb02765.x.

Jackson, J.B.C., M.X. Kirby, W.H. Berger, K.A. Bjorndal, L.W. Botsford, B.J. Bourque et al. 2001. Historical overfishing and the recent collapse of coast. Science 293: 629–637. doi: 10.1126/science.1059199.

Jiménez, S., A. Domingo, A. Brazeiro, O. Defeo, A.G. Wood, H. Froy et al. 2016. Sex-related variation in the vulnerability of wandering albatrosses to pelagic longline fleets. Anim. Conserv. 19: 281–295. doi.org/10.1111/acv.12245.

Kelleher, K. 2005. Discards in the world's marine fisheries. An update. FAO Fisheries Technical Paper No. 470, Rome, Italy.

Laneri, K.F., M. Louzao, A. Martínez-Abraín, J.M. Arcos, E. Belda and E. Guallart. 2010. Trawling regime influences longline seabird bycatch in the Mediterranean: new insights from a small-scale fishery. Mar. Ecol. Progr. Ser. 420: 241–252. doi: 10.3354/meps08847.

Le Bot, T., A. Lescroël and D. Grémillet. 2018. A toolkit to study seabird–fishery interactions. ICES J. Mar. Sci. 75: 1513–1525. doi: 10.1093/icesjms/fsy038.

Le Bot, T., A. Lescroël, J. Fort, C. Péron, O. Gimenez, P. Provost et al. 2019. Fishery discards do not compensate natural prey shortage in Northern gannets from the English Channel. Biol. Conserv. 236: 375–384. doi: 10.1016/j.biocon.2019.05.040.

Leitão, F. and V. Baptista. 2017. The discard ban policy, economic trends and opportunities for the Portuguese fisheries sector. Mar. Policy 75: 75–82. doi: 10.1016/j.marpol.2016.10.012.

Lokkeborg, S. and G. Robertson. 2002. Seabird and longline interactions: Effects of a bird-scaring streamer line and line shooter on the incidental capture of northern fulmars *Fulmarus glacialis*. Biol. Conserv. 106: 359–364. doi: 10.1016/S0006-3207(01)00262-2.

Mangel, J.C., J. Wang, J. Alfaro-Shigueto, S. Pingo, A. Jimenez, F. Carvalho et al. 2018. Illuminating gillnets to save seabirds and the potential for multi-taxa bycatch mitigation. Roy. Soc. Open Sci. 5: 4–7. doi: 10.1098/rsos.180254.

Marçalo, A., T. Marques, J. Araújo, P. Pousão-Ferreira, K. Erzini and Y. Stratoudakis. 2010. Fishing simulation experiments for predicting effects of purse seine capture on sardines (*Sardina pilchardus*). ICES J. Mar. Sci. 67: 334–344.doi: 10.1093/icesjms/fsp244.

Marçalo, A., J. Araújo, P. Pousão-Ferreira, G.J. Pierce, Y. Stratoudakis and K. Erzini. 2013. Behavioural responses of sardines *Sardina pilchardus* to simulated purse-seine capture and slipping. J. Fish Biol. 83: 480–500. doi: 10.1111/jfb.12184.

Mariano-Jelicich, R., S. Copello, J.P. Seco Pon and M. Favero. 2014. Contribution of fishery discards to the diet of the Black-browed albatross (*Thalassarche melanophris*) during the non-breeding season: An assessment through stable isotope analysis. Mar. Biol. 161: 119–129. doi: 10.1007/s00227-013-2320-7.

Martin, G.R. and R. Crawford. 2015. Reducing bycatch in gillnets: A sensory ecology perspective. Global Ecol. Conserv. 3: 28–50. doi: 10.1016/j.gecco.2014.11.004.

Martínez-Abraín, A., J. González-solis, V. Pedrocchi, M. Genovart, J.C. Abella, X. Ruiz et al. 2003. Kleptoparasitism, disturbance and predation of yellow-legged gulls on Audouin's gulls in three colonies of the western Mediterranean. Sci. Mar. 67: 89–94. doi: 10.3989/scimar.2003.67s289.

Matos, D.M., J.A. Ramos, J.G. Calado, F.R. Ceia, J. Hey and V.H. Paiva. 2018. How fishing intensity affects the spatial and trophic ecology of two gull species breeding in sympatry. ICES J. Mar. Sci. 75: 1949–1964. doi: 10.1093/icesjms/fsy137.

McCauley, D.J., M.L. Pinsky, S.R. Palumbi, J.A. Estes, F.H. Joyce and R.R. Warner. 2015. Marine defaunation: Animal loss in the global ocean. Science 347: 1255641. doi: 10.1126/science.1255641.

McInnes, J.C., S.N. Jarman, M.-A. Lea, B. Raymond, B.E. Deagle, R.A. Phillips et al. 2017. DNA metabarcoding as a marine conservation and management tool: A circumpolar examination of fishery discards in the diet of threatened albatrosses. Front. Mar. Sci. 4: 277. doi: 10.3389/fmars.2017.00277.

Montevecchi, W. 2002. Interactions between fisheries and seabirds. pp. 527–557. *In*: Schreiber, E. and J. Burger [eds.]. Biology of Marine Birds. CRC Press, Boca Raton, FL, USA. doi: 10.1201/9781420036305.ch16.

Navarro, J., D. Oro, A. Bertolero, M. Genovart, A. Delgado and M.G. Forero. 2010. Age and sexual differences in the exploitation of two anthropogenic food resources for an opportunistic seabird. Mar. Biol. 157: 2453–2459. doi: 10.1007/s00227-010-1509-2.

Norriss, J.V., E.A. Fisher and A.M. Denham. 2020. Seabird bycatch in a sardine purse seine fishery. ICES J. Mar. Sci. 77: 2971–2983. doi: 10.1093/icesjms/fsaa179.

Oliveira, N., A. Henriques, J. Miodonski, J. Pereira, D. Marujo, A. Almeida et al. 2015. Seabird bycatch in Portuguese mainland coastal fisheries: An assessment through on-board observations and fishermen interviews. Glob. Ecol. Conserv. 3: 51–61.doi: 10.1016/j.gecco.2014.11.006.

Oliveira, N., A. Almeida, H. Alonso, E. Constantino, A. Ferreira, I. Gutiérrez et al. 2020. A contribution to reducing bycatch in a high priority area for seabird conservation in Portugal. Bird Conserv. Int. 1–20. doi: 10.1017/S0959270920000489.

Oro, D., M. Bosch and X. Ruiz. 1995. Effects of a trawling moratorium on the breeding success of the Yellow-legged Gull *Larus cachinnans*. Ibis 137: 547–549. doi: 10.1111/j.1474-919X.1995.tb03265.x.

Oro, D. 1996. Effects of trawler discard availability on egg laying and breeding success in the lesser black-backed gull Larus fuscus in the western Mediterranean. Mar. Ecol. Prog. Ser. 132: 43–46. doi: 10.3354/meps132043.

Oro, D. and X. Ruiz. 1997. Seabirds and trawler fisheries in the northwestern Mediterranean: Differences between the Ebro Delta and the Balearic Islands areas. ICES J. Mar. Sci. 54: 695–707. doi: 10.1006/jmsc.1997.0246.

Pauly, D. and V. Christensen. 1995. Primary production required to sustain global fisheries. Nature 374: 255–257. doi: 10.1038/374255a0.

Pauly, D., J. Alder, E. Bennett, V. Christensen, P. Tyedmers and R. Watson. 2003. The future for fisheries. Science 302: 1359–1361. doi: 10.1126/science.1088667.

Pauly, D., R. Watson and J. Alder. 2005. Global trends in world fisheries: Impacts on marine ecosystems and food security. Philos. Trans. R. Soc. B 360: 5–12. doi: 10.1098/rstb.2004.1574.

Pereira, J.M., V.H. Paiva, L. Kruger and S.C. Votier. 2022. Tracking seabirds for conservation and marine spatial planning. pp. 59–74. *In*: Ramos, J.A. and L. Pereira [eds.]. Seabird Biodiversity and Human Activities. CRC Press, Boca Raton, FL.

Phillips, R.A., R. Gales, G.B. Baker, M.C. Double, M.G. Favero, F. Quintana et al. 2016. The conservation status and priorities for albatrosses and large petrels. Biol. Conserv. 201: 169–183. doi: 10.1016/j.biocon.2016.06.017.

Pierre, J., M. Gerner and L. Penrose. 2014. Assessing the effectiveness of seabird mitigation devices in the trawl sectors of the Southern and Eastern Scalefish and Shark Fishery in Australia. Australian Fisheries Management Authority, Canberra, Australia.

Regular, P., W. Montevecchi, A. Hedd, G. Robertson and S. Wilhelm. 2013. Canadian fishery closures provide a large-scale test of the impact of gillnet bycatch on seabird populations. Biol. Lett. 9: 20130088.doi: 10.1098/rsbl.2013. 0088.

Rodríguez, A., J.M. Arcos, V. Bretagnolle, M.P. Dias, N.D. Holmes, M. Louzao et al. 2019. Future directions in conservation research on petrels and shearwaters. Front. Mar. Sci. 6: 1–27. doi: 10.3389/fmars.2019.00094.

Romano, M.D., J.F. Piatt and D.D. Roby. 2006. Testing the junk-food hypothesis on Marine Birds: Effects of prey type on growth and development. Waterbirds 29: 407–414. doi: 10.1675/1524-4695(2006)29%5B407:TTJHOM%5D2.0.CO;2.

Rousseau, Y., R.A. Watson, J.L. Blanchard and E.A. Fulton. 2019. Evolution of global marine fishing fleets and the response of fished resources. Proc. Natl. Acad. Sci. U. S. A. 116: 12238–12243. doi: 10.1073/pnas.1820344116.

Salas, S., R. Chuenpagdee, J.C. Seijo and A. Charles. 2007. Challenges in the assessment and management of small-scale fisheries in Latin America and the Caribbean. Fish. Res. 87: 5–16. doi: 10.1016/j.fishres.2007.06.015.

Sánchez, A. and E. Belda. 2003. Bait loss caused by seabirds on longline fisheries in the northwestern mediterranean: Is night setting an effective mitigation measure? Fish. Res. 60: 99–106. doi: 10.1016/S0165-7836(02)00055-3.

Senko, J., E.R. White, S.S. Heppell and L.R. Gerber. 2013. Comparing bycatch mitigation strategies for vulnerable marine megafauna. Anim. Conserv. 17: 5–18. doi: 10.1111/acv.12051.

Sherley, R.B., H. Ladd-Jones, S. Garthe, O. Stevenson and S.C. Votier. 2020. Scavenger communities and fisheries waste: North Sea discards support 3 million seabirds, 2 million fewer than in 1990. Fish Fish. 21: 132–145. doi: 10.1111/faf.12422.

Smith, A.D.M., C.J. Brown, C.M. Bulman, E.A. Fulton, P. Johnson, I.C. Kaplan et al. 2011. Impacts of fishing low-trophic level species on marine ecosystems. Science 333: 1147–1150. doi: 10.1126/science.1209395.

Soriano-Redondo, A., V. Cortéz, J.M. Reyes González, S. Guallar, J. Bécares, B. Rodríguez et al. 2016. Relative abundance and distribution of fisheries influence risk of seabird bycatch. Sci. Rep. 6: 37373. doi: 10.1038/srep37373.

Stratoudakis, Y. and A. Marçalo. 2002. Sardine slipping during purse-seining off northern Portugal. ICES J. Mar. Sci. 59: 1256–1262. doi: 10.1006/jmsc.2002.1314.

Swartz, W., E. Sala, S. Tracey, R. Watson and D. Pauly. 2010. The spatial expansion and ecological footprint of fisheries (1950 to Present). PLoS ONE 5(12): e15143. doi: 10.1371/journal.pone.0015143.

Sydeman, W.J., S.A. Thompson, T. Anker-Nilssen, M. Arimitsu, A. Bennison and S. Bertrand. 2017. Best practices for assessing forage fish fisheries-seabird resource competition. Fish. Res. 194: 209–221. doi: 10.1016/j.fishres.2017.05.018.

Tasker, M.L., C.J. Camphuysen, J. Cooper, S. Garthe, W.A. Montevecchi and S.J.M. Blaber. 2000. The impacts of fishing on marine birds. ICES J. Mar. Sci. 57: 531–547. doi: 10.1006/jmsc.2000.00714.

Tegner, M.J. and P.K. Dayton. 1999. Ecosystem effects of fishing. Trends Ecol. Evol. 14: 261–262. doi: 10.1016/S0169-5347(99)01650-X.

Thurstan, R.H., S. Brockington and C.M. Roberts. 2010. The effects of 118 years of industrial fishing on UK bottom trawl fisheries. Nat. Commun. 1: 1–15. doi: 10.1038/ncomms1013.

Uhlmann, S.S., C. Ulrich and S.J. Kennelly. 2019. The European Landing Obligation. Reducing discards in complex, multi-species and multi-jurisdictional fisheries. Springer International Publishing, Cham, Switzerland.

Veit, R.R. and N.M. Harrison. 2017. Positive Interactions among Foraging Seabirds, Marine Mammals and Fishes and Implications for Their Conservation. Front. Ecol. Evol. 5: 121. doi: 10.3389/fevo.2017.00121.

Votier, S.C., R.W. Furness, S. Bearhop, J.E. Crane, R.W. Caldow, P. Catry et al. 2004. Changes in fisheries discard rates and seabird communities. Nature 427: 727–730. doi: 10.1038/nature02315.

Weimerskirch, H., M. Le Corre, E.T. Kai and F. Marsac. 2010. Foraging movements of great frigatebirds from Aldabra Island: Relationship with environmental variables and interactions with fisheries. Prog. Oceanogr. 86: 204–213. doi: 10.1016/j.pocean.2010.04.003.

Weimerskirch, H., D.P. Filippi, J. Collet, S.M. Waugh and S.C. Patrick. 2017. Use of radar detectors to track attendance of albatrosses at fishing vessels. Conserv. Biol. 32: 240–245. doi: 10.1111/cobi.12965.

Williams, N. 1998. Overfishing disrupts entire ecosystems. Science 279: 809. doi: 10.1126/science.279.5352.809.

Williams, D., R.G. Pople, D.A. Showler, L.V. Dicks, M.F. Child, E.K.H.J. zu Ermgassen et al. 2012. Bird Conservation—Global evidence for the effects of interventions. Pelagic Publishing, Exeter, UK. doi: 10.13140/2.1.1927.3924.

Žydelis, R., C. Small and G. French. 2013. The incidental catch of seabirds in gillnet fisheries: A global review. Biol. Conserv. 162: 76–88. doi:10.1016/j.biocon.2013.04.002.

CHAPTER 2

Urban Gulls Living with Humans

Joana Pais de Faria,[1,*] *Catarina S. Lopes*,[1] *Edward Kroc*,[2] *Louise K. Blight*[3] and *Ruedi G. Nager*[4]

Introduction

The increase of the human population and consequent growing urbanization have been progressively leading to the reduction and fragmentation of natural habitats, and amplifying anthropogenic pressures on ecosystems (e.g., Gosling et al. 2017). Growing urban areas, however, also offer new ecological opportunities for wildlife to modify their behaviour and successfully colonize urban environments (Lowry et al. 2013). The habitat changes induced by urbanization, in combination with novel biotic interactions, result in several selection pressures that favour certain traits in urban populations (Alberti et al. 2017, Ouyang et al. 2018). For example, birds have adjusted to urban environments and changed migratory patterns, breeding phenologies and phenotypic traits such as song (Garcia et al. 2017, Sepp et al. 2017, Hensley et al. 2019). Certain traits may have facilitated a species' colonization of urban environments, including cognitive and problem-solving performance (Snell-Rood and Wick 2013, Audet et al. 2016, Castano et al. 2020), heightened tolerance and habituation (Lowry et al. 2013, Sol et al. 2013), and ability to shift their dietary niche (Pagani-Núñez et al. 2019, Murray et al. 2020), all of which could affect survival and reproductive success in the urban environment. These adjustments are particularly evident in generalist species with higher phenotypic and behavioural plasticity such as gulls, allowing them to overcome the challenges of a novel environment by, for example, adapting and exploiting novel food and nesting site opportunities (Belant 1997, Sayol et al. 2020, Carmona et al. 2021). Ultimately, expanding urbanization and the colonization of urban areas by opportunistic species will increase instances of human-wildlife

[1] University of Coimbra, MARE - Marine and Environmental Sciences Centre, Department of Life Sciences, 3000-456 Coimbra, Portugal.
Email: catarina.s.lopes90@gmail.com
[2] University of British Columbia, Measurement, Evaluation, and Research Methodology Unit, 2125 Main Mall, Vancouver, BC V6T 1Z4, Canada.
Email: ed.kroc@ubc.ca
[3] Procellaria Research & Consulting, 944 Dunsmuir Road, Victoria, BC V9A 5C3, Canada and University of Victoria, School of Environmental Studies, Victoria, BC V8W 2Y2, Canada.
Email: louise.blight@gmail.com
[4] University of Glasgow, Institute of Biodiversity, Animal Health and Comparative Medicine, College of Medical, Veterinary & Life Sciences, Glasgow G12 8QQ, Scotland, UK.
Email: ruedi.nager@glasgow.ac.uk
* Corresponding author: paisdefaria.joana@gmail.com

encounters, causing a variety of interactions and conflicts in the cities. This chapter provides an overview of the relationships between urban gulls and humans. To address patterns of urban gull population expansion by different gull species, and the challenges faced when studying the ecology of urban gulls, we give an overview of the history of gull colonization of urban areas in the British Isles and North America. Then, we review the urban features influencing breeding success of urban gull populations, and summarize the main human-gull interactions within cities. Lastly, we consider the importance of management actions to minimize conflicts related to urban gulls.

History of urban gull populations in the British Isles and North America

Studying urban gull populations comes with a variety of challenges, particularly when counting and predicting population size, because urban areas rarely provide good vantage points with an unobstructed view to all surrounding nests. This results in a likely underestimation of the true size of urban gull populations worldwide (Coulson and Coulson 2015). Counting efforts of urban-nesting gulls seem to have been more consistent in the British Isles than elsewhere, perhaps because it was one of the first places where gulls started nesting on roofs outside of the Black Sea region, where roof-nesting European Herring Gulls (*Larus argentatus*; Herring Gull hereafter) were first recorded in the early 1890s (Goethe 1960, Nankinov 1992). 'Urban nesting' can be defined as nesting on buildings or other man-made structures, such as bridges, in areas frequented by humans (Cramp 1971), including factory roofs and other industrial buildings outside urban centres. This definition was used in all the censuses in the British Isles. The first known records of roof-nesting Herring Gulls in the British Isles dates from 1910 at two sites in Cornwall, southwest England (Cramp 1971). Herring Gulls nesting on buildings remained very rare before the second World War, with just six known sites in 1939 concentrated in the southwest of England, and all with less than 10 breeding pairs per site (Witherby et al. 1938–1941, Cramp 1971). It is unclear what drove Herring Gulls to construct their first nest on roofs and several reasons have been considered. For example, Took (1955) speculated that long-range artillery fire during the war drove Herring Gulls off their usual nesting sites on chalk cliffs around the city of Dover, and onto roof-tops, particularly into the city centre. It is possible that the architecture of traditional British houses, with several separate chimney stacks and small ledges and niches on the roof, provided attractive nesting sites that resembled the Herring Gulls' natural nesting sites on cliffs. This is in contrast to Lesser Black-backed Gulls (*Larus fuscus*), which nest more on flat grounds (Goethe 1960). From the beginning of the 20th century until the early 1970s the Herring Gull population in the British Isles increased at a rate of 12–13% per annum, possibly due to reduced persecution and increased availability of anthropogenic food resources from human activities (Cramp et al. 1974). During that time, the roof-nesting habit of Herring Gulls also increased, spreading from southwest England to many coastal towns along the southern parts of the British Isles, and then expanded inland and to the north (Cramp 1971, Monaghan and Coulson 1977). In Scotland, the first roof-nesting pair of Herring Gulls was recorded in Inverness in 1965 (Parslow 1967).

During 1969–70, the first census of roof-nesting gulls in the British Isles was organized as part of a national census of seabirds (Cramp et al. 1974), and by then roof-nesting Herring Gulls had spread to at least 61 sites and numbered 1,252 nests (Fig. B2.1a), with five sites occupied by over 100 breeding pairs each, and the largest colony of roof-nesting gulls being at Dover, with 225 nests (Cramp 1971, updated by Monaghan and Coulson 1977). Although the breeding productivity of roof-nesting Herring Gulls was high, possibly due to the local availability of food from anthropogenic sources and reduced nest predation (Monaghan 1979, Rock 2005), it is unlikely that the reproductive output from local roof-top breeders alone could have sustained the observed large increase in numbers (Chabrzyk and Coulson 1976). Hence, roof-nesting colonies were likely recruiting from natural sites nearby, where large Herring Gull colonies were possibly reaching

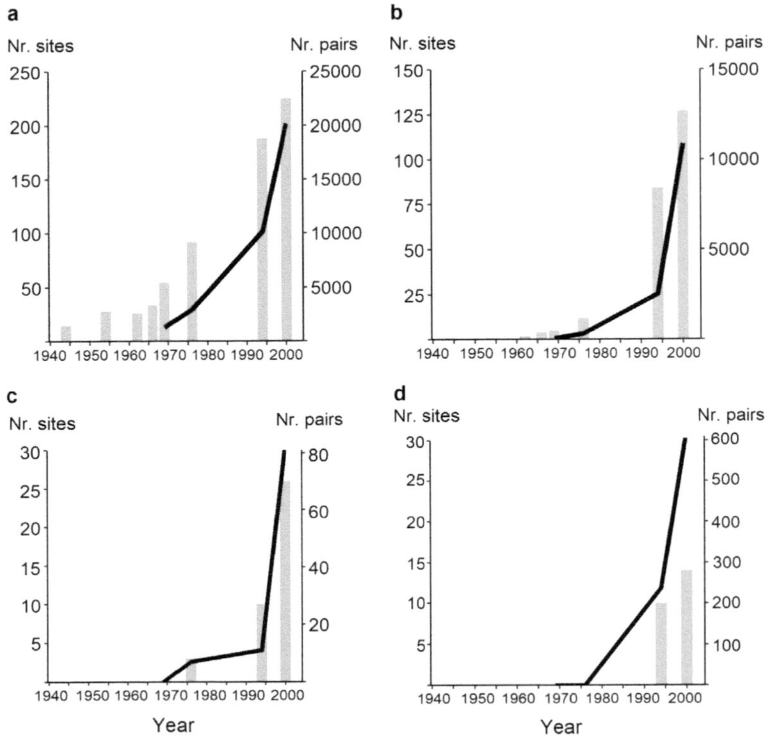

Figure B2.1. Increase in the number of sites (histogram and left axis) and number of pairs (black line and right axis) of the four *Larus* species that have been recorded to nest on roofs in built-up areas in the British Isles, from their earliest records until the census of 1998–2002: (a) European Herring Gulls (*Larus argentatus*), (b) Lesser Black-backed Gulls (*Larus fuscus*), (c) Great Black-backed Gulls (*Larus marinus*) and (d) Common Gulls (*Larus canus*). Note that the scale differs between vertical axes due to the large differences in numbers of roof-nests between the species. Data from Cramp (1971), Monaghan and Coulson (1977), Raven and Coulson (1997) and Mitchell et al. (2004).

saturation levels in the 1970s (Monaghan and Coulson 1977). Therefore, the increasing number of roof-nesting Herring Gulls was likely a consequence of the species' rapid population growth in the 20th century, when young breeders were forced to find alternative nesting grounds (Monaghan and Coulson 1977). In addition, culling and disturbance by human activities at natural sites could have dispersed Herring Gulls away from those sites and into built-up areas (Raven and Coulson 1997).

At that time, the only other gull species recorded breeding on roofs in the British Isles was the Lesser Black-backed Gull, first recorded nesting on an inland factory-roof in Glamorgan, in 1945 (Salmon 1958). By 1969–70 the species expanded to a total roof-nesting population of at least 61 pairs in seven locations mainly confined to south Wales (Cramp 1971). The smaller numbers and slower increase of the roof-nesting Lesser Black-backed Gulls was likely due to their slower rate of increase of the total British population compared to Herring Gulls at that time. Further censuses of roof-nesting gulls in Britain, using the same methods as the 1969 census, were organized in 1976 (Monaghan and Coulson 1977), in 1994 (Raven and Coulson 1997) and in 1998–2002 (Mitchell et al. 2004); the next one is expected to be concluded in 2021.

Since the 1970s, the population of Herring Gulls nesting at natural sites in the British Isles fell by 43% in 1985 (Lloyd et al. 1991) and a further 13% in 2000 (Mitchell et al. 2004). However, the numbers of roof-nesting gulls kept increasing (Fig. B2.1a), but at a slower rate than the declines at natural sites, so that the overall population of Herring Gulls in the British Isles decreased. Roof-nesting Herring Gulls had spread to at least 92 locations with a minimum of 2,968 breeding pairs in 1976, corresponding to an increase of 17% per annum (Monaghan and Coulson 1977), growing faster than populations nesting at natural sites had been growing before the 1970s (13% per annum; Chabrzyk and Coulson 1976). By 1994, at least 188 sites with 10,184 breeding pairs across most of Britain were occupied by roof-nesting Herring Gulls (Raven and Coulson 1997), and by the most recent census in 1998–2002 this further grew to at least 225 sites with 20,170 breeding pairs (Fig. B2.1a; Mitchell et al. 2004). Another study estimated more than 100,000 gull pairs nesting on roofs in 2004 (Rock 2005), but this was an extrapolation from a stronghold of roof-nesting gulls to the rest of the British Isles. Not only did roof-nesting Herring Gulls colonize new sites, including sites further away from the coast, but also the average number of nests at those sites increased. The mean size of roof-nesting colonies increased from 49 pairs in 1976 to 75 in 1994 (Raven and Coulson 1997), and to 90 pairs in 2000 (Mitchell et al. 2004). Although most urban colonies were modest in size, a few colonies became very large, such as in the city of Aberdeen, Scotland, with 3350 Herring Gull roof-nests in 2001 (Mitchell et al. 2004). Roof-nesting gulls increased from 0.6% to 8.2% and 13.7% of the total British Herring Gull population in 1976, 1994 and 2000, respectively. The rate of increase of roof-nesting Herring Gulls slowed down after the 1970s with 10% per annum between 1976 and 1994 (Raven and Coulson 1997), and 3% per annum between 1994 and 2000 (Mitchell et al. 2004). Increased breeding density in built-up areas might be one of the factors explaining the slowing of the expansion of roof-nesting Herring Gulls towards the end of the 20th century (Raven and Coulson 1997). The continued increase in roof-nesting Herring Gulls from 1970s onwards, despite a large decline in numbers at natural breeding sites, could possibly be attributed to the deterioration of foraging conditions and increased predation and persecution at natural sites, leading to birds looking for refuge in built-up areas (Nager and O'Hanlon 2016), where the availability of food and nesting sites safe from predation is favourable (Monaghan and Coulson 1977, Raven and Coulson 1997, Rock 2005).

Numbers of roof-nesting Lesser Black-backed Gulls and the number of colonized sites in built-up areas also increased after 1990 (Fig. B2.1b) while the overall population also kept increasing, with a higher rate of increase than in Herring Gulls (Raven and Coulson 1997). In most coastal regions of the British Isles, roof-nesting Lesser Black-backed Gulls joined existing Herring Gull colonies and were outnumbered by them. However, some inland areas, for example the Forth-Clyde region of Scotland, were first colonized by Lesser Black-backed Gulls and then Herring Gulls followed later, with roof-nesting Lesser Black-backed Gulls being more numerous than Herring Gulls (Raven and Coulson 1997). Between 1970s and 2000s, Herring Gull and Lesser Black-back Gull populations showed a different relationship between numbers nesting at natural sites and on roof-tops (Nager and O'Hanlon 2016). For Herring Gulls the number of roof-nests increased as the number of nests in natural sites decreased, showing an apparent negative correlation between numbers at natural sites and built-up areas, while Lesser Black-backed Gull populations were still expanding in natural sites, as the numbers of roof-nests also increased, showing an apparent positive correlation between the number of breeding pairs at natural and urban sites. This suggests that as with observations for Herring Gulls before 1970s, young Lesser Black-backed Gulls might have been forced to disperse from saturated natural sites into built-up areas. In both cases the changes in numbers at natural sites were larger than in built-up areas.

Other gull species were slower to turn to roof-nesting in Britain and Ireland, probably due to their lower numbers and slower population increases at the time. The first roof-nesting pairs of Great Black-backed Gulls (*Larus marinus*) were recorded in 1970 for a single pair in Cornwall, and by the turn of the century this had expanded to at least 83 pairs at 26 sites (Fig. B2.1c; Mitchell et al. 2004), mostly in small colonies of 1–2 pairs with other gulls on surrounding roofs (Raven and Coulson 1997). In 1971 a breeding pair of Common Gulls (*Larus canus*) was recorded on a roof in Inverness (Cramp 1971), and by 2000 there were at least 621 roof-nesting pairs at 14 locations in Scotland (Fig. B2.1d; Mitchell et al. 2004). In addition to the *Larus* spp. gulls, it is worth noting that Black-legged Kittiwakes (*Rissa tridactyla*) were also recorded nesting on man-made structures (mainly piers and window ledges of warehouses) from the early 1930s in southeast Scotland and spread along the British North Sea coast to seven colonies, with a total of at least 410 pairs in 1969–70 (Cramp 1971). This number increased to between 664 to 755 breeding pairs in 2007–2009 in one area of northeast England (Turner 2010), but more recent counts in other areas are not known.

Over time, large gulls have also colonized cities across all of Europe, North America, Australia and Japan (Cramp 1971, Rock 2005), mostly during the late 20th century, as shown by the first records from 1971 for Italy by Yellow-legged Gulls (*Larus michahellis*; Cignini and Zapparoli 1996), and those from the 1980s for Japan by Slaty-backed Gulls (*Larus schistisagus*; Artyukhin 2002 in Zelenskaya 2019). However, information on the history of urban-nesting gulls in regions outside Europe is scarce, with some records from North America and isolated accounts of their behaviour in other parts of the world (e.g., Turbott 1969, Chávez-Villavicencio 2014, Yorio et al. 2016). North American records of roof-nesting gulls are far less detailed than those from the UK; although some jurisdictions appear to survey urban gull populations periodically (e.g., Roby et al. 2007), most urban counts result from location- or species-specific research rather than initiatives at the national scale as in the British Isles (cf. Hooper 1998, Blight et al. 2019).

The first North American record of urban-nesting gulls appears to have been of Western Gulls (*Larus occidentalis*) breeding in San Francisco, California, circa 1920 (Fisk 1978). This same species first colonized Seattle, in Washington state further north on the US Pacific coast, in 1946, nesting on buildings near the port (Eddy 1982). Further north still, the first roof-top nesting record for the Pacific coast of Canada was in 1962, in Vancouver, for a single Glaucous-winged Gull (*Larus glaucescens*) nest near the port (Oldaker 1963). The apparently rapid initial expansion of this species in the city of Vancouver was reasonably well-documented until the 1980s, with four nests described in one city neighbourhood in 1972–73 (Sanford 1974), and a total of 88 nests found at two other localities later in this same decade (Campbell 1975, Poynter 1976). By 1986, Vermeer et al. (1988) estimated that 500 pairs nested in the northeast quadrant of the city on buildings near the waterfront, including the downtown core. About 30 years later, in 2017, this northeast quadrant population was estimated at 1231 pairs (95% CI: 1182, 1279; E. Kroc, L. Blight and W. Cao, unpublished data). The same researchers have estimated that the total breeding population in Vancouver was 1690 pairs (95% CI: 1626, 1755), with an additional 200 nests at least occupying the city's adjacent coastline to the south in the bordering city of Richmond. Their estimation for the downtown core breeding population (449 pairs, 95% CI: 441, 457) is essentially unchanged from 1988; the current Vancouver population occurs at a low density over much of the city, and no longer clusters near the harbour (E. Kroc, L. Blight and W. Cao, unpublished data). In the absence of control measures it is likely that this population will continue to grow (Kroc et al. 2018). For this same species in the nearby maritime city of Victoria, Hooper (1988) reported that in 1986, 114 pairs nested on various structures (mostly on roof-tops) near the waterfront. In 2017–2018, Blight et al. (2019) used drones to estimate a population of 346 pairs in their survey area, with an additional 102 pairs at a warehouse colony excluded from the drone survey, for a total of about 448 pairs through the downtown core and adjacent neighbourhoods. As with Vancouver, the growing population of nesting gulls in Victoria maintained a low density and spread farther into the urban area over time, rather than aggregating at increasing densities near the waterfront. This same pattern seems to be repeating itself in the nearby city of Nanaimo, British Columbia as well (Kroc et al. 2018). Like the European Herring Gull in the British Isles after the 1970s, these two urban population increases have occurred at a time when the overall regional population was in decline, implying a shift in preferred habitats by nesting Glaucous-winged Gulls in this region of Pacific Canada (Blight et al. 2015, Blight et al. 2019).

Elsewhere on the west coast of North America, urban nesting habitats have facilitated a range expansion of Heermann's Gulls (*Larus heermanni*) from their nesting areas in Mexico. In 1979 the species began nesting on the central California coast, when a single pair nested on Alcatraz Island in San Francisco Bay for three consecutive years. In the 1980s and 1990s, 1–2 pairs nested at natural offshore sites in this same region, then in 1999 three pairs successfully nested at an artificial island on a lake in the city of Seaside in the Monterey Bay area. As this island eroded, the population subsequently relocated to the region's roof-tops, and has since increased to about 100 pairs (Howell et al. 1983, Roberson et al. 2001, Golden Gate Audubon Society 2021). This location is seen as facilitating a potential northwards range shift in the face of climate change (Herrera 2019). Western Gulls have also continued to colonize urban centres along the California coast. As far back as 1935, Western Gulls nested on roof-tops throughout San Diego County in extreme southern California (Unitt 2004), while gulls in San Francisco have been recorded nesting along the downtown shoreline and nearby Alcatraz Island (Pierotti and Annett 2001).

Western Gulls have also colonized urban centres around Monterey Bay and will sometimes nest on roof-tops of structures that are immediately adjacent to nesting Heermann's Gulls (Kroc and Blight 2019). In the extreme north of the continent, small populations of both Herring and Common Gulls (10–20 pairs each) have been documented nesting on roof-tops in and around the downtown core of Anchorage, Alaska (Kroc and Blight 2019). Interestingly, Common Gulls have also been documented nesting atop lampposts in the small, subarctic city of Whitehorse in Canada's Yukon Territory (J. Helmer pers. comm.; Kroc and Blight 2019), more than 250 km from the nearest (Pacific) ocean.

Somewhat surprisingly, the first record of urban-nesting birds for the East Coast of North America (for American Herring Gulls, *Larus smithsonianus*, in Boston, USA) does not appear in the literature until 1961, although 150 nests were thought to have become established some years earlier; these subsequently expanded by the 1970s (Paynter 1963, Fisk 1978). In the Great Lakes region, first roof-top nests were documented in Ontario, Canada, in the early 1970s, for Ring-billed Gulls (*Larus delawarensis*) and American Herring Gulls (Blokpoel et al. 1990), and in 1978 on the US side of the Great Lakes for these same two species (Dwyer et al. 1996). In the Canadian side of the Great Lakes, American Herring Gulls grew in number from 440 to 1,300 pairs from 1976 to 1990 (about one-third were urban colonies), while Ring-billed Gulls increased from 56,000 to 283,000 pairs (13 of 27 colonies were urban) during this same period, with concerns about health and economic effects of the urban portion of these gull populations increasing apace (Blokpoel and Tessier 1991). By the mid-1990s, 4% of the US Great Lakes' growing American Herring Gull population was nesting on roof-tops, as were 2% of Ring-billed Gulls in the region, for a total of 7,922 pairs of both species (Dwyer et al. 1996). Although these two gull species have been declining in the Great Lakes region since about 1990 (Morris et al. 2003, 2011, Hebert et al. 2008), the urban-nesting habit, gained in their population expansion stage, has been retained and they continue to breed in urban habitats (Morris et al. 2011; L. Blight and E. Kroc, unpublished data). On the Atlantic coast, American Herring Gulls and Great Black-Backed Gulls have been documented nesting atop waterfront roof-tops around New York City harbour for the past 35 years (Elbin et al. 2019, Kolodzinski et al. 2019). Urban breeding populations reached peak numbers of 2516 and 792 pairs, respectively, in 1990, but have since declined (Elbin et al. 2019). These species have also been observed nesting in the downtown cores of the Canadian coastal cities of Saint John, Halifax, and St. John's, with an aggregated estimate of more than 200 pairs (Kroc and Blight 2019).

It seems based on disparate studies to date that gulls colonized urban habitats for nesting either while their regional population was expanding, with possibly surplus individuals seeking new nesting opportunities, or while their regional populations declined (individuals finding a refuge when conditions at natural sites deteriorated). Once gulls have expanded into an urban habitat, they remain, regardless of the status of their original source populations. The reasons for this are unclear, though may be related to the greater breeding success experienced by some urban colonies (below); the geographic structuring of gull metapopulations comprised of both urban and natural colonies is clearly a direction for future research.

Factors affecting the success of urban gull populations

Several studies have reported a higher breeding success for urban gull colonies than gulls breeding at natural sites (Monaghan 1979, Kroc 2018a, Zelenskaya 2019), which may be linked to the lower nest density and consequently lower intraspecific aggression from other nesting adults in the neighbourhood. This suggests that roof-top habitats may represent suitable breeding grounds of equal or higher quality than those offered by natural coastal or insular habitats at that time (Monaghan 1979). However, other studies found no differences in breeding parameters between gulls nesting at natural and urban sites. Hooper (1988) found a high egg and chick mortality in urban-nesting Glaucous-winged Gulls in Victoria, British Columbia, which was mostly caused by the predation of Northwestern Crows (*Corvus caurinus*) and other adult gulls, leading to hatching (72%) and fledgling success (51%) that were similar to those observed at a natural site on nearby Mandarte Island (70–83% and 36–58%, respectively; Vermeer 1963). More recently, Perlut et al. (2016) studied the breeding parameters of a roof-top breeding population of American Herring Gulls in Portland, Maine, USA, and found a lower clutch size (mean of 2.3 eggs per clutch) and hatching success (46–48%), but higher chick survival up to 30 days of age (62–73%), than for a population nesting at a natural site on Appledore Island (2.6 eggs per clutch, 56–71% hatching success and 49–53% chick survival). Nest failure in Portland was mostly caused by predation (38%), weather conditions (27%) and management interventions of nest removal during incubation (23%). There was also no evidence of a difference in breeding success between natural and urban colonies in Yellow-legged Gulls in the Venice region, Italy, possibly because these were recently established colonies assumed to be mostly composed of younger birds with a poorer breeding performance (Soldatini et al. 2008). Vermeer et al. (1988) found great variation in the breeding success of roof-nesting Glaucous-winged Gulls in downtown Vancouver, Canada, with colonial breeders having lower (35%) fledgling success than solitary and more experienced breeding pairs (73%), although human disturbance may have caused some chick mortality among colonial breeders.

Numerous other factors might influence differential breeding success between natural and urban sites, such as the quality of the diet of adult breeders. Some studies have reported a higher breeding output in gulls foraging on natural marine resources (Hunt 1972, Annett and Pierotti 1999, O'Hanlon et al. 2017), which commonly include prey with high energetic and nutritional value, particularly rich in essential fatty acids and micronutrients (Gladyshev et al. 2009, EFSA NDA Panel 2014). The view that a marine diet is beneficial to growing chicks is further supported by the observation that breeding gulls often switch their diet during the chick-rearing period to more marine prey (Annett and Pierotti 1989, Isaksson et al. 2016, Pais de Faria et al. 2021a, but see Sotillo et al. 2019 for an alternative view). However, some studies found that gulls feeding on anthropogenic food did better compared to gulls feeding on natural resources (Pons and Migot 1995), such as the high hatching success in Lesser Black-backed Gulls (Gyimesi et al. 2016) and the higher fledgling rates in European Herring Gulls, when compared to gulls feeding on natural marine prey (van Donk et al. 2017), both in the Netherlands. It is likely, however, that urban-nesting gulls use the same food resources as the birds nesting in the surrounding natural sites (e.g., Huig et al. 2016, Rock et al. 2016, Spelt et al. 2019). The higher breeding success observed in gulls with an anthropogenic diet, from both natural and urban colonies, might be a result of the higher predictability of these food recourses (Oro et al. 2013, Martínez-Abrain and Jiménez 2016). Some studies have recently shown the ability of gulls to adapt their foraging behaviour to follow predictable human habits. Spelt et al. (2021) showed that gulls breeding in Bristol, UK, matched their foraging patterns to the times when anthropogenic food became available at school breaks and opening hours of a waste centre. Other than matching their foraging patterns to predictable human activities, in a highly touristic area in the centre of Porto, Portugal, Yellow-legged Gulls were reported defending favourable vantage spots on the top of traffic lights near food outlets, from where they could swoop down and steal the food carried by humans coming out of the shops and waiting to cross the street (Pais de Faria et al. 2021b). Overall, the increase in the availability of anthropogenic food sources in cities and landfills, coupled with fishery discards in nearby marine areas and ports, have likely contributed to the growth of urban gull populations.

Urban habitats may also provide a range of other potentially favourable conditions for gulls. Urban areas typically provide a milder microclimate than surrounding areas (Chown and Duffy 2015), with warmer temperatures (heat island effect), and shelter from extreme weather conditions provided by roof-top structures (e.g., chimneys or walls), both potentially promoting the success of urban gull populations. General conditions in urban areas may also be favourable to gulls all year-round, with some gulls no longer migrating and continuing to occupy their nesting grounds over the entire year (Rock 2005, Kroc 2018b, Pais de Faria et al. 2021b). Lower disturbance levels at uninhabited buildings within cities seem to provide suitable breeding opportunities for a range of avian species (Reynolds et al. 2019), but the effect of microclimate and availability of uninhabited buildings on roof-nesting pairs has not yet been studied in gulls.

Proximity to water bodies is another feature of urban areas that can make them attractive to gulls. Colonization of built-up areas by gulls initially started in coastal or estuarine areas and progressively expanded to more distant inland areas (e.g., see the expansion of the British roof-nesting gull populations). For example, Zelensky (2019) studied the growth of the urban population of Slaty-backed Gulls in Magadan, Russia, and reported an increasing distance of nest sites to the coast as the population expanded. The same pattern of expansion away from the immediate shoreline was observed in Seattle, USA, from 1961 to 1982, and in Victoria, British Columbia, between the 1980s and 2010s (Eddy 1982, Blight et al. 2019). Freshwater bodies are commonly used by gulls as roosting locations (Clark et al. 2016) and their availability in or near built-up areas may also attract gulls to urban areas (Vermeer et al. 1988, Washburn et al. 2016). Although exploiting predictable anthropogenic resources would allow gulls to potentially reduce the energetic costs of foraging, urban environments are also commonly associated with an increased risk of disturbance (Møller 2008) or exposure to contaminants (Chen et al. 2012, Isaksson 2018, Zapata et al. 2018, Sorais et al. 2020) and pathogens (Alm et al. 2018, Smith et al. 2020), potentially jeopardizing their health condition and producing long-term negative effects for gulls.

Human-gull interactions within the cities

Gulls have always been part of coastal landscapes, where positive relationships and culturally positive connections were common between humans and gulls. Recently, however, human-gull relationships have been changing, with gulls increasingly becoming a constant part of urban landscapes worldwide (Belant 1997) and causing conflicts with the human population, such as damaging urban structures, general nuisance behaviour and spilling waste when foraging in trash containers (Fig. B2.2). Negative human-gull interactions seem to intensify during the gulls' breeding season, when they are constrained by breeding duties and show higher aggressive territorial behaviour near their roof-top nests, and take higher risks to obtain food (e.g., stealing food from humans; Belant 1997, Huig et al. 2016). Huig et al. (2016) reported human-gull interactions for the Lesser Black-backed and Herring Gulls breeding at natural sites, but visiting the city of The Hague, Netherlands, during the breeding season, causing nuisance events such as raiding rubbish bags or searching for food scraps in residential areas, especially during the chick-rearing period. However, the number of urban gulls in some cities may be larger outside the breeding season, as observed in areas of southern Europe when migratory gulls join the local resident populations (Pais de Faria et al. 2021b).

Figure B2.2. Gulls exploring the city: adult and immature gulls *Larus* spp. spilling waste when foraging in trash containers in the street (top left) and in a small city park (top right), in the city of Porto, Portugal. Gulls resting in an urban pond (bottom) within the city of Matosinhos, Portugal, at the end of the day.

The common use of landfills and large water bodies by foraging and roosting gulls raises several human public health concerns, as gulls might end up serving as pathogen vectors (Alm et al. 2018), contributing to the dissemination of pathogens from landfills into roosting locations commonly shared with humans (e.g., beaches, ponds, lakes, rivers and reservoirs; Fig. B2.2; Clark et al. 2016). In fact, several studies have reported the presence of antimicrobial-resistant (AMR) bacteria and Avian Influenza viruses in gull species from different countries (Antilles et al. 2015, Arnal et al. 2015). For example, in France, researchers found identical populations of *Escherichia coli* on Yellow-legged Gull and humans, suggesting that gulls may be contaminated in landfills and then act as an environmental reservoir of AMR pathogens (Bonnedahl et al. 2009). In Berlenga Island, Portugal, gulls are considered a major source of faecal pollution in coastal waters near a beach that is largely visited by tourists (Alves et al. 2014, Araujo et al. 2014), and known to carry AMR *E. coli* (Radhouani et al. 2009).

Human-nature interactions and associated dynamics can be highly complex (Gaston et al. 2018, Soga and Gaston 2020), exhibiting spatial and temporal variation and being commonly associated with socio-economic factors (Soga and Gaston 2020). The same complexity is apparent when it comes to human-gull interactions. A great part of direct human-gull interactions is associated with the intentional and unintentional provisioning of food subsidies in urban areas (Oro et al. 2013, Newsome et al. 2015, Cox and Gaston 2018). The intentional provision of food is commonly a response to the 'extinction of experience', where people feed wildlife to experience a relationship with nature in urban areas, resulting in a positive interaction with several psychological benefits (Cox and Gaston 2016, Soga and Gaston 2016). However, these positive interactions can rapidly progress to negative ones, with opportunistic species increasing their human tolerance through a habituation process, and humans losing control over the interactions that they initially initiated (e.g., Kumar et al. 2019).

Several aspects may influence direct human-gull interactions (Soga and Gaston 2020), such as social and educational factors (e.g., lack of knowledge about the ecological consequences of wildlife feeding) or socio-economic factors, with gulls in some cases preferring to build their nests on older buildings with lower human disturbance levels (Reynolds et al. 2019, J. Pais de Faria pers. obser.). Humans intentionally feeding wildlife may often be elderly and lonely citizens sharing their

Figure B2.3. Citizen sharing his food with a group of gulls, *Larus* sp., in a city park within the city of Porto, Portugal.

food (Fig. B2.3). This could be related to a particular orientation and/or predisposition to feed and, ultimately, take care of the 'most vulnerable ones' (Cox and Gaston 2016, Soga and Gaston 2020), or a consequence of both being from social groups that suffer the most from loneliness, and will most frequently and subconsciously seek psychological benefits of interacting with nearby wild animals, such as gulls (Cox and Gaston 2016, Cox et al. 2017, Barnes et al. 2019).

Managing human-gull conflicts

The complexity of human-gull interactions underlines the need to consider the activities, routines and behaviour of humans when designing management measures aiming to decrease human-gull conflicts within cities. Pais de Faria et al. (2021b) found the number of people feeding Yellow-legged Gulls within the city of Porto increased during the winter months, attracting large numbers of foraging gulls, but gull-related complaints to the City Council peaked during the breeding season, when gulls' territorial aggressive behaviour tended to increase near their roof-top nests. Therefore, efficient management measures should consider local dynamics and seasonality of human-gull interactions, while including environmental education and social awareness campaigns. Although feeding by individual humans alone is probably not enough to sustain entire urban populations, this surely contributes to human-gull conflicts driven by the increasing habituation of gulls to humans. But changing human behaviour can be extremely challenging. Clark et al. (2015) observed people repeatedly feeding gulls in parking lots after being approached during awareness campaigns, and suggested the use of community-based social marketing techniques as a possible and more efficient approach. Contrarily, the Litter Free Coast and Sea "Don't Feed the Locals" Campaign (DFLC 2017) that occurred in Weymouth, UK, seemed to be very successful in engaging the local community, tourists and local business on several activities that aimed to educate people about the reasons why they should not feed gulls.

Several management measures that have been attempted aim to: disperse gulls (e.g., pyrotechnics, broadcast of gull distress calls, blank ammunitions, hazing by trained raptors; Cook et al. 2008, Rock 2013, Thieriot et al. 2015); prevent gulls from roosting in important water bodies (Clark et al. 2013); prevent gulls from constructing their nests on roof-tops (e.g., application of wires, spikes and roof netting); and reduce the availability of anthropogenic food subsidies (e.g., the use of gull-proof litter bins or closing open-air landfills; Payo-Payo et al. 2015, Steigerwald et al. 2015). However, due to gulls' extreme resilience, most of the applied measures were inefficient in the long-term, and in some cases simply relocated gulls, and their related conflicts, to new locations (Rock 2005, 2013). The general use of scaring devices is probably one of the least efficient techniques when it comes to dispersing gulls from roosting or nesting roof-top locations, with gulls commonly observed to nest next to these devices (Rock 2013, Kroc 2018a). Roof netting or wiring, on the other hand, may deter gulls from nesting on roof-tops, but only if mesh size and wire spacing is correctly considered to match the gulls' size and behaviour (Rock 2005, 2013; L. Blight, personal observ.). The use of a similar technique was shown to be highly effective at preventing Ring-billed Gulls from roosting in wastewater treatment tanks in Millbury, USA, where single-strand stainless-steel piano wires (1 mm in diameter) were spaced at 0.9–3.3 m with a 136 kg tensile strength (Clark et al. 2013). On landfills, the use of trained falcons to scare and disperse foraging gulls also seems to be effective (Thieriot et al. 2015), especially when combined with other techniques to reduce gulls' habituation to a single method (Cook et al. 2008). But again, all these measures aiming to disperse gulls, even if highly effective, could just be moving the problem elsewhere. Measures aimed at reducing breeding success, such as egg oiling, do not seem to lead to the abandonment of the colony or the movement to new locations (DeVault et al. 2014), but the management effort required is high, with campaigns being conducted during several consecutive breeding seasons, as gulls are long-lived and highly philopatric species, meaning that the loss of a couple of seasons in productivity does not affect population size.

Overall, changes in the availability of predictable anthropogenic food resources can affect major population trends at breeding sites (Oro et al. 2004, Duhem et al. 2008, Camphuysen 2013). In some cases, these changes may affect the gulls' average body mass, egg volume and clutch size (Pons and Migot 1995, Steigerwald et al. 2015). Although information on worldwide urban-nesting gull population trends is generally lacking, it has been shown for Europe that gulls nesting in roof-top areas forage both in landfills (Spelt et al. 2019) and fishing ports (Méndez et al. 2020). Therefore, combining measures that reduce the availability of predictable anthropogenic subsidies and those that aim to reduce urban gulls' breeding output, along with environmental education and social awareness campaigns to minimize direct human-gull conflict within cities, could potentially increase effectiveness of managing increasing urban gull populations.

References cited

Alberti, M., C. Correa, J.M. Marzluff, A.P. Hendry, E.P. Palkovacs and K.M. Gotanda. 2017. Global urban signatures of phenotypic change in animal and plant populations. PNAS. 114: 8951–8956. doi: 10.1073/pnas.1606034114.

Alm, E.W., Q.R. Daniels-Witt, D.R. Learman, H. Ryu, D.W. Jordan, T.M. Gehring et al. 2018. Potential for gulls to transport bacteria from human waste sites to beaches. Sci. Total Environ. 615: 123–130. doi: 10.1016/j.scitotenv.2017.09.232.

Alves, M.S., A. Pereira, S.M. Araújo, B.B. Castro, A.C.M. Correia and I. Henriques. 2014. Seawater is a reservoir of multi–resistant *Escherichia coli*, including strains hosting plasmid–mediated quinolones resistance and extended–spectrum beta-lactamases genes. Front. Microbiol. 5: 426. doi: 10.3389/fmicb.2014.00426.

Annett, C. and R. Pierotti. 1989. Chick hatching as a trigger for dietary switching in the western gull. Colon. Waterbirds 12: 4–11. doi: 10.2307/1521306.

Annett, C.A. and R. Pierotti. 1999. Long-term reprodutive output in western gulls: consequences of alternate tactis in diet choice. Ecology 80: 288–297. doi: 10.1890/0012-9658(1999)080[0288:LTROIW]2.0.CO;2.

Antilles, N., L. Garcia-Migura, K.G. Joensen, P. Leekitcharoenphon, F.M. Aarestrup, M. Cerda–Cuellar et al. 2015. Audouin's gull, a potential vehicle of an extended spectrum beta-lactamase producing *Salmonella* Agona. FEMS Microbiol. Lett. 362: 4. doi: 10.1093/femsle/fnu039.

Araujo, S., I.S. Henriques, S.M. Leandro, A. Alves, A. Pereira and A. Correia. 2014. Gulls identified as major source of fecal pollution in coastal waters: A microbial source tracking study. Sci. Total Environ. 470: 84–91. doi: 10.1016/j.scitotenv.2013.09.075.

Arnal, A., M. Vittecoq, J. Pearce–Duvet, M. Gauthier–Clerc, T. Boulinier and E. Jourdain. 2015. Laridae: A neglected reservoir that could play a major role in avian influenza virus epidemiological dynamics. Crit. Rev. Microbiol. 41: 508–519. doi: 10.3109/1040841X.2013.870967.

Artyukhin, Y.B. 2002. Unusual nesting of Japanese cormorant *Phalacrocorax filamentosus* and slaty-backed gull *Larus schistisagus* in the south of the Far East. *In:* Yu. B. Artyukhin and Yu. N. Gerasimov [eds.]. Biologiya i Okhrana Ptits Kamchatki (Biology and Conservation of Birds of Kamchatka). Moscow, Russia

Audet, J.-N., S. Ducatez and L. Lefebvre. 2016. The town bird and the country bird: Problem solving and immunocompetence vary with urbanization. Behav. Ecol. 27: 637–644. doi: 10.1093/beheco/arv201.

Barnes, M.R., M.L. Donahue, B.L. Keeler, C.M. Shorb, T.Z. Mohtadi and L.J. Shelby. 2019. Characterizing nature and participant experience in studies of nature exposure for positive mental health: An integrative review. Front. Psychol. 9: 2617. doi: 10.3389/fpsyg.2018.02617.

Belant, J.L. 1997. Gulls in urban environments: landscape-level management to reduce conflict. Landsc. Urban Plan. 38: 245–258. doi: 10.1016/S0169-2046(97)00037-6.

Blight, L.K., M.C. Drever and P. Arcese. 2015. A century of change in Glaucous-winged Gull (*Larus glaucescens*) populations in a dynamic coastal environment. Condor 117: 108–120. doi: 10.1650/CONDOR-14-113.1.

Blight, L.K., D.F. Bertram and E. Kroc. 2019. Evaluating UAV-based techniques to census an urban-nesting gull population on Canada's Pacific coast. J. Unmanned Veh. Syst. 7: 312–324. doi: 10.1139/juvs-2019–0005.

Blokpoel, H., W.F. Weller, G.D. Tessier, B. Smith, H. Gulls and R.B. Gulls. 1990. Roof-nesting by Ring-billed Gulls and Herring Gulls in Ontario in 1989. Ont. Birds 8: 55–60.

Blokpoel, H. and G.D. Tessier. 1991. Distribution and abundance of colonial waterbirds nesting in the Canadian portions of the lower Great Lakes System in 1990. Can. Wildl. Serv. Tech. Rep. Ser. 117.

Bonnedahl, J., M. Drobni, M. Gauthier-Clerc, J. Hernandez, S. Granholm, Y. Kayser et al. 2009. Dissemination of *Escherichia coli* with CTX-M Type ESBL between Humans and Yellow-Legged Gulls in the South of France. Plos One 4: 6. doi: 10.1371/journal.pone.0005958.

Campbell, R.W. 1975. Marginal habitat used by Glaucous-winged Gulls for nesting. Syesis. 8: 395–396.

Camphuysen, C.J. 2013. A historical ecology of two closely related gull species (Laridae): Multiple adaptations to a man-made environment. PhD. Thesis, University of Groningen, Groningen, Netherlands.

Carmona, M., R. Aymí and J. Navarro. 2021. Importance of predictable anthropogenic food subsidies for an opportunistic gull inhabiting urban ecosystems. Eur. J. Wildl. Res. 67: 9. doi: 10.1007/s10344-020-01446–2.

Castano, M.V., L.M. Biondi, F. Zumpano, M. Favero and G.O. García. 2020. Behavional responses to a novel feeding problem in the Olrog's Gull *Larus atlanticus*. Mar. Ornithol. 48: 149–155.

Chabrzyk, G. and J.C. Coulson. 1976. Survival and recruitment in the Herring Gull *Larus argentatus*. J. Anim. Ecol. 45: 187–203. doi: 10.2307/3774.

Chávez-Villavicencio, C. 2014. Aproximación a la selección de sitios de nidificación de la Gaviota Dominicana (*Larus Dominicanus* Lichtenstein 1823) en un área urbana de la región de Coquimbo (Chile) y un nuevo sustrato de nidificación. The Biologist 12: 33–44. doi: 10.24039/rtb2014121384.

Chen, D., R.J. Letcher, N.M. Burgess, L. Champoux, J.E. Elliott, C.E. Hebert et al. 2012. Flame retardants in eggs of four gull species (Laridae) from breeding sites spanning Atlantic to Pacific Canada. Environ. Pollut. 168: 1–9. doi: 10.1016/j.envpol.2012.03.040.

Chown, S.L. and G.A. Duffy. 2015. Thermal physiology and urbanization: Perspectives on exit, entry and transformation rules. Funct. Ecol. 29: 902–912. doi: 10.1111/1365–2435.12478.

Cignini, B. and M. Zapparoli. 1996. Atlante degli uccelli nidificanti a Roma. Palombi Editori, Roma, Italy.

Clark, D.E., K.K.G. Koenen, K.G. Mackenzie, J.W. Pereira and S. Destefano. 2013. Stainless-steel wires exclude gulls from a wastewater treatment plant. J. Am. Water Work. Assoc. 105: E609–E618. doi: 10.5942/jawwa.2013.105.0118.

Clark, D.E., J.J. Whitney, K.G. MacKenzie, K.K.G. Koenen and S. DeStefano. 2015. Assessing gull abundance and food availability in urban parking lots. Hum. Wildl. Interact. 9: 180–190. doi: 10.26077/m3ts-7d08.

Clark, D.E., S. DeStefano, K.G. MacKenzie, K.K.G. Koenen and J.J. Whitney. 2016. Roost site selection by ring-billed and herring gulls. J. Wildl. Manage. 80: 708–719. doi: 10.1002/jwmg.1066.

Cook, A., S. Rushton, J. Allan and A. Baxter. 2008. An evaluation of techniques to control problem bird species on landfill sites. Environ. Manage 41: 834–843. doi: 10.1007/s00267-008-9077-7.

Coulson, J.C. and B.A. Coulson. 2015. The accuracy of urban nesting gull censuses. Bird Study 62: 170–176. doi: 10.1080/00063657.2015.1013523.

Cox, D.T.C. and K.J. Gaston. 2016. Urban bird feeding: Connecting people with nature. PLOS ONE 11: e0158717. doi: 10.1371/journal.pone.0158717.

Cox, D.T.C., D.F. Shanahan, H.L. Hudson, R.A. Fuller, K. Anderson, S. Hancock et al. 2017. Doses of nearby nature simultaneously associated with multiple health benefits. Int. J. Environ. Res. Public Health 14: 172. doi: 10.3390/ijerph14020172.

Cox, D.T.C. and K.J. Gaston. 2018. Human-nature interactions and the consequences and drivers of provisioning wildlife. Philos. Trans. Royal Soc. B: Biol. Sci. 373(1745): 20170092. doi: 10.1098/rstb.2017.0092.

Cramp, S. 1971. Gulls nesting on buildings in Britain and Ireland. Br. Birds 64: 476–487.

Cramp, S., W.R.P. Bourne and D. Saunders. 1974. The Seabirds of Britain and Ireland. Collins, London, UK.

DeVault, T.L., P.M. Schmidt, F.E. Pogmore, J. Gobeille, J.L. Belant and T.W. Seamans. 2014. Influence of egg oiling on colony presence of ring-billed gulls. Hum. Wildl. Interact. 8: 22–30. doi: 10.26077/674e-qn50.

DFLC – Don't Feed the Locals Post Campaign Report. 2017. Weymouth, UK. http://www.litterfreecoastandsea.co.uk/wp–content/uploads/2017/05/Dont–Feed–the–Locals–Post–campign–report.pdf, accessed 15 Frebruary 2021.

Duhem, C., P. Roche, E. Vidal and T. Tatoni. 2008. Effects of anthropogenic food resources on yellow-legged gull colony size on Mediterranean islands. Popul. Ecol. 50: 91–100. doi: 10.1007/s10144-007-0059–z.

Dwyer, C.P., J.L. Belant and R.A. Dolbeer. 1996. Distribution and abundance of roof-nesting gulls in the Great Lakes region of the United States. Ohio J. Sci. 96: 9–12.

Eddy, G. 1982. Glaucous-winged Gulls nesting on buildings in Seattle, Washington. The Murrelet 63: 27–28.

EFSA NDA Panel (EFSA Panel on Dietetic Products, Nutrition and Allergies) (2014) Scientific Opinion on health benefits of seafood (fish and shellfish) consumption in relation to health risks associated with exposure to methylmercury. EFSA J. 12: 3761. doi: 10.2903/j.efsa.2014.3761.

Elbin, S., K.L. Parkins and E. Tobón. 2019. Urban gulls: A case study from New York City. Presented at the 43rd Annual Meeting of the Waterbird Society, Maryland, USA.

Fisk, E.J. 1978. The growing use of roofs by nesting birds. Bird-banding 49: 134–141.

Garcia, C.M., M. Suárez–Rodríguez and I. López–Rull. 2017. Becoming citizens: Avian adaptations to urban life. pp. 91–112. *In:* Murgui, E. and M. Hedblom [eds.]. Ecology and Conservation of Birds in Urban Environments. Springer, Cham, Switzerland. doi: 10.1007/978-3-319-43314-1_6.

Gaston, K.J., M. Soga, J.P. Duffy, J.K. Garrett, S. Gaston and D.T.C. Cox. 2018. Personalised Ecology. Trends Ecol. Evol. 33: 916–925. doi: 10.1016/j.tree.2018.09.012.

Gladyshev, M., M. Arts and N. Sushchik. 2009. Preliminary estimates of the export of omega-3 Highly Unsaturated Fatty Acids (EPA + DHA) from Aquatic to Terrestrial Ecosystems. pp. 179–210. *In:* Kainz, M., M. Brett and M. Arts [eds.]. Lipids in Aquatic Ecosystems. Springer, New York, USA.

Goethe, F. 1960. Felsbrütertum und weitere beachtenswerte Tendenzen bei der Silbermöwe. Proc. Int. Ornithol. Congress. 12: 252–258.

Golden Gate Audubon Society. 2021. Heermann's gulls nesting in California. https://goldengateaudubon.org/speaker_series/heermanns-gulls-nesting-in-california/. Accessed 14 April 2021.

Gosling, W.D., J. de Kruif, S.J. Norder, E.J. de Boer, H. Hooghiemstra, K.F. Rijsdijk et al. 2017. Mauritius on fire: Tracking historical human impacts on biodiversity loss. Biotropica 49: 778–783. doi: 10.1111/btp.12490.

Gyimesi, A., T.J. Boudewijn, R.J. Buijs, J.Z. Shamoun-Baranes, J.W. de Jong, R.C. Fijn et al. 2016. Lesser Black-backed Gulls *Larus fuscus* thriving on a non-marine diet. Bird Study 63: 241–249. doi: 10.1080/00063657.2016.1180341.

Hebert, C.E., D.V.C. Weseloh, A. Idrissi, M.T. Arts, R. O'Gorman. O.T. Gorman et al. 2008. Restoring piscivorous fish populations in the Laurentian Great Lakes causes seabird dietary change. Ecology 89: 891–897. doi: 10.1890/07-1603.1.

Hensley, C.B., C.H. Trisos, P.S. Warren, J. MacFarland, S. Blumenshine, J. Reece et al. 2019. Effects of urbanization on native bird species in three southwestern US Cities. Front. Ecol. Evol. 7: 71. doi: 10.3389/fevo.2019.00071.

Herrera, J. 2019. Nesting island to launch Saturday on Roberts Lake in Seaside. Monterey Herald.https://www.montereyherald.com/2019/04/09/nesting-island-to-launch-saturday-on-roberts-lake-in-seaside. Accessed 20 April 2021.

Hooper, T.D. 1988. Habitat, reproductive parameters, and nest-site tenacity of urban–nesting glaucous–winged gulls at Victoria, British Columbia. The Murrelet 69: 10–14. doi: 10.2307/3534880.

Howell, J., D. Laclergue, S. Paris, W.I. Boarman, A.R. Degange and L.C. Binford. 1983. First nests of Heermann's Gull in the United States. West. Birds 14: 39–46.

Huig, N., R.-J. Buijs and E. Kleyheeg. 2016. Summer in the city: Behaviour of large gulls visiting an urban area during the breeding season. Bird Study: 1–9. doi: 10.1080/00063657.2016.1159179.

Hunt Jr., G.L. 1972. Influence of food distribution and human disturbance on the reproductive success of herring gulls. Ecology 53: 1051–1061. doi: 10.2307/1935417.

Isaksson, C. 2018. Impact of urbanization on birds. pp. 235–257. *In*: Tietze, D.T. [ed.]. Bird Species: How They Arise, Modify and Vanish. Springer International Publishing, Cham, Switzerland.

Isaksson, N., T.J. Evans, J. Shamoun-Baranes and S. Akesson. 2016. Land or sea? Foraging area choice during breeding by an omnivorous gull. Mov. Ecol. 4: 14. doi: 10.1186/s40462–016–0078–5.

Joint Nature Conservation Committee (JNCC). 2014. UK regional seas. JNCC, Peterborough, U.K. http://jncc. defra.gov.uk/page–1612. Accessed 5 February 2021.

Kolodzinski, J., S. Elbin and K. Parkins. 2019. Jailbirds: The history and future of Herring Gulls on Rikers Island in New York City. Presented at the 43rd Annual Meeting of the Waterbird Society, Maryland, USA.

Kroc, E., L.K. Blight, and W. Cao. 2018. Effects of urban density and structure on the spatial distribution of urban-nesting Glaucous-winged Gulls (*Larus glaucescens*) of the Salish Sea. Poster presentation. Int. Ornithol. Congress, Vancouver, BC.

Kroc, E. 2018a. Reproductive Ecology of Urban-nesting Glaucous-winged gulls *Larus glaucescens* in Vancouver, BC, Canada. Mar. Ornithol. 46: 155–164.

Kroc, E. 2018b. An example of year-round nest fidelity among urban-resident Glaucous-winged Gulls (*Larus glaucescens*) in Vancouver, B.C., Canada. British Columbia Birds 18: 28–31.

Kroc, E. and L.K. Blight. 2019. Differences in urban nesting behaviours and nest site selection of seven common North American species of gull. Presented at the 43rd Annual Meeting of the Waterbird Society, Maryland, USA.

Kumar, N., Y.V. Jhala, Q. Qureshi, A.G. Gosier and F. Sergio. 2019. Human-attacks by an urban raptor are tied to human subsidies and religious practices. Sci. Rep. 9: 10. doi: 10.1038/s41598-019-38662–z.

Lloyd, C., M.L. Tasker and K. Partridge. 1991. The Status of Seabirds in Britain and Ireland. T and AD Poyser, London, UK.

Lowry, H., A. Lill and B.B. Wong. 2013. Behavioural responses of wildlife to urban environments. Biol. Rev. 88: 537–549. doi: 10.1111/brv.12012.

Martínez-Abraín, A. and J. Jiménez. 2016. Anthropogenic areas as incidental substitutes for original habitat. Conserv. Biol. 30: 593–598. doi: 10.1111/cobi.12644.

Méndez, A., T. Montalvo, R. Aymí, M. Carmona, J. Figuerola and J. Navarro. 2020. Adapting to urban ecosystems: Unravelling the foraging ecology of an opportunistic predator living in cities. Urban Ecosyst. 23: 1117–1126. doi: 10.1007/s11252-020-00995-3.

Mitchell, I.P., S. Newton, N. Ratcliffe and T.E. Dunn. 2004. Seabird populations of Britain and Ireland. T and AD Poyser, London, UK.

Møller, A.P. 2008. Flight distance of urban birds, predation, and selection for urban life. Behav. Ecol. Sociobiol. 63(63). doi: 10.1007/s00265-008-0636-y.

Monaghan, P. and J.C.C. Coulson. 1977. Status of large gulls nesting on buildings. Bird Study 24: 89–104. doi: 10.1080/00063657709476538.

Monaghan, P. 1979. Aspects of the breeding biology of Herring Gulls *Larus argentatus* in urban colonies. Ibis 121: 475–481. doi: 10.1111/j.1474-919X.1979.tb06687.x.

Morris, R.D., D.V. Weseloh and J.L. Shutt. 2003. Distribution and abundance of nesting pairs of Herring Gulls (*Larus argentatus*) on the North American Great Lakes, 1976 to 2000. J. Great Lakes Res. 29: 400–426. doi: 10.1016/s0380-1330(03)70447–x.

Morris, R.D., D.V. Weseloh, L.R. Wires, C. Pekarik, F.J. Cuthbert and D.J. Moore. 2011. Population trends of ring-billed gulls breeding on the North American Great Lakes, 1976 to 2009. Waterbirds 34: 202–212. doi: 10.1675/063.034.0209.

Murray, M.H., E.W. Lankau, A.D. Kidd, C.N. Welch, T. Ellison, H.C. Adams et al. 2020. Gut microbiome shifts with urbanization and potentially facilitates a zoonotic pathogen in a wading bird. PLOS ONE15: e0220926. doi: 10.1371/journal.pone.0220926.

Nager, R.G. and N.J. O'Hanlon. 2016. Changing numbers of three gull species in the British Isles. Waterbirds 39: 15–28. doi: 10.1675/063.039.sp108.

Nankinov, D.N. 1992. The nesting by the Herring Gull (*Larus argentatus*) in the towns and villages of Bulgaria. Avocetta 16: 93–94.

Newsome, T.M., J.A. Dellinger, C.R. Pavey, W.J. Ripple, C.R. Shores, A.J. Wirsing et al. 2015. The ecological effects of providing resource subsidies to predators. Glob. Ecol. Biogeogr. 24: 1–11. doi: 10.1111/geb.12236.

O'Hanlon, N., R. McGill and R. Nager. 2017. Increased use of intertidal resources benefits breeding success in a generalist gull species. Mar. Ecol. Prog. Ser. 574: 193–210. doi: 10.3354/meps12189.

Oldaker, R.F. 1963. Unusual nest site of the Glaucous-winged Gull. Can. Field-Nat. 77: 65–66.

Oro, D., E. Cam, R. Pradel and A. Martínez-Abraín. 2004. Influence of food availability on demography and local population dynamics in a long-lived seabird. Proc. R. Soc. Lond. B. 271: 387–396. doi: 10.1098/rspb.2003.2609.

Oro, D., M. Genovart, G. Tavecchia, M.S. Fowler and A. Martínez-Abraín. 2013. Ecological and evolutionary implications of food subsidies from humans. Ecol. Lett. 16: 1501–1514. doi: 10.1111/ele.12187.

Ouyang, J.Q., C. Isaksson, C. Schmidt, P. Hutton, F. Bonier and D. Dominoni. 2018. A new framework for urban ecology: an integration of proximate and ultimate responses to anthropogenic change. Integr. Comp. Biol. 58: 915–928. doi: 10.1093/icb/icy110.

Pagani–Núñez, E., D. Liang, C. He, X. Zhou, X. Luo, Y. Liu et al. 2019. Niches in the Anthropocene: passerine assemblages show niche expansion from natural to urban habitats. Ecography 42: 1360–1369. doi: 10.1111/ecog.0420.

Pais de Faria, J., P.T. Vaz, C.S. Lopes, J.G. Calado, J.M. Pereira, S.N. Veríssimo et al. 2021a. The importance of marine resources in the diet of urban gulls. Mar. Ecol. Prog. Ser. 660: 189–201. doi: 10.3354/meps13599.

Pais de Faria, J., V.H. Paiva, S.N. Veríssimo, A.M.M. Gonçalves and J.A. Ramos. 2021b. Seasonal variation in habitat use, daily routines and interactions with humans by urban-dwelling gulls. Urban Ecosyst. doi: 10.1007/s11252-021-01101-x.

Parslow, J. 1967. Changes in status among breeding birds in Britain and Ireland: part 3. Br. Birds 60: 177–202.

Paynter, R.A. 1963. North American Herring Gulls nesting on a building. The Wilson Bulletin, 75: 88–88.

Payo-Payo, A., D. Oro, J.M. Igual, L. Jover, C. Sanpera and G. Tavecchia. 2015. Population control of an overabundant species achieved through consecutive anthropogenic perturbations. Ecol. Appl.25: 2228–2239. doi: 10.1890/14-2090.1.sm.

Perlut, N.G., D.N. Bonter, J.C. Ellis and M.S. Friar. 2016. Roof-top nesting in a declining population of herring gulls (*Larus argentatus*) in Portland, Maine, USA. Waterbirds 39: 68–73. doi: 10.1675/063.039.sp113.

Pierotti, R. and C. Annett. 2001. The ecology of Western Gulls in habitats varying in degree of urban influence. pp. 307–329. *In:* J. Marzluff, R. Bowman and R. Donnelly [eds.]. Avian Ecology and Conservation in an Urbanizing World. Springer Science and Business Media, New York, USA.

Pons, J.M. and P. Migot. 1995. Life-History strategy of the herring gull: Changes in survival and fecundity in a population subjected to various feeding conditions. J. Anim. Ecol. 64: 592–599. doi: 10.2307/5802.

Poynter, A. 1976. Glaucous-winged Gulls—unusual nesting sites. Vancouver Natural History Society Discovery, New Series 5: 8–10.

Radhouani, H., P. Poeta, G. Igrejas, A. Goncalves, L. Vinue and C. Torres. 2009. Antimicrobial resistance and phylogenetic groups in isolates of *Escherichia coli* from seagulls at the Berlengas nature reserve. Vet. Record 165: 138–142. doi: 10.1136/vr.165.5.138.

Raven, S.J. and J.C. Coulson. 1997. The distribution and abundance of *Larus* gulls nesting on buildings in Britain and Ireland. Bird Study 44: 13–34. doi: 10.1080/00063659709461035.

Reynolds, S.J., J.D. Ibáñez-Álamo, P. Sumasgutner and M.C. Mainwaring. 2019. Urbanisation and nest building in birds: A review of threats and opportunities. J. Ornithol. 160: 841–860. doi: 10.1007/s10336-019-01657-8.

Roberson, D., S.F. Bailey and R. Carratello. 2001. First successful united states nesting of Heermann's Gull: Challenges of an Urban Colonist. N. Am. Birds 55: 375–378.

Roby, D., J.Y. Adkins, K. Bixler, K. Collins and A. Evans. 2007. Surveys for nesting Glaucous-winged/Western Gulls and Caspian Terns in Puget Sound, Washington. 2007 Final Annual Report. Submitted to US Fish and Wildlife Service, Region 1, Migratory Birds and Habitat Programs, Portland, Ore., USA.

Rock, P. 2005. Urban gulls: problems and solutions. Br. Birds 98: 338–355.

Rock, P. 2013. Urban gulls. Why current control methods always fail. Riv. Ital. Orn. 82. doi: 10.4081/rio.2012.112.

Rock, P., C.J. Camphuysen, J. Shamoun-Baranes, V.H. Ross-Smith and I.P. Vaughan. 2016. Results from the first GPS tracking of roof-nesting Herring Gulls *Larus argentatus* in the UK. Ringing Migr. 31: 47–62. doi: 10.1080/03078698.2016.1197698.

Salmon, H.M. 1958. Lesser Black-backed and Herring Gulls nesting on a factory-roof inland in Glamorgan. Br. Birds 51: 399–401.

Sanford, F.J. 1974. Gulls nesting on water street - 1972–1973. Vancouver Nat. Hist. Soc. Discovery, New Ser. 5: 8–10.

Sayol, F., D. Sol and A.L. Pigot. 2020. Brain size and life history interact to predict urban tolerance in birds. Front. Ecol. Evol. 8. doi: 10.3389/fevo.2020.00058.

Sepp, T., K.J. McGraw, A. Kaasik and M. Giraudeau. 2017. A review of urban impacts on avian life-history evolution: Does city living lead to slower pace of life? Glob. Change Biol. 24: 1452–1469. doi:10.1111/gcb.13969.

Smith, O.M., W.E. Snyder and J.P. Owen. 2020. Are we overestimating risk of enteric pathogen spillover from wild birds to humans? Biol. Rev. 95: 652–679. doi:10.1111/brv.12581.

Snell-Rood, E.C. and N. Wick. 2013. Anthropogenic environments exert variable selection on cranial capacity in mammals. Proc. R. Soc. B. 280: 20131384. doi: 10.1098/rspb.2013.1384.

Soga, M. and K.J. Gaston. 2016. Extinction of experience: the loss of human-nature interactions. Front. Ecol. Environ. 14: 94–101. doi: 10.1002/fee.1225.

Soga, M. and K.J. Gaston. 2020. The ecology of human-nature interactions. Proc. R. Soc. B. 287(1918): 20191882. doi: 10.1098/rspb.2019.1882.

Sol, D., O. Lapiedra and C. González-Lagos. 2013. Behavioural adjustments for a life in the city. Anim. Behav. 85: 1101–1112. doi: 10.1016/j.anbehav.2013.01.023.

Soldatini, C., Y.V. Albores-Barajas, D. Mainardi and P. Monaghan. 2008. Roof nesting by gulls for better or worse? Ital. J. Zool. 75: 295–303. doi: 10.1080/11250000701884805.

Sorais, M., M.J. Mazerolle, J.F. Giroux and J. Verreault. 2020. Landfills represent significant atmospheric sources of exposure to halogenated flame retardants for urban-adapted gulls. Environ. Int. 135: 10. doi: 10.1016/j.envint.2019.105387.

Sotillo, A., J.M. Baert, W. Muller, E.W.M. Stienen, A. Soares and L. Lens. 2019. Time and energy costs of different foraging choices in an avian generalist species. Mov. Ecol. 7: 11. doi: 10.1186/s40462–019–0188–y.

Spelt, A., C. Williamson, J. Shamoun-Baranes, E. Shepard, P. Rock and S. Windsor. 2019. Habitat use of urban-nesting lesser black-backed gulls during the breeding season. Sci. Rep. 9: 10527. doi: 10.1038/s41598-019-46890–6.

Spelt, A., O. Soutar, C. Williamson, J. Memmott, J. Shamoun-Baranes, P. Rock et al. 2021. Urban gulls adapt foraging schedule to human-activity patterns. Ibis 163: 274–282. doi: 10.1111/ibi.12892.

Stedt, J., J. Bonnedahl, J. Hernandez, B.J. McMahon, B. Hasan, B. Olsen et al. 2014. Antibiotic resistance patterns in *Escherichia coli* from gulls in nine European countries. Infect. Ecol. Epidemiology 4: 21565. doi: 10.3402/iee.v4.21565.

Steigerwald, E.C., J.M. Igual, A. Payo-Payo and G. Tavecchia. 2015. Effects of decreased anthropogenic food availability on an opportunistic gull: evidence for a size-mediated response in breeding females. Ibis 157: 439–448. doi: 10.1111/ibi.12252.

Thieriot, E., M. Patenaude-Monette, P. Molina and J.F. Giroux. 2015. The Efficiency of an integrated program using falconry to deter gulls from landfills. Animals 5: 214–225. doi: 10.3390/ani5020214.

Took, G.E. 1995. Nesting of Herring Gulls on roof-tops in Dover. Br. Birds 48: 88–89.

Turbott, E.G. 1969. Roof nesting of black-backed gulls. Notornis 16: 187–89.

Turner, D.M. 2010. Counts and breeding success of Blacklegged Kittiwakes *Rissa tridactyla* nesting on man-made structures along the River Tyne, northeast England, 1994–2009. Seabird 23: 111–126.

Unitt, P. 2004. San Diego County Bird Atlas. San Diego Natural History Museum and Sunbelt Publications, San Diego, USA.

van Donk, S., K.C.J. Camphuysen, J. Shamoun-Baranes and J. van der Meer. 2017. The most common diet results in low reproduction in a generalist seabird. Ecol. Evol. 7: 4620–4629. doi: 10.1002/ece3.3018.

Vermeer, K. 1963. The breeding ecology of the Glaucous-winged Gull (*Larus glaucescens*) darte Island, B.C. British Columbia. Prov. Mus. Occ. 13: 92.

Vermeer, K., D. Power and G.E.J. Smith. 1988. Habitat selection and nesting biology of roof-nesting glaucous-winged Gulls. Colon. Waterbirds 11: 189–201.

Washburn, B.E., S.B. Elbin and C. Davis. 2016. Historical and current population trends of herring gulls (*Larus argentatus*) and Great Black-backed Gulls (*Larus marinus*) in the New York Bight, USA. Waterbirds 39: 74–86. doi: 10.1675/063.039.sp114.

Witherby, H.F., F.C.R.R. Jourdain, N.F. Ticehurst and B.W. Tucker. 1938–1941. The Handbook of British Birds. H. F. and G. Witherby, London, UK.

Yorio, P., J.O. Branco, J. Lenzi, G. Luna-Jorquera, and C. Zavalaga. 2016. Distribution and trends in kelp gull (*Larus dominicanus*) coastal breeding populations in South America. Waterbirds 39: 114–135. doi: 10.1675/063.039.sp103.

Zapata, P., R. Ballesteros-Cano, P. Colomer, A. Bertolero, P. Viana, S. Lacorte et al. 2018. Presence and impact of stockholm convention POPs in gull eggs from Spanish and Portuguese natural and national parks. Sci. Total Environ. 633: 704–715. doi: 10.1016/j.scitotenv.2018.03.081.

Zelenskaya, L.A. 2019. Ecology of an Urban Population of the Slaty-Backed Gull (*Larus schistisagus*) in Comparison with Natural Colonies: 1. Features of Nest Location and Productivity. Biol. Bull. 46: 1108–1123. doi: 10.1134/S106235901909019X.

Chapter 3

Seabirds and Marine Renewable Energy Sources

Andrew J.P. Harwood[1,*] and *Sue King*[2]

Introduction

The world's oceans are becoming increasingly crowded, and a variety of marine users compete for space with seabirds. Many seabird species are already under pressure from a wide range of anthropogenic stressors including commercial fisheries, pollution and climate change. The Marine Renewable Energy (MRE) industry represents a relatively new, but quickly expanding, developer of the coastal and offshore waters that are so important to many seabird species and other wildlife. While MRE projects contribute to decarbonisation, climate change mitigation, delivery of energy security and provide socio-economic benefits, we do not fully understand their effects on seabirds (Snyder and Kaiser 2009, Graziano et al. 2016, Copping et al. 2020). As the industry accelerates globally and moves further offshore, research and monitoring are needed to fill knowledge gaps and ensure growth can continue without significantly impacting wildlife (Borthwick 2016).

MRE projects may have wide-ranging effects on seabirds that could result in negative (e.g., collisions and disturbance) or positive (e.g., increases in biodiversity and prey resources) population-level impacts on species of conservation concern (Drewitt and Langston 2006, Gibson et al. 2017, Perrow 2019, Copping et al. 2020). Understanding these effects is therefore crucial for the protection of important seabird populations through consenting, mitigation and compensation. Further, as projects move into deeper waters and exploit new geographic areas there may be a paucity of information about the seabirds present that will need to be addressed.

This chapter provides a brief overview of MRE devices currently in use or development and their possible effects on seabirds. The roles of spatial planning, impact assessment and seabird monitoring methods are reviewed. Lastly, we consider the importance of mitigation, compensation and restoration for the conservation of seabirds and the continued growth of the industry.

[1] Natural England, Mail Hub, County Hall, Spetchley Road, Worcester, WR5 2NP, UK.

[2] Sue King Consulting Ltd, The Coach House, Hampsfell Road, Grange over Sands, Cumbria, LA11 6BG, UK. Email: sue@kingconsulting-ltd.com

* Corresponding author: Andrew.Harwood@naturalengland.org.uk

Harvesting energy in the marine environment

Extracting energy from the marine environment is challenging. The continuous attack of saltwater, harsh weather conditions, biofouling, sedimentation and problems associated with isolation (e.g., maintenance and electrical transmission) pose technical and economic issues for developers (Borthwick 2016, Melikoglu 2018). However, where these problems can be overcome, the world's oceans present a significant and sustainable renewable energy source (Khan et al. 2017). Here we define MRE devices as those which may harness wind or water movements and exploit gradients in temperature or salinity to generate energy (see Fig. B3.1 for examples).

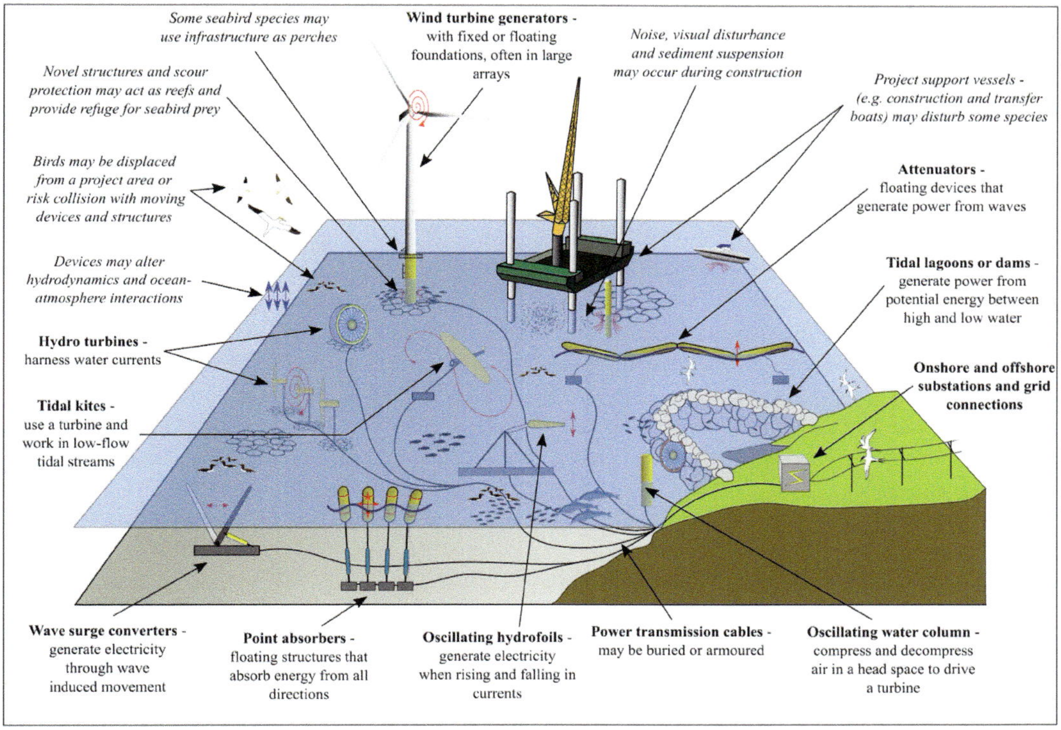

Figure B3.1. Diversity of marine renewable energy devices, associated infrastructure, and potential effects on seabirds. Impacts may occur throughout the life of a project due to the presence of vessels, the device and associated infrastructure influencing bird behaviour and distribution (displacement) and presenting a collision hazard. Indirect and food web effects may occur through impacts on lower trophic levels and prey availability.

Offshore wind farms

Offshore Wind Farms (hereafter OWFs) currently dominate the MRE industry, with full-scale commercial projects currently providing a total estimated installed global capacity of 29.1 GW in 2019 (GWEC 2020, IEA 2020). Despite this, offshore wind only accounted for 0.3% of global electricity generation in 2018 (IEA 2020). Current global predictions estimate the addition of 205 GW of capacity by 2030, and countries in Europe alone are targeting 450 GW of capacity by 2050 (GWEC 2020). Indeed, Europe currently remains the largest offshore market, comprising 75% of all global offshore wind assets (GWEC 2020). However, North America, China, Taiwan, Vietnam, Japan and South Korea are all accelerating investment in OWFs (GWEC 2020, IEA 2020).

OWFs are comprised of an array of turbines, a network of cables to transport electricity to shore and often include electrical substations. Turbines are typically comprised of a foundation (e.g., piled, gravity or suction), tower and nacelle with rotor blades that turn a drive shaft through a gearbox that then spins a generator. Whilst piled, horizontal axis three-bladed designs are most common, horizontal axis two-bladed designs have been deployed and vertical axis designs are being explored. As development areas move further offshore into deeper water various floating designs, anchored to the seabed, are likely to become more common (Castro-Santos and Diaz-Casas 2016).

The first OWFs were small. Vindeby OWF, Denmark, commissioned in 1991 with 11 turbines, was the first true OWF, but was decommissioned in 2017 after 25 years of service (Topham and McMillan 2017). Vindeby covered less than 0.5 km^2 and had a capacity of just 4.95 MW. Each 450 kW turbine employed rotors with a diameter of 35 m. In comparison, at the end of 2020, the Hornsea Project Three OWF, with up to 300 turbines and a capacity of up to 2.4 GW, was consented. By 2025, individual turbines may generate upwards of 10 MW of electricity from rotors with diameters of over 200 m (GWEC 2020). The first OWFs were situated close to shore around Europe, where there was already a lot of information about a limited range of well-studied seabird species. However, as the geographic spread of projects increases and there is a move offshore, a project may affect a wider range of species about which less is known.

Wet marine renewable energy devices

Wet MRE devices generate electricity from tides, waves, ocean currents, thermal and salinity gradients. According to Khan et al. (2017), wet MRE devices alone have the potential to fulfil the total global electricity demand. However, the commercialisation of wet MRE devices has been slower than for other renewables because of the challenges of operating in the marine environment (Melikoglu 2018, Copping et al. 2020). Of the 243 European wave, tidal and salinity gradient projects recorded in the EMODnet spatial database (EMODnet 2020), only 26 (11%) are classed as operational or under construction; of these, 12 are tidal, 13 are wave and one exploits salinity gradients.

Tidal range energy can be harvested using a dam or barrage along an estuary or river, or via construction of a tidal lagoon (Neill et al. 2018, Chowdhury et al. 2020). The first tidal power plant, the La Rance Tidal Power Station (240 MW capacity), was commissioned in the Rance river mouth, France, in 1966 (Retiere 1994). Other operating tidal barrages include the Sihwa Lake Tidal Power Station in South Korea (254 MW capacity) and smaller plants, with less than 20 MW of capacity, in Canada, China, Russia and South Korea (Neill et al. 2018). Such developments are likely to have high construction costs and could have considerable environmental impacts on the dynamics of an estuary and associated habitat (Neill et al. 2018). Energy can also be harvested from strong tidal currents using tidal turbines of varying designs (Figs. B3.1 and B3.2). Such devices may vary in capacity from a few kW to multi-megawatt (MW), can be deployed in arrays, and are beginning to move toward commercial deployment (Fox et al. 2017).

Wave energy is also a highly abundant, clean and predictable energy resource in certain areas of the world (e.g., Alaska, northern Canada, western coast of Scotland, southern Africa, and Australia) where there are consistently high energy waves (Aderinto and Li 2018, Ocean Energy Systems 2019). Various Wave Energy Converter (WEC) designs exist and work on the principle of converting wave energy into mechanical energy and then into electricity (Aderinto and Li 2018). WECs can be installed on the seabed, sea surface or land and numerous designs include attenuators, point absorbers, oscillating wave surge converters, oscillating water column, over topping or terminator, submerged pressure differential, bulge wave and rotating mass (see Fig. B3.1 for examples). However, few have been deployed or used commercially (Aderinto and Li 2018). Currently, tidal and wave projects are largely constrained to coastal areas but may overlap with important seabird, wildfowl and wader populations that exploit similar areas.

Figure B3.2. Examples of structures and vessels involved with MRE projects. From top-left clockwise: an Open Hydro test tidal turbine on a barge (photo courtesy of Iain Hill, Specialised Marine Support Ltd.), a jack-up vessel installing blades (photo courtesy of Richard Berridge, ECON Ecological Consultancy Ltd.), a crew transfer vessel and wind farm cardinal buoy, here used by Sandwich Terns (*Thalasseus sandvicensis*).

The exploitation of Ocean Thermal Energy Conversion (OTEC) and salinity gradients (osmotic power) are less well developed than tidal and wave devices, but could provide a significant resource in the future (Khan et al. 2017, Tufa et al. 2018, Zhang et al. 2018).

Effects on seabirds

MRE projects may affect seabirds through direct or indirect pathways during construction (e.g., piling), operation (e.g., active devices and maintenance) and decommissioning (e.g., use of explosives or cutting) of below- and above-water structures and additional infrastructure (e.g., Fig. B3.2). Seabirds could collide with devices, be entangled in mooring systems and be displaced from, or attracted to, the project area due to disturbance and barrier effects or provision of novel habitat (Drewitt and Langston 2006, Masden et al. 2009, Furness et al. 2012, Furness et al. 2013a, Masden et al. 2013, King 2019, Vanermen and Stienen 2019, Copping et al. 2020). Furthermore, indirect effects could result in changes in seabird prey resources, resulting in negative or positive effects (Linley et al. 2007, Grecian et al. 2010, Lindeboom et al. 2011, Perrow et al. 2011, Scott et al. 2014, Goodale and Milman 2016, Perrow 2019). Effects on seabirds may be short-or long-term, but could reduce local biodiversity or, more importantly, impact wider seabird populations.

Collision risk and entanglement

Direct mortality, resulting from collisions with devices, or entanglement with mooring systems, could have significant implications for k-selected species such as seabirds, of which many are already in decline (Desholm and Kahlert 2005, Drewitt and Langston 2006, Furness et al. 2012, Paleczny et al. 2015, Bowgen and Cook 2018). Collisions are mainly assumed to occur with moving blades, but barotrauma (i.e., injury due to pressure changes around moving blades) and collision with fixed structures cannot be ruled out (Barclay et al. 2017, May et al. 2020). Until relatively recently, assessment of collision risk has generally been predictive based on pre-construction behavioural data. The quantification of collisions and avoidance at sea remains a challenge (e.g., carcass recovery is difficult), though our understanding is improving and in some cases changing our pre-conceptions (Furness et al. 2013a, King 2019, Copping et al. 2020).

To date, there is limited evidence of collisions of seabirds with OWFs. For example, in Kalmar Sound, Sweden, four Common Eider (*Somateriamollissima*) from a migrating flock collided with a turbine blade, or its wake turbulence, with only three flying away (Pettersson 2005). Six collisions were also recorded at Thanet OWF, UK, during the ORJIP study (Skov et al. 2018). Collisions of seabirds with coastal and onshore wind farms are better documented and mainly include gulls, Northern Gannet (*Morus bassanus*), tern species and Northern Fulmar (*Fulmarus glacialis*) (Percival 2003, Dierschke and Garthe 2006, MacArthur Green et al. 2013, Cook et al. 2014, Cook et al. 2018). Petrels and shearwaters may also be susceptible to collisions where they nest in high numbers near terrestrial wind farms in the southern hemisphere (Hull et al. 2013, Bull et al. 2013). Whilst there are currently no detailed records of seabird collisions with wet MRE devices, they still pose a risk to diving birds (Waggitt and Scott 2014, Isaksson et al. 2020). Similarly, there have been no observations of seabirds becoming entangled and this seems unlikely to pose a significant threat (Copping et al. 2020, Garavelli 2020).

Collision risk is affected by species-specific factors (e.g., morphology, ecology and behaviour), project-specific factors (e.g., device design and project layout) and location-specific factors (e.g., weather and hydrodynamic conditions) (Marques et al. 2014, King 2019, Copping et al. 2020, Isaksson et al. 2020). To be at risk of collision, a species must overlap with a project in space and time and demonstrate a lack of avoidance behaviour (Waggitt and Scott 2014, Isaksson et al. 2020).An understanding of avoidance behaviour is therefore key to assessing collision effects. Box B3.1 provides a brief introduction to the different scales of avoidance behaviour that may occur in relation to OWFs. Further, habituation or attraction to a MRE project may increase exposure to collision risk, though behavioural adaptations may counter the risks.

Beyond avoidance, flight height and behaviour (e.g., speed and sinuosity) are likely to be the strongest determinants of collision risk (Cook et al. 2014, Johnston et al. 2014, King 2019). Furness et al. (2013a) identified large gulls and Northern Gannet as likely to be the most vulnerable seabird species to collision impacts with OWFs around the British Isles. Other 'risk' species included Common Gull (*Larus canus*) and Black-legged Kittiwake. The major driver for these scores was the amount of time spent at rotor blade height, though this was based on earlier, smaller wind turbine models. Indeed, post-construction studies have shown that Northern Gannets avoid OWFs and are therefore less likely to be subject to high levels of collision risk. Some species (e.g., Northern Gannet, Black-headed Gull *Chroicocephalus ridibundus* and Black-legged Kittiwake) may also adjust flight altitudes and behaviour within an OWF (Krijgsveld et al. 2011, Skov et al. 2012, Skov et al. 2018). Such observations may further alter initial perspectives on collision vulnerability.

Exposure to collision risk may be highly seasonal due to increased local abundance during breeding, passage and over-wintering periods (Krijgsveld et al. 2011, Thaxter et al. 2019). Breeding birds may be more likely to take greater risks due to their reproductive status. For example, at Zeebrugge, Belgium, a wind farm built on a breakwater close to a breeding colony caused significant collision mortality to tern species that undertook regular foraging flights across the turbine row.

BOX B3.1: Avoidance behaviour and offshore wind farms.

For OWFs, overall project avoidance is usually considered as a composite of avoidance of the whole site (macro-avoidance), avoidance within the site up to a small buffer (~ 10 m) around the device (meso-avoidance) or last-second responses (Cook et al. 2014, Skov et al. 2018). Understanding of avoidance behaviour is important for Collision Risk Modelling (CRM), used in impact assessment, and is inherently linked to displacement and barrier effects.

Some studies have provided macro-avoidance rates for Northern Gannet, ranging between 64 and 85% in Dutch and Belgian wind farms and 79.7% at Thanet, UK (Krijgsveld et al. 2011, Vanerman et al. 2015, Garthe et al. 2017, Skov et al. 2018). At Thanet, macro-avoidance was found to be lower for European Herring Gull (*Larus argentatus*) and Great Black-backed Gull (*Larus marinus*) (44.2%), Black-legged Kittiwake (*Rissa tridactyla*) (56.6%) and Lesser Black-backed Gull (*Larus fuscus*) (61.9%) (Skov et al. 2018). Using visual tracking, Harwood et al. (2017) estimated macro-avoidance to be 30% for Sandwich Terns (*Thalasseus sandvicensis*) at Sheringham Shoal OWF during its construction.

Meso-avoidance has also been shown clearly, in several studies, where birds fly down the middle of rows of turbines and may avoid operational turbines at greater distances (Desholm and Kahlert 2005, Pettersson 2005, Petersen et al. 2006, Krijgsveld et al. 2011, Hill et al. 2014). Micro-avoidance is more difficult to quantify due to the preceding levels of macro- and meso-avoidance resulting in what should be very few last-second encounters, with several studies confirming this (Krijgsveld et al. 2011, Cook et al. 2014, 2018, Skov et al. 2018).

Overall empirical avoidance rates, however, are likely to be most sensitive to micro-avoidance rates that may be difficult to quantify due to small sample sizes. The study at Thanet suggested a high chance of collision for birds that crossed perpendicular to the rotor blade rather than orientating parallel to it (Skov et al. 2018). Further large-scale studies are required to broaden our understanding of avoidance behaviour and collision risk for different species at different OWFs and sites around the world. However, such studies require significant funding and may need investment from multiple developers and other stakeholders.

However, subsequent repowering, with fewer larger turbines and greater spacing, likely contributed to reductions in collision risk at Zeebrugge (Everaert and Steinen 2007, Everaert 2014). During winter, collision risk may increase if wind farms intersect heavily used routes between foraging areas and roosts (Thaxter et al. 2019). Sensory perception of risk, and therefore avoidance, will also be influenced by factors including project configuration, turbine size, lighting and weather conditions that could drive variations in collision rates (Anderson et al. 1999, Drewitt and Langston 2006, Cook et al. 2011, Masden et al. 2012, May 2017, King 2019, May et al. 2020).

Our knowledge of collision risk associated with wet MRE devices is generally theoretical (Sparling et al. 2020). In the UK, Furness et al. (2012) identified Razorbill (*Alca torda*), European Shag (*Gulosus aristotelis*), Black Guillemot (*Cepphus grylle*), Common Guillemot (*Uria aalge*) and Great Cormorant (*Phalacrocorax carbo*) as being most vulnerable to effects of tidal energy devices. Auks, gulls and terns that utilise tidal stream environments or dive to depths where tidal turbines could be deployed may also be vulnerable to collisions, whilst sea ducks and divers avoid such environments (Langton et al. 2011, Furness et al. 2012, Robbins 2017, MBIEG 2019, Isaksson et al. 2020). Specific micro-habitat use within these tidal stream environments is likely to be highly variable and species- and site-specific, making it difficult to accurately quantify collision risk (Waggitt and Scott 2014, Isaksson et al. 2020). For diving birds, collision risk may be affected by factors such as seasonal and diel variations in site use, tidal cycles, prey distribution, foraging behaviour, agility, visual acuity and characteristics of the device itself (Wilson et al. 2006, Waggitt and Scott 2014, Copping et al. 2020, Isaksson et al. 2020, Sparling et al. 2020). Untangling the relative importance of these factors remains a significant challenge. Wave energy devices are likely to present less risk as they are generally low to the surface with parts that move in unison with natural wave motions (Grecian et al. 2010). However, if the device projects above from the sea, but is relatively inconspicuous, low flying seabirds that shear close to the water and waves (e.g., petrels and shearwaters) may be at risk.

Displacement and attraction

Species or individuals that avoid MRE projects may suffer from habitat fragmentation, increased travel times and energetic costs (barrier effects) and reduced access to potential foraging habitat (displacement), with implications for adult survival and colony productivity (Fox et al. 2006, Masden et al. 2010, Busch and Garthe 2016, Vanermen and Stienen 2019). Conversely, above water structures may attract some species as they provide roosts and/or access to new prey resources (Furness et al. 2012, Dierschke et al. 2016, Harwood et al. 2017, Vanermen and Stienen 2019). Indirect or food web effects and fishing activity may also lessen or compound avoidance and attraction (Nilsson and Green 2011, Dierschke et al. 2016).

For species that may be displaced, the effects and resultant impacts could be varied. Breeding birds may be more vulnerable to displacement effects where daily commutes intersect local MRE projects. This could result in reductions in adult fitness, reductions in provisioning rates and nest attendance which could result in chick abandonment or influence chick survival (Masden et al. 2010, Furness et al. 2012, Furness et al. 2013a, Searle et al. 2014, Searle et al. 2018). Displacement during non-breeding periods could compound higher winter mortality rates and lead to reduced adult fitness at the start of the next breeding season, with consequences for productivity (Dierschke et al. 2016, Furness et al. 2013a).

Furness et al. (2013a) identified diver species and Common Scoter (*Melanitta nigra*) as being most likely to be at risk of disturbance and displacement from OWFs. Scaup (*Aythya marila*), Common Eider, Black Guillemot and Slavonian Grebe (*Podiceps auritus*) were also considered as species of concern. In contrast, they suggest petrels, shearwaters, skuas, gulls, Little Auk (*Alle alle*), Northern Fulmar and Northern Gannet were unlikely to be seriously affected by disturbance and displacement.

Post-construction studies at European OWF sites have identified a range of avoidance responses and, whilst there was variability between sites, some species consistently showed strong avoidance or attraction. Dierschke et al. (2016) found that divers strongly avoid OWFs, but also found consistent avoidance by Northern Gannet. They also found some evidence for avoidance by Great-crested Grebe (*Podiceps cristatus*) and Northern Fulmar. In contrast, cormorant generally displayed a strong attraction to OWFs. Common Guillemot, Razorbill, Long-tailed Duck (*Clangula hyemalis*), Common Scoter, Little Gull (*Hydrocoloeus minutus*) and Sandwich Tern provided less conclusive results. Vanerman and Stienen (2019) came to similar conclusions but also found evidence for Great Black-backed Gull attraction. Large gulls and cormorants appear to routinely use turbine bases and other infrastructure at OWFs (Petersen et al. 2006, Leopold et al. 2013, Vanerman and Stienen 2019, Vanermen et al. 2020). Sandwich Tern and Common Tern (*Sterna hirundo*) have also been seen using wind turbine bases before tower installation, or project marker buoys (Walls et al. 2008, Vanermen et al. 2013, Harwood et al. 2017, Perrow 2019).

The magnitude of any displacement effect, and the subsequent impact on seabird populations, will again depend on a variety of species-, project- and location-specific factors that may contribute to avoidance. For example, Heinänen et al. (2020) found that, for Red-throated Diver (*Gavia stellata*), the displacement effect was detectable at 10 km from OWFs but increased with proximity. Further, they found the displacement effect lessened when visibility was reduced, but increased at night, possibly due to wind farm lighting. The layout of a project could also influence individual risk perception and therefore avoidance. Densely packed arrays or lines of wind turbines are more likely to appear as a barrier than widely spaced turbines with larger corridors oriented to important flyways (Masden et al. 2012, Harwood and Perrow 2019).

Displacement may also vary due to changes in vessel traffic, timing and type of construction events and possibly due to habituation (Fox and Petersen 2006, Fox et al. 2006, Petersen and Fox 2007). Where seabird surveys have continued during construction, studies suggest varying responses. For example, wintering and passing razorbills were displaced from the Sheringham Shoal OWF, UK, during construction (Perrow 2019). In contrast, Common Guillemot at Robin Rigg, UK, where the birds are present during the breeding season, were not (Vallejo et al. 2017). Furthermore, the sensitivity of seabirds to vessel and helicopter activity associated with offshore projects is likely to be species-specific (Garthe and Hüppop 2004, Furness et al. 2012, Furness et al. 2013a). For example, some gulls might be attracted to vessels, fulmars and shearwaters can be indifferent, auks may be disturbed at up to several hundred metres and divers and Common Scoter can flush at over a kilometre (Platteeuw and Beekman 1994, Austin et al. 2000, Ronconi and Clair 2002, Bellefleur et al. 2009, Schwemmer et al. 2011, Topping and Petersen 2011, Mendel et al. 2019). Nevertheless, apart from in high traffic areas, disturbance by vessels is likely to be short-lived.

In contrast to OWFs, there is currently a lack of studies investigating the displacement effects of wet MRE devices on seabirds. However, it may be expected that there is less risk of displacement in comparison to OWFs with large arrays of imposing turbines. Furness et al. (2012) suggested that some species (Black Guillemot, Razorbill, European Shag, Common Guillemot and Great Cormorant) may be particularly vulnerable to impacts associated with tidal turbines in UK waters and that wave energy is generally likely to represent a lesser hazard. However, disturbance or attraction may still occur during construction, because of service vessels or provision of novel structures on which birds may roost or prey might aggregate around (Grecian et al. 2010, Isaksson et al. 2020). A study at Strangford Lough, Northern Ireland, found multiple tidal energy devices in a row may form a barrier to movements for diving seabirds (Savidge et al. 2014). Diving birds may also lose available foraging habitat underwater or indirect effects may alter prey resources influencing foraging behaviour (Grecian et al. 2010, Scott et al. 2014). The impacts for diving birds will ultimately depend on energetic constraints, their reliance on effected areas and ability to adapt foraging behaviour.

Indirect or food web effects

Indirect or food web effects can influence seabird prey distributions and abundance through various ecological pathways (Perrow 2019). Seabird prey resources around MRE projects may be affected by noise, creation of suitable new habitat, destruction of pre-existing habitat, changes in primary productivity resources and shifting competition and predation pressure (Lindeboom et al. 2015, Broström et al. 2019, Rees and Judd 2019, Degraer et al. 2020).

MRE projects may create 'reef-effects', due to the introduction of new hard structures, which can result in new communities developing, supporting filter feeders, crustaceans and fish (Petersen and Malm 2006, Langhamer 2012, Gill and Wilhelmsson 2019, Degraer et al. 2020). These communities may influence primary productivity (e.g., Slavik et al. 2019) and provide important prey resources for top predators including seals and seabirds (Bergström et al. 2013, Reubens et al. 2014, Russell et al. 2014, Mavraki 2020, Wilber et al. 2020). For example, European Herring Gull and Lesser Black-backed Gulls have been seen feeding on the epifauna on foundations (May 2008, Vanermen et al. 2015). Great Cormorant, European Shag, Sandwich Tern, Common Guillemot, Common Scoter and Long-tailed Duck have also been observed foraging inside OWFs, with other exceptional records of Red-throated Diver and Northern Gannet (Kahlert et al. 2004, Petersen et al. 2006, Petersen and Fox 2007, Leopold and Camphuysen 2008, Walls et al. 2008, Krijgsveld et al. 2010, Krijgsveld et al. 2011, Leopold et al. 2013, Petersen et al. 2014). MRE projects may also cause changes in localised hydrodynamics, ocean-atmosphere dynamics and other physical processes (Scott et al. 2014, Haverson et al. 2018, Broström et al. 2019, Rees and Judd 2019). Such processes can influence primary productivity or modify habitats and thus the spatial distribution of seabird prey, with possible implications for collision risk and displacement.

Indirect effects are difficult to predict and evidence without holistic monitoring across trophic levels, however, such monitoring is rarely undertaken (Scott et al. 2014, Perrow 2019). Where studies do address this, the results indicate indirect and food web effects are likely to occur and could impact seabirds (Lindeboom et al. 2011, BSH and BMU 2014). For example, Perrow et al. (2011) found that Little Tern (*Sternula albifrons*) foraging success and productivity declined during the construction of the nearby Scroby Sands OWF, UK. These reductions coincided with a significant decline in young Atlantic herring (*Clupea harengus*) that corresponded with pile driving activity during the winter spawning period. Given the increasing numbers of MRE projects around the world, the potential for such impacts in isolation, or due to cumulative effects of multiple developments, warrants concern and dedicated investigation.

Marine spatial planning, impact assessment and seabird monitoring

Spatial planning for MRE projects

Marine spatial planning for MRE projects is a complex process that needs to satisfy relevant policies and legislation (including wildlife conservation), keep pace with industry developments, and tackle competition or conflicts with other marine users (Köppel et al. 2019, Flannery and McAteer 2020). However, in the North Sea, adjacent jurisdictions are collaborating to develop a common framework for environmental impact assessment (SEANSE 2020). In addition, seabird sensitivity maps have been developed for transnational sea areas such as the North East Atlantic (Waggitt et al. 2019). Similar exercises have also been carried out to predict more local distributions based on tracking data (e.g., Cleasby et al. 2020). Such initiatives are particularly important at a time when the industry is looking toward ever larger projects further offshore. Strategic Environmental Assessments (SEAs), for regional or national MRE plans, should draw on sensitivity mapping exercises when identifying possible development areas. However, in the short-term, such assessments are likely to remain qualitative due to a lack of suitable data for relevant seabird species across large areas.

Where suitable MRE sites are identified and leased to developers, Environmental Impact Assessments (EIAs), or similar, are required to quantify project-specific effects on seabirds. These must provide sufficient information to allow competent authorities to make appropriate consenting decisions. Further assessments (e.g., Habitats Regulations Assessment (HRA) in Europe) may also be required if a project is likely to have significant effects on designated nature conservation sites that can include seabird features (e.g., a Special Area of Conservation (SAC) or Special Protection Area (SPA)). However, even where significant adverse effects from a project, or cumulative projects, are predicted, consent may still be granted if there is a case for imperative reasons of overriding public interest (IROPI) and appropriate compensation can be secured (Jameson et al. 2019, ABPmer 2020).

Assessing impacts on seabirds

Impact assessments generally begin with screening and scoping (identifying potential impacts), then progress to baseline characterisation, assessment of likely impacts and mitigation measures (Masden et al. 2010, Busch et al. 2013). Project-specific ornithological monitoring programmes are usually required to inform assessments of primary effects (collision risk and displacement/barrier effects) but should also consider indirect or food web effects (Perrow 2019).

Baseline characterisation informs the selection of sensitive seabird receptors based on the abundance in the project area, conservation status and connectivity to important colonies. The latter can be established from foraging ranges (e.g., Thaxter et al. 2012, Woodward et al. 2019) or tracking data where available. The impacts of the effects on these species and colonies can then be assessed using the best available data and assessment tools. For example, a 'Matrix Approach' may be used for the assessment of displacement impacts (Busch and Garthe 2016). Here, the displacement rate is considered against the mortality rate. Collision risk is generally assessed using CRMs that generate collision mortality estimates based on densities of flying birds and various parameters (see Cook and Masden (2019) for a review). Mortality estimates can be derived for specific seasons, consider sabbatical rates and be split across adults and sub-adults. These estimates can then be assessed against the background mortality of relevant populations. Where colony-specific assessments are required, additional mortality may need to be apportioned accordingly using suitable methods (e.g., Butler et al. 2020). The population-scale impacts of additional mortality, or changes in demographic rates for the designated feature, are then projected using models. For example, this may involve estimation of potential biological removal or population viability analysis, to predict population-level impacts over the lifetime of the project (Searle et al. 2019a).

Further, impacts from different effects can be integrated within cumulative assessments, though such assessments may include large sources of uncertainty (Busch and Garthe 2016, Searle et al. 2019b). Cumulative effects of displacement or collision risk, within a specific area across multiple projects, may also be hard to assess accurately given variations in individual approaches to EIA and associated uncertainties (Willsteed et al. 2018, King 2019). Cumulative assessments could represent a significant consenting risk for developers. For example, the proposed Docking Shoal OWF, UK, was cumulatively assessed with other proposed sites in the area, but ultimately had consent refused due to potential impacts on Sandwich Tern within a single SPA (DECC 2012).

Seabird monitoring for MRE projects

Ornithological monitoring programmes for MRE projects are critical to impact assessments, validation of predictions and provide a wider contribution to our knowledge of seabird ecology. Characterisation of a project area should identify spatio-temporal trends in seabird abundance and behaviour (e.g., flight heights, flight directions and foraging behaviour) within the site and buffer and/or control area. Following consent, construction and post-construction surveys should provide comparative data for appropriate statistical analyses (Vanermen et al. 2015, Mackenzie et al. 2017, Webb and Nehls 2019). Baseline surveys are typically carried out once a month over two years, though a higher frequency could be required to capture fine-scale patterns in space use (Webb and Nehls 2019). Digital aerial strip transect or grid surveys, boat-based (European Seabirds at Sea (ESAS)) line transects and vantage point surveys can be employed depending on the project scale and survey area (Komdeur et al. 1992, Camphuysen et al. 2004). Each method has associated benefits and limitations but may provide complementary data (Vanermen and Stienen 2019, Webb and Nehls 2019). Boat surveys uniquely offer the opportunity to simultaneously monitor environmental parameters and prey abundance, allowing further investigation of seabird distributions (e.g., Scott et al. 2010, Embling et al. 2012). Monitoring methods may need to adapt to provide suitable spatio-temporal coverage of the larger offshore areas currently being proposed, particularly in relation to migratory bird movements which may be relatively short-lived.

Seabird tracking, ringing and other bespoke studies can deliver valuable additional data on site use, connectivity to colonies and demographic rates. Studies may involve visual tracking of birds (Harwood et al. 2017, Skov et al. 2018), use of radar systems to track bird movements (Krijgsveld et al. 2011, Brabant et al. 2012, Plonczkier and Simms 2012, Skov et al. 2018) tag telemetry to assess individual interactions and behaviour (Thaxter et al. 2018, Thaxter et al. 2019, Vanermen et al. 2020) or camera systems and other devices to detect possible collisions (Desholm et al. 2006, Skov et al. 2018, Matzner et al. 2020, Largey et al. 2021). For diving birds, a better understanding of specific foraging behaviour and potential interactions with wet MRE devices is required. This can be addressed through GPS tracking using tags that detect acceleration and dive depths (e.g., Garthe et al. 2017) or through the use of multisensory acoustic systems to monitor local currents and space use by birds, fish and marine mammals around underwater structures (e.g., Williamson et al. 2017).

Mitigation, compensation and restoration

Lowering the risk of significant impacts to 'acceptable' levels constitutes a key environmental issue for consenting of MRE projects (Fox et al. 2006, Langston 2013, Schuster et al. 2015, Goodale and Milman 2016, Green et al. 2016). Given the global expansion of the MRE industry and other anthropogenic pressures on seabird populations, mitigation and compensation measures are already becoming a focal point of the consenting process (Harwood and Perrow 2019, ABPmer 2020).

Mitigating potential impacts

Mitigation provides a means of reducing negative impacts to ensure development is compliant with conservation legislation and should be an integral part of the planning process (Gartman et al. 2016a, Harwood and Perrow 2019, Jameson et al. 2019). Even where predicted impacts are unlikely to be significant, practicable mitigation can reduce risk and provide 'headroom' for other projects. However, mitigation may be costly and difficult to validate. There are several reviews on the topic of mitigation in relation to OWFs effects on seabirds that provide different perspectives (Hüppop et al. 2006, Drewitt and Langston 2006, Edkins 2008, Cook et al. 2011, Marques et al. 2014, May et al. 2015, Harwood and Perrow 2019). However, most studies agree that mitigation needs to be project-specific.

Assuming SEAs have informed site selection, the process should already look to mitigate the risk of potential impacts on seabirds by excluding high-risk sites (Köppel et al. 2019). Such assessments should also identify mitigation measures that should be considered (e.g., Scottish Government 2020). Seabird sensitivity mapping can also be applied at the project level to scope important bird areas out of development (Bradbury et al. 2014, Loukogeorgaki et al. 2018, Best and Halpin 2019, Harwood and Perrow 2019). Consideration should also be given to project design, including micro-siting of devices to modify permeability to seabirds and associated risk of collision and displacement effects (Masden et al. 2012). Harwood and Perrow (2019) provide an overview of site-level planning approaches that could be adopted for OWFs and maybe applicable to wet MRE projects.

MRE devices can also be modified to reduce collision risk. For example, employing fewer, larger, turbines, with a suitable clearance beneath the rotor swept area, is a strategy that is regularly used to reduce theoretical collision risk for some seabird species, but this could increase the risk for migratory birds (Johnston et al. 2014, Harwood and Perrow 2019, Hüppop et al. 2019). Similar principles could be applied to tidal turbines by installing them at depths that preclude interactions with diving birds and employing fewer larger devices in an array. Less impactful construction methods should be employed where possible (e.g., floating wind, gravity bases or suction caissons) and timing of construction should ideally avoid breeding periods (Peire et al. 2009, Houlsby et al. 2005, Perrow et al. 2011, Gartman et al. 2016b). Best-practices (e.g., avoiding aggregations of rafting birds) for associated project vessels should also be followed to reduce the potential effects of high traffic on sensitive species.

Targeted temporary shut-down, or curtailment, may also provide effective mitigation for collision risk (de Lucas et al. 2012, Fijn et al. 2015, Gartman et al. 2016b). Modifications to offshore infrastructure lighting and use of auditory or visual deterrents could also reduce the attraction of nocturnal migrants and seabird species (e.g., cormorants and some gulls) (Poot et al. 2008, Gilsdorf et al. 2002, Cook et al. 2011). Finally, the painting of turbine blades, to increase visibility and therefore avoidance rates, appears to provide a promising method for reducing collision risk, though further trials at OWFs are required (Hodos 2003, May et al. 2020).

Despite the array of possible measures, few have been trialled properly in the offshore environment to date. Whilst measuring the efficacy of mitigation measures remains challenging, appropriate studies could benefit the industry moving forward.

Mitigation may be incorporated into consent conditions or through statutory requirements for Environmental Management Plans. Furthermore, adaptive management and regular reviews, based on emerging evidence, should also be incorporated into planning to deliver reactive mitigation and project improvements (Hanna et al. 2016).

The role of compensation

Where mitigation is insufficient, and no suitable alternative solutions exist, compensation is required to offset residual impacts on important seabird populations (Langston and Pullan 2003, Gardner et al. 2013, ABPmer 2020). Compensation must go beyond normal conservation measures already in place for a population and account for uncertainty in impact assessment and efficacy of proposed measures. Compensation to offset additional mortality involves enhancing populations, manipulating any limiting biological factors or reducing other anthropogenic impacts (Furness et al. 2013b, Marques et al. 2014).

Potential compensatory solutions could include: measures to enhance colony productivity (e.g., wardening, predator control, habitat enhancement and extension of protection to important offshore areas), creation of new habitat where this is limiting, and enhancement of prey resources (e.g., improve prey species habitat and reduce impacts of commercial fisheries). Furness et al. (2013b) provided a review of possible mitigation and compensation measures for MREs that could be applied outside project areas for species around the British Isles. Their suggestion for the most cost-effective measures were: provision of new nest platforms (red-throated divers), eradication of predators (Manx Shearwater *Puffinus puffinus* and Common Tern), supplementary feeding (Arctic Skuas *Stercorarius parasiticus*), cessation of culls (large gulls), fisheries closures (Black-legged Kittiwakes and auks), protection from predators and flooding (Sandwich Tern), and reduction in seabird by-catch or harvesting of chicks (Northern Gannet). Commercial fisheries have had significant global impacts on seabirds and reducing these effects (e.g., reduction in prey and seabird by-catch) will likely confer benefits across seabird species and geographic regions (Tasker et al. 2000, Daunt et al. 2008, Žydelis et al. 2013). The growth of the MRE sector offers an opportunity to address these impacts on seabirds whilst benefiting a much wider range of wildlife (Jameson et al. 2019). However, the process of implementing measures may be complex and require consideration of timing, siting, planning regulations, stakeholder consultations, collaborations, incentives or subsidies and monitoring. Recently, the proposed Hornsea Project Three OWF was required to produce a compensation plan for Black-legged Kittiwake to support a case for IROPI (GoBe Consultants and Ørsted 2020). The plan focused on the provision of four artificial nesting towers in suitable areas and this contributed to consent being granted in 2020. However, the plan has received criticism from non-governmental organisations who have questioned the certainty in the suggested measures. Indeed, determining whether measures will be fit-for-purpose may be difficult due to underlying uncertainties in assessments and lack of current validation. This requires further dedicated studies to provide confidence in solutions and reduce consenting risk.

Site restoration

When a project reaches the end of life, decommissioning or repowering must be considered. Decommissioning may be a costly and destructive process, particularly given the potential for reef effects associated with underwater structures that increase local biodiversity (Smyth et al. 2015, Topham and McMillan 2017, Dannheim et al. 2020, Gill and Wilhelmsson 2019). Thus, it may be that sites are only partially decommissioned and may even become important habitats for seabirds and their prey, though the specific value of this for many seabirds remains uncertain (Perrow 2019). Further, reductions in available development space may make the repowering of sites with more efficient turbines, including new mitigation, even more desirable.

Summary and conclusions

The MRE industry is expanding globally and represents an important, albeit partial, solution to the climate crisis. However, projects can conflict with wildlife conservation and affect seabirds through a variety of pathways, at varying spatio-temporal scales, compounding pressure from other anthropogenic stressors. Our understanding of the possible effects of OWFs on certain seabird species, and their general ecology, in coastal areas has developed with the industry due to the studies that have been carried out to inform impact assessment and validate predictions. However, OWFs are beginning to move further offshore and are expanding into geographic areas where seabird populations and behaviour are less well described. Wet MRE devices, though currently less likely to affect seabirds seriously, place further pressure on seabird populations. As the development of wet MRE devices continues, they may become more commercially viable, leading to growth in the size and number of projects and thus the potential effects on seabirds.

Future research needs to focus on investigating variations in displacement and associated impacts, species- and site-specific avoidance behaviour and complex indirect effects that are inherently difficult to detect. Whilst, predictive models and tools are constantly evolving, post-construction monitoring (at varying scales) is needed to provide empirical evidence to inform and validate models and predictions. Seabird tracking studies and MRE project monitoring systems are important tools in the study of effects on seabirds. However, such studies need to target a wider range of projects, geographic areas and species to inform future development. Moreover, a holistic ecosystem approach to monitoring should be adopted so we can start to understand the complex indirect effects of projects on seabirds. In all cases, studies should look to reduce the uncertainty around our understanding of effects, appropriate mitigation and, where required, compensation.

References cited

ABPmer. 2020. UK Offshore Wind Expansion, Meeting the challenges of Article 6(4) of the Habitats Directive, ABPmer White Paper, January 2020. Southampton, UK.

Aderinto, T. and H. Li. 2018. Ocean wave energy converters: Status and challenges. Energies 11: 1250. doi: 10.3390/en11051250.

Anderson, R., M. Morrison, K. Sinclair and D. Strickland. 1999. Studying wind energy/bird interactions: A guidance document. Metrics and methods for determining or monitoring potential impacts on birds at existing and proposed wind energy sites. National Wind Coordinating Committee, Washington, DC.

Austin, J.E., A.D. Afton, M.G. Anderson, R.G. Clark, C.M. Custer, J.S. Lawrence et al. 2000. Declining scaup populations: Issues, hypotheses, and research needs. Wildl. Soc. Bull. 28: 254–263.

Barclay, R.M.R., E.F. Baerwald and J. Rydell. 2017. Bats. pp. 191–221. *In*: Perrow, M.R. [ed.]. Wildlife and Wind Farms, Conflicts and Solutions. Volume 1. Onshore: Potential effects. Pelagic Publishing, Exeter, UK.

Bellefleur, D., P. Lee and R.A. Ronconi. 2009. The impact of recreational boat traffic on marbled murrelets (*Brachyramphus marmoratus*). J. Environ. Manage. 90: 531–538. doi: 10.1016/j.jenvman.2007.12.002.

Bergström, L., F. Sundqvist and U. Bergström. 2013. Effects of an offshore wind farm on temporal and spatial patterns in the demersal fish community. Mar. Ecol. Prog. Ser. 485: 199–210. doi: 10.3354/meps10344.

Best, B.D. and P.N. Halpin. 2019. Minimizing wildlife impacts for offshore wind energy development: Winning tradeoffs for seabirds in space and cetaceans in time. PLoS One 14(5): e0215722. doi: 10.1371/journal.pone.0215722.

Borthwick, A.G.L. 2016. Marine Renewable Energy Seascape. Engineering 2: 69–78.doi: 10.1016/J.ENG.2016.01.011.

Bowgen, K. and A. Cook. 2018. Bird Collision Avoidance: Empirical evidence and impact assessments. JNCC Report No. 614, JNCC, Peterborough, ISSN 0963–8091.

Brabant, R., L. Vigin, E.W.M. Stienen, N. Vanermen and S. Degraer. 2012. Radar research on the impact of offshore wind farms on birds: Preparing to go offshore. pp. 111–126. *In*: Degraer, S., R. Brabant and B. Rumes [eds.]. Offshore windfarms in the Belgian part of the North Sea: Heading for an understanding of environmental impacts. Royal

Belgian Institute of Natural Sciences, Management Unit of the North Sea Mathematical Models, Marine Ecosystem Management Unit.

Bradbury, G., M. Trinder, R. Furness, A.N. Banks, R.W.G. Caldow and D. Hume. 2014. Mapping seabird sensitivity to offshore wind farms. PloS ONE 9(9): e106366. doi: 10.1371/journal.pone.0106366.

Broström, G., E. Ludewig, A. Schneehorst and T. Pohlmann. 2019. Atmosphere and ocean dynamics. pp. 47–63. *In*: Perrow, M.R. [ed.]. Wildlife and Wind Farms, Conflicts and Solutions. Volume 3. Offshore: Potential effects. Pelagic Publishing, Exeter, UK.

BSH and BMU 2014. Ecological Research at the Offshore Windfarm alpha ventus: Challenges, results and perspectives. Federal Maritime and Hydrographic Society of Germany (BSH) and Federal Ministry for the Environment, Nature Conservation and Nuclear Safety (BMU). Weisbaden: Springer Spektrum. doi: 10.1007/978-3-658-02462-8.

Bull, L.S., S. Fuller and D. Sim. 2013. Post-construction avian mortality monitoring at Project West Wind. N. Z. J. Zool. 40: 28–46. doi: 10.1080/03014223.2012.757242.

Busch, M., A. Kannen, S. Garthe and M. Jessopp. 2013. Consequences of a cumulative perspective on marine environmental impacts: Offshore wind farming and seabirds at North Sea scale in context of the EU Marine Strategy Framework Directive. Ocean Coast. Manag. 71: 213–224. doi: 10.1016/j.ocecoaman.2012.10.016.

Busch, M. and S. Garthe. 2016. Approaching population thresholds in presence of uncertainty: assessing displacement of seabirds from offshore wind farms. Environ. Impact Assess. Rev. 56: 31–42. doi: 10.1016/j.eiar.2015.08.007.

Butler, A., M. Carroll, K. Searle, M. Bolton, J. Waggitt, P. Evans et al. 2020. Attributing seabirds at sea to appropriate breeding colonies and populations (CR/2015/18). Scottish Marine and Freshwater Science 11(8). doi: 10.7489/2006-1.

Camphuysen, K.J., A.D. Fox, M.F. Leopold and I.K. Petersen. 2004. Towards Standardised Seabirds at Sea Census Techniques in Connection with Environmental Impact Assessments for Offshore Wind Farms in the UK (Report No. COWRIE–BAM–02–2002). Report by Royal Netherlands Institute for Sea Research (NIOZ). Report for Collaborative Offshore Wind Research into the Environment (COWRIE).

Castro-Santos, L. and V. Diaz-Casas [eds.]. 2016. Floating offshore wind farms. Springer, Cham, Switzerland. doi: 10.1007/978-3-319-27972-5.

Chowdhury, M.S., K.S. Rahman, V. Selvanathan, N. Nuthammachot, M. Suklueng, A. Mostafaeipour et al. 2020. Current trends and prospects of tidal energy technology. Environ. Dev. Sustain. doi: 10.1007/s10668-020-01013-4.

Cleasby, I.R., E. Owen, L. Wilson, E.D. Wakefield, P. O'Connell and M. Bolton. 2020. Identifying important at-sea areas for seabirds using species distribution models and hotspot mapping. Biol. Conserv. 241: 108375. doi: 10.1016/j.biocon.2019.108375.

Cook, A.S.C.P., V.H. Ross-Smith, S. Roos, N.H.K. Burton, N. Beale, C. Coleman et al. 2011. Identifying a range of options to prevent or reduce avian collision with offshore wind farms, using a UK-based case study. BTO Research Report No. 580, BTO, Thetford.

Cook, A.S.C.P., N.H.K. Burton, E.M. Humphreys and E.A. Masden. 2014. The Avoidance Rates of Collision Between Birds and Offshore Turbines. Scottish Marine and Freshwater Science Vol. 5 No 16. Scottish Government, Edinburgh. doi: 10.7489/1553–1.

Cook, A.S.C.P., E.M. Humphreys, F. Bennet, E.A. Masden and N.H.K. Burton. 2018. Quantifying avian avoidance of offshore wind turbines: Current evidence and key knowledge gaps. Mar. Environ. Res. 140: 278–288. doi: 10.1016/j.marenvres.2018.06.017.

Cook, A.S.C.P. and E.A. Masden. 2019. Modelling collision risk and predicting population-level consequences. pp. 135–166. *In*: Perrow, M.R. [ed.]. Wildlife and Wind Farms, Conflicts and Solutions Volume 4 Offshore: Monitoring and mitigation. Pelagic Publishing, Exeter, UK.

Copping, A.E., L.G. Hemery, D.M. Overhus, L. Garavelli, M.C. Freeman, J.M. Whiting et al. 2020. Potential environmental effects of marine renewable energy development-the state of the science. J. Mar. Sci. Eng. 8: 18. doi: 10.3390/jmse8110879.

Dannheim, J., L. Bergström, S.N.R. Birchenough, R. Brzana, A.R. Boon, J.W.P. Coolen et al. 2020. Benthic effects of offshore renewables: Identification of knowledge gaps and urgently needed research. ICES J. Mar. Sci. 77: 1092–1108. doi: 10.1093/ icesjms/fsz018.

Daunt, F., S. Wanless, S.P.R. Greenstreet, H. Jensen, K.C. Hamer and M.P. Harris. 2008. The impact of the sandeel fishery closure in the northwestern North Sea on seabird food consumption, distribution and productivity. Can. J. Fish. Aquat. Sci. 65: 362–381. doi: 10.1139/f07-164.

Degraer, S., D.A. Carey, J.W.P. Coolen, Z.L. Hutchison, F. Kerckhof, B. Rumes et al. 2020. Offshore wind farm artificial reefs affect ecosystem structure and functioning-A Synthesis. Oceanogr. 33: 48–57. doi: 10.5670/oceanog.2020.405.

de Lucas, M., M. Ferrer, M.J. Bechard and A.R. Muñoz. 2012. Griffon vulture mortality at wind farms in southern Spain: Distribution of fatalities and active mitigation measures. Biol. Conserv. 147: 184–189. doi: 10.1016/j.biocon.2011.12.029.

Department of Energy and Climate Change (DECC) 2012. Record of the Appropriate Assessment Undertaken for Applications Under Section 36 of the Electricity Act 1989. Projects: Docking Shoal offshore wind farm (as amended), Race Bank offshore wind farm (as amended), Dudgeon offshore wind farm, December 2011, updated June 2012.

Desholm, M. and J. Kahlert. 2005. Avian collision risk at an offshore wind farm. Biol. Lett. 1: 296–298. doi: 10.1098/rsbl.2005.0336.

Desholm, M., A.D. Fox, P.D.L. Beasley and J. Kahlert. 2006. Remote techniques for counting and estimating the number of bird-wind turbine collisions at sea: A review. Ibis 148: 76–89. doi: 10.1111/j.1474-919X.2006.00509.x.

Dierschke, V. and S. Garthe. 2006. Literature review of offshore wind farms with regard to seabirds. Bundesamt Naturschutz–Skripten 186: 131–198.

Dierschke, V., Furness, R.W. and S. Garthe. 2016. Seabirds and offshore wind farms in European waters: Avoidance and attraction. Biol. Conserv. 202: 59–68. doi: 10.1016/j.biocon.2016.08.016.

Drewitt, A.L. and R.H.W. Langston. 2006. Assessing the impacts of wind farms on birds. Ibis 148: 29–42. doi: 10.1111/j.1474–919X.2006.00516.x.

Edkins, M.T. 2008. Impacts of wind energy development on birds and bats: Looking into the problem. FPL Energy, Juno Beach, Florida, USA.

Embling, C.B., J. Illian, E. Armstrong, J. Van der Kooij, J. Sharples, C.J. Camphuysen et al. 2012. Investigating fine-scale spatio-temporal predator-prey patterns in dynamic marine ecosystems: A functional data analysis approach. J. Appl. Ecol. 49: 481–492. doi: 10.1111/j.1365–2664.2012.02114.x.

EMODnet 2020. EMODnet Human Activities: Ocean Energy Projects. Spatial dataset. Accessed 25th March 2021 from: https://www.emodnet-humanactivities.eu/search-results.php?dataname=Project+Locations.

Everaert, J. and E.W.M. Steinen. 2007. Impact of wind turbines on birds in Zeebrugge (Belgium). Biodivers. Conserv. 16: 3345–3359. doi: 10.1007/s10531–006–9082–1.

Everaert, J. 2014. Collision risk and micro-avoidance rates of birds with wind turbines in Flanders. Bird Study 61: 220–230. doi: 10.1080/00063657.2014.894492.

Fijn, R.C., K.L. Krijgsveld, M.J.M. Poot and S. Dirksen. 2015. Bird movements at rotor heights measured continuously with vertical radar at a Dutch offshore wind farm. Ibis 157: 558–566. doi: 10.1111/ibi.12259.

Flannery, W. and B. McAteer. 2020. Assessing marine spatial planning governmentality. Marit. Stud. 19: 269–284. doi: 10.1007/s40152–020–00174–2.

Fox, A.D. and I.K. Petersen. 2006. Assessing the degree of habitat loss to marine birds from the development of offshore wind farms. pp. 801–804. *In*: Boere, G.C., C.A. Galbraith and D.A. Stroud [eds.]. Waterbirds Around the World. A Global Overview of the Conservation, Management and Research of the World's Waterbird Flyways. The Stationery Office, Edinburgh, UK.

Fox, A.D., M. Desholm, J. Kahlert, T.K. Christensen and I.B.K. Petersen. 2006. Information needs to support environmental impact assessment of the effects of European marine offshore wind farms on birds. Ibis 148: 129–144. doi: 10.1111/j.1474–919X.2006.00510.x.

Fox, C., S. Benjamins, E. Masden and R. Miller. 2017. Challenges and opportunities in monitoring the impacts of tidal-stream energy devices on marine vertebrates. Renew. Sust. Energ. Rev. 81: 1926–1938. doi: 10.1016/j.rser.2017.06.004.

Furness, R.W., H.M. Wade, A.M.C. Robbins and E.A. Masden. 2012. Assessing the sensitivity of seabird populations to adverse effects from tidal stream turbines and wave energy devices, ICES J. Mar. Sci. 69: 1466–1479.doi: 10.1093/icesjms/fss131.

Furness, R.W., H.M. Wade and E.A. Masden. 2013a. Assessing vulnerability of marine bird populations to offshore wind farms. J. Environ. Manag. 119: 56–66. doi: 10.1016/j.jenvman.2013.01.025.

Furness, B., D. MacArthur, M. Trinder and K. MacArthur. 2013b. Evidence review to support the identification of potential conservation measures for selected species of seabird. MacArthur Green Technical Report for Cefas. doi: 10.13140/RG.2.1.4666.3200.

Garavelli, L. 2020. Encounters of Marine Animals with Marine Renewable Energy Device Mooring Systems and Subsea Cables. pp. 147–153. *In*: Copping, A.E. and L.G. Hemery [eds.]. OES-Environmental 2020 State of the Science Report: Environmental Effects of Marine Renewable Energy Development Around the World. Report for Ocean Energy Systems (OES). doi: 10.2172/1633184.

Gardner, T.A., A. Von Hase, S. Brownlie, J.M.N. Ekstrom, J.D. Pilgrim, C.E. Savy et al. 2013. Biodiversity offsets and the challenge of achieving no net loss. Conserv. Biol. 27: 1254–1264. doi: 10.1111/cobi.12118.

Garthe, S. and O. Hüppop. 2004. Scaling possible adverse effects of marine wind farms on seabirds: Developing and applying a vulnerability index. J. Appl. Ecol. 41: 724–734. doi: 10.1111/j.0021–8901.2004.00918.x.

Garthe, S., N. Markones and A. Corman. 2017. Possible impacts of offshore wind farms on seabirds: A pilot study in Northern Gannets in the Southern North Sea. J. Ornithol. 158: 345–349. doi: 10.1007/s10336-016-1402-y.

Gartman, V., L. Bulling, M. Dahmen, G. Geißler and J. Köppel. 2016a. Mitigation measures for wildlife in wind energy development, consolidating the state of knowledge-Part 1: Planning and Siting, Construction. J. Environ. Assess. Policy Manag. 18: 1650013. doi: 10.1142/S1464333216500137.

Gartman, V., L. Bulling, M. Dahmen, G. Geißler and J. Köppel. 2016b. Mitigation measures for wildlife in wind energy development, consolidating the state of knowledge-Part 2: Operation, Decommissioning. J. Environ. Assess. Policy Manag. 18: 1650014-1-31. doi: 10.1142/S1464333216500149.

Gibson, L., E.N. Wilman and W.F. Laurance. 2017. How green is "green" energy? Trends Ecol. Evol. 17: 922–935. doi: 10.1016/j.tree.2017.09.007.

Gill, A.B. and D. Wilhelmsson. 2019. Fish. pp. 86–111. *In*: Perrow, M.R. [ed.]. Wildlife and Wind Farms, Conflicts and Solutions. Volume 3. Offshore: Potential Effects. Pelagic Publishing, Exeter, UK.

Gilsdorf, J.M., S.E. Hygnstrom and K.C. VerCauteren. 2002. Use of frightening devices in wildlife damage management. Integrated Pest Manag. Rev. 7: 29–45. doi: 10.1023/A:1025760032566.

GoBe Consultants and Ørsted. 2020. Response to the Secretary of State's Minded to Approve Letter, Appendix 2: Kittiwake Compensation Plan. September 2020. Accessed 25th March 2021 from: https://infrastructure.planninginspectorate.gov.uk/wp–content/ipc/uploads/projects/EN010080/EN010080–003246–HOW03–30Sep_Appendix%202%20Kittiwake%20Compensation%20Plan%20(06543754_A).pdf.

Goodale, M.W. and A. Milman. 2016. Cumulative adverse effects of offshore wind energy development on wildlife. J. Environ. Plan. Manag. 59: 1–21. doi: 10.1080/09640568.2014.973483.

Graziano, M., S.-L. Billing, J.O. Kenter and L. Greenhill. 2016. A transformational paradigm for marine renewable energy development. Energy Res. Soc. Sci. 23: 136–147. doi: 10.1016/j.erss.2016.10.008.

Grecian, W.J., R. Inger, M.J. Attrill, S. Bearhop, B.J. Godley, M.J. Witt et al. 2010. Potential impacts of wave-powered marine renewable energy installations on marine birds. Ibis 152: 683–697. doi: 10.1111/j.1474-919X.2010.01048.x.

Green, R.E., R.H.W. Langston, A. McCluskie, R. Sutherland and J.D. Wilson. 2016. Lack of sound science in assessing wind farm impacts on seabirds. J. Appl. Ecol. 53: 1635–1641. doi: 10.1111/1365-2664.12731.

GWEC. 2020. Global Offshore Wind Report 2020. Global Wind Energy Council. Accessed 25th March 2021 from: https://gwec.net/global–offshore–wind–report–2020/.

Hanna, L., A. Copping, S. Geerlofs, L. Feinberg, J. Brown-Saracino, P. Gilman et al. 2016. Assessing Environmental Effects (WREN): Adaptive Management White Paper. Report by Bureau of Ocean Energy Management (BOEM), Marine Scotland Science, Norwegian Institute for Nature Research (NINA), Pacific Northwest National Laboratory (PNNL), TechnischeUniversität Berlin, and US Department of Energy (DOE).

Harwood, A.J.P., M.R. Perrow, R.J. Berridge, M.L. Tomlinson and E.R. Skeate. 2017. Unforeseen responses of a breeding seabird to the construction of an offshore wind farm. pp 19–41. *In*: Köppel, J. [ed.]. Wind Energy and Wildlife Interactions - Presentations from the CWW2015 Conference. Springer, Cham, Switzerland.

Harwood, A.J.P. and M.R. Perrow. 2019. Mitigation for birds. pp. 242–280. *In*: Perrow, M.R. [ed.]. Wildlife and Wind Farms, Conflicts and Solutions Volume 4 Offshore: Monitoring and mitigation. Pelagic Publishing, Exeter, UK.

Haverson, D., J. Bacon, H. Smith, V. Venugopal and Q. Xiao. 2018. Modelling the hydrodynamic and morphological impacts of a tidal stream development in Ramsey Sound. Renew. Energ. 126: 876–887. doi: 10.1016/j.renene.2018.03.084.

Heinänen, S., R. Žydelis, B. Kleinschmidt, M. Dorsch, C. Burger, J. Morkūnas et al. 2020. Satellite telemetry and digital aerial surveys show strong displacement of red-throated divers (*Gavia stellata*) from offshore wind farms. Mar. Environ. Res. 160: 104989. doi: 10.1016/j.marenvres.2020.104989.

Hill, R., K. Hill, R. Aumuller, A. Schulz, T. Dittmann, C. Kulemeyer et al. 2014. Of birds, blades and barriers: Detecting and analysing mass migration events at alpha ventus. pp. 111–131. *In*: Beiersdorf, A. and A. Radecke [eds.]. Ecological Research at the Offshore Windfarm Alpha Ventus: Challenges, Results and Perspectives. Springer Spektrum Wiesbaden, Germany. doi: 10.1007/978-3-658-02462-8.

Hodos, W. 2003. Minimization of motion smear: Reducing avian collisions with wind turbines. National Renewable Energy Laboratory, Golden, Colorado USA.

Houlsby, G.T., R.B. Kelly, J. Huxtable and B.W. Byrne. 2005. Field trials of suction caissons in clay for offshore wind turbine foundations. Géotechnique 55: 287–296. doi: 10.1680/geot.2005.55.4.287.

Hull, C.L., E.M. Stark, S. Peruzzo and C.C. Sims. 2013. Avian collisions at two wind farms in Tasmania, Australia: taxonomic and ecological characteristics of colliders versus non-colliders, N. Z. J. Zool. 40: 47–62. doi: 10.1080/03014223.2012.757243.

Hüppop, O., J. Dierschke, K.-M. Exo, E. Fredrich and R. Hill. 2006. Bird migration studies and potential collision risk with offshore wind turbines. Ibis 148: 90–109. doi: 10.1111/j.1474-919X.2006.00536.x.

Hüppop, O., B. Michalik, L. Back, R. Hill and S.K. Pelletier. 2019. Migratory birds and bats. pp. 142–173. *In*: Perrow, M.R. [ed.]. Wildlife and Wind Farms, Conflicts and Solutions. Volume 3. Offshore: Potential Effects. Pelagic Publishing, Exeter, UK.

IEA 2020. Global Energy Review 2019. IEA, Paris. Accessed 25th March 2021 from: https://www.iea.org/reports/global–energy–review–2019.

Isaksson, N., E. Masden, B. Williamson, M. Costagliola-Ray, J. Slingsby, J. Houghton et al. 2020. Assessing the effects of tidal stream marine renewable energy on seabirds: A conceptual framework. Mar. Pollut. Bull. 157: 111314. doi: 10.1016/j.marpolbul.2020.111314.

Jameson, H., E. Reeve, B. Laubek and H. Sittel. 2019. The nature of offshore wind farms. pp. 1–29. *In*: Perrow, M.R. [ed.]. Wildlife and Wind Farms, Conflicts and Solutions. Volume 3 Offshore: Potential Effects. Pelagic Publishing, Exeter, UK.

Johnston, A. A.S.C.P. Cook, L.J. Wright, E.M. Humphreys and N.H.K. Burton. 2014. Modelling flight heights of marine birds to more accurately assess collision risk with offshore wind turbines. J. Appl. Ecol. 51: 3–10. doi: 10.1111/1365–2664.12191.

Kahlert, J., I.K. Petersen, A.D. Fox, M. Desholm and I. Clausager. 2004. Investigations of birds during construction and operation of Nysted offshore wind farm at Rødsand. Annual status report 2003. NERI Report.

Khan, N., A. Kalair, N. Abas and A. Haider. 2017. Review of ocean tidal, wave and thermal energy technologies. Renew. Sust. Energ. Rev. 72: 590–604. doi: 10.1016/j.rser.2017.01.079.

King, S. 2019. Seabirds: collision. pp. 206–234. *In*: Perrow, M.R. [ed.]. Wildlife and Wind Farms, Conflicts and Solutions Volume 3 Offshore: Potential Effects. Pelagic Publishing, Exeter, UK.

Komdeur J., J. Bertelsen and G. Cracknell [eds.]. 1992. Manual for Aeroplane and Ship Surveys of Waterfowl and Seabirds. IWRB Special Publ. No. 19, National Environmental Research Institute Kalø.

Köppel, J., J. Biehl, M. Dahmen, G. Geissler and M.E. Portman. 2019. Perspectives on marine spatial planning. pp. 281–317. *In*: Perrow, M.R. [ed.]. Wildlife and Wind Farms, Conflicts and Solutions Volume 4 Offshore: Monitoring and mitigation. Pelagic Publishing, Exeter, UK.

Krijgsveld, K.L., R.C. Fijn, C. Heunks, P.W. van Horssen, J. de Fouw, M. Collier et al. 2010. Effect studies Offshore Wind Farm Egmond aan Zee. Progress report on fluxes and behaviour of flying birds covering 2007 & 2008. Noordzeewind report OWEZ R 231 T1 20100810. Bureau Waardenburg, Culemborg The Netherlands.

Krijgsveld, K.L., R.C. Fijn, M. Japink, P.W. van Horssen, C. Heunks, M.P. Collier et al. 2011. Effect studies Offshore Wind Farm Egmond aan Zee: Final report on fluxes, flight altitudes and behaviour of flying birds. Bureau Waardenburg report no. 10–219. Commissioned by NordzeeWind The Netherlands.

Langhamer, O. 2012. Artificial reef effect in relation to offshore renewable energy conversion: State of the art. Sci. World J. 386713. doi: 10.1100/2012/386713.

Langton, R., I.M. Davies and B.E. Scott. 2011. Seabird conservation and tidal stream and wave power generation: Information needs for predicting and managing potential impacts. Mar. Policy 35: 623–630. doi: 10.1016/j.marpol.2011.02.002.

Langston, R.H.W. and J.D. Pullan. 2003. Wind farms and birds: an analysis of the effects of wind farms on birds, and guidance on Environmental Assessment criteria and site selection issues. Birdlife International report on behalf of the Bern Convention. Accessed 25th March 2021 from: http://www.birdlife.org/eu/pdfs/BirdLife_Bern_windfarms.pdf

Langston, R.H.W. 2013. Birds and wind projects across the pond: A UK perspective. Wildl. Soc. Bull. 37: 5–18. doi: 10.1002/wsb.262.

Largey, N., A.S.C.P. Cook, C.B. Thaxter, A. McCluskie, B.G. Stokke, B. Wilson et al. 2021. Methods to quantify avian airspace use in relation to wind energy development. Ibis. doi: 10.1111/ibi.12913.

Leopold, M.F. and C.J. Camphuysen. 2008. Local birds in and around the offshore Wind Park Egmond aan Zee (OWEZ) (T1). Noordzeewind Rapport OWEZ R 221 T1 20080201. Imares, Wageningen, The Netherlands.

Leopold, M.F., R.S.A. van Bemmelen and A.F. Zuur. 2013. Responses of local birds to the offshore wind farms PAWP and OWEZ off the Dutch mainland coast. Report C151/12. Imares, Wageningen, The Netherlands.

Lindeboom, H., H. Kouwenhoven, M. Bergman, S. Bouma, S. Brasseur, R. Daan et al. 2011. Short-term ecological effects of an offshore wind farm in the dutch coastal zone: A compilation. Environ. Res. Lett. 6: 035101. doi: 10.1088/1748-9326/6/3/035101.

Lindeboom, H., S. Degraer, J. Dannheim, A.B. Gill and D. Wilhelmsson. 2015. Offshore wind park monitoring programmes, lessons learned and recommendations for the future. Hydrobiologia 756: 169–180. doi: 10.1007/s10750-015-2267-4.

Linley, E.A.S., T.A. Wilding, K. Black, A.J.S. Hawkins and S. Mangi. 2007. Review of reef effects of offshore wind farm structures and their potential for enhancement and mitigation. Report from PML Applications Ltd. and the Scottish Association for Marine Science to the Department for Business, Enterprise and Regulatory Reform (BERR), Contract No: RFCA/005/0029P. BERR/DEFRA, London, UK.

Loukogeorgaki, E., D.G. Vagiona and M. Vasileiou. 2018. Site selection of hybrid offshore wind and wave energy systems in Greece incorporating environmental impact assessment. Energies 11: 2095. doi: 10.3390/en11082095.

MacArthur Green, British Trust for Ornithology, NIRAS and Royal Haskining DHV. 2013. Review of avoidance rates in seabirds at offshore wind farms and applicability of use in the Band collision risk model. Report for Smart-Wind and Forewind UK.

Mackenzie, M.L., L.A.S. Scott-Hayward, C.G. Paxton and M.L. Burt. 2017. Quantifying the Power to Detect Change: methodological development and implementation using the R package MRSeaPower. CREEM-13804-2016-1. Provided to the Scottish Government and Marine Scotland Science (USA/012/15).

Marques, A.T., H. Batalha, S. Rodrigues, H. Costa, M.J.R. Pereira, C. Fonseca et al. 2014. Understanding bird collisions at wind farms: An updated review on the causes and possible mitigation strategies. Biol. Conserv. 179: 40–52. doi: 10.1016/j.biocon.2014.08.017.

Masden, E.A., D.T. Haydon, A.D. Fox, R.W. Furness, R. Bullman and M. Desholm. 2009. Barriers to movement: Impacts of wind farms on migrating birds. ICES J. Mar. Sci. 66: 746–753. doi: 10.1093/icesjms/fsp031.

Masden, E.A., A.D. Fox, R.W. Furness, R. Bullman and D.T. Haydon. 2010. Cumulative impact assessments and bird/wind farm interactions: developing a conceptual framework. Environ. Impact Assess. Rev. 30: 1–7. doi: 10.1016/j.eiar.2009.05.002.

Masden, E.A., R. Reeve, M. Desholm, A.D. Fox, R.W. Furness and D.T. Haydon. 2012. Assessing the impact of marine wind farms on birds through movement modelling. J. R. Soc. Interface 9: 2120–2130. doi: 10.1098/rsif.2012.0121.

Masden, E.A., S. Foster and A.C. Jackson. 2013. Diving behaviour of Black Guillemots *Cepphus grylle* in the Pentland Firth, UK: Potential for interactions with tidal stream energy developments. Bird Study 60: 547–549. doi: 10.1080/00063657.2013.842538.

Matzner, S., T. Warfel and R.E. Hull. 2020. ThermalTracker-3D: A thermal stereo vision system for quantifying bird and bat activity at offshore wind energy sites. Ecol. Inform. 57: 101069. doi: 10.1016/j.ecoinf.2020.101069.

Mavraki, N. 2020. On the Food-Web Ecology of Offshore Wind Farms, the Kingdom of Suspension Feeders. PhD Dissertation, Ghent University, Belgium.

May, J. 2008. North Hoyle Offshore Wind Farm. Final annual FEPA monitoring report (2006–7) & Five year monitoring programme summary. NWP Offshore Ltd.

May, R., O. Reitan, K. Bevanger, S.H. Lorentsen and T. Nygard. 2015. Mitigating wind-turbine induced avian mortality: Sensory, aerodynamic and cognitive constraints and options. Renew. Sust. Energ. Rev. 42: 170–181.doi: 10.1016/j.rser.2014.10.002.

May, R. 2017. Mitigation options for birds. pp. 124–145. *In*: Perrow, M.R. [ed.]. Wildlife and Wind Farms: Conflicts and Solutions. Volume 2: Onshore Monitoring and Mitigation. Pelagic Publishing, Exeter, UK.

May, R., T. Nygård, U. Falkdalen, J. Åström, Ø. Hamre and B.G. Stokke. 2020. Paint it black: Efficacy of increased wind-turbine rotor blade visibility to reduce avian fatalities. Ecol. Evol. 10: 8927–8935. doi: 10.1002/ece3.6592.

MBIEG. 2019. Improved understanding of underwater collision risks of marine renewable energy devices for birds while diving. A report produced by British Trust for Ornithology for the Marine Management Organisation on behalf of the Marine Biodiversity Impacts Evidence Group, MMO Project No: 1139, May 2019.

Melikoglu, M. 2018. Current status and future of ocean energy sources: A global review. Ocean Eng. 148: 563–573. doi: 10.1016/j.oceaneng.2017.11.045.

Mendel, B., P. Schwemmer, V. Peschko, S. Müller, H. Schwemmer, M. Mercker et al. 2019. Operational offshore wind farms and associated ship traffic cause profound changes in distribution patterns of Loons (*Gavia* spp.). J. Environ. Manag. 231: 429–438. doi: 10.1016/j.jenvman.2018.10.053.

Neill, A.P., A. Angeloudis, P.E. Robins, I. Walkington, S.L. Ward, I. Masters et al. 2018. Tidal range energy resource and optimization past perspectives and future challenges. Renew. Energ. 127: 763–778. doi: 10.1016/j.renene.2018.05.007.

Nilsson, L. and M. Green. 2011. Birds in southern Öresund in relation to the wind farm at Lillgrund. Final report of the monitoring program 2001–2011. University of Lund, Lund, Sweden.

Ocean Energy Systems. 2019. Annual Report, An Overview of Ocean Energy Activities in 2019; The Executive Committee of Ocean Energy Systems: Lisbon, Portugal. Accessed 25th March 2021 from: https://www.ocean–energy–systems.org/publications/oes-annual-reports/.

Paleczny, M., E. Hammill, V. Karpouzi and D. Pauly. 2015. Population Trend of the World's Monitored Seabirds, 1950–2010. PLoS ONE 10(6): e0129342. doi: 10.1371/journal.pone.0129342.

Peire, K., H. Nonneman and E. Bosschem. 2009. Gravity Base Foundations for the Thornton Bank Offshore Wind Farm. Terra et Aqua 115: 19–29. Accessed 25th March 2021 from: https://www.iadc–dredging.com/article/thornton-offshore-wind-farm/.

Percival, S. 2003. Birds and wind farms in Ireland: A Review of Potential Issues and Impact Assessment. Ecology Consulting, Durham, UK.

Perrow, M.R., J.J. Gilroy, E.R. Skeate and M.L. Tomlinson. 2011. Effects of the construction of Scroby Sands offshore wind farm on the prey base of Little tern *Sternula albifrons* at its most important UK colony. Mar. Pollut. Bull. 62: 1661–1670. doi: 10.1016/j.marpolbul.2011.06.010.

Perrow, M.R. 2019. A synthesis of effects and impacts. pp. 235–277. *In*: Perrow, M.R. [ed.]. Wildlife and Wind Farms, Conflicts and Solutions Volume 3 Offshore: Potential Effects. Pelagic Publishing, Exeter, UK.

Petersen, I.K., T.K. Christensen, J. Kahlert, M. Desholm and A.D. Fox. 2006. Final Results of Bird Studies at the Offshore Wind Farms at Nysted and Horns Rev, Denmark. National Environmental Research Institute, Kalø Denmark.

Petersen, I.K. and A.D. Fox. 2007. Changes in Bird Habitat Utilisation around the Horns Rev 1 offshore Wind Farm, with Particular Emphasis on Common Scoter. NERI Report, Århus.

Petersen, I.K., R.D. Nielsen and M.L. Mackenzie. 2014. Post-Construction Evaluation of Bird Abundances and Distributions in the Horns Rev 2 offshore Wind Farm Area, 2011 and 2012. Aarhus University, Aarhus Denmark.

Petersen, J.K. and T. Malm. 2006. Offshore windmill farms: Threats to or possibilities for the marine environment. AMBIO: A J. of the Human Environ. 35: 75–80. doi: 10.1579/0044-7447(2006)35[75:OWFTTO]2.0.CO;2.

Pettersson, J. 2005. The Impact of Offshore Wind Farms on Bird Life in Southern Kalmar Sound, Sweden. Report requested by Swedish Energy Agency.

Platteeuw, M. and J.H. Beekman. 1994. Disturbance of waterbirds by ships on lakes Ketelmeer and Ijsselmeer. Limosa 67: 27–33.

Plonczkier, P. and I.C. Simms. 2012. Radar monitoring of migrating Pink-footed Geese: behavioural responses to offshore wind farm development. J. Appl. Ecol. 49: 1187–1194. doi: 10.1111/j.1365–2664.2012.02181.x.

Poot, H., B.J. Ens, H. de Vries, M.A.H. Donners, M.R. Wernand and J.M. Marquenie. 2008. Green light for nocturnally migrating birds. Ecol. Soc. 13: 47.

Rees, J.M. and A.D. Judd. 2019. Physical and chemical effects. pp. 30–46. *In*: Perrow, M.R. [ed.]. Wildlife and Wind Farms, Conflicts and Solutions. Volume 3. Offshore: Potential effects. Pelagic Publishing, Exeter, UK.

Retiere, C. 1994. Tidal power and the aquatic environment of La Rance. Biol. J. Linn. Soc. 51: 25–36. doi: 10.1006/bijl.1994.1004.

Reubens, J.T., S. Degraer and M. Vincx. 2014. The ecology of benthopelagic fishes at offshore wind farms: A synthesis of 4 years of research. Hydrobiologia 727: 121–136.doi: 10.1007/s10750-013-1793-1.

Robbins, A. 2017. Seabird Ecology in High-Energy Environments: Approaches to Assessing Impacts of Marine Renewables (Doctoral Dissertation), Institute of Biodiversity Animal Health and Comparative Medicine, University of Glasgow UK.

Ronconi, R.A. and C.C.S. Clair. 2002. Management options to reduce boat disturbance on foraging black guillemots (*Cepphus grylle*) in the Bay of Fundy. Biol. Conserv. 108: 265–271. doi: 10.1016/S0006-3207(02)00126-X.

Russell, D.F., S.M.J.M. Brasseur, D. Thompson, G.D. Hastie, V.M. Janik, G. Aarts et al. 2014. Marine mammals trace anthropogenic structures at sea. Curr. Biol. 24: R638–R639. doi: 10.1016/j.cub.2014.06.033.

Savidge, G., D. Ainsworth, S. Bearhop, N. Christen, B. Elsäßer, F. Fortune et al. 2014. Strangford Lough and the SeaGen tidal turbine. pp. 153–172. *In*: Shields, M.A. and A.I.L. Payne [eds.]. Marine Renewable Energy Technology and Environmental Interactions. Humanity and the Sea. Springer, Dordrecht. doi: 10.1007/978-94-017-8002-5_12.

Schuster, E., L. Bulling and J. Köppel. 2015. Consolidating the state of knowledge: A synoptical review of wind energy's wildlife effects. Environ. Manag. 56: 300–331. doi: 10.1007/s00267-015-0501-5.

Schwemmer, P., B. Mendel, N. Sonntag, V. Dierschke and S. Garthe. 2011. Effects of ship traffic on seabirds in offshore waters: implications for marine conservation and spatial planning. Ecol. Appl. 21: 1851–1860. doi: 10.1890/10-0615.1.

Scott, B.E., J. Sharples, O.N. Ross, J. Wang, G.J. Pierce and C.J. Camphuysen. 2010. Sub-surface hotspots in shallow seas: Fine-scale limited locations of top predator foraging habitat indicated by tidal mixing and sub-surface chlorophyll. Mar. Ecol. Prog. Ser. 408: 207–226. doi: 10.3354/meps08552.

Scott, B.E., R. Langton, E. Philpott and J.J. Waggitt. 2014. Seabirds and marine renewables: Are we asking the right questions? pp 81–92. *In*: Shields, M.A. and A.I.L. Payne [eds.]. Marine Renewable Energy Technology and Environmental Interactions. Humanity and the Sea. Springer, Dordrecht The Netherlands. doi 10.1007/978-94-017-8002-5_7.

Scottish Government 2020. Sectoral Marine Plan for Offshore Wind Energy. October 2020. 79 pp. Accessed 25th March 2021 from: https://www.gov.scot/publications/sectoral-marine-plan-offshore-wind-energy/.

SEANSE 2020. Strategic Environmental Assessment on North Sea Energy as an aid for Maritime Spatial Planning, Summary Report. Accessed 25th March 2021 from: https://northseaportal.eu/downloads/.

Searle, K., D. Mobbs, A. Butler, M. Bogdanova, S. Freeman, S. Wanless et al. 2014. Population consequences of displacement from proposed offshore wind energy developments for seabirds breeding at Scottish SPAs (CR/2012/03) report to Marine Scotland Science.

Searle, K.R., D.C. Mobbs, A. Butler, R.W. Furness, M.N. Trinder and F. Daunt. 2018. Finding out the Fate of Displaced Birds. Scottish Marine and Freshwater Science 9(8). doi: 10.7489/12118–1.

Searle, K., D. Mobbs, F. Daunt and A. Butler. 2019a. A population viability analysis modelling tool for seabird species. Natural England Commissioned Reports, Number 274.

Searle, K.R., A. Butler, D.C. Mobbs, M. Trinder, J. Waggitt, P. Evans et al. 2019b. Scottish Waters East Region Regional Sectoral Marine Plan Strategic Ornithology Study: Final report. Report for MS via SEANSE. Accessed 25th March 2021 from: https://northseaportal.eu/downloads/.

Skov, H., S.B. Leonhard, S. Heinänen, R. Zydelis, N.E. Jensen, J. Durinck et al. 2012. Horns Rev 2 Monitoring 2010–2012. Migrating Birds. Orbicon, DHI, Marine Observers and Biola. Report commissioned by DONG Energy.

Skov, H., S. Heinänen, T. Norman, R.M. Ward, S. Mendez-Roldan and I. Ellis. 2018. ORJIP bird collision and avoidance study. The Carbon Trust, London, UK.

Slavik, K., C. Lemmen, W. Zhang, O. Kerimoglu, K. Klingbeil and K.W. Wirtz. 2019. The large-scale impact of offshore wind farm structures on pelagic primary productivity in the southern North Sea. Hydrobiologia 845: 35–53.doi: 10.1007/s10750-018-3653-5.

Smyth, K., N. Christie, D. Burdon, J.P. Atkins, R. Barnes and M. Elliott. 2015. Renewables-to-reefs?—Decommissioning options for the offshore wind power industry. Mar. Pollut. Bull. 90: 247–258. doi: 10.1016/j.marpolbul.2014.10.045.

Snyder, B. and M.J. Kaiser. 2009. A comparison of offshore wind power development in europe and the U.S.: Patterns and drivers of development. Appl. Energy 86: 1845–1856. doi: 10.1016/j.apenergy.2009.02.013.

Sparling, C.E., A.C. Seitz, E. Masden and K. Smith. 2020. Collision risk for animals around Turbines. pp. 29–65. *In*: Copping, A.E. and L.G. Hemery [eds.]. OES-Environmental 2020 State of the Science Report: Environmental Effects of Marine Renewable Energy Development Around the World. Report for Ocean Energy Systems (OES). doi: 10.2172/1632881.

Tasker, M.L., C.J. Camphuysen, J. Cooper, S. Garthe, W.A. Montevecchi and S.J.M. Blaber. 2000. The impacts of fishing on marine birds. ICES J. Mar. Sci. 57: 531–547. doi: 10.1006/jmsc.2000.00714.

Thaxter, C.B., B. Lascelles, K. Sugar, A.S.C.P. Cook, S. Roos, M. Bolton et al. 2012. Seabird foraging ranges as a tool for identifying candidate marine protected areas. Biol. Conserv. 156: 53–61. doi: 10.1016/j.biocon.2011.12.009.

Thaxter, C.B., V.H. Ross-Smith, W. Bouten, E.A. Masden, N.A. Clark, G.J. Conway et al. 2018. Dodging the blades: New insights into three-dimensional area use of offshore wind farms by lesser black-backed gulls *Larus fuscus*. Mar. Ecol. Prog. Ser. 587: 247–253. doi: 10.3354/meps12415.

Thaxter, C.B., V.H. Ross-Smith, W. Bouten, N.A. Clark, G.J. Conway, E.A. Masden et al. 2019. Avian vulnerability to wind farm collision through the year: Insights from lesser black-backed gulls (*Larus fuscus*) tracked from multiple breeding colonies. J. Appl. Ecol. 56: 2410–2422. doi: 10.1111/1365-2664.13488.

Topham, E. and D. McMillan. 2017. Sustainable decommissioning of an offshore wind farm. Renew. Energy 102: 470–480. doi: 10.1016/j.renene.2016.10.066.

Topping, C. and I.K. Petersen. 2011. Report on a Red-throated Diver Agent-Based Model to assess the cumulative impact from offshore wind farms. Report commissioned by the Environmental Group. Aarhus University, DCE – Danish Centre for Environment and Energy.

Tufa, R.A., S. Pawlowski, J. Veerman, K. Bouzek, E. Fontananova, G. di Profio et al. 2018. Progress and prospects in reverse electrodialysis for salinity gradient energy conversion and storage. Appl. Energy 225: 290–331. doi: 10.1016/j.apenergy.2018.04.111.

Vallejo, G.C., K. Grellier, E.J. Nelson, R.M. McGregor, S.J. Canning, F.M. Caryl et al. 2017. Responses of two marine top predators to an offshore wind farm. Ecol. Evol. 7: 8698–8708. doi: 10.1002/ece3.3389.

Vanermen, N., E.W.M. Stienen, W. Courtens, T. Onkelinx, M. Van de walle and H. Verstraete. 2013. Bird monitoring at offshore wind farms in the Belgian part of the North Sea. Assessing seabird displacement effects. Rapporten van het Instituut voor Natuur- en Bosonderzoek 2013 (INBO.R.2013.755887). Instituut voor Natuur- en Bosonderzoek, Brussels, Belgium.

Vanermen, N., T. Onkelinx, W. Courtens, M. Van de walle, H. Verstraete and E.W.M. Stienen. 2015. Seabird avoidance and attraction at an offshore wind farm in the Belgian part of the North Sea. Hydrobiologia 756: 51–61. doi: 10.1007/s10750-014-2088-x.

Vanermen, N. and E.W.M. Stienen. 2019. Seabirds: displacement. pp. 174–205. *In*: Perrow, M.R. [ed.]. Wildlife and Wind Farms, Conflicts and Solutions. Volume 3. Offshore: Potential effects. Pelagic Publishing, Exeter, UK.

Vanermen, N., W. Courtens, R. Daelemans, L. Lens, W. Müller, M. Van de walle et al. 2020. Attracted to the outside: A mesoscale response pattern of lesser black-backed gulls at an offshore wind farm revealed by GPS telemetry. ICES J. Mar. Sci. 77: 701–710. doi: 10.1093/icesjms/fsz199.

Waggitt, J.J. and B.E. Scott. 2014. Using a spatial overlap approach to estimate the risk of collisions between deep diving seabirds and tidal stream turbines: A review of potential methods and approaches. Mar. Policy 44: 90–97. doi: 10.1016/j.marpol.2013.07.007.

Waggitt, J.J., P.G.H. Evans, J. Andrade, A. Banks, O. Boisseau, M. Bolton et al. 2019. Distribution maps of cetacean and seabird populations in the North-East Atlantic. J. Appl. Ecol. 57: 253–269. doi: 10.1111/1365-2664.13525.

Walls, R., C. Pendlebury and L. Turner. 2008. Lynn & Inner Dowsing Offshore Wind Farm. Boat-based Ornithological Monitoring Report July–December 2007. RPS, Glasgow, UK.

Webb, A. and G. Nehls. 2019. Surveying seabirds. pp. 60–95. *In*: Perrow, M.R. [ed.]. Wildlife and Wind Farms, Conflicts and Solutions Volume 4 Offshore: Monitoring and mitigation. Pelagic Publishing, Exeter, UK.

Wilber, D., L. Read, M. Griffin and D. Carey. 2020. Block Island Wind Farm Demersal Fish Trawl Survey Synthesis Report-Years 1 to 6, October 2012 through September 2018. Technical report prepared for Deepwater Wind, INSPIRE Environmental, Newport, RI, USA.

Williamson, B.J., S. Fraser, P. Blondel, P.S. Bell, J.J. Waggitt and B.E. Scott. 2017. Multisensor acoustic tracking of fish and seabird behavior around tidal turbine structures in Scotland. IEEE J. Ocean. Eng. 42: 948–965. doi: 10.1109/JOE.2016.2637179.

Willsteed, E.A., S. Jude, A.B. Gill and S.N.R. Birchenough. 2018. Obligations and aspirations: A critical evaluation of offshore wind farm cumulative impact assessments. Renew. Sust. Energ. Rev. 82: 2332–2345. doi: 10.1016/j.rser.2017.08.079.

Wilson, B., R.S. Batty, F. Daunt and C. Carter. 2006. Collision Risks between Marine Renewable Energy Devices and Mammals, Fish and Diving Birds. Report to the Scottish Executive. Scottish Association for Marine Science, Oban, Scotland, PA37 1QA.

Woodward, I., C.B. Thaxter, E. Owen and A.S.C.P. Cook. 2019. Desk-based revision of seabird foraging ranges used for HRA screening. BTO research report number 724.

Zhang, W., Y. Li, X. Wu and S. Guo. 2018. Review of the applied mechanical problems in ocean thermal energy conversion. Renew. Sust. Energ. Rev. 93: 231–244. doi: 10.1016/j.rser.2018.05.048.

Zydelis, R., C. Small and G. French. 2013. The incidental catch of seabirds in gillnet fisheries: a global review. Biol. Conserv. 162: 76–88. doi: 10.1016/j. biocon.2013.04.002.

CHAPTER 4

Seabirds and Biotoxins

María Victoria M. Casero,[1,*] *Jaime A. Ramos*[2] and *Leonel Pereira*[2]

Introduction

Wildlife biologists are now reporting an increasing number of seabird casualties (Dobson and Foufopoulos 2001, Newman et al. 2007) and wildlife rehabilitation centres are receiving an increasing number of seabirds to rehabilitate (Montesdeoca et al. 2017). The media is also reporting large numbers of moribund or dead seabirds, particularly in coastal areas, and affecting mostly opportunistic species such as gulls (*Larus* spp.) (SGO 2019). Large scale seabird wrecks, i.e., a vast number of dead, injured or exhausted seabirds wash up ashore, are a well-known phenomenon to seabird biologists. Often, the cause of death is difficult to determine, but strong storms, starvation or major pollution events are often implicated (Newman et al. 2007, Morley et al. 2016, Tavares et al. 2016). However, the particularly smaller scale seabird wreck events may be associated with other environmentally natural causes of death such as toxins produced by: (1) marine algae, mainly during strong increases in microalgae biomass (algal blooms), which affect mostly pelagic seabirds, and (2) bacteria such as *Clostridium botulinum* affecting mostly coastal seabird species such as gulls (Shumway et al. 2003). The possible impact of algal toxins on seabird deaths has been described since the Late Pliocene, because toxic red tides may have been responsible for the death of seabirds preserved in marine sediments from the late Pliocene of Florida (Emslie et al. 1996). One of the earliest and best documented events of a large-scale seabird mortality by toxins produced by marine algae (in this case by the dinoflagellate *Alexandrium tamarense*, formerly *Gonyaulax tamarensis*) occurred in 1968 on the Farne Islands, Northumberland, UK, where 80% of the Shag (*Phalacrocorax aristotelis*) breeding population was estimated to be killed (Coulson et al. 1968). Large numbers of dead sandeels (*Ammodytes* spp.), the main diet of Shags, were washed ashore dead and birds were observed with loss of equilibrium and vomiting sandeels.

This chapter provides a brief overview of the biotoxins affecting seabirds, the symptons they show once intoxicated and treatment procedures. We also describe the harmful algae blooms and toxin-producing species and discuss the main interactions between seabirds and biotoxins, given the current climate change that promotes such interactions. To address the symptoms of ill and

[1] RIAS - Wildlife Rehabilitation and Research Center of Ria Formosa, Rua do Parque Natural da Ria Formosa, Centro Ambiental de Marim, 8700-194 Olhão, Portugal.

[2] MARE - Marine and Environmental Sciences Centre, Department of Life Sciences, University of Coimbra, 3004-517 Coimbra, Portugal.

Emails: jramos@uc.pt; Leonel.pereira@uc.pt

* Corresponding author: mariavmcasero@gmail.com

moribund birds as well as their treatment we will use the case of gulls with Paretic Syndrome received at the Wildlife Rehabilitation and Research Centre—RIAS, a wildlife hospital located in Olhão, at the Ria Formosa Natural Park headquarters in the Algarve, Portugal.

Harmful algae blooms and toxin-producing species

Algal blooms occur naturally in the oceans associated with changes, mainly seasonal, in the sunlight regime, availability of nutrients and hydrodynamic conditions, which modify the stability of the water column, particularly the depth of the mixing layer (Solé et al. 2005). In fact, algal blooms can be very common and linked to the physical, chemical and biological interactions particular to each location (Solé et al. 2005). In particular, Masó and Garcés (2006) highlighted the internal control that microalgae themselves exercise on the dynamics of flowering, through some adaptive behavioral strategies such as phototaxis, the ability to migrate vertically, formation of aggregates and cysts, in addition to the production of substances toxic to other organisms.

Several studies have shown how the formation of algal blooms can be harmful to the development of other species, due to the large accumulation of biomass or toxic compounds, capable of causing deleterious effects on competitors or acting as defense mechanisms against predators. These blooms are called Harmful Algal Blooms (HABs), the term first coined in 1974 during the "1st International Conference on Toxic Dinoflagellate Blooms" (Hallegraeff et al. 1995). In this context, it is necessary to make it clear that the term HABs should be used to describe increases in a short period of time in the microalgae population that has found favorable conditions for their development. It is also essential to highlight that the term is used for as different classes of microalgae as Dinophyceae, Cyanophyceae, Bacillariophyceae, Raphidophyceae and Coccolithophyceae (Hallegraeff 2010).

According to Anderson et al. (2008) the observed increase in HABs is a function of both the natural dispersion of species by currents or storms, and the influence of human activities, such as the increase in coastal eutrophication or via transport of microalgae by ballast water and by the increase in temperature. In particular, the availability of nutrients is essential for the growth of the biomass of marine microalgae, and situations of alteration of nutrients are mostly described as the main causes for the increase in the development, persistence and expansion of many algal blooms (Granéli and Johansson 2003, Dyhrman 2008, Granéli et al. 2008, Heisler et al. 2008). The excessive increase in the concentration of nitrogen and phosphorus in the water column, caused by the increase of the river supply and the nutrient load of terrestrial origin or due to the anthropic influence (mainly from agricultural activities), is the main factor responsible for the increase in primary productivity, mainly in coastal regions (Van Der Bergh et al. 2002, Granéli et al. 2008). The association between biotoxins and warmer water temperature has been described by several authors (Van Egmond et al. 2004, Vidal et al. 2013, Gibble and Hoover 2018, Barbosa et al. 2019, Li et al. 2020). A relationship between botulism outbreaks and hotter air temperature (> 26ºC) was also reported by Vidal et al. (2013) for Spanish wetlands.

Worldwide, many cases of morbidity and mortality among marine wildlife, including seabirds, have been associated with the ingestion of algae toxins (Burek et al. 2008, Lefebvre et al. 2016), with HAB-related incidents becoming increasingly common (Landsberg et al. 2014). Numerous seabird species have also experienced die-offs attributed to HABs (Shumway et al. 2003), and these mortality events, presumed to result from consumption of toxic prey, have occurred in various geographic locations and affected taxa with diverse foraging strategies and habitat preferences, suggesting that impacts of HABs occur broadly throughout the marine environment (Van Hemert et al. 2020). Three main marine toxins, saxitoxin, brevetoxin and domoic acid (Table B4.1) are associated with seabird mortalities. Shellfish harbour toxins from weeks to months, but the vector to piscivorous seabirds such as terns and cormorants should be fish. Pelagic fish species such as sardines and anchovies may accumulate these marine toxins in their viscera during blooms and appear to excrete the toxin rapidly (Costa and Garrido 2004, Busse et al. 2006).

Table B4.1. Main species of toxin-producing algae associated with seabird mortalities.

Class/Species	Toxin produced	Reference
Bacillariophyceae *Halamphora perpusilla* *Nitzschia bizertensis* *Nitzschia navis-varingica* *Pseudo-nitzschia* spp.	Domoic acid (DA)	Zabaglo et al. 2016
Dinophyceae *Alexandrium tamarense*	Paralytic shellfish poisoning (PSP)	Nisbet 1983
Dinophyceae *Akashiwo sanguinea*	Surfactant-like protein exudates	Jessup et al. 2009
Dinophyceae *Alexandrium catenella*	Saxitoxin (STX)	Van Hemert et al. 2020
Dinophyceae *Karenia brevis* (also known as *Gymnodinium breve* or *Ptychodiscus brevis*)	Brevetoxin (BTX)	Turner et al. 2015

Symptoms developed by seabirds intoxicated with biotoxins

Marine biotoxins, freshwater biotoxins and botulism can affect seabirds. The symptoms developed by intoxicated seabirds are usually similar for these three types of neurobiotoxins and are known as Paretic Syndrome. The most common clinical symptoms are loss of equilibrium (inability to stand and keep the head up), moderate to severe paralysis, motor incoordination, vomiting and diarrhoea (Fire and van Dolah 2012).

Between 2010 and 2020, 6455 waterbird admittances with paretic syndrome symptoms belonging to 39 species occurred at RIAS, Olhão, Portugal (72% were seabirds, 27% were aquatic birds and 1% were waders). Gulls were the most common group, with 2046 individuals of Yellow-legged Gull (*Larus michahellis*) and 1542 individuals of Lesser Black-backed Gull (*L. fuscus*). The annual number of gull admissions varied from 53 to 164 between 2010 and 2015, but increased strongly thereafter (2016 = 324, 2017 = 262, 2018 = 248, 2019 = 989, 2020 = 1156). The monthly distribution of all the gulls admitted from 2010 to 2020 presents two annual peaks, a smaller peak in April and a larger peak in October (Fig. B4.1).

Of all the gull admissions, 565 were dead upon entrance and 3023 were alive. Alive gulls were weak, with drooping wings, incapable of flying and showed signs of diarrhoea (Fig. B4.2). Dead animals were considered as paretic syndrome victims due to the presence of diarrhoea, absence of injuries, necropsies and also because the date of arrival concurred with a paretic syndrome outbreak.

Clinical signs of paretic syndrome include different degrees of ascendant flaccid paresis, weakness, third eyelid paralysis, diarrhoea and dyspnoea. Due to the prostration, paralysis and weakness, secondary conditions like fungal pneumonia, decubitus sores at the tarsus and *keratoconjunctivitis sicca* can appear. Clinical procedures are based on supportive treatment, including fluid therapy, feeding assistance and management of secondary conditions. The most common necropsy findings in dead seabirds with paretic syndrome were the absence of traumatic injuries or other diseases signs, diarrhoea, poor body condition, damaged kidneys and fungus infections (Table B4.2).

Most studies (reviewed by Shumway et al. 2003, Sonne et al. 2012), thesis and reports usually document paretic syndrome in seabirds, but the presence of the toxin is often not confirmed by analyses (Shumway et al. 2003, Soares 2014, Gibble and Hoover 2018). However, laboratory analyses of seabird tissues are the only way to clearly establish the exact toxin affecting the birds. Differential diagnosis includes marine biotoxins, freshwater toxins and botulism. Due to the strong evidence of neuro-biotoxins as an aetiology of paretic syndrome in RIAS gulls, laboratory analysis

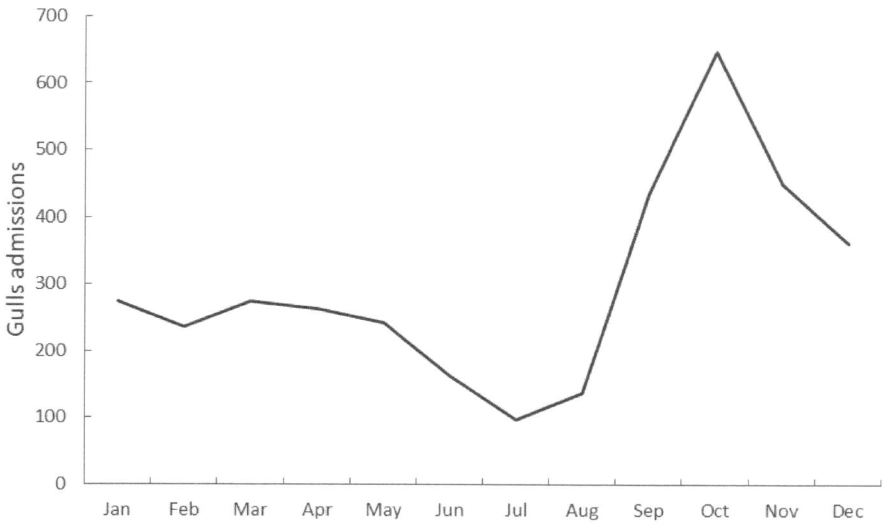

Figure B4.1. Gulls (Yellow-legged Gull *Larus michahellis* and Lesser Black-backed Gull *L. fuscus*) monthly admittances between 2010 and 2020 at the RIAS Wildlife Rehabilitation and Research Centre.

Figure B4.2. Yellow-legged Gull (*Larus michahellis*) with symptons of paretic syndrome: with drooping wings, full leg paralysis and unable to stand.

for botulinum toxin, saxitoxin and domoic acid were performed. Botulinum toxin was analysed in two pools of five livers and five intestines of gulls admitted during February 2020 at the National Reference Center for Anaerobic Bacteria and Botulism (Centre National de Référence des Bactéries Anaérobies et du Botulisme) at the Pasteur Institute in Paris, France. All samples were positive for botulism, and the toxic character was confirmed by a lethality test on mice injected intraperitoneally with filtered supernatant of the culture. Saxitoxin and domoic acid analysis were carried out at the Spanish Oceanographic Institute (Instituto Español de Oceanografía) in Vigo, Spain. Twenty-three gull samples belonging to ten gulls admitted during February 2020 were analysed by the High-Performance Liquid Chromatography. All the samples were negative for the presence of saxitoxin and domoic acid.

The results of laboratory analyses conducted by the RIAS Wildlife Rehabilitation and Research Centre suggest a high importance of botulism to explain the large numbers of coastal seabirds with paretic syndrome. Newman et al (2007) also used diagnoses from necropsy data of 3619 mortality events, involving 633708 dead individuals of aquatic birds (seabirds, shorebirds, waders and sea ducks) collected in the United States from 1971 to 2005, and reported that gulls represented the largest percentage (27.7%) of mortality events, and that 54.3% of all deaths were attributed to *C. botulinum* exposure, mostly in nearshore/coastal and freshwater areas. Again, such large datasets suggest that botulism may be the most widespread cause of coastal seabird mortality, although the potential effects of other biotoxins cannot be ruled out because no biotoxin analyses were reported by Newman et al. (2007). Most of the mortality due to botulism was apparently caused by ingestion of forms of botulinum toxin C/D (distinct from those that affect people when tinned food or in a poor

Table B4.2. Necropsy findings of Yellow-legged Gull (*Larus michahellis*) and Lesser Black-backed Gull (*L. fuscus*) with paretic syndrome.

Findings	Description	Figure
Diarrhoea	Dirty feathers around the cloaca	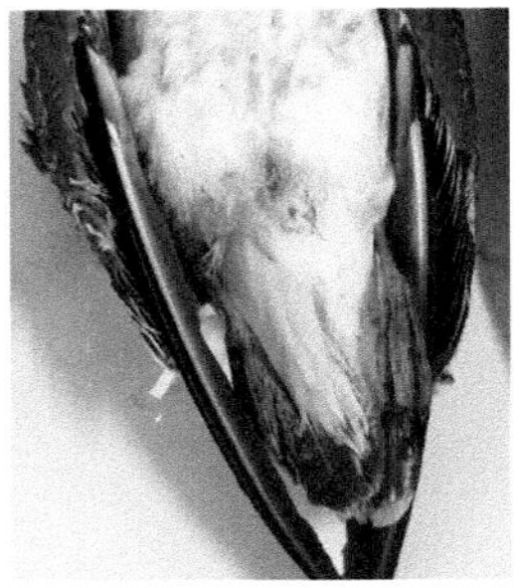
Poor body condition	Loss of pectoral muscle mass due to starvation	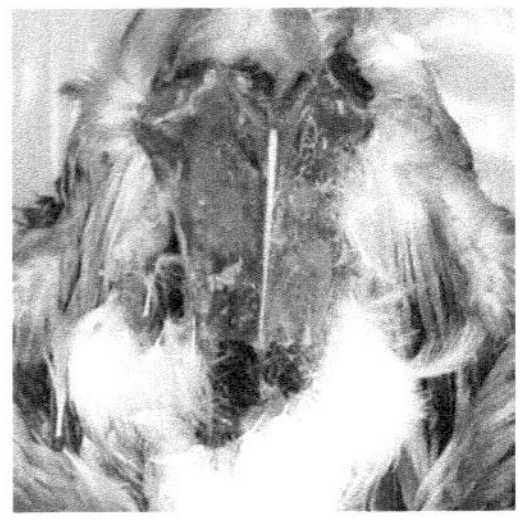
Damaged kidneys	Pale kidneys with marbled pattern compatible with kidney failure lesions due to dehydration	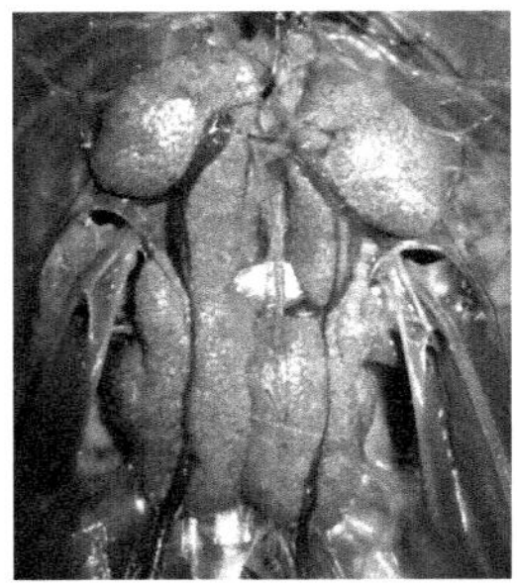
Fungus infections	Green-white masses with different sizes compatible with aspergilloma caused by the fungus *Aspergiullus* spp.	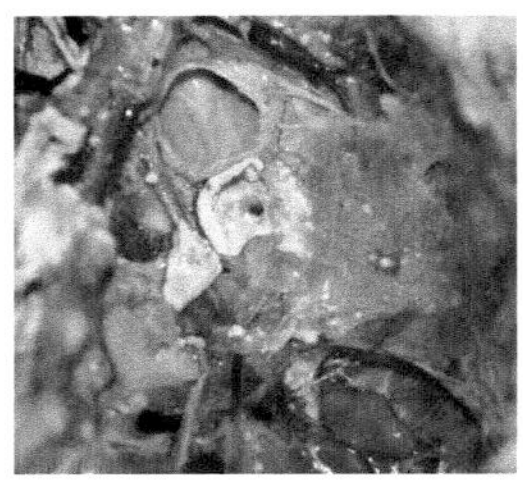

condition are consumed), rather than primary infection of the gastrointestinal tract by *C. botulinum* bacteria. In the case of the gulls admitted at the RIAS Wildlife Rehabilitation and Research Centre, GPS tracking indicates that birds might ingest the botulinum toxin in wastewater treatment plants (Matos et al. 2018).

Interactions between seabirds and algal toxins

Shumway et al. (2003) presented a review of seabird and toxic algal interactions and showed that: (1) the impacts appear to be stronger on coastal species such as terns and cormorants, but the impact on offshore species such as shearwaters and petrels may be overlooked because these species may be affected offshore and not detected, (2) large coastal birds such as shags and cormorants more often appear on the records of affected birds, presumably because of their coastal habits, easiness of detection and ingestion of relatively large fish (which will bioaccumulate the toxins), and (3) symptoms developed by seabirds vary according to seabird species and the algal toxin involved, but inability to stand and keep the head up and paralysis are often involved. (4) Seabirds may learn to avoid consumption of marine organisms with high levels of algal toxins, such as Common Eider (*Somateria mollissima*) avoiding large mussels with more toxins, and terns (*Sterna* spp.) apparently vomiting fish with a higher toxin content.

It appears that paralytic shellfish poisoning caused by saxitoxin may be responsible for large scale mortalities of coastal seabird species such as gulls, shags, pelicans, gannets and black ducks (Adams et al. 1968, Amstrong et al. 1978, Work et al. 1993), but reports of pelagic seabirds exposed to algal biotoxins are much less common. Soliño et al. (2019) analysed the concentration of domoic acid (produced by blooms of *Pseudo-nitzschia* spp., the main amnesic neurotoxin-producer microalgae) in the blood of 61 Cory's Shearwater (*Calonectris borealis*) and 87 Scopoli's Shearwater (*C. diomedea*) individuals foraging in the Canary current upwelling system and around Menorca Island, Mediterranean Sea, respectively. They found that 50.8% of Cory's Shearwaters and 32.2% of Scopoli's Shearwaters had detectable levels of domoic acid (1.0–10.6 ng ml^{-1}). This was the first study showing that pelagic seabird species appear to show a chronic exposure to marine biotoxins, and recaptures of the same individuals suggest that these shearwater species may excrete or degrade the toxin within 2 to 4 weeks.

Since 1998 domoic acid has been associated with large marine mammal and seabird mortalities along the coast of California: (a) from 1998 to 2006 1,335 cases of confirmed or suspected of domoic acid poisoning of California Sea Lions (*Zalophus californianus*) were reported, and about 50% of them died (Bejarano et al. 2008, Goldstein et al. 2009), (b) in 1991, 43 Brown Pelicans (*Pelecanus occidentalis*) and 95 Brandt's Cormorants (*Phalacrocorax penicillatus*) were reported dead on the west coast of the United States after ingesting anchovies with *Pseudo-nitzschia* frustules, and in 1996, 50% of the Brown pelican colony in Baja California (Mexico) was wiped out, with over 150 deaths attributed to domoic acid poisoning (Sierra-Beltran et al. 1997, Fire and van Dolah 2012). In addition, the effects on seabirds may depend on sex and/or age: Nisbet (1983) described a detailed poisoning event in 1978 by paralytic shellfish poisoning produced by the dinoflagellate *Alexandrium tamarense* (formerly *Gonyaulax excavata*) on a Common Tern (*Sterna hirundo*) colony in Massachusetts, where mortality affected mostly females and 3-years old birds. Although the toxin was present for about 3 weeks, terns died only on two days; it seems that birds might have developed an aversion to infected fish because many were observed vomiting. Overall, intoxication of marine top predators such as seabirds usually occurs via the consumption of contaminated marine invertebrates and fish, but there are also reports of mass fish mortalities, indicating that mid-trophic level prey is also affected. Pelagic fish may excrete the toxins rapidly (Costa and Garrido 2004, Busse et al. 2006), which may contribute to explain why seabird poisoning events are usually constrained to a narrow time window, presumably the initial phase of the bloom period (Nisbet 1983) when the toxin accumulates rapidly in fish.

Seabirds may also be affected via a proteinaceous foam surfactant that coats seabird plumage in a similar way to petroleum oil and causes the loss of waterproofness. Such foam may originate from the cellular breakdown of dinoflagellates such as *Akashiwo sanguinea*, with the first cases documented in 2007 in the coast of California (Jessup et al. 2009), and in 2009 in the coast of Oregon (Phillips et al. 2011). These two events affected hundreds of seabirds of several species, including Common Loons (*Gavia immer*), Common Murres (*Uria aalge*), Rhinoceros Auklets (*Cerorhinca monocerata*), Pacific Loons (*G. pacifica*), Red-throated Loons (*G. stellata*), Surf Scoters (*Melanitta perspicillata*), Double-crested Cormorants (*Phalacrocorax auritus*), Sooty Shearwaters (*Ardenna grisea*) and Northern Fulmars (*Fulmarus glacialis*). Some of these species had just finished their southward migration, and their poor body condition possibly exacerbated their mortality (Jessup et al. 2009).

Laboratory studies about the effect of algal toxins on seabirds are rare (but see Gochfeld and Burger 1998), but needed to determine the level of toxins that are harmful to seabirds. Harmful Algal Blooms may be a threat to human health (Gratan et al. 2016). The oral LD50 of saxitoxin for humans is 5.7 μg/kg (Nguyen et al. 2020), and the toxicity levels reported for several seabird species in the review by Shumway et al. (2003) are usually much higher, meaning that biotoxins can reach very high concentrations in marine ecosystems. The proliferation of HABs in coastal environments is expected to increase in intensity and frequency due to global warming and various anthropic influences (Glibert et al. 2014). With the increase in the incidence of HABs coupled with greater global awareness of their potential impacts on bird populations, it is highly likely that more incidences of bird deaths from HABs will be confirmed in the future. But the difficulty in this process involves several causes. First of all, HABs includes a complex aetiology involving different neuro-biotoxins that can be synthetized by numerous and diverse microorganisms. Secondly, the lack of means and financial resources have been identified as the main obstacle to ascertain the definitive diagnosis of the outbreaks. Furthermore, outbreaks in generalist species such as gulls are not treated as concerning events because these species are very abundant and there is no concern about their population trend. However, the decline in gulls' populations in Europe reported in the last years (Nager and O'Hanlon 2016, BirdLife International 2020) and its relationship with paretic syndrome as the cause (SGO 2019) should not be ignored. It is necessary to develop deeper studies about massive mortality and disease events, its link with biotoxins and its impact in seabird populations. Therefore, wildlife hospitals should make the effort to collect complete and detailed information in their data bases, perform necropsies and collect samples, and the authorities should collect dead animals in the field more efficiently. Environmental education programs are also necessary to train researchers and the public to recognize and help to observe the potential impacts of HABs on seabird health. It would be very useful to test archived bird samples using current, and more sophisticated and sensitive methods of detecting biotoxins (Shumway et al. 2003), and regular analysis for biotoxins could be implemented in routine seabird monitoring programmes.

Acknowledgments

To the RIAS team for the support. Laboratory tests were conducted at The Pasteur Institute and the Spanish Institute of Oceanography at Vigo. The study benefitted from the strategic program of MARE, financed by FCT (UIDB/04292/2020). We thank Joel Bried for comments on a draft.

References cited

Adams, J.A., D.D. Seaton, J.B. Buchanan and M.R. Longbottom. 1968. Biological observations associated with the toxic phytoplankton bloom off the east coast. Nature 220: 24–25. doi: 10.1038/220024a0.

Anderson, D.M., J.M. Burkholder, W.P. Cochlan, P.M. Glibert, C.J. Gobler, C.A. Heil et al. 2008. Harmful algal blooms and eutrophication: Examining linkages from selected coastal regions of the United States. Harmful Algae 8: 39–53. doi: 10.1016/j.hal.2008.08.017.

Armstrong, I.H., J.C. Coulson, P. Hawkey and M.J. Hudson. 1978. Further mass seabirds' deaths from paralytic shellfish poisoning. Br. Birds 71: 58–68.

Barbosa, V., M. Santos, P. Anacleto, A.L. Maulvault, P. Pousão-Ferreira, P.R. Costa et al. 2019. Paralytic shellfish toxins and ocean warming: Bioaccumulation and ecotoxicological responses in juvenile gilthead seabream (*Sparus aurata*). Toxins 11: 408. doi: 10.3390/toxins11070408.

Bejarano, A.C., E.M. Gulland, T. Goldstein, J.S. Leger, M. Hunter, L.H. Schwacke et al. 2008. Demographics and spatio-temporal signature of the biotoxin domoic acid in California sea lion *Zalophus californianus* stranding records. Mar. Mamm. Sci. 24: 899–912. doi: 10.1111/j.1748-7692.2008.00224.x.

BirdLife International. 2020. IUCN Red List for birds. Birdlife International, Cambridge. Retrieved May 2021 from http://www.birdlife.org.

Burek, K.A., F.M. Gulland and T.M. O'Hara. 2008. Effects of climate change on Arctic marine mammal health. Ecol. Appl. 18: 126–134. doi: 10.1890/06-0553.1.

Busse, L.B., E.L. Venrick, R. Antrobus, P.E. Miller, V. Vigilant, M.W. Silver et al. 2006. Demoic acid in phytoplankton and fish in San Diego, CA, USA. Harmful Algae 5: 91–101. doi: 10.1016/j.hal.2005.06.005.

Costa, P.R. and S. Garrido. 2004. Domoic acid accumulation in the sardine *Sardina pilchardus* and its relationships to *Pseudo-nitzschia* diatom ingestion. Mar. Ecol. Prog. Ser. 284: 261–268. doi: 10.3354/meps284261.

Coulson, J.C., G.R. Potts, I.R. Deans and S.M. Fraser. 1968. Mortality of shags and other sea birds caused by paralytic shellfish poison. Nature 220: 23–24. doi: 10.1038/220023a0.

Dobson, A. and J. Foufopoulos. 2001. Emerging infectious pathogens of wildlife. Philos. Trans. R. Soc. Lond. B 356: 1001–1012. doi: 10.1098/rstb.2001.0900.

Dyhrman, S.T. 2008. Molecular approaches to diagnosing nutritional physiology in harmful algae: Implications for studying the effects of eutrophication. Harmful Algae 8: 167–174. doi: 10.1016/j.hal.2008.08.016.

Emslie, S.D., W.D. Allmon, F.J. Rich, J.H. Wrenn and S.D. de France. 1996. Integrated taphonomy of an avian death assemblage in marine sediments from the late Pliocene of Florida. Palaeogeogr. Palaeocl. 124: 107–136. doi: 10.1016/0031-0182(96)00005-3.

Fire, S.E. and F.M. van Dolah. 2012. Algal blooms as health threats to marine wildlife. pp. 374–389. *In*: Aguirre, A.A., R.S. Ostfiels and P. Daszak [eds.]. New Directions in Conservation Medicine: Applied Cases in Ecological Health. Oxford University Press, New York, USA.

Gibble, C.M. and B.A. Hover. 2018. Seabird and harmful algal bloom interactions. pp. 223–242. *In*: Shumway, S.E., J.M. Burkholder and S. Morton [eds.]. Harmful Algal Blooms: A Compendium Desk Reference. Wiley-Blackwell, New York, USA. doi: 10.1002/9781118994672.ch6.

Glibert, P.M., J. Icarus Allen, Y. Artioli, A. Beusen, L. Bouwman, J. Harle et al. 2014. Vulnerability of coastal ecosystems to changes in harmful algal bloom distribution in response to climate change: projections based on model analysis. Glob. Change Biol. 20: 3845–3858. doi: 10.1111/gcb.12662.

Gochfeld, M. and J. Burger. 1998. Apparent paralytic shellfish poisoning in captive herring gulls fed commercial scallops. Toxicon 36: 411–412. doi: 10.1016/s0041-0101(97)00106-2.

Goldstein, T., T.S. Zabka, R.L. Delong, E.A. Wheeler, G. Yiitalo, S. Bargu et al. 2009. The role of domoic acid in abortion and premature parturition of California sea lions (*Zalophus californianus*) on San Miguel Island, California. J. Wildl. Dis. 45: 91–108. doi: 10.7589/0090-3558-45.1.91.

Granéli, E. and N. Johansson. 2003. Increase in the production of allelopathic substances by *Prymnesium parvum* cells grown under N- or P-deficient conditions. Harmful Algae 2: 135–145. doi: 10.1016/S1568-9883(03)00006-4.

Granéli, E., M. Weberg and P.S. Salomon. 2008. Harmful algal blooms of allelopathic microalgal species: The role of eutrophication. Harmful Algae 8: 94–102. doi: 10.1016/j.hal.2008.08.011.

Grattan, L.M., S. Holobaugh and J.G. Morris, Jr. 2016. Harmful algal blooms and public health. Harmful Algae 57: 2–8. doi: 10.1016/j.hal.2016.05.003.

Hallegraeff, G.M. 2010. Ocean climate change, phytoplankton community responses, and harmful algal blooms: A formidable predictive challenge. J. Phycol. 46: 220–235. doi: 10.1111/j.1529-8817.2010.00815.x.

Hallegraeff, G.M., D.M. Anderson, A.D. Cembella and H.O. Enevoldsen. 1995. IOC Manual on harmful marine microalgae. Manuals and Guides 33, UNESCO, Paris, FR. Available on: https://unesdoc.unesco.org/ark:/48223/pf0000122021.

Heisler, J., P.M. Glibert, J.M. Burkholder, D.M. Anderson, W. Cochlan, W.C. Dennison et al. 2008. Eutrophication and harmful algal blooms: A scientific consensus. Harmful Algae 8: 3–13. doi: 10.1016/j.hal.2008.08.006.

Jessup, D.A., M.A. Miller, J.P. Ryan, H.M. Nevis, H.A. Kerkering, A. Mekebri et al. 2009. Mass stranding of marine birds caused by a surfactant-producing red tide. PLoS ONE 4(2): e4550. doi: 10.1371/journal. pone.0004550.

Landsberg, J.H., K.A. Lefebvre and L.J. Flewelling. 2014. Effects of toxic microalgae on marine organisms. pp. 379–449. *In*: Rossini, G.P. [eds.]. Toxins and Biologically Active Compounds from Microalgae. CRC Press, Boca Raton, FL, USA. doi: 10.1201/b16806-17.

Lefebvre, K.A., L. Quakenbush, E. Frame, K.B. Huntington, G. Sheffield, R. Stimmelmayr et al. 2016. Prevalence of algal toxins in Alaskan marine mammals foraging in a changing arctic and subarctic environment. Harmful Algae 55: 13–24. doi: 10.1016/j.hal.2016.01.007.

Li, X., M. Yan, J. Gu, V.T. Lam, T. Wai, D.M. Baker, P.T. Leung. 2020. The effect of temperature on physiology, toxicity and toxin content of the benthic dinoflagellate Coolia malayensis from a seasonal tropical region. Water Res. 185: 116264. doi: 10.1016/j.watres.2020.116264.

Masó, M. and E. Garcés. 2006. Harmful microalgae blooms (HAB); problematic and conditions that induce them. Mar. Pollut. Bull. 53: 620–630. doi: 10.1016/j.marpolbul.2006.08.006.

Matos, D., J.A. Ramos, J.G. Calado, F.R. Ceia, J. Hey and V.H. Paiva. 2018. How fishing intensity affects the spatial and trophic ecology of two gull species breeding in sympatry. ICES J. Mar. Sci. 75: 1949–1964. doi:10.1093/icesjms/fsy096.

Montesdeoca N., P. Calabuig, J.A. Corbera and J. Orós. 2017. A long-term retrospective study on rehabilitation of seabirds in Gran Canaria Island, Spain (2003–2013). PLoS ONE 12: e0177366. doi: 10.1371/journal.pone.0177366.

Morley, T.I., A.L. Fayet, H. Jessop, P. Veron, M. Veron, J. Clark et al. 2016. The seabird wreck in the Bay of Biscay and Southwest Approaches in 2014: A review of reported mortality. Seabird 29: 22–38.

Nager R.G. and N.J. O'Hanlon. 2016. Changing numbers of three gull species in the British Isles. Waterbirds, 39(sp1): 15–28. doi: 10.1675/063.039.sp108.

Newman, S.H., A. Chmura, K. Converse, A.M. Kilpatrick, N. Patel, E. Lammers et al. 2007. Aquatic bird disease and mortality as an indicator of changing ecosystem health. Mar. Ecol. Prog. Ser. 352: 299–309. doi: 10.3354/meps07076.

Nguyen, H.V.N., M.E. Smith and H.D. Swoboda. 2020. Shellfish toxicity. (Updated 221 July 2020). *In*: StatPearls (Internet). StatPearls Publishing, Treasure Island, Florida. Available from: https://www.ncbi.nlm.nih.gov/books/NBK470225/.

Nisbet, I.C.T. 1983. Paralytic shellfish poisoning effects on breeding terns. Condor 85: 338–345. doi: 10.2307/1367071.

Phillips, E.M., J. E. Zamon, H.M. Nevis, C.M. Gibble, R.S. Duerr and L.H. Kerr. 2011. Summary of birds killed by a harmful algal bloom along the South Washington and North Oregon coasts during October 2009. Northwest. Nat. 92: 120–126. doi: 10.1898/10-32.1.

Shumway, S.E., S.M. Allen and P. Dee Boersma. 2003. Marine birds and harmful algal blooms: Sporadic victims or under-reported events? Harmful Algae 2: 1–17. doi: 10.1016/S1568-9883(03)00002-7.

Sierra-Beltran, A., M. Palafox-Uribe, J. Grajales-Montiel, A. Cruz-Villacorta and J.L. Ochoa. 1997. Sea bird mortality at Cabo San Lucas, Mexico: Evidence that toxic diatom blooms are spreading. Toxicon 35: 447–453. doi: 10.1016/S0041-0101(96)00140-7.

Soares, S.P.V. 2014. Parectic syndrome in gulls (Laridae) in the south of Portugal. Master thesis, University of Lisbon, Lisbon, Portugal.

Sociedade Galega de Ornitoloxía (SGO) (2019, June 16). Cada vez menos gaivotas en Galicia [Press release]. Retrieved March 2020, from http://www.sgosgo.org/index.php/prensa/16-notas-de-prensa/161-cada-vez-menos-gaivotas-en-galicia.

Solé, J., E. García-Ladona, P. Ruardij and M. Estrada. 2005. Modelling allelopathy among marine algae. Ecol. Model. 183: 373–384. doi: 10.1016/j.ecolmodel.2004.08.021.

Soliño, L., J. Ferrer-Obiol, L. Navarro-Herrero, J. González-Solís and P.R. Costa. 2019. Are pelagic seabirds exposed to amnesic shellfish poisoning toxins. Harmful Algae 84: 172–180. doi: 10.1016/j.hal.2019.03.014.

Sonne, C., A.K. Alstrup and O.R. Therkildsen. 2012. A review of the factors causing paralysis in wild birds: Implications for the paralytic syndrome observed in the Baltic Sea. Sci. Total Environ. 416: 32–39. doi: 10.1016/j.scitotenv.2011.12.023.

Tavares, D.C., J. Fulgencio de Moura and S. Siciliano. 2016. Environmental predictors of seabird wrecks in a tropical coastal area. PLoS ONE 11: e0168717. doi: 10.1371/journal. pone.0168717.

Turner, A.D., C. Higgins, K. Davidson, A. Veszelovszki, D. Payne, J. Hungerford et al. 2015. Potential threats posed by new or emerging marine biotoxins in UK waters and examination of detection methodology used in their control: brevetoxins. Mar. Drugs 13: 1224–1254. doi: 10.3390/md13031224.

Van den Bergh, J.C.J.M., P.A.L.D. Nunes, H.M. Dotinga, W.H.C.F. Kooistra, E.G. Vrieling and L. Peperzak. 2002. Exotic harmful algae in marine ecosystems: an integrated biological-economic-legal analysis of impacts and policies. Mar. Policy 26: 59–74. doi: 10.1016/S0308-597X(01)00032-X.

Van Egmond, H.P., M.E. Van Apeldoorn and G.J.A. Speijers. 2004. Marine biotoxins. Food and nutrition Technical papers. Food and Agriculture Organization of the United Nations (FAO), Rome, Italy. ISBN 95-5-105129-1.

Van Hemert, C., S.K. Schoen, R.W. Litaker, M.M Smith, M.L. Arimitsu, J.F. Piatt et al. 2020. Algal toxins in Alaskan seabirds: Evaluating the role of saxitoxin and domoic acid in a large-scale die-off of Common Murres. Harmful Algae 92: 101730. doi: 10.1016/j.hal.2019.101730.

Vidal, D., I. Anza, M.A. Taggart, E. Perez-Ramirez, E. Crespo, U. Hofle et al. 2013. Environmental factors influencing the prevalence of a *Clostridium botulinum* type C/D mosaic strain in non-permanent mediterranean wetlands. Applied and Environ. Microbiol. 79: 4264–4271. doi: 10.1128/aem.01191-13.

Work, T.M., B. Barr, M.B. Allison, L. Fritz, M.A. Quilliam and J.L.C. Wright. 1993. Epidemiology of domoic acid poisoning in brown Pelicans (*Pelecanus occidentalis*) and Brandt's Cormorants (*Phalacrocorax penicillatus*) in California. J. Zoo Wildl. Med. 24: 54–62.

Zabaglo, K., E. Chrapusta, B. Bober, A. Kaminski, M. Adamski and J. Bialczyk. 2016. Environmental roles and biological activity of domoic acid: A review. Algal Res. 13: 94–101. doi: 10.1016/j.algal.2015.11.020.

Part C
Seabirds as Ecological Indicators

CHAPTER 1

Seabirds as Indicators of Forage Fish Stocks

Jaime A. Ramos[1,*] and *Robert W. Furness*[2]

Introduction

Professional and sports fishermen around the world know that a flock of seabirds hovering above the water or plunge diving means a school of small fish underneath, possibly being chased to the surface by large predatory fish or marine mammals. Such inter-specific associations between seabirds and other marine top predators are present in all oceans, influencing population parameters and the structure of these marine communities (Veit and Harrison 2017), and have been used to navigate, locate and capture fish since historical times (Steadman 2006). The roots of using seabirds as indicators of changes in fish populations lie in the presence of these strong feeding associations.

Ecological indicators are appealing for scientists, managers, politicians and the general public (Niemi and MacDonald 2004), because they convey (often complex) information about the state of the environment in a relatively easy way, and may provide an early warning signal for ecological problems or be used as monitors for trends in ecological resources. However, to be effective, they should capture the complexity of the ecosystem concerned, and at the same time, be simple enough to be understood and cost effective to be routinely monitored (Dale and Beyler 2001). Seabirds may provide ecological indicators to show the general condition or 'health' of aquatic ecosystems, in which case they provide a qualitative indicator and are often referred as sentinel species (Furness and Camphuysen 1997), or they may show a specific functional relationship with a specific stressor, in which case they may be used as a quantitative ecological indicator (Piatt et al. 2007). Seabirds are suitable candidates to act as ecological indicators, because they are conspicuous top predators likely integrating ecosystems processes occurring at lower trophic levels (Frederiksen et al. 2006), easy to monitor on land and there are many long-term studies of seabird populations worldwide. However, to be ideal indicators of changes in fish stocks seabird measures should show linear relationships along the full range of fish stock variation encountered by seabirds (Dale and Beyler 2001). A first word of caution is needed because different seabird reproductive and behavioural measures are likely to respond differently to variations in forage fish, and responses vary among species (Montevecchi 1993, Furness and Camphuysen 1997).

[1] University of Coimbra, MARE - Marine and Environmental Sciences Centre, Department of Life Sciences, 3004-517 Coimbra, Portugal.

[2] MacArthur Green, 93 South Woodside Road, Glasgow G20 6NT, U.K.
Email: bob.furness@macarthurgreen.com

* Corresponding author: jramos@uc.pt

Seabirds are long-lived, and life history predicts that they respond to a reduction in food abundance by altering their behaviour or breeding effort, in order to buffer adult survival (Stearns 1992). Their main prey, mostly small pelagic fish, are short-lived animals with strong variations in recruitment rates, which overall abundance is highly influenced by stochastic changes in oceanographic conditions (Houde 2009). Therefore natural selection provided seabirds with the ability to cope with changes in food abundance (Furness 2007). All seabird parameters, from breeding numbers to laying dates, breeding success and demographic rates are influenced by environmental variation. During years of very low food abundance, such as El Niño years in tropical areas, most of these parameters from different species may show very strong responses, to the extent that in some years seabirds may skip breeding altogether (Catry et al. 2013, Ramos et al. 2002, Cubaynes 2011), and can thus be used as sentinels of changes in marine ecosystems. However, in order to provide an informative decision as an indicator of changes in fish stocks, seabird parameters must include a wide range of environmental conditions apart from such extreme years, and the detail of the relationship between the seabird parameter and a measure of fish stock abundance is very important to consider (Cairns 1987, Piatt et al. 2007).

Since the 1990s hundreds of publications showed close relationships between seabird parameters and changes in marine environments (often in food supplies), including several reviews between 2007 and 2010 (Piatt et al. 2007, Parsons et al. 2008, Einoder, 2009, Durant et al. 2009). Einoder (2009) provided a comprehensive review on the use of seabirds as ecological indicators, important within the framework of ecosystem-based management. The traditional approach of assessing fish populations as single-stock management, i.e., considering solely the impact of fisheries, led to overfishing and the ecological deterioration of many marine ecosystems (Jackson et al. 2001). Modern views advocate that fish stocks should now be managed in relation to the structure and functionality of marine food webs (Pikitch et al. 2004). Einoder et al. (2009) outlined the limitations and uncertainties that should be considered before a seabird ecological indicator is used as a management tool: (1) the need to distinguish between the effects of fisheries and environmental change on fish stocks; (2) when selecting the most appropriate indicator species consider the spatial scale (the species should reflect the same scale that the fishery operates), life history parameters (small bodied species commonly forage at their maximum rate during good environmental conditions and are less resilient to perturbations in food conditions than large bodied species), and species foraging characteristics (low food conditions will affect more dietary specialists such as surface feeders than wide ranging and more generalist species); (3) when selecting the most important parameter as an indicator consider the temporal scale that the parameter reflects: reproductive measures such as chick growth or breeding success reflect a scale of months and are relatively easy to obtain, whereas population parameters such as adult survival reflect a scale of years to decades, but are very labourious to obtain.

The vast literature on seabird ecology clearly shows links between changes in prey abundance and behavioural, reproductive and demographic seabird measures, but as predicted by life history theory, breeding parameters such as breeding success or chick growth rate show a close relationship with food abundance, whereas other variables such as adult body condition, adult survival or number of breeding birds may not be closely related with such short-term fluctuations in food abundance (Boyd et al. 2006). Some measures such as adult body condition, evaluated as an index of body mass relative to structural body size, are easy to measure in seabird studies, but often do not show direct and clear relationships with direct or surrogate measures of food abundance (Reid et al. 2005, Piatt et al. 2007), except for years with extreme low food conditions (Paiva et al. 2013, Pereira et al. 2020). Other measures such as adult survival rates are time consuming and much more difficult to measure, and the relationship with food abundance is less clear (Davis et al. 2005, Frederiksen et al. 2005).

Whereas Einoder (2009) provided a general overview of the main factors to consider when assessing how seabirds respond to changes in the composition and availability of their prey, here we concentrate on the details of the relationship between breeding season parameters of (mostly) coastal feeding seabirds and measures of fish abundance (i.e., the functional response curve, Fig. C1.1), outlining and discussing the implications of these relationships for the monitoring of fish stocks.

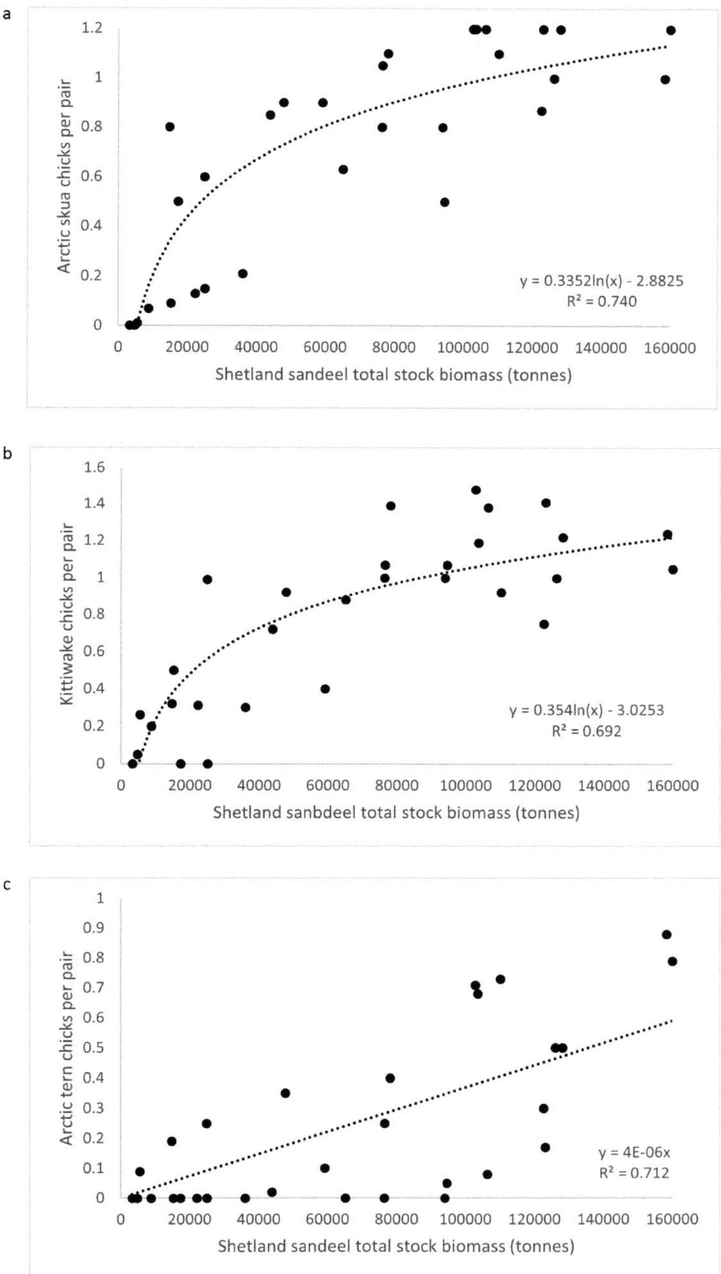

Figure C1.1. Relationships between Shetland sandeel stock total biomass each year and breeding success (mean chicks per pair) of (a) Arctic Skua (*Stercorarius parasiticus*), (b) Black-legged Kittiwake (*Rissa tridactyla*) and (c) Arctic Tern (*Sterna paradisaea*) at Foula, Shetland, for the years when fish stock biomass and seabird breeding success were measured (1976 to 2005).

The use of seabird measures as indicators of forage fish stocks

Cairns (1987) made three important predictions for the use of seabird reproductive and behavioural measures as indicators of marine food supplies: (1) these measures would show a curvilinear relationship with variations in food supply, (2) the threshold (half-way or inflexion point in the sigmoid functional response curve) in prey abundance variation would differ for each parameter, and (3) different species should respond differently to changes in food supply, depending on their foraging behaviour and capacity to adjust time budgets. Piatt et al. (2007) examined these three predictions using up to 12 reproductive and behavioural measures of Common Murres (*Uria aalge*) and Black-legged Kittiwake (*Rissa tridactyla*) for three colonies varying in prey density over a five-year period and found that predictions (1) and (3) were supported, but prediction (2) was not. (1) Apart from linearly related relationships over a small range of prey densities, the overall functional relationships were mostly sigmoidal, i.e., measures fluctuated rapidly from low to high values over small ranges of prey abundance (Fig. C1.1). (2) A relatively similar inflexion point, along a narrow range of prey densities, was found for all reproductive and behavioural measures. The authors attributed this to the fact that the main physical threshold for seabirds should be the density of fish school above which seabirds can forage successfully to maintain their daily energy budgets and provision for their chicks. (3) The breeding success of Black-legged Kittiwake was highly influenced by food supply, but that of the common murres was not.

Many studies point to a wide spectrum in the sensitivity of a species to variations in food abundance, which is ultimately related with its ecological niche. Seabird species with a wider foraging amplitude, either in terms of foraging range such as shearwaters, or diving at different depths of the water column such as alcids (murres, guillemots, auklets, puffins and murrelets), exhibit a higher foraging plasticity and may change their foraging behaviour or diet relatively easily (Harding et al. 2007). However, small, surface feeding seabird species, with a specialised diet, chiefly small pelagic fish species, and with a small foraging range such as tern species are very sensitive to changes in food abundance (Furness and Tasker 2000). The main reason is that surface feeding species foraging by plunge diving, target prey at a limited depth and perform close to the limits of their energetic budget, and thus have little leeway to buffer against a decrease in forage fish (Pearson 1968). Therefore, these species should provide a much clearer and direct signal of variations in food abundance, providing more consistent and easily interpretable outcomes to be used as indicators of changes in fish stocks. In fact, several long-term studies that examined a functional relationship between prey abundance and breeding success for plunge diving terns obtained a linear relationship: this was the case for: (a) Common Tern (*Sterna hirundo*) foraging on Atlantic herring (*Clupea harengus*) in the North Sea, and on sprat (*Sprattus sprattus*) in the Wadden Sea (Danhardt and Becker 2011, Vedder et al. 2019) and Scotland (Jennings et al. 2012), (b) Arctic Tern (*Sterna paradisaea*) foraging on lesser sandeels (*Ammodytes marinus*) in the North Sea, Shetland stock (Furness 2007, Fig. C1.1). Cury et al. (2011) related seabird breeding success with prey abundance using a comprehensive global dataset of long-term studies for 14 seabird species across seven ecosystems. They found a generality for an asymptotic relationship and identified a common threshold in prey abundance for all species (=one third of the maximum prey biomass in each ecosystem), below which breeding success was consistently reduced and more variable. However, the only plunge diving species in the analysis, the arctic tern, showed no threshold, meaning that the relationship was linear.

It appears that food abundance is always a limiting factor for the breeding success of plunge diving terns. Even for less responsive reproductive measures such as the number of breeding birds, long-term studies show strong correlations with fish stock estimates. This is the case for Swift Terns (*Sterna bergii*) foraging on anchovy (*Engraulis capensis*) and sardines (*Sardinops sagax*) in South Africa Western Cape (Crawford 2003). Obviously, this does not mean that such relationships can be used directly to estimate fish stocks, but shows that the high linear responsiveness of plunge diving terns to food abundance may be strong enough to develop quantitative indicators of fish stocks. This is particularly noticed when certain tern populations rely on a specific prey.

Changes in the diet of plunge diving terns can also be used as indicators of changes in fish stocks. The catch per unit effort of Pacific sardines (*Sardinops caeruleus*) was predicted with an accuracy of 73% with a model using the proportion of Pacific sardines in the diet of Elegant Terns (*Sterna elegans*), the reproductive success of Heermann's Gulls (*Larus heermanni*) and the spring sea surface temperature anomaly in the Sea of Corte's, Baja California (Velarde et al. 2004). There are numerous studies of changes in the diet of plunge diving terns worldwide, apparently related with changes in fish stocks, such as Crested Terns (*Sterna bergii*) feeding on sardines (*Sardinops sagax*) in the Gulf of St Vincent, South Australia (McLeay et al. 2009). Overall, breeding success of seabird species heavily dependent on the abundance of specific prey species, have been used in models to inform fisheries' management: this is the case of Brown Pelicans (*Pelecanus occidentalis californicus*) in Southern California Bight for northern anchovy (*Engraulis mordax*) (Anderson et al. 1980). Changes in the diet of seabirds may also be particularly useful in providing information about the status of fish stocks at different spatial and temporal scales (Montevecchi and Myers 1995). For instance, the diet composition of Puffin (*Fratercula arctica*), Black-legged Kittiwake and Common Guillemot may indicate the stock condition of Atlantic herring in the Southern Barents Sea (Barrett and Krasnov 1996), and Common Tern diet in the Gulf of Maine, Canada, provide information for Atlantic herring (Scopel et al. 2018). A detailed case study about the use of seabirds in the North Sea as indicators of forage fish stocks is presented below.

Seabirds in the North Sea as indicators of forage fish stocks

Several studies have shown correlations between abundance of forage fish stocks in the North Sea and seabird diet composition (Anderson et al. 2014, Church et al. 2018, Wanless et al. 2018). Many more show correlations with breeding success (Furness and Tasker 2000, Frederiksen et al. 2005, Wanless et al. 2007, Danhardt and Becker 2011, Cook et al. 2014, Carroll et al. 2017, Vedder et al. 2019, JNCC 2020, Olin et al. 2020). Few show correlations with adult survival (Oro and Furness 2002, Davis et al. 2005), or breeding numbers (Jennings et al. 2012, Perkins et al. 2018, Miller et al. 2019, JNCC 2020). There are well-established methods for monitoring breeding numbers and breeding success of seabirds. Long-term studies have provided extensive data over many decades at some colonies, some based on professional research, but often based on expert citizen science (e.g., JNCC 2020). In contrast, seabird diets and adult survival can be much more difficult to monitor.

Studies listed above have all focused on the effect of variations in forage fish abundance on seabirds. Relatively few studies have used these relationships between seabird ecology and fish abundance to predict fish abundance from seabird ecology (but some good examples from other parts of the world can be seen, Lorentsen et al. 2018 and Scopel et al. 2018). In many parts of the world, forage fish stocks are not harvested by fisheries, and so there are no fishery-based data on forage fish abundance. Without any commercial fishery, monitoring would be expensive (requiring dedicated acoustic surveys or *in situ* sampling). In such circumstances, seabirds may provide useful indications of the status and dynamics of forage fish stocks. However, some forage fish are especially difficult to monitor even where there is a desire to do so. The lesser sandeel (hereafter sandeel) is a good example of this because it spends long periods of the year buried in sandy sediment. Sandeels come out of the sand and move up to the surface to feed on the plankton bloom at the appropriate time of year. However, their behaviour may vary according to predation risk and the amount of lipid they have stored. Their complex and erratic behaviour makes the species difficult to monitor. In the North Sea, the sandeel has been the focus of a very large industrial fishery (catching sandeels to manufacture fish meal and fish oil). Sandeel is also the main forage fish eaten by most breeding seabirds in most areas of the North Sea during spring and summer (Furness and Tasker 2000).

Furness and Tasker (2000) used ecological features of the different seabird species breeding in the North Sea to compute an index of their vulnerability to reductions in sandeel abundance. Terns, Black-legged Kittiwake and Arctic Skua (*Stercorarius parasiticus*) ranked highest in this vulnerability index. Data from Shetland, where sandeel abundance had fallen between two time periods, showed that these species were also the ones with the largest decreases in breeding success between periods with high and low sandeel abundance (Fig. C1.1). Furness and Tasker therefore proposed that breeding success of these seabirds could be used to indicate the state of the sandeel stock (whether sandeel abundance was above or below some threshold that determined seabird breeding success). Subsequent research has supported this general conclusion (JNCC 2020). Perkins et al. (2018) identified Arctic Skua as highly vulnerable to reductions in food availability in terms of breeding success and breeding numbers. On the other hand, Kittiwakes show strong relationships between sandeel abundance and breeding success (Oro and Furness 2002, Frederiksen et al. 2005, Cury et al. 2011, Carroll et al. 2017, Miller et al. 2019, Olin et al. 2020).

From the start of the sandeel fishery in the 1950s to the 1970s, the North Sea was treated as a single fishery management area, effectively assuming that there was a single stock throughout this area. However, with the development of a small fishery at Shetland in the early 1970s it became apparent that sandeels at Shetland were a separate stock from those elsewhere in the North Sea. Shetland sandeels showed independent dynamics. The Shetland sandeel stock collapsed in the mid-1980s, and the fishery was closed from 1989 to 1995. It reopened in 1996 but ended in 2005, when stock biomass had fallen to a few percent of the original unfished biomass. Since then, the stock has not been monitored because the cost of fishery-independent stock assessment is high. However, seabird numbers and breeding success at Shetland have been monitored throughout the period of the sandeel fishery and after its closure. Kittiwake, Arctic skua and Arctic tern breeding success at Foula, Shetland all show clear and statistically robust correlations with the biomass of Shetland sandeel stock over the period when that was monitored. All these species had higher breeding success in years before the sandeel stock collapsed, recovered slightly during the period of closure from 1989–1995, but fell to very low levels by 2005 and have generally remained low since then. The strong relationship between seabird breeding success and Shetland sandeel total stock biomass suggests that seabird breeding success can be used as an indicator of sandeel abundance around Shetland over the period for which no sandeel assessment data are available (after 2005). It is possible that the indicator is affected by other forage fish species and not just by sandeel, so it must be treated with caution.

Kittiwake breeding success is monitored at many colonies along the UK North Sea coast, and it soon became apparent that the temporal patterns in breeding success showed strong consistency between neighbouring colonies, but differences between regions. Since breeding kittiwakes at all North Sea colonies seem to feed mainly on sandeels while breeding (Furness and Tasker 2000), this suggested that the North Sea holds several distinct stocks of sandeels in different regions, with independent dynamics among regions. That hypothesis was proposed by Furness et al. (1996) based on kittiwake breeding success data as a proxy indicator for sandeel stock dynamics. Independent evidence from sandeel biologists subsequently supported such interpretation. Proctor et al. (1998) suggested, on the basis of hydrographic modelling, that the limited larval transport of sandeels within the North Sea would result in separate stocks of sandeels in different regions. The International Council for the Exploration of the Sea (ICES) then began to model and derive management recommendations for North Sea sandeel stocks on a regional basis rather than treating the North Sea as a single management unit (Pedersen et al. 1999). There are now considered to be seven separate stocks of sandeels in the North Sea (ICES 2017), and these map onto regionally distinct kittiwake productivity areas (Frederiksen et al. 2005). Further analysis of Black-legged Kittiwake breeding success at UK North Sea colonies (Olin et al. 2020) is consistent with the revised management of sandeels as regional stocks. Olin et al. (2020) reported that "the scale of kittiwake synchrony, just over 200 km, is similar to the scale of sandeel synchrony (Wright et al. 2019), where similar scales of synchrony in the hypothesised driver and the target population can be used to infer a causal relationship" and "in many colonies, in particular in Shetland and on the Scottish east coast, breeding success fluctuated on a biannual time-scale, which fits with similar biannual fluctuations in sandeel recruitment (e.g., Arnott and Ruxton 2002, van Deurs et al. 2009), and thus provides additional support for sandeels as a potential driver of synchrony".

Olin et al. (2020) also pointed out that the strong relationship between kittiwake breeding success and sandeel abundance in the local stock is likely to be clearest in the northern North Sea, where sandeel is by far the most important forage fish. In the southern North Sea there are stocks of sprat and juvenile herring, and kittiwakes can feed on those species (Furness and Tasker 2000), so the relationship between kittiwake breeding success and sandeel abundance may become less clear if alternative forage fish prey are available to support kittiwake breeding success. It is therefore important to bear in mind possible long-term changes in fish distribution that could confound simple correlations between productivity and sandeel abundance.

While the relationship between kittiwake breeding success and sandeel stock biomass is strong, it can also be affected by changes in sandeel fishing and depletion of sandeel stock biomass by the fishery. Frederiksen et al. (2004) showed that breeding success of kittiwakes at the Isle of May in east Scotland was correlated with an environmental driver of sandeel abundance (sea surface temperature) and with the presence or absence of a sandeel fishery in the region. Carroll et al. (2017) showed that at Flamborough and Filey Coast Special Protection Area (SPA), the largest kittiwake colony in England, lower temperatures and lower fishing mortality on the Dogger Bank sandeel stock were positively associated with sandeel biomass, and higher sandeel biomass and lower fishing mortality were positively associated with kittiwake breeding success.

During the early 1990s, a sandeel fishery developed off the Firth of Forth, east Scotland. The landings from this fishery peaked at over 100,000 tonnes in 1993 but then subsequently fell (ICES 2017). The Firth of Forth area is important for breeding seabirds and the removal of such large quantities of sandeels within their foraging range became a matter of concern. In 1999, the UK called for a moratorium on sandeel fishing adjacent to seabird colonies along the UK coast and in response the EU requested advice from an ICES study group. In the absence of detailed sandeel stock assessment data for this part of the North Sea (which at the time was managed as part of the whole North Sea rather than as a separate stock), ICES noted that there was suggestion of a negative effect of the Firth of Forth fishery on the local sandeel abundance which coincided with a particularly low breeding success of seabirds, especially kittiwakes. The ICES study group concluded that there were reasons for continued concern about this area that provided the basis for a precautionary closure. Given available information, the study group proposed that kittiwake breeding success was the best practical indicator of sandeel availability, at least to seabirds, and threshold levels of the breeding success of this species should be used to guide future decisions. After the ICES acceptance of the study group's advice, the EU advised that the fishery should be closed from 2000 whilst maintaining a monitoring catch in the area. It should be stressed that this closure was based on kittiwake breeding success as a proxy for the condition of the local sandeel stock, so was to protect the sandeel stock and ecosystem as a whole. It was not specifically to protect breeding kittiwakes. Research (acoustic, trawl and dredge surveys) and commercial abundance indices indicated an increase in sandeel abundance during the period of the closure (ICES 2017). Dredge surveys indicated a significant decrease of total mortality of 1+ sandeels following the closure, and increasing catches taken from just outside the closed area in years when sandeel abundance in other parts of the North Sea was in decline, also indicated that the closure was effective in allowing sandeel abundance to recover (ICES 2017). Black-legged Kittiwake breeding success at the Isle of May and adjacent colonies benefited from the reduction in mortality of age 1+ sandeels, although benefits for other seabird species were less strong (Daunt et al. 2008). A further development of seabirds as indicators would be the establishment of reference points for predators in the management of North Sea sandeel stocks and other forage-fish fisheries (Hill et al. 2020).

Analysis of the remains in over 20,000 pellets regurgitated by Great Skuas (*Stercorarius skua*) at Foula, Shetland, from 1973 to 2017 showed that sandeels were the most common dietary item during the 1970s, but became scarce in the diet in the 1980s, increased slightly in the 1990s then became virtually absent from the 2000s onward (Church et al. 2018). The reduction in presence of sandeel otoliths in Great Skua pellets fits with the fishery-based assessment of a severe decline in Shetland sandeel stock abundance in the 1980s, slight recovery in the 1990s then even stronger reduction in the 2000s. The amount of sandeel in the great skua pellets correlates with the sandeel total stock biomass estimates, although with considerable scatter around the regression, which is likely to reflect high annual variability in discard availability which also influences Great Skua diet (Fig. C1.2).

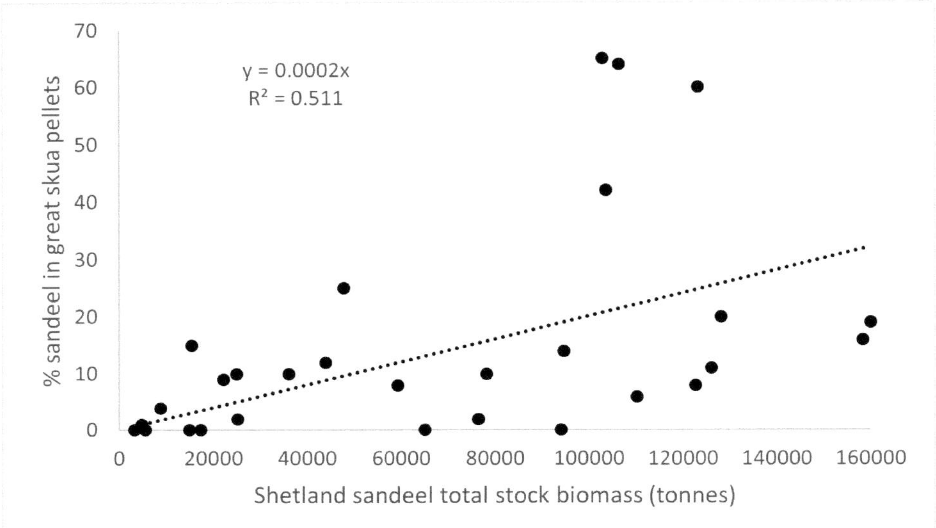

Figure C1.2. Percentage of pellets comprising sandeel bones and otoliths in Great Skua (*Stercorarius skua*) pellet samples from Foula, Shetland, in relation to Shetland sandeel total stock biomass estimates for 1976 to 2005.

Seabirds as indicators of fish stocks: the way forward

The abundant literature about seabirds as indicators of fish stocks decreased after 2010. The papers of Durant et al. (2009) and Grémillet and Charmantier (2010) suggest the main reasons for some disappointment with the concept of seabirds as quantitative indicators of fish stocks: (1) most studies advocating the use of seabirds as bioindicators were based on correlations. It is clear that prey abundance is important in explaining the productivity of seabirds and other animals, but the opposite is not necessarily true, and independent data to examine this is very scarce, particularly in addressing the variability and uncertainty of how seabird parameters respond to changes in food abundance. (2) Seabirds may show strong plasticity in responding to environmental changes such as changes in food supply, which may obscure the functional relationship between seabird parameters and fish stocks. Such plasticity may be particularly relevant in wide ranging seabirds such as the Cory's Shearwater (*Calonectris borealis*) in the North Atlantic, which change their foraging strategies in relation to prevailing oceanographic conditions (Paiva et al. 2010). Diet plasticity may also explain why a seabird species may be a good bioindicator for a certain area, and not for others. In the North Sea, Black-legged Kittiwake rely almost exclusively on sandeels, and their breeding success is a good indicator of the sandeel stock (Wanless et al. 2007); however in the Celtic Sea other prey may be available, making this species less reliable as a quantitative indicator of sandeel abundance for this area (Cook et al. 2014).

Therefore, the main advances in using seabird parameters as bioindicators have been along the lines of ecosystem-based marine management (Cury et al. 2008). The principle of "one-third for the birds" was advocated by the authors as an important management guideline in terms of ecosystem management, because it would preserve the integrity of predator-prey relationships (Cury et al. 2011). It is now well established that seabirds are more powerful as qualitative indicators to indicate a binary response (that signal changes from good to poor environmental conditions and vice versa) such as the changes in diet shown by analysis of Northern Gannet (*Morus bassanus*) regurgitations from 1977 to 2006 from temperate and sub-tropical prey such as Atlantic mackerel (*Scomber scombrus*) in 1977–1989, to low arctic prey such as capelin (*Mallotus villosus*) after 1990, which show a major shift in the pelagic food webs in Newfoundland, Canada (Montevecchi 2007).

Other important aspects have been the development of composed indices integrating information of different species and/or different parameters. Frederiksen et al. (2007) entered the breeding success (nº of fledged chicks per nest) of four sandeel-dependent seabird species on the Isle of May (European Shag *Phalacrocorax aristotelis*, Black-legged Kittiwake, Razorbill *Alca torda*, Atlantic Puffin *Fratercula arctica*) into a principal component analysis, and the first principal component (PC1) showed a significant correlation with an abundance index of larval sandeel in the previous spring: however PC1 explained only 29% of this variation, which was similar to the variation explained solely by the breeding success of the European Shag (32%) or the Black-legged Kittiwake (31%). This shows that these two species were the most important for the indicator. Frederiksen et al. (2007) also added the mean mass of near-fledged chicks of Common Guillemot and Atlantic Puffin to the principal component analysis and proposed the PC1 of these seven variables as a seabird indicator for the NW North Sea. Combining several species, often those that belong to the same ecological guild (Parsons 2008), into a single indicator may provide a better caption of the seabird relationships with forage fish for a larger spatial scale, and should be a good approach within the framework of ecosystem-based resource management (Jennings 2005). Nevertheless, the Black-legged Kittiwake is a very good example of a bioindicator of sandeel stocks. In fact, the breeding success of black-legged kittiwake in the North Sea is now considered a North Sea Ecological Quality Element by OSPAR (Heslenfeld and Enserink 2008).

Analysis of Great Skua pellets demonstrated major changes in diets of Great Skuas related with the availability of discarded ground fish from the trawl fishery (bottom-living fish species, such as haddock and whiting that are not otherwise available to surface-feeding seabirds such as Great Skuas). Proportions of different discarded fish species in Great Skua diet reflect very closely the amounts of those species being discarded in different years by the fishing fleet (Votier et al. 2004). However, in recent years, otoliths in pellets have included some of fish species that are no longer allowed to be discarded under the EU common fisheries policy landings obligation (Church et al. 2018). This suggests that analysis of pellets regurgitated by gulls and skuas could be used to monitor compliance (or lack of compliance) with fisheries regulations, as was first suggested by Maitland and Furness (2013).

The use of seabirds as quantitative indicators of fish stocks is virtually non-existent. However, information on diet, chick growth and breeding success of tern species is relatively easy and cost-effective to obtain and may provide indicators of juvenile fish abundance in real-time (McLeay et al. 2009) given the high linear responsiveness of plunge diving terns to food abundance. This means that measures of tern breeding success could be used to indicate pre-recruitment rates of forage fish. However, the limited foraging range of plunge diving tern species during the breeding season means they represent only a small-scale, but fish abundance within and beyond this small scale may be inter-related, in which case the surface-feeding terns would be indicators for a wider oceanic area (Danhardt and Becker 2011). Long-term data series on pelagic prey fish abundance and tern reproductive traits in several study systems throughout the world should provide strong evidence for the fact that reproductive traits (including multivariate indices considering several reproductive parameters) of surface feeding terns can be used as indicators of pelagic fish populations. This would be relevant in terms of fisheries management because the abundance of juvenile fish, the main prey of coastal seabird species, is usually not known and very expensive to obtain, except for specific important fishery stocks.

Acknowledgments

This work benefitted from the strategic program of MARE, financed by FCT (UIDB/04292/2020) through national funds. Vitor H. Paiva commented on an earlier version.

References cited

Anderson, D.W., F. Gress, K.F. Mais and P.R. Kelly. 1980. Brown pelicans as anchovy stock indicators and their relationships to commercial fishing. CalCOFI [California Cooperative Oceanic Fisheries Investigations] Report 21: 54– 61.

Anderson, H.B., P.G.H. Evans, J.M. Potts, M.P. Harris and S. Wanless. 2014. The diet of common guillemot *Uria aalge* chicks provides evidence of changing prey communities in the North Sea. Ibis 156: 23–34. doi: 10.1111/ibi.12099.

Arnott, S.A. and G.D. Ruxton. 2002. Sandeel recruitment in the North Sea: demographic, climatic and trophic effects. Mar. Ecol. Prog. Ser. 238: 199–210. doi: 10.3354/meps238199.

Barrett, R.T. and J.V. Krasnov. 1996. Recent responses to changes in stocks of prey species by seabirds breeding in the southern Barents Sea. ICES J. Mar. Sci. 53: 713–722. doi: 10.1006/jmsc.1996.0090.

Barrett, R.T., K.C.J. Camphuysen, T. Anker-Nilssen, J.W. Chardine, R.W. Furness, S. Garthe et al. 2007. Diet studies of seabirds: a review and recommendations. ICES J. Mar. Sci. 64: 1675–1691. doi: 10.1093/icesjms/fsm152.

Boyd, I.L., S. Wanless and C.J. Camphuysen (eds.). 2006. Top predators in marine ecosystems: Their role in monitoring and management. Cambridge University Press, Cambridge. doi: 10.1017/cbo9780511541964.

Cairns, D.K. 1987. Seabirds as indicators of marine food supplies. Biol. Oceanogr. 5: 261–271.

Carroll, M.J., M. Bolton, E. Owen, G.Q.A. Anderson, E.K. Mackley, E.K. Dunn et al. 2017. Kittiwake breeding success in the southern North Sea correlates with prior sandeel fishing mortality. Aquat. Conserv. 27: 1164–1175. doi: 10.1002/aqc.2780.

Catry, T., J.A. Ramos, I. Catry, D. Monticelli and J.P. Granadeiro. 2013. Inter-annual variability in the breeding performance of six tropical seabird species: Influence of life-history traits and relationship with oceanographic parameters. Mar. Biol. 160: 1189–1201. doi: 10.1007/s00227-013-2171-213.

Church, G.E., R.W. Furness, G. Tyler, L. Gilbert and S.C. Votier. 2018. Change in the North Sea ecosystem from the 1970s to the 2010s: Great skua diets reflect changing forage fish, seabirds and fisheries. ICES J. Mar. Sci. 76: 925–937. doi: 10.1093/icesjms/fsy165.

Cook, A.S.C.P., M. Dadam, I. Mitchell, V.H. Ross-Smith and R.A. Robinson. 2014. Indicators of seabird reproductive performance demonstrate the impact of commercial fisheries on seabird populations in the North Sea. Ecological Indicators 38: 1–11. doi: 10.1016/j.ecolind.2013.10.027.

Crawford, R.J.M. 2003. Influence of food on numbers breeding, colony size and fidelity to localities of swift terns in South Africa's Western Cape, 1987–2000. Waterbirds 26: 44–53. doi: 10.1675/1524-4695(2003)026[0044:iofonb]2.0.co;2.

Cubaynes, S., P.F. Doherty, Jr., E.A. Schreiber and O. Gimenez. 2011. To breed or not to breed: A seabird's response to extreme climatic events. Biol. Lett. 7: 303–306 doi: 10.1098/rsbl.2010.0778.

Cury, P.M., Y.-J. Shin, B. Planque, J.M. Durant, J.-M. Fromentin, S. Kramer-Schadt et al. 2008. Ecosystem oceanography for global change in fisheries. Trends Ecol. Evol. 23: 338–346. doi: 10.1016/j.tree.2008.02.005.

Cury, P.M., I.L. Boyd, S. Bonhommeau, T. Anker-Nilssen, R.J.M. Crawford, R.W. Furness et al. 2011. Global Seabird Response to Forage Fish Depletion-One-Third for the Birds. Science 334: 1703–1706. doi: 10.1126/science.1212928.

Dale, V.H. and S.C. Beyler. 2001. Challenges in the development and use of ecological indicators. Ecol. Indic. 1: 3–10. doi: 10.1016/s1470-160x(01)00003-6.

Danhardt, A. and P.H. Becker. 2011. Herring and sprat abundance predict chick growth and reproductive performance of Common Terns in the Wadden Sea. Ecosystems 14: 791–803. doi: 10.1007/s10021-011-9445-7.

Daunt, F., S. Wanless, S.P.R. Greenstreet, H. Jensen, K.C. Hamer and M.P. Harris. 2008. The impact of the sandeel fishery closure on seabird food consumption, distribution and productivity in the northwestern North Sea. Can. J. Fish Aquat. Sci. 65: 362–381. doi: 10.1139/f07-164.

Davis S.E., R.G. Nager and R.W. Furness. 2005. Food availability affects adult survival as well as breeding success of parasitic jaegers. Ecology 86: 1047–1056. doi: 10.1890/04-0989.

Durant, J.M., D.Ø. Hjermann, M. Frederiksen, J.B. Charrassin, Y. Le Maho, P.S. Sabarros et al. 2009. Pros and cons of using seabirds as ecological indicators. Mar. Ecol. Prog. Ser. 39: 115–129. doi: 10.3354/cr00798.

Einoder, L.D. 2009. A review of the use of seabirds as indicators in fisheries and ecosystem management. Fish. Res. 95: 6–13. doi: 10.1016/j.fishres.2008.09.024.

Frederiksen, M., M.P. Harris, F. Daunt, P. Rothery and S. Wanless. 2004. The role of industrial fisheries and oceanographic change in the decline of North Sea black-legged kittiwakes. J. Appl. Ecol. 41: 1129–1139. doi: 10.1111/j.0021-8901.2004.00966.x.

Frederiksen, M., P.J. Wright, M. Heubeck, M.P. Harris, R.A. Mavor and S. Wanless. 2005. Regional patterns of kittiwake *Rissa tridactyla* breeding success are related to variability in sandeel recruitment. Mar. Ecol. Prog. Ser. 300: 201–211. doi: 10.3354/meps300201.

Frederiksen, M., M. Edwards, A.J. Richardson, N.C. Halliday and S. Wanless. 2006. From plankton to top predators: Bottom up control of a marine food web across four trophic levels. J. Anim. Ecol. 75: 1259–1268. doi: 10.1111/j.1365-2656.2006.01148.x.

Frederiksen, M., R.A. Mavor and S. Wanless. 2007. Seabirds as environmental indicators: The advantages of combining data sets. Mar. Ecol. Prog. Ser. 352: 205–211. doi: 10.3354/meps07071.

Furness, R.W., S.P.R. Greenstreet and P.M. Walsh. 1996. Spatial and temporal variability in the breeding success of seabirds around the British Isles: Evidence for distinct sandeel stocks? ICES Coop. Res. Rep. 216: 63−65.

Furness, R.W. and K.C.J. Camphuysen. 1997. Seabirds as monitors of the marine environment. ICES J. Mar. Sci. 54: 726–737. doi: 10.1006/jmsc.1997.0243.

Furness, R.W. and M.L. Tasker. 2000. Seabird-fishery interactions: Quantifying the sensitivity of seabirds to reductions in sandeel abundance, and identification of key areas for sensitive seabirds in the North Sea. Mar. Ecol. Prog. Ser. 202: 253−264. doi: 10.3354/meps202253.

Furness, R.W. 2007. Responses of seabirds to depletion of food fish stocks. J. Ornithol. 148(Suppl 2): S247–S252. doi: 10.1007/s10336-007-0152-2.

Grémillet, D. and A. Charmantier. 2010. Shifts in phenotypic plasticity constrain the value of seabirds as ecological indicators of marine ecosystems. Ecol. Appl. 20: 1498–1503. doi: 10.1890/09-1586.1.

Harding, A.M.A., J.F. Piatt, J.A. Schmutz, M.T. Shultz, T.I. van Pelt, A.B. Kettle et al. 2007. Prey density and the behavioral flexibility of a marine predator: the common murre (*Uria aalge*). Ecology 88: 2024–2033. doi: 10.1890/06-1695.1.

Heslenfeld, P. and E.L. Enserink. 2008. OSPAR Ecological Quality Objectives: The utility of health indicators for the North Sea. ICES J. Mar. Sci. 65: 1392–1397. doi: 10.1093/icesjms/fsn159.

Hill, S.L., J. Hinke, S. Bertrand, L. Fritz, R.W. Furness, J.N. Ianelli et al. 2020. Reference points for predators will progress ecosystem-based management of fisheries. FishFish. 21: 368–378. doi: 10.1111/faf.12434.

Houde, E.D. 2009. Recruitment variability. pp. 91–171. *In*: Jakobsen, T., M.J. Fogarty, B.A. Megrey and E. Moksness [eds.]. Fish Reproductive Biology. Wiley-Blackwell Scientific Publications, Chichester, UK. doi: 10.1002/9781444312133.ch3.

ICES. 2017. Report of the benchmark workshop on sandeel (WKSand 2016). ICES CM 2016/ACOM: 3331. October - 4 November 2016, Bergen, Norway.

Jackson, J.B.C., M.X. Kirby, W.H. Berger, K.A. Bjorndal, L.W. Botsford, B.J. Bourque et al. 2001. Historical overfishing and the recent collapse of coastal ecosystems. Science 293: 629–637. doi: 10.1126/science.1059199.

Jennings, G., D.J. McGlashan and R.W. Furness. 2012. Responses to changes in sprat abundance of common tern breeding numbers at 12 colonies in the Firth of Forth, east Scotland. ICES J. Mar. Sci. 69: 572–577. doi: 10.1093/icesjms/fss022.

Jennings, S. 2005. Indicators to support an ecosystem approach to fisheries. Fish and Fisheries 6: 212–232. doi: 10.1111/j.1467-2979.2005.00189.x.

JNCC. 2020. Seabird population trends and causes of change: 1986–2018 report. JNCC, Peterborough, UK.https://jncc.gov.uk/our-work/smp-report-1986-2018.

Lorentsen, S.H., T. Anker-Nilssen and K.E. Erikstad. 2018. Seabirds as guides for fisheries management: European shag *Phalacrocorax aristotelis* diet as indicator of saithe *Pollachius virens* recruitment. Mar. Ecol. Prog. Ser. 586: 193–201. doi: 10.3354/meps12440.

Maitland, S. and R.W. Furness. 2013. Her bonxie boy. pp. 1–21. Chapter 1 in Moss Witch and other stories. Comma Press, Manchester UK.

McLeay, L.J., B. Page, S.D. Goldsworthy, T.M. Ward, D.C. Paton, M. Waterman et al. 2009. Demographic and morphological responses to prey depletion in a crested tern (*Sterna bergii*) population: Can fish mortality events highlight performance indicators for fisheries management? ICES J. Mar. Sci. 66: 237–247. doi: 10.1093/icesjms/fsn195.

Miller, J.A.O., R.W. Furness, M. Trinder and J. Matthiopoulos. 2019. The sensitivity of seabird populations to density-dependence, environmental stochasticity and anthropogenic mortality. J. Appl. Ecol. 56: 2118–2130. doi: 10.1111/1365-2664.13448.

Montevecchi, W.A. 1993. Birds as indicators of change in marine prey stocks. pp. 217–266. *In*: Birds as Monitors of Environmental Change. R.W. Furness and J.J.D. Greenwood [eds.]. Chapman and Hall, London. doi: 10.1007/978-94-015-1322-7_6.

Montevecchi, W.A. and R.A. Myers. 1995. Prey harvests of seabirds reflect pelagic fish and squid abundance on multiple spatial and temporal scales. Mar. Ecol. Prog. Ser. 117: 1–9. doi: 10.3354/meps117001.

Montevecchi, W.A. 2007. Binary dietary responses of northern gannets *Sula bassana* indicate changing food web and oceanographic conditions. Mar. Ecol. Prog. Ser. 352: 213–220. doi: 10.3354/meps07075.

Niemi, G.J. and M.E. MacDonald. 2004. Application of ecological indicators. Annu. Rev. Ecol. Evol. Syst. 35: 89–111. doi: 10.1146/annurev.ecolsys.35.112202.130132.

Olin, A.B., N.S. Banas, P.J. Wright, M.R. Heath and R.G. Nager. 2020. Spatial synchrony of breeding success in the black-legged kittiwake *Rissa tridactyla* reflects the spatial dynamics of its sandeel prey. Mar. Ecol. Prog. Ser. 638: 177–190. doi: 10.3354/meps13252.

Oro, D. and R.W. Furness. 2002. Influences of food availability and predation on survival of Kittiwakes. Ecology 83: 2516–2528. doi: 10.1890/0012-9658(2002)083[2516:iofaap]2.0.co;2.

Parsons, M., I. Mitchell, A. Butler, N. Ratcliffe, M. Frederiksen, S. Foster and J.B. Reid. 2008. Seabirds as indicators of the marine environment. ICES J. Mar. Sci. 65: 1520–1526. doi: 10.1093/icesjms/fsn155.

Paiva, V.H., P. Geraldes, I. Ramirez, A. Meirinho, S. Garthe and J.A. Ramos. 2010. Foraging plasticity of a pelagic seabird species foraging along a marine productivity gradient. Mar. Ecol. Prog. Ser. 398: 259–274. doi: 10.3354/meps08319.

Paiva, V.H., P. Geraldes, V. Marques, R. Rodríguez, S. Garthe and J.A. Ramos. 2013. Effects of environmental variability on different trophic levels of the North Atlantic food web. Mar. Ecol. Prog. Ser. 477: 15–28. doi: 10.3354/meps10180.

Pearson, T.H. 1968. The feeding biology of seabird species breeding on the Fame islands, Northumberland. J. Anim. Ecol. 37: 521–552. doi: 10.2307/3073.

Pedersen, S.A., P. Lewy and P.J. Wright. 1999. Assessments of the lesser sandeel (*Ammodytes marinus*) in the North Sea based on revised stock divisions. Fish. Res. 41: 2210241. doi: 10.1016/s0165-7836(99)00026-0.

Pereira, J.M., V.H. Paiva, FR. Ceia and J.A. Ramos. 2020. Facing extremes: Cory's shearwaters adjust their foraging behaviour differently in response to contrasting phases of North Atlantic Oscillation. Reg. Environ. Change 20: 77. doi: 10.1007/s10113-020-01662-1.

Perkins, A., N. Ratcliffe, D. Suddaby, B. Ribbands, C. Smith, P. Ellis et al. 2018. Combined bottom-up and top-down pressures drive catastrophic population declines of Arctic skuas in Scotland. J. Anim. Ecol. 87: 1573–1586. doi: 10.1111/1365-2656.12890.

Piatt, J.F., A.M.A. Harding, M. Schultz, S.G. Speckman, T.I. van Pelt, G.S. Drew et al. 2007. Seabirds as indicators of marine food supplies: Cairns revisited. Mar. Ecol. Prog. Ser. 352: 221–234. doi: 10.3354/meps07078.

Pikitch, E.K., C. Santora, E.A. Babcock, A. Bakun, R. Bonfil, D.O. Conover et al. 2004. Ecosystem-based fishery management. Science 305: 346–347. doi: 10.1126/science.1098222.

Proctor, R., P.J. Wright and A. Everitt. 1998. Modelling the transport of larval sandeels on the north-west European shelf. Fish Oceanogr. 7: 347–354. doi: 10.1046/j.1365-2419.1998.00077.x.

Ramos J.A., A.M. Maul, V. Ayrton, I. Bullock, J. Hunter, J. Bowler et al. 2002. The influence of local and large scale weather events and timing of breeding on tropical Roseate Tern reproductive parameters. Mar. Ecol. Prog. Ser. 243: 271–279. doi: 10.3354/meps243271.

Reid, K., J.P. Croxall, D.R. Briggs and E.J. Murphy. 2005. Antarctic ecosystem monitoring: Quantifying the response of ecosystem indicators to variability in Antarctic krill. ICES J. Mar. Sci. 62: 366–373. doi: 10.1016/j.icesjms.2004.11.003.

Stearns, S.C. 1992. The evolution of life histories. Oxford University Press, Oxford, UK.

Scopel, L.C., A.W. Diamond, S.W. Kress, A.R. Hards and P. Shannon. 2018. Seabird diets as bioindicators of Atlantic herring recruitment and stock size: A new tool for ecosystem-based fisheries management. Can. J. Fish. Aquat. Sci. 75: 1215–1229. doi: 10.1139/cjfas-2017-0140.

Steadman, DW. 2006. Extinction and biogeography of tropical Pacific birds. The University of Chicago Press, Chicago, USA.

vanDeurs, M., R. van Hal, M.T. Tomczak, S.H. Jónasdóttir and P. Dolmer. 2009. Recruitment of lesser sandeel *Ammodytes marinus* in relation to density dependence and zooplankton composition. Mar. Ecol. Prog. Ser. 381: 249–258. doi: 10.3354/meps07960.

Vedder, O., H. Zhang, A. Danhardt and S. Bouwhuis. 2019. Age-specific offspring mortality economically tracks food abundance in a piscivorous seabird. Am. Nat. 193: 588–597. doi: 10.1086/702304.

Veit, R.R. and N.M. Harrison. 2017. Positive interactions among foraging seabirds, marine mammals and fishes and implications for their conservation. Front. Ecol. Evol. 5: 121. doi: 10.3389/fevo.2017.00121.

Velarde, E., E. Ezcurra, M.A. Cisneros-Mata and M.F. Lavin. 2004. Seabird ecology, El Nin˜o anomalies, and prediction of sardine fisheries in the Gulf of California. Ecol. Appl. 14: 607–615. doi: 10.1890/02-5320.

Votier, S.C., R.W. Furness, S. Bearhop, J.E. Crane, R.W.G. Caldow, P. Catry et al. 2004. Changes in fisheries discard rates and seabird communities. Nature 427: 727–730. doi: 10.1038/nature02315.

Wanless, S., M. Frederiksen, F. Daunt, B.E. Scott and M.P. Harris. 2007. Black-legged Kittiwake as indicators of environmental change in the North Sea: evidence from long-term studies. Prog. Oceanogr. 72: 30–38. doi: 10.1016/j.pocean.2006.07.007.

Wanless, S., M.P. Harris, M.A. Newell, J.R. Speakman and F. Daunt. 2018. Community-wide decline in the occurrence of lesser sandeels *Ammodytes marinus* in seabird chick diets at a North Sea colony. Mar. Ecol. Prog. Ser. 600: 193–206. doi: 10.3354/meps12679.

Wright, P.J., A. Christensen, T. Régnier, A. Rindorf and M. van Deurs. 2019. Integrating the scale of population processes into fisheries management, as illustrated in the sandeel, *Ammodytes marinus*. ICES J. Mar. Sci. 76: 1453–1463. doi: 10.1093/icesjms/fsz013.

CHAPTER 2
Seabirds as Indicators of Oceanographic Changes

Vitor H. Paiva

Introduction

In a time of unprecedented environmental change, the identification of ecological parameters for ecosystem monitoring is important for the early detection of disruptive phenomena in marine food-webs. This is especially relevant, given the fast climate-driven changes in marine communities currently in place (Poloczanska et al. 2013, Poloczanska et al. 2016). Much of the increase in heat projected by the Intergovernmental Panel on Climate Change (IPCC), predictably around 3ºC more by 2040s, will be assimilated by the world's oceans, which will probably exert added stress on an already disturbed environment, because of many persistent human-related stressors (Gissi et al. 2020). Cascade bottom-up effects and potential disruptions on marine food webs can then be predicted with measurable effects at all levels, including top-predators such as seabirds (Lynam et al. 2017). Thus, seabirds can be studied to deepen our understanding of the effects of climate change on the marine environment, and be used as ecological indicators and sentinels of marine ecosystems' health (Parsons et al. 2008, Hazen et al. 2019). During the last decade, an impressive number of manuscripts have built evidence for the great potential of seabird ecological parameters as indicators of oceanographic changes (Grémillet and Boulinier 2009, Oro 2014). These include, but are not restricted to, shifts in distribution (Louzao et al. 2013), reproductive parameters (Divoky et al. 2015), phenology (Descamps et al. 2019) or demographic traits (Jenouvrier et al. 2018) (Table C1.1).

Using seabirds as ecological indicators of oceanographic changes implies the accurate identification of specific oceanographic processes, to which there is a 'functional response' by seabirds' measures such as reproductive parameters (Grémillet and Charmantier 2010, Ramos and Furness 2022). Changes in seabird ecology parameters can be (1) an indirect response to changes in prey distribution and/or availability, which was conditioned by shifts in oceanographic processes or (2) a direct response to climate or oceanographic shifts, which altered the conditions for optimal foraging at-sea and may ultimately affect their survival. Such processes might be framed by shifts in large-scale climatic indices or local-scale oceanographic variables, with the former influencing the

University of Coimbra, MARE—Marine and Environmental Sciences Centre, Department of Life Sciences, 3004-517 Coimbra, Portugal.
Email: vitorpaiva@ci.uc.pt

later. Most immediate seabird responses to severe oceanographic conditions and local food scarcity will necessarily be shifts in foraging distribution in search of alternative prey patches, possibly breeding failure or skipping breeding on a particular year with very poor resource conditions. Responses in phenology (e.g., timing of breeding) or demographic traits (e.g., population growth) will necessarily occur later in time (Fig. C2.1).

Because seabirds are long-lived organisms, they can accommodate a certain extent of inter-annual environmental stochasticity without a direct and linear response, which precludes their immediate use as quantitative ecological indicators (Durant et al. 2009, Grémillet and Charmantier 2010). Yet, when the environmental pressure is too severe or perseverant, it might cross the seabirds' resilience tipping point and induce a response. Therefore, seabirds will at least be sentinels of ecological changes. Seabird responses will always aim at maximising the individual probability of survival, which ultimately translates into increased population survival. Because oceanographic changes are likely to have a more direct impact on prey availability to seabirds, they should have a stronger influence on reproductive parameters than on adult survival (Sandvik et al. 2005).

In the following paragraphs, the reader will gain a broad perspective of how different seabird ecological parameters can act as biological indicators of large- and small-scale oceanographic changes. He will also learn about the non-linear and sometimes site-specific response of seabird species/populations to oceanographic shifts and how individual traits, such as age or sex, can act as drivers of variation in seabird responses to those shifts (Fig. C2.1). Studies used to illustrate such ecological functional responses relied on the last ten years of publications and on individual studies

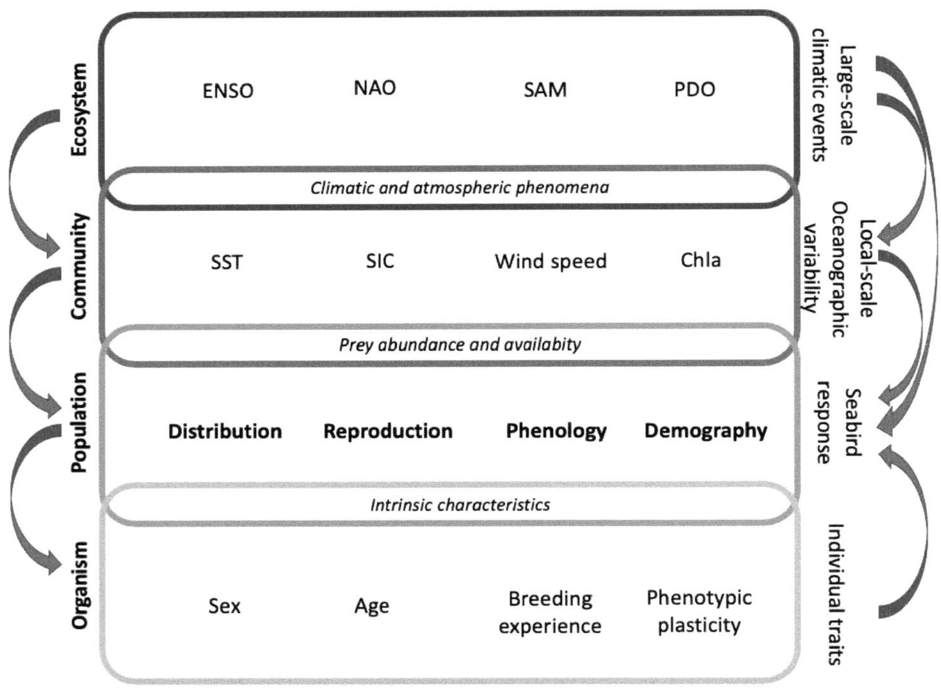

Figure C2.1. Diagram of relationships between the different hierarchical levels through which pelagic seabirds' ecological responses can be used as indicators of oceanographic changes. Large-scale climatic events occurring at the ecosystem level of ecological systems, may directly influence seabirds and are drivers of local-scale oceanographic variability, occurring at the community level. These influence seabird responses at the population level. Individual traits can influence seabird responses at the level of the 'organism'. ENSO—El Niño Southern Oscillation index; NAO—North Atlantic Oscillation index; SAM—Southern Annular Mode; PDO—Pacific Decadal Oscillation; SST—Sea Surface Temperature; Chla—Chlorophyll a concentration; SIC—Sea Ice Concentration.

based on at least ten years of data collection, to reduce biases introduced through inter-annual variability, covering a wide range of pelagic seabird study species inhabiting diverse ocean basins.

Indirect influence of environmental variability

Seabird evolution took place in a mostly stochastic realm, where environmental and oceanographic processes are often unpredictable. There are multiple examples in the literature of indirect associations between seabird ecological responses and changes in the marine environment (Poloczanska et al. 2013, Sydeman et al. 2015). Shifts in oceanographic patterns can create systemic effects through marine food webs with amplified responses at each trophic level (Möllmann et al. 2008). A key mechanistic way in which climate stochasticity may affect seabirds is through spatio-temporal mismatch with prey availability, conditioned by oceanographic processes (Champion et al. 2021) and energy demands at the peak of their breeding season (Shultz et al. 2009, Hipfner et al. 2020). While some pelagic seabird species might have the phenotypical plasticity to adapt to these shifts in prey availability, others have less capacity to buffer such mismatches, by advancing or delaying their reproductive period accordingly (Descamps et al. 2019, Merkel et al. 2019).

Large-scale climatic events

El Niño Southern Oscillation (ENSO)

The well-known El Niño Southern Oscillation (ENSO) frames the intermittent shifts in oceanographic and atmospheric conditions responsible for setting large-scale atmospheric pressure, wind and currents intensities and sea level height (Timmermann et al. 2018). The ENSO cycle develops first in the Central and East-Central Equatorial Pacific Ocean and causes global changes in temperature and precipitation. A positive ENSO value (El Niño), characterized by high air pressure in the western Pacific and low air pressure in the eastern Pacific, depicts oceanographic conditions characterizsed by Sea Surface Temperatures (SST) warmer than usual, sometimes leading to the disappearance of prey-fish and ultimately the collapse of the entire food-chain (Barber and Chavez 1983, Ballance et al. 2006). Negative ENSO values (La Niña) frame oceanographic regimes characterised by SST colder than usual and thus very productive years for marine ecosystems and food webs. Yet, unusually colder water regimes can alter the timing, movement and availability of pelagic fish and squid to seabirds. Until recently, El Niño events had a periodicity of two to seven years, but because of global climate change they are becoming more frequent and aperiodic. Thus, marine species might not have the ability to buffer and adapt to the impacts of these severe climatic events. An increasing ENSO value depicts climatic effects on the ecosystem which can go from mild to catastrophic, with many seabirds dying from starvation because of prey scarcity (Brainard et al. 2018).

El Niño events have been reported to affect the breeding performance of tropical seabirds in the Indian Ocean (Surman et al. 2012), Red-footed Boobies (*Sula sula*) in the tropical Atlantic Ocean (Cubaynes et al. 2011) or to influence the demographic parameters of Northern (*Macronectes halli*) and Southern (*M. giganteus*) Giant Petrels in the Southern Ocean (Gianuca et al. 2019) and European Storm Petrels (*Hydrobates pelagicus*) in the Atlantic Ocean (Matović et al. 2017). In years of higher ENSO values, Laysan (*Phoebastria immutabilis*) and Black-footed (*P. nigripes*) Albatrosses engaged in longer and farther foraging excursion over the North Pacific Ocean. Energetic costs of such increased foraging effort translated into a significant decrease in reproductive success (Thorne et al. 2015). Still in the North Pacific, Cassin's Auklets (*Ptychoramphus aleuticus*) shifted from an energetically richer copepod-plus-euphausiids diet to an impoverished larval fish diet in response to a decrease in ENSO. Such dietary shift was reflected on their chick's condition, which attained a comparably lower mass at 25 days of age (Hipfner et al. 2020). In the Southern Atlantic, the intensification of the ENSO, increased the mortality of Manx Shearwaters (*Puffinus puffinus*), through the enhancement of local storm activity. Adverse climatic conditions can be highly deleterious for seabirds and even sporadic storm events can cause seabird wrecks, with exacerbated negative effects on population trends (Tavares et al. 2020).

Although the ENSO phenomenon seems to have a more pronounced effect in tropical regions, especially in the South Pacific Ocean, it is also experienced in other ocean basins, such as the South Atlantic, Indian or Southern Oceans (Doney et al. 2012). As oceans are globally inter-connected, the occurrence of El Niño events at the southern hemisphere may be, to some extent, depicted at the northern hemisphere by other climatic indices, such as the Pacific Oscillation (PDO) index, Artic Oscillation (AO) index or the North Atlantic Oscillation (NAO) index.

The North Atlantic Oscillation (NAO)

The NAO index frames a north-south shift in atmospheric mass (pressure) between the Arctic and the subtropical Atlantic. This is known to influence patterns of wind intensity and direction, precipitation, air and sea temperature and current strength and upwelling intensity (Stenseth et al. 2003). In the North Atlantic region, a positive NAO index is characterised by prevailing westerly winds moving northwards, with increased frequency of stormy weather and precipitation both at-sea and inland. Contemporaneously, the southern area of the North Atlantic and Mediterranean Sea should experience the opposite climatological and oceanographic conditions, with fewer storms and dry weather. During a negative phase of NAO, there is a shift in these patterns, with the northern section of the Atlantic and Arctic Oceans experiencing fewer storms (Stenseth et al. 2002). Variability in NAO has been shown to be correlated with changes in local oceanographic regimes, such as values of SST, patterns of marine productivity, abundance and availability of prey, and subsequently with various seabird measures (Oro 2014). Yet, there is also some evidence supporting the idea that the NAO might not always be correlated with trends in SST and for instance, SST explained most of the variation in adult survival of Common Guillemot (*Uria aalge*) in the Arctic Ocean (Sandvik et al. 2012).

The NAO index has been increasing and in a positive phase since the 1970s, although with some years of negative values. For instance, in the winter of 2009–2010, the NAO reached a record-breaking negative value (Osborn 2011), which within the Portuguese current was characterised by low SST and an abnormally intense upwelling. These conditions forced plankton and planktivorous fish larvae to offshore areas, inducing high mortality and very low recruitment into prey-fish stocks a few months later (Santos et al. 2007). As a reflexion of those drastic changes and a likely decrease in prey-fish availability in the colony surroundings, chick provisioning Cory's Shearwaters (*Calonectris borealis*) responded by increasing their foraging range and effort. This in time, decreased chicks' body condition and overall fledging and breeding success in that particular year (Paiva et al. 2013). A similar distributional response was reported for a Cory's shearwater population breeding at Madeira archipelago (Avalos et al. 2017), while a disparate response observed in Cory's Shearwaters breeding in the Azores, where birds decreased their foraging range and increased their breeding success (Pereira et al. 2020). This highlights the importance of understanding the mechanistic effect of large-scale indices taking place at regional and local scales, because they might be influenced locally by different oceanographic processes and thus induce contrasting behavioural responses by diverse populations of the same species (Parsons et al. 2008).

In literature, studies on the response of pelagic seabirds to large-scale climate indices are dominated by the El Niño Southern Oscillation (ENSO) and the North Atlantic Oscillation (NAO) indices, although several other more regional indices have been used to frame large-scale oceanographic conditions driving seabird responses (Table C2.1). Often, large-scale climatic indices do not seem to have a direct relation with seabird measures. In the Barents Sea region, Arctic Ocean, researchers found that inter-annual variations (30-year period) in population numbers of Common

Guillemots did not correlate with NAO index values. Instead, demographic trends seemed to respond to low-pressure systems in place at the region, which were a source of heat transport into the region and forcing upwelling mixing in the ocean. These oceanographic changes created favourable conditions for successful breeding and survival of the study species (Mesquita et al. 2015).

Local-scale oceanographic variability

Changes in SST have been largely used as a local proxy of marine productivity in investigating seabird-environmental interactions (Tremblay et al. 2009). Increasing SST has been shown to have a negative effect on the foraging distribution and abundance of murres (*Uria* spp.) (Kuletz et al. 2020) and Laysan Albatrosses (*Phoebastria immutabilis*) (Thorne et al. 2015) in the Pacific Ocean and demographic parameters of Wandering Albatrosses (*Diomedea exulans*) (Fay et al. 2015) and Snow Petrels (*Pagodroma nivea*) (Barbraud et al. 2011) in the Southern Ocean, or on the breeding success of Black-legged Kittiwake (*Rissa tridactyla*) (Carroll et al. 2015) and Red-footed Booby (*Sula sula*) (Cubaynes et al. 2011) in the Atlantic Ocean (Table C2.1). Nevertheless, a recent global meta-analysis of 209 phenological time series from 145 seabird breeding populations showed that seabird populations worldwide have not adjusted their breeding seasons since the 1950s, neither to global warming nor to shifts in SST. Still, resident seabird species (e.g., boobies) exhibited a marked between-year variation in timing of breeding. Limited plasticity of seabirds' reproductive timing in response to changes in SST, may turn these marine top predators susceptible to potential future mismatches with lower-trophic-level resources (Keogan et al. 2018).

Ocean warming seems also to influence the seabirds' trophic ecology, with Antarctic (*Pachyptila desolata*) and Broad-billed (*P. vittata*) Prions markedly decreasing the nitrogen isotopic signature in their feathers over the last 87 years. This decrease in trophic level seems to depict long-term decreases in marine productivity (Grecian et al. 2016). In the same line, Sooty Terns (*Onychoprion fuscatus*) from Ascension Island showed a steep decrease in both carbon and nitrogen isotopic values of their feathers over the past 125 years. This being indicative of a profound shift to forage at more pelagic environments, and on comparatively lower trophic level prey. Both shifts were found to be related to a (1) a notorious trend in ocean warming in the tropical Atlantic region, and (2) tuna's overfishing by industrial fishing fleets in the post-industrial era (Reynolds et al. 2019).

Yet, some studies report that the influence of increasing SST in adult survival of a particular seabird species, can either be positive or negative depending on the location of the breeding colony (Sandvik et al. 2005). Contrasting with results from polar and temperate regions, increasing SST was found to be beneficial for adult survival of Audubon's Shearwater (*Puffinus lherminieri*) in the tropical Atlantic Ocean. Indeed, in that area a higher discharge from the Amazon River is responsible for the enhancement of both SST and sea surface chlorophyll *a* concentration (Chla), which enriches marine productivity and prey availability for seabirds. This highlights the need for more studies on the effect of oceanographic changes on seabirds from tropical regions (Precheur et al. 2016).

Other local-scale oceanographic variables being used to interpret changes in seabird measures include, but are not restricted to, Chla (Krüger et al. 2017), Sea Surface Height (SSH; Louzao et al. 2013) or sea ice concentration (SIC; Ramírez et al. 2017) (see Table C2.1). In the Arctic Ocean, increasing SIC was found to positively influence Thick-billed Murres (*Uria lomvia*), which laid energy-denser eggs. In the same area Northern Fulmars (*Fulmarus glacialis*) seemed not to be influenced by changes in SIC (Cusset et al. 2019). However, the retreat in ice extension may also create novel foraging opportunities, by providing prey that would otherwise be inaccessible to seabirds, under the ice sheets (Grémillet et al. 2015, Amélineau et al. 2019). A decrease in SIC was found to drive an increase in body condition and chick growth and an anticipation of the hatching date of Little Auks (*Alle alle*) (Amélineau et al. 2019).

These array of seabird responses to the same oceanographic trend, highlights the relevance of understanding the mechanisms behind the response, and the caution when selecting a seabird species

Table C2.1. Selected last ten years publications (since 2010), reporting seabird ecological responses (i.e., bioindicator parameters) to oceanographic changes. We selected subsets of observations which span ten years or more, reducing biases introduced through inter-annual variability and studies in which the authors clearly/directly describe a relationship between an oceanographic process and a seabird functional response. The omission of an oceanographic process/variable on the driver column, implies that there was no reported effect of such variables on a seabird-related parameter. SSTA—Sea Surface Temperature Anomaly; Chla—Sea Surface Chlorophyll *a* concentration; SIOD—Subtropical Indian Ocean Dipole; SAM—Southern Annular Mode; ENSO—El Niño-Southern Oscillation; NAO—North Atlantic Oscillation index; PDO—Pacific Decadal Oscillation; NPGO—North Pacific Gyre Oscillation; SAM—Southern Annular Mode; SIC—sea ice concentration, MER—meridional wind speed; NPI—North Pacific Index; PDO—Pacific Decadal Oscillation; AGMT—annual global mean temperature; SGI—subpolar gyre index; WeMOi—Western Mediterranean Oscillation index; SOI—Southern Oscillation Index; AO—Arctic Oscillation index; ALPI—Winter Aleutian Low Pressure Index; ONI—Oceanic Niño Index.

Species	Ocean basin	Study Years (Period)	Driver	Response in at-sea distribution and abundance	Reference
Little Auks (*Alle alle*)	Arctic	34 (1979–2013)	Decrease in SIC	Shorter foraging distances, low foraging effort, equal chick growth rates, slight decrease in body condition	(Grémillet et al. 2015)
Thirty-six seabird species	Arctic	37 (1975–1981 and 2007–2012)	Long-term earlier SIC retreat	Historical, dominated by Black-legged Kittiwakes (*Rissa tridactyla*) and murres (*Uria* spp.); Current, dominated by Crested Auklets (*Aethia cristatella*) and Short-tailed Shearwaters (*Puffinus tenuirostris*)	(Gall et al. 2017)
Northern Gannets (*Morus bassanus*)	North Atlantic	11 (2006–2017)	Decrease in SST and Chla, and increase in thermal fronts strength	Increase in foraging probability; females foraged further and for longer than males	(Clark et al. 2021)
Six Alcid species	North Atlantic	57 (1954–2011)	Higher NAO	Higher abundance of Atlantic Puffin (*Fratercula arctica*) and Black Guillemot (*Cepphus grylle*); Lower abundance of Little Auk (*Alle alle*) and Thick-billed Murre (*Uria lomvia*)	(Veit and Manne 2015)
Murres (*Uria* spp.), Auklets (*Aethia* spp.), Short-tailed Shearwaters (*Ardenna tenuirostris*)	North Pacific	12 (2007–2019)	Higher SST, lower SIC	Abundance of murres decreased; Planktivorous auklet density increased; Short-tailed shearwaters expanded, with nearly twice the density of the previous decade	(Kuletz et al. 2020)
Thirty-four seabird species	North Pacific	39 (1975–2014)	Lower SST and higher (current) velocity	Higher species richness and total abundance	(Santora et al. 2018)
Laysan (*Phoebastria immutabilis*) and Black-footed (*P. nigripes*) Albatrosses	North Pacific	10 (2003–2012)	Higher SST, ENSO and NPGO	Longer and farther foraging excursions; Lower reproductive success	(Thorne et al. 2015)
Seven albatross and petrel species	Southern	15 (1996–2011)	Projected increasing SST, decreasing Chla, wind speed and SIC (2050 and 2100)	General poleward shifts and range contraction, increased under the projected scenarios of increasing SST	(Krüger et al. 2017)

Wandering Albatross (*Diomedea exulans*)	Southern	43 1958–2001	Higher sea surface height	Increasingly poleward foraging distribution	(Louzao et al. 2013)
Wandering Albatross (*Diomedea exulans*)	Southern	20 (1990–2010)	Increasing wind speed and poleward direction	Decrease in trip duration decreased, increase in breeding success and adult body mass	(Weimerskirch et al. 2012)
King Penguin (*Aptenodytes patagonicus*)	Southern	16 (1992–2008) Projected until 2100	Increasing SST	A twice fold increase in foraging distances; Poleward shift	(Péron et al. 2012)
Twelve pelagic seabird species	Southern	26 (1981–2007)	Increasing SST	Southward shift of wandering albatross and prions;White-chinned Petrel distribution shifted northward - more complex mechanisms.	(Péron et al. 2010)
Species	**Ocean basin**	**Study Years (Period)**	**Driver**	**Response in breeding phenology and reproductive parameters**	**Reference**
Thirty seabird species	Arctic	35 (1982–2016)	Higher SST in the colony surroundings	Earlier hatching date for surface-feeders from populations at the Pacific Arctic region	(Descamps et al. 2019)
Little Auks (*Alle alle*)	Arctic	12 (2004–2015)	Decrease in SIC	Increase in adult body condition and chick growth rate; Decrease in hatching date	(Amélineau et al. 2019)
Brünnich's Guillemot (*Uria lomvia*), Black-legged Kittiwake (*Rissa tridactyla*), Little Auk (*Alle alle*)	Arctic	11 (2006–2016)	Increasing temporal lag between Chla and max SIC	Decrease in chick survival at 15 days; Hatching success not affected by the lag	(Ramírez et al. 2017)
Black Guillemot (*Cepphus gryllemandtii*)	Arctic	37 (1975–2012)	Increasing SST, decreasing SIC	Five-fold increase in nestling starvation, reduction in nestling growth and fledging mass	(Divoky et al. 2015)
Black-legged Kittiwakes (*Rissa tridactyla*) and Thick-billed Murres (*Uria lomvia*)	North Pacific	36 (1975–2010)	Decreasing SST	Increase in laying and overall breeding success	(Renner et al. 2014)
Rhinoceros Auklet (*Cerorhinca monocerata*), Japanese Cormorant (*Phalacrocorax filamentosus*), Black-tailed Gulls (*Larus crassirostris*)	North Pacific	25 (1984–2009)	Increasing SST	No change in the timing of breeding, breeding success and diet composition	(Watanuki and Ito 2012)
Least, Whiskered and Crested Auklets (*Aethia pusilla, A. pygmaea and A. cristatella)*	North Pacific	18 (1990–2008)	a) Increasing ALTI; b) Increasing SST; c) Increasing sea level pressure	a) Decrease in reproductive success; b) Increase in auklet productivity at the western North Pacific and Bering Sea and decrease in the Gulf of Alaska; c) Decrease in productivity	(Bond et al. 2011)

Table C2.1 contd. ...

...Table C2.1 contd.

Leach's Storm Petrels (*Hydrobates leucorhous*)	North Atlantic	55 (1955–2010)	Increase in AGMT	Increase in hatching success until a tipping AGMT value, decrease from that value onwards; Impact of AGMT higher on inexperienced breeders	(Mauc et al. 2018)
Black-legged Kittiwake (*Rissa tridactyla*)	North Atlantic	19 (1986–2004)	Higher SST	Lower breeding success in 11 colonies	(Carroll et al. 2015)
Black-legged Kittiwake (*Rissa tridactyla*)	North Atlantic	17 (1991–2007)	Increasing SST	No change in seabird productivity	(Lauria et al. 2013)
Lesser Noddy (*Anous tenuirostris)*,Brown Noddy (*Anous stolidus*), Sooty Tern (*Onychoprion fuscata*), Wedge-tailed Shearwater (*Ardenna pacifica*)	Tropical Indian	19 (1991–2010)	a) Higher SST; b) Higher sea level and SOI	a) Earlier first eggs laid; b) higher breeding success	(Surman et al. 2012)
Red-footed Booby (*Sula sula*)	Tropical Atlantic	19 (1983–2002)	Lower ENSO, higher SST	a) Experienced breeders skipping breeding; b) higher likelihood of pre-breeders to engage in first breeding attempts	(Cubaynes et al. 2011)
Fairy Prions (*Pachyptila turtur*)	South Pacific	11 (2001–2011)	Higher SOI and SST	a) Higher wing length at fledging; b) no change in breeding performance	(Manno et al. 2014)
Antarctic Petrel (*Thalassoica antarctica*)	Southern	30 (1985–2015)	Decrease in SOI	Decrease in breeding productivity and adult survival and delayed timing of hatching	(Descamps et al. 2016)
Seven Antarctic seabird species	Southern	31 (1982–2013)	Drastic decrease in SIC	Decrease in breeding success of all species, with a total failure for two of them	(Barbraud et al.2015)
Species	**Ocean basin**	**Study Years (Period)**	**Driver**	**Response in demography**	**Reference**
Common Guillemot (*Uria aalge*)	Arctic	31 (1980–2011)	Higher mean sea level pressure	Increase in population numbers	(Mesquita et al. 2015)
Northern Rockhopper Penguins (*Eudyptes moseleyi*)	South Indian	22 (1994–2016)	Increasing SAM, SIOD and SSTa	Females decreased body condition	(Delord et al. 2021)
Northern (NGP, *Macronectes halli*) and Southern (SGP, *M. giganteus*) Giant Petrels	South Atlantic	13 (2001–2014)	a) Increase in wind speed; b) Increase in ENSO and decrease in SIC; c) Increase in SAM	a) Decrease in survival for male NGP but increase for female NGP; b) Increase in female survival; c) increase in male SGPs survival	(Gianuca et al. 2019)

Black-legged (*Rissa tridactyla*) and Red-legged Kittiwakes (*R. brevirostris*), Common (*Uria aalge*) and Thick-billed Murres (*U. lomvia*)	North Pacific	40 (1974–2014)	Increase in SST and NPI, decrease in PDO	Decrease in population numbers; Kittiwakes, best indicator of oceanographic shifts; Murres, resilient to environmental change	(Goyert et al. 2018)
Cassin's Auklet (*Ptychoramphus aleuticus*)	North Pacific	26 (1980–2005)	Increasing SST	Decrease in survival probability and breeding success	(Wolf et al. 2010)
Brünnich's Guillemot (*Uria lomvia*)	North Atlantic	25 (1986–2011)	Higher SGI	Higher adult survival	(Fluhr et al. 2017)
European Shag (*Phalacrocorax aristotelis*)	North Atlantic	20 (1992–2012)	a) Positive NAO; b) Negative NAO	a) Fledgling sex ratio female skewed;b) fledgling sex ratio male skewed	(Barros et al. 2013)
European Storm Petrels (*Hydrobates pelagicus*)	Mediterranean Sea and Atlantic	12 (2002–2014)	Higher WeMOi and ENSO	Higher mortality of Mediterranean breeders' due to higher costs of migration, especially during La Niña years	(Matović et al. 2017)
Blue-footed Booby (*Sula nebouxii*)	Tropical Pacific	23 (1991–2013)	Occurrence of Hurricanes	Increase in adult survival and recruitment after hurricane events	(Ancona et al. 2017)
Round Island Petrel (*Pterodroma arminjoniana*)	Tropical Indian	20 (1993–2012)	Occurrence of tropical cyclones in a) colony surroundings or b) non-breeding areas	a) Negative impact on juvenile (first year) individuals; b) positive impact on juveniles	(Nicoll et al. 2017)
Audubon's Shearwater (*Puffinus lherminieri*)	Tropical Atlantic	19 (1995–2014)	Higher SST and Chla	Higher adult survival	(Precheur et al. 2016)
Westland Petrel (*Procellaria westlandica*)	Tropical Pacific	44 (1970–2014)	Higher SSTA	a) Lower adult survival during the breeding stage; b) higher adult survival during the non-breeding phase	(Waugh et al. 2015)
Black-browed Albatross (*Thalassarche melanophris*)	South Indian	33 (1982–2015)	Increasing SST	Lower juvenile survival causing lower population growth; little effect on migratory schedule and breeding performance	(Jenouvrier et al. 2018)
Wandering Albatross (*Diomedea exulans*)	Southern	50 (1962–2012)	a) Increasing SSTA; b) increasing SAM	a) Decrease in juvenile survival; b) increase in early recruitment, respectively	(Fay et al. 2015, Fay et al. 2017)
Amsterdam (*Diomedea amsterdamensis*) and Black-browed Albatrosses (*Thalassarche melanophris*) and Snow Petrels (*Pagodroma nivea*)	Southern	39 (1968–2007)	a) Increasing SST, b) decreasing SIC	a) Steep projected declines for southern populations; b) northern populations little affected by projected ocean warming	(Barbraud et al. 2011)

and/or life history parameter as an ecological indicator of changes in a specific oceanographic variable (Oro 2014, Grémillet and Charmantier 2010). Besides, the multiple trophic links between a specific oceanographic condition and the response of a given seabird measure, means it may be challenging to establish a direct and strong connection between the two extremities of the ecological response string. Changes in oceanographic conditions might also add-up to the low prey availability, by enhancing the energetic costs and difficulty of locating suitable alternative prey, for instance when facing stormy winds or oceanic turbulence.

Direct effects of environmental variability

Direct impacts of environmental variability on seabirds have been greatly under-studied when compared with indirect effects. For instance, extreme weather/oceanographic events at-sea may cause death of large seabird numbers (Tavares et al. 2016, Louzao et al. 2019). Increasing periodicity of such events is a motive of foremost concern for the survival of seabird populations, and especially for the conservation of threatened species. In the winter of 2013/2014 the storms hitting the Atlantic coastal area were unusually harsh. As a result, a total of ~ 55000 seabirds were washed ashore, most found dead with signs of starvation and exhaustion (Morley et al. 2016). During the winter of 2014/2015, thousands of Cassin's Auklets (*Ptychoramphus aleuticus*) were found dead on beaches of the Pacific coast with clear signs of starvation. This was attributed to a drastic change in oceanographic regimes (sea water heat wave) causing a shift in zooplankton availability (Jones et al. 2018). Although intermittent, these seabird wreck events give us a glimpse of the exacerbated impact they may have on seabird populations, given the predictable increase in periodicity of extreme oceanographic events (Lorenzo and Mantua 2016). Many seabird species can be negatively affected by the increased periodicity of extreme wind conditions (Louzao et al. 2019), with some species, like the endangered Black-capped Petrel (*Pterodroma hasitata*), at risk of extinction because of enhanced hurricane activity (Hass et al. 2012). Prevalent stormy conditions can induce changes in optimal nesting sites, with measurable effects on the breeding success of seabird species. For example, 22% of the monitored Herring Gull (*Larus argentatus*) nests in a study in Maine, USA, were lost because of a single storm in June 2012 (Bonter et al. 2014).

Sea-level rise, caused by global warming, may flood breeding colonies at coastal sites, such as those of terns, boobies or tropicbirds (Hatfield et al. 2012), thus affecting their breeding success and sometimes adult survival. This can be particularly negative for populated seabird colonies in tropical regions, which frequently breed in shallow islets, overcrowded with multiple seabird species and occupying all available space until the high tide line. Projected sea-level rise can thus create ecological traps, especially for tropical seabirds. An anticipated planning of potential climate change effects and installation of higher elevation seabird refugia should help mitigate the issue (Reynolds et al. 2015).

Seabird responses to oceanographic changes

Shifts in at-sea distribution/abundance and migration patterns

The interest in the role of oceanographic changes in driving distribution changes of pelagic seabirds has increased recently (Sydeman et al. 2015, Sherley et al. 2017, Kuletz et al. 2020). In fact, this is unsurprisingly one of the most immediate seabird responses to a decrease in food availability, driven by oceanographic shifts. Between 1990–2010, westerly winds of the Southern Ocean have increased in intensity and moved poleward. In response to this, Wandering Albatrosses (*Diomedea exulans*) shifted their foraging distribution polewards, increasing their travelling distance and flight speeds. Curiously, the use of such less energetically costly wind fields, led to a decrease in trip duration, increases in breeding success and body mass by over 1 kg (Weimerskirch et al. 2012). The use of favourable wind conditions for a facilitated low-cost flight has also been observed in seven other albatross and petrel species of the Southern Ocean (Krüger et al. 2017), as well as on Balearic Shearwaters (*Puffinus mauretanicus*) foraging over the Mediterranean Sea (Afán et al. 2021).

Poleward distributional shifts, driven by wind shifts, warming sea waters or changes in marine productivity have been observed in other seabird species (Péron et al. 2012, Quillfeldt et al. 2010). In the North Pacific, a predicted increase in SST of 1 to 6ºC by 2100 should lead to a northward 35% change in core habitat of some albatross species (and other marine top predators). For species already under environmental stress, increased migration duration and loss of suitable pelagic foraging habitat may accelerate population declines in the future (Hazen et al. 2012). Twenty years of at-sea census surveys of 12 albatross and petrel species in the Southern Indian Ocean, documented that those changes in distribution and decreases in abundance were higher for species using subtropical waters, where ocean warming was more pronounced, while species using the southern region remained more constant over time, despite the tendency for poleward movements (Péron et al. 2010).

In a recent study about the effect of ocean warming on the foraging distribution of seven seabird species of the North Pacific Ocean, researchers found disparate species-specific responses of the seabird community inhabiting the area. For instance, while piscivorous murres (*Uria* spp.) decreased in abundance, planktivorous auklets (*Arthia* spp.) increased by 70% in some regions, and Short-tailed Shearwaters (*Ardaenna tenuirostris*) had an overall expansion and increase in density over the area (Kuletz et al. 2020). Both studies support the idea that the response of different seabird species to changes in the marine environment is not linear nor uniform, and may be driven by several ecological factors (Russell et al. 2015a). Under the current context of rapidly changing environmental conditions suitable habitats might shift. Hence, species showing high site fidelity might not alter their distribution fast enough to avoid negative effects at the population level (Merkel et al. 2020).

For long-distant migrants, impacts of oceanographic variability might occur in regions far beyond their breeding grounds. This is the case for Sabine's Gull (*Xema sabini*) a trans-equatorial migrant from the Artic Ocean, for which adult mortality was higher in years of extreme oceanographic conditions at its non-breeding areas (Fife et al. 2018). Ultimately, seabirds might even change their breeding colonies in response to oceanographic changes in the surroundings of their usual breeding range (Bonter et al. 2014).

Phenology and reproduction

One of the most studied responses of seabirds to oceanographic variability has been their changes in phenology, i.e., the timing of seasonally recurring life-history events (Sauve et al. 2019). Climate and oceanographic changes are causing shifts in the typical pattern of oceans' primary productivity, with spring arriving earlier and summer ending later (Moore et al. 2018). Thus, we can expect an advancement in the timing of spring migration and a delay in autumn migration. It has been reported that globally, breeding phonologies' of pelagic seabird species have advanced by around 4 days per decade since mid-20th century (Sydeman et al. 2015, Poloczanska et al. 2013). Yet, in the Pacific Ocean, Rhinoceros Auklet (*Cerorhinca monocerata*), Japanese Cormorant (*Phalacrocorax filamentosus*) and Black-tailed Gulls (*Larus crassirostris*) exhibited no change in the timing of breeding, despite a clear trend in increasing SST (Watanuki and Ito 2012). In the Indian Ocean, researchers gathered 50 years (1968–2009) of laying date for 24 seabird species. They found that mean laying date might be a more sensible bioindicator of local oceanographic conditions when compared with population size. Hence, monitoring seabird breeding chronology should permit an earlier detection of temporal trends in monitoring programmes, when compared to demographic measures which might buffer the immediate effects of environmental stochasticity, or at least exhibit a response much later in time (e.g., a decrease in population numbers years later) (Chambers et al. 2015).

There are several examples in literature of seabirds' reproductive parameters being used as indicators of oceanographic changes (Surman et al. 2012, Mauck et al. 2018). When facing a decrease in food availability triggered by changes in oceanographic processes, seabirds should prioritize their own survival (Nevoux et al. 2010). This will obviously have consequences on reproductive parameters, such as a decrease in chick growth (Divoky et al. 2015) or productivity (Descamps et al. 2016). Broadly, breeding success and productivity are reported to be negatively correlated with ocean warming for Black-legged Kittiwakes in the North Atlantic (Carroll et al. 2015), Black Guillemots (*Cepphus gryllemandtii*) in the Arctic (Divoky et al. 2015) or Thick-billed Murres (*Uria lomvia*) in the Pacific Ocean (Renner et al. 2014). Yet, this is not a linear relationship, and, in some regions, seabirds seem to improve their reproductive performance under increasing SST in their foraging areas. In the Indian Ocean, resident Lesser Noddies (*Anous tenuirostris*) anticipated their egg-laying period and had an increase in breeding success in years of higher SST, sea level and SOI. On the contrary, migratory Wedge-tailed Shearwaters (*Ardenna pacifica*), had lower breeding participation and success under the same oceanographic trends (Surman et al. 2012). It also seems evident that the response to the same array of oceanographic processes changes among species, or sometimes among populations, inhabiting the same region. In the North Pacific, Least (*Aethia pusilla*), Whiskered (*A. pygmaea*) and Crested (*A. cristatella*) Auklets decreased their reproductive success and productivity in response to increasing Aleutian Low Pressure Index (ALPI) and sea level pressure. In the same 18-year study, auklet productivity increased for populations inhabiting the western North Pacific and Bering Sea, and decreased for those populations in the Gulf of Alaska in response to increasing SST (Bond et al. 2011).

In some occasions the breeding performance of certain seabird populations seem insensitive to changes in oceanographic parameters. Black-legged Kittiwakes in the North Atlantic and Fairy Prions (*Pachyptila turtur*) in the Pacific Ocean did not noticeably change breeding performance through 11 years of increasing SST (Lauria et al. 2013, Renner et al. 2014). However in the Pacific Ocean, Rhinoceros Auklet, Japanese Cormorant and Black-tailed Gulls exhibited no change in breeding success, despite a clear trend of increasing SST in the region (Watanuki and Ito 2012). This emphasises that there might be other extrinsic and intrinsic factors acting as stronger drivers of breeding performance for those populations.

Demographic parameters

Given their conspicuousness and predictable annual rendezvous at their breeding colonies, a wealth of seabird monitoring data has been gathered, allowing to study how seabird demographic parameters respond to oceanographic changes (Goyert et al. 2018, Jenouvrier et al. 2018, Fluhr et al. 2017). Several of those studies make use of powerful Capture-Mark-Recapture (CMR) models of interannual changes, to get a more reliable prediction of demographic trends, and incorporate the effect of oceanographic changes in such population modelling exercises (Pardo et al. 2013). For instance, survival of Cassin's Auklets in the Pacific Ocean (Wolf et al. 2010), Wandering Albatrosses in the Southern Ocean (Fay et al. 2015, Fay et al. 2017) or Black-browed Albatross in the Indian Ocean (Jenouvrier et al. 2018) was shown to respond negatively to an increase in SST. In the Atlantic region, adult survival of Cory's Shearwater was reported to increase under higher SOI and NAO (Genovart et al. 2013). In the Southern Ocean, 52% of variation in recruitment age of Wandering Albatrosses was explained by the Average Southern Annular Mode over the previous 5 years (Fay et al. 2017). Population numbers of Common Guillemots were seen to increase under higher mean sea level pressure (Mesquita et al. 2015).

It is also worth noting that any short-term oceanographically driven change in demographic parameters, such as juvenile or adult survival, will likely have worsened effects in the long-term in long-lived seabirds (Frederiksen et al. 2008). In fact, long-term demographic data also allows for modelling historical and current population trends, and predict their responses to future predicted oceanographic changes (Sydeman et al. 2015). For instance, steep population declines until 2060 were projected for Snow Petrels of the Southern Ocean, under scenarios of increase in SST and decrease in SIC (Barbraud et al. 2011). In the North Pacific, an 11–45% decrease in population growth of Cassin's Auklets by the end of the century is expected to occur under projections of increasing SST and upwelling intensity generated by a regional climate model (RCM) (Wolf et al. 2010).

Nevertheless, for demographic parameters such as population numbers, it might be difficult to realise which percentage of the inter-annual variation is explained by oceanographic-driven factors, compared to the concurrent effects of other life-history traits, such as intra- and inter-species competition (Almaraz and Oro 2011, Ahola et al. 2007) or competition with and bycatch from fisheries (Barbraud et al. 2012, Grémillet et al. 2018). As an example, 21 years of CMR of Mediterranean Storm Petrels (*Hydrobates pelagicus*) revealed that juvenile survival rate was not affected by shifts in values of NAO, SST nor Chla (Soldatini et al. 2014). In a similar perspective, 13 years of data on adult survival of Northern Giant Petrels have shown an increase in male survival and a decrease in female survival with increasing wind speed. Larger male Northern Giant Petrels are more reliant on terrestrial foraging (i.e., carrion) than the more pelagic females, which may thus be more impacted by wind shifts, as drivers of foraging conditions and marine prey availability (Gianuca et al. 2019).

Individual traits as sources of variation

Life history traits

A diverse array of individual traits can also influence the population response to oceanographic changes. For instance, in a migration study conducted in the Indian Ocean between 2006–2013, higher SST in late winter, reflecting poor winter conditions, resulted in Black-browed Albatrosses departing earlier and spending more time on their pre-breeding period. Yet, an earlier spring migration and return to the breeding grounds were associated with a lower breeding success, especially for the male when compared to female parents (Desprez et al. 2018). In the North Atlantic, 20-years of data collection revealed that fledgling sex ratio was female skewed in years of positive NAO, and male skewed in years of negative NAO. This represents a rare evidence that climate/oceanographic conditions can influence sex-specific density in population dynamics (Barros et al. 2013). In the North Atlantic region also, inter-sexual differences in the foraging behaviour, trophic ecology and body condition of Cory's Shearwaters were only noticeable in years of negative NAO (Paiva et al. 2017). Overall, ecological differences between sexes can translate into disparate responses to environmental drivers. Ignoring these sex-related differences may camouflage the impact of oceanographic changes on demography (Gianuca et al. 2019).

Age, a major factor influencing demography in long-lived seabirds, may also influence population-level responses to oceanographic drivers. During periods of higher SST, survival of old and young individuals increased, while it decreased for middle-aged individuals. In colder years and while breeding, old and young birds may face higher intraspecific competition for space and resources than middle-aged counterparts, and this may shape future population dynamics (Pardo et al. 2013). In the Indian Ocean, changes in winter SST affected the survival of juvenile Black-browed Albatross (*Thalassarche melanophris*) but showed none or little effect on adult survival (Jenouvrier et al. 2018).

Breeding experience of individual birds can also be a source of variation in a population-level response to environmental drivers. During El Niño years, Red-footed Boobies (*Sula sula*) from the tropical Atlantic were more likely to skip breeding, due to an increase in local SST. This might be an adaptive strategy of birds to face the trade-off between reproduction and their own survival owing to environmental constrains. Curiously, the likelihood of first breeding attempts by inexperienced breeders was higher during El Niño years, perhaps because of a lowered competition with experienced breeders in such years of higher probability in skip breeding (Cubaynes et al. 2011). In years of negative NAO, inexperienced Cory's Shearwaters breeders exhibited higher foraging effort and fed on lower trophic-level prey when compared to experienced breeders. However, such differences disappeared in years of positive NAO (Haug et al. 2015) (Table C2.1).

Individual plasticity and time lags

Within seabird populations, individuals can have disparate responses to a given oceanographic driver. As a result, the population-level response to an oceanographic stressor might differ from site to site, between populations, depending on the degree of individual plasticity within each seabird population (Grémillet and Charmantier 2010). Behavioural plasticity should be one of the first seabird reactions to short-term oceanographic shifts, and seabird tracking has helped in understanding how these individual reactions may shape the population-level response to changes in their environment (Studholme et al. 2019). At their breeding grounds, Little Auks were seen to respond to variations in SST with individual plasticity in their foraging behaviour, while maintaining their body condition (Grémillet et al. 2012). Besides, in response to a decrease in SIC, the same species engaged in shorter foraging trips (in distance and duration) and low foraging effort, while upholding chick growth rates on historically similar levels (Grémillet et al. 2015). Yet, individual plasticity has its limits, and, considering the rapidly changing environment and the persistence of stochastic events, the change may be too fast for seabirds to adapt and prevent population extinctions (Sauve et al. 2019). As an example, a drastic and sudden decrease in SIC at the southern ocean, led to a sharp decrease in breeding success of seven seabird species, including a total breeding failure for two of them (Barbraud et al. 2015).

Temporal lags between oceanographic changes (e.g., ENSO) and seabird responses (e.g., survival) may also challenge our perception on the responsiveness of specific seabird measures as ecological indicators. In addition to the lagged responses spanning many trophic levels, seabirds as long-lived organisms, may respond to the change in the environment months or years later. For instance, for a current negative effect of increasing SST on juvenile survival (Pardo et al. 2017), one should expect a repercussion on lower recruitment numbers into the breeding population years later, and thus a potential impairment on population dynamics and adult survival (Nevoux et al. 2010, Jenouvrier et al. 2018). On the other hand, seabirds may have the capacity to buffer immediate impacts of oceanographic changes, overcoming nutritional deficits and reproducing successfully (Grémillet et al. 2012), but this might impact their future survival (Sydeman et al. 2012).

Final considerations and further knowledge needed

Most studies dealing with the effects of oceanographic changes on an array of pelagic seabird life history traits, have been conducted in seabird populations from higher latitudes and are based on a few model species per study (Sydeman et al. 2012). In addition most studies are concentrated on particular oceanic regions (e.g., the Arctic Ocean), hindering the perception on how different populations of the same species inhabiting diverse regions react to changes in a given oceanographic driver. Besides, multi-species studies are usually much more informative, of the community-level response of seabirds to oceanographic drivers. In the Arctic, the at-sea census of 26 seabird species for 37 years, revealed a shift in the seabird community structure in response to long-term earlier SIC retreat, which increased the availability of large zooplankton prey. Historically dominated by piscivorous murres, the seabird community at the region is currently dominated by planktivorous Crested Auklets (*Aethia cristatella*) (Gall et al. 2017). Thus, the latter species should be currently the ideal bioindicator of changes in SIC. More representative and scientifically sound oceanographic—

pelagic seabird relationships should be attained if future research is increasingly focussed on marine tropical ecosystems and encompass multiple species in each research study (Sydeman et al. 2012).

When studying oceanographic-seabird relationships, there is also an imperative need to disentangle the mechanisms and processes in which a certain oceanographic change alters marine productivity patterns, which markedly leads to shifts in prey and ultimately generate an ecological response on seabird populations (Paiva et al. 2013). Yet, in most studies the possibility for a mechanistic comprehension of this succession of effects is hindered by the lack of concurrent information on prey abundance and availability. The comprehension that effects of oceanographic changes are non-linear, may have long-lasting consequences (e.g., demographic traits induced by incidence on juvenile survival (Pardo et al. 2013)) and when interacting with additional stressors (e.g., fisheries bycatch) may be even exacerbated.

Chronically limited funding for long-term empirical studies can peril the collection of key data to ascertain the validity of seabirds as indicators of climate variability. Without such long-term monitoring of the seabirds' response to environmental changes, one may miss the degree of phenotypical plasticity in response to inter-annual oceanographic changes. This individual phenotypic plasticity can be high in certain species, which has led some researchers to argue that seabirds should be more powerful as qualitative (binary response) rather than quantitative (gradual response) indicators of oceanographic changes (Piatt et al. 2007, Grémillet and Charmantier 2010, Ramos and Furness 2022).

New and more powerful Species Distribution Models (SDMs) are emerging, ensuring a more holistic perception of the mechanistic processes behind a seabird ecological response to environmental changes (Krüger et al. 2017). These also allow predictions of seabird responses to, for instance, global warming scenarios (Péron et al. 2012). A recent study based on SDMs applied to ~ 1500 tracked individuals from five seabird species in the North Atlantic, used a global warming scenario exceeding 2°C, and showed that future winter distributions were predicted to shift in response to climate change. When applying the Paris Agreement of limiting global warming to < 2°C, shifts in the at-sea distribution would be comparably more limited for this seabird community of the North Atlantic Ocean during the 21st century (Clairbaux et al. 2020). Similar behaviour was observed for Little Auks tracked from five breeding colonies in the Arctic Ocean. SDMs predicted that under scenarios of 2°C increase in SST, most Little Auk populations would lose suitable foraging habitat. Response of this species to future conditions will depend on its behavioural plasticity and pace of oceanographic changes (Jakubas et al. 2017). In this respect, the increasing use of Climate Envelope Models (CEM) to create local climatic suitability indices (Russell et al. 2015b), or simulate prey changes in accordance to oceanographic conditions (Clairbaux et al. 2020), will certainly deepen our understanding of seabirds' responsiveness to oceanographic changes, particularly when investigating future projected responses under climate change scenarios.

Finally, the three-step methodology described by (Jenouvrier 2013) is a useful tool to ascertain if a specific seabird ecological response can be a suitable bioindicator and monitor of oceanographic changes occurring locally or more globally. Namely, the need to select and measure the proper response variable. This meaning the variable must be sensitive and responsive to a particular small- or large-scale oceanographic parameter. The need to understand the mechanism by which an oceanographic parameter is driving a direct or indirect response by a seabird species. And the need to use this knowledge to predict future seabird responses under scenarios of enhanced oceanographic stochasticity. This framework might also be applied when using a specific seabird response in environmental monitoring programmes, for instance, to the early detection of potential ruptures in commercial fish stocks.

References cited

Afán, I., J.M. Arcos, F. Ramírez, D. García, B. Rodríguez, K. Delord et al. 2021. Where to head: Environmental conditions shape foraging destinations in a critically endangered seabird. Mar. Biol. 168: 23. doi: 10.1007/s00227-021-03830-1.

Ahola, M.P., T. Laaksonen, T. Eeva and E. Lehikoinen. 2007. Climate change can alter competitive relationships between resident and migratory birds. J. Anim. Ecol. 76: 1045–1052. doi: 10.1111/j.1365-2656.2007.01294.x.

Almaraz, P. and D. Oro. 2011. Size-mediated non-trophic interactions and stochastic predation drive assembly and dynamics in a seabird community. Ecology 92: 1948–1958. doi: 10.1890/11-0181.1.

Amélineau, F., D. Grémillet, A.M.A. Harding, W. Walkusz, R. Choquet and J. Fort. 2019. Arctic climate change and pollution impact little auk foraging and fitness across a decade. Sci. Rep. 9: 1014. doi: 10.1038/s41598-018-38042-z.

Ancona, S., H. Drummond, C. Rodríguez and J.J. Zúñiga-Vega. 2017. Long-term population dynamics reveal that survival and recruitment of tropical boobies improve after a hurricane. J. Avian Biol. 48: 320–332. doi: 10.1111/jav.01133.

Avalos, M.R., J.A. Ramos, M. Soares, F.R. Ceia, A.I. Fagundes, C. Gouveia et al. 2017. Comparing the foraging strategies of a seabird predator when recovering from a drastic climatic event. Mar. Biol. 164: 48. doi: 10.1007/s00227-017-3082-4.

Ballance, L.T., R.L. Pitman and P.C. Fiedler. 2006. Oceanographic influences on seabirds and cetaceans of the eastern tropical pacific: A review. Progr. in Oceanogr. 69: 360–390. doi:10.1016/j.pocean.2006.03.013.

Barber, R.T. and F.P. Chavez. 1983. Biological Consequences of El Niño. Science 222: 1203–1210. doi: 10.1126/science.222.4629.1203.

Barbraud, C., P. Rivalan, P. Inchausti, M. Nevoux, V. Rolland and H. Weimerskirch. 2011. Contrasted demographic responses facing future climate change in southern ocean seabirds. J. Anim. Ecol. 80: 89–100. doi: 10.1111/j.1365-2656.2010.01752.x.

Barbraud, C., V. Rolland, S. Jenouvrier, M. Nevoux, K. Delord and H. Weimerskirch. 2012. Effects of climate change and fisheries bycatch on southern ocean seabirds: A review. Mar. Ecol. Prog. Ser. 454: 285–307. doi: 10.3354/meps09616.

Barbraud, C., K. Delord and H. Weimerskirch. 2015. Extreme ecological response of a seabird community to unprecedented sea ice cover. Royal Soc. Open Sci. 2: 140456–140456. doi: 10.1098/rsos.140456.

Barros, A., D. Álvarez and A. Velando. 2013. Climate influences fledgling sex ratio and sex-specific dispersal in a seabird. PLoS ONE 8: e71358. doi: 10.1371/journal.pone.0071358.

Bond, A.L., I.L. Jones, W.J. Sydeman, H.L. Major, S. Minobe, J.C. Williams et al. 2011. Reproductive success of planktivorous seabirds in the north pacific is related to ocean climate on decadal scales. Mar. Ecol. Prog. Ser. 424: 205–218. doi: 10.3354/meps08975.

Bonter, D.N., S.A. MacLean, S.S. Shah and M.C. Moglia. 2014. Storm-induced shifts in optimal nesting sites: A potential effect of climate change. J. Ornith. 155: 631–638. doi: 10.1007/s10336-014-1045-9.

Brainard, R.E., T. Oliver, M.J. McPhaden, A. Cohen, R. Venegas, A. Heenan et al. 2018. Ecological impacts of the 2015/16 El Niño in the central equatorial pacific. Bull. Amer. Meteo. Soc. 99: S21–S26. doi: 10.1175/bams-d-17-0128.1.

Carroll, M.J., A. Butler, E. Owen, S.R. Ewing, T. Cole, J.A. Green et al. 2015. Effects of sea temperature and stratification changes on seabird breeding success. Clim. Res. 66: 75–89. doi: 10.3354/cr01332.

Chambers, L.E., T. Patterson, A.J. Hobday, J.P.Y. Arnould, G.N. Tuck, C. Wilcox et al. 2015. Determining trends and environmental drivers from long-term marine mammal and seabird data: Examples from Southern Australia. Reg. Env Change 15: 197–209. doi: 10.1007/s10113-014-0634-8.

Champion, C., S. Brodie and M.A. Coleman. 2021. Climate-driven range shifts are rapid yet variable among recreationally important coastal-pelagic fishes. Front. Mar. Sci. 8: 622299. doi: 10.3389/fmars.2021.622299.

Clairbaux, M., W. Cheung, P. Mathewson, W. Porter, N. Courbin, J. Fort et al. 2020. Meeting paris agreement objectives will temper seabird winter distribution shifts in the North Atlantic Ocean. Glob. Change Biol. 27: 1457–1469. doi: 10.1111/gcb.15497.

Clark, B.L., S.L. Cox, K.M. Atkins, S. Bearhop, A.W.J. Bicknell, T.W. Bodey et al. 2021. Sexual segregation of gannet foraging over 11 years: Movements vary but isotopic differences remain stable. Mar Ecol. Prog. Ser. 661: 1–16. doi: 10.3354/meps13636.

Cubaynes, S., P.F. Doherty, E.A. Schreiber and O. Gimenez. 2011. To breed or not to breed: A seabird's response to extreme climatic events. Biol. Lett. 7: 303–306. doi: 10.1098/rsbl.2010.0778.

Cusset, F., J. Fort, M. Mallory, B. Braune, P. Massicotte and G. Massé. 2019. Arctic seabirds and shrinking sea ice: Egg analyses reveal the importance of ice-derived resources. Sci. Rep. 9: 15405. doi: 10.1038/s41598-019-51788-4.

Delord, K., C. Cotté, P. Terray, C.-A. Bost, H. Weimerskirch and C. Barbraud. 2021. Factors affecting adult body condition in the endangered Northern Rockhopper Penguin. Mar. Biol. 168: 27. doi: 10.1007/s00227-021-03832-z.

Descamps, S., A. Tarroux, S.-H. Lorentsen, O.P. Love, Ø. Varpe and N.G. Yoccoz. 2016. Large-scale oceanographic fluctuations drive antarctic petrel survival and reproduction. Ecography 39: 496–505. doi: 10.1111/ecog.01659.

Descamps, S., F. Ramírez, S. Benjaminsen, T. Anker-Nilssen, R.T. Barrett, Z. Burr et al. 2019. Diverging phenological responses of arctic seabirds to an earlier spring. Glob. Chang. Biol. 25: 4081–4091. doi: 10.1111/gcb.14780.

Desprez, M., S. Jenouvrier, C. Barbraud, K. Delord and H. Weimerskirch. 2018. Linking oceanographic conditions, migratory schedules and foraging behaviour during the non-breeding season to reproductive performance in a long-lived seabird. Func. Ecol. 32: 2040–2053. doi: 10.1111/1365-2435.13117.

Divoky, G.J., P.M. Lukacs and M.L. Druckenmiller. 2015. Effects of recent decreases in arctic sea ice on an ice-associated marine bird. Prog. Oceanogr. 136: 151–161. doi: 10.1016/j.pocean.2015.05.010.

Doney, S.C., M. Ruckelshaus, J.E. Duffy, J.P. Barry, F. Chan, C.A. English et al. 2012. Climate change impacts on marine ecosystems. Annu. Rev. Mar. Sci. 4: 11–37. doi: 10.1146/annurev-marine-041911-111611.

Durant, J.M., D.Ø. Hjermann, M. Frederiksen, J.B. Charrassin, Y. Le Maho, P.S. Sabarros et al. 2009. Pros and cons of using seabirds as ecological indicators. Clim. Res. 39: 115–129. doi: 10.3354/cr00798.

Fay, R., H. Weimerskirch, K. Delord and C. Barbraud. 2015. Population density and climate shape early-life survival and recruitment in a long-lived pelagic seabird. J. Anim Ecol. 84: 1423–1433. doi: 10.1111/1365-2656.12390.

Fay, R., C. Barbraud, K. Delord and H. Weimerskirch. 2017. Contrasting effects of climate and population density over time and life stages in a long-lived seabird. Func. Ecol. 31: 1275–1284. doi: 10.1111/1365-2435.12831.

Fife, D.T., S.E. Davis, G.J. Robertson, H.G. Gilchrist, I.J. Stenhouse, D. Shutler et al. 2018. Correlating tropical climate with survival of an arctic-breeding, trans-equatorial migrant seabird. Arct. Sci. 4: 656–668. doi: 10.1139/as-2017-0018.

Fluhr, J., H. Strøm, R. Pradel, O. Duriez, G. Beaugrand and S. Descamps. 2017. Weakening of the subpolar gyre as a key driver of north atlantic seabird demography: A case study with Brünnich's Guillemots in svalbard. Mar. Ecol. Prog. Ser. 563: 1–11. doi: 10.3354/meps11982.

Frederiksen, M., F. Daunt, M.P. Harris and S. Wanless. 2008. The demographic impact of extreme events: stochastic weather drives survival and population dynamics in a long-lived seabird. J. Anim. Ecol. 77: 1020–1029. doi: 10.1111/j.1365-2656.2008.01422.x.

Gall, A.E., T.C. Morgan, R.H. Day and K.J. Kuletz. 2017. Ecological shift from piscivorous to planktivorous seabirds in the Chukchi Sea, 1975–2012. Pol. Biol. 40: 61–78. doi: 10.1007/s00300-016-1924-z.

Genovart, M., A. Sanz-Aguilar, A. Fernández-Chacón, J.M. Igual, R. Pradel, M.G. Forero et al. 2013. Contrasting effects of climatic variability on the demography of a trans-equatorial migratory seabird. J. Anim. Ecol. 82: 121–130. doi: 10.1111/j.1365-2656.2012.02015.x.

Gianuca, D., S.C. Votier, D. Pardo, A.G. Wood, R.B. Sherley, L. Ireland et al. 2019. Sex-specific effects of fisheries and climate on the demography of sexually dimorphic seabirds. J. Anim. Ecol. 88: 1366–1378. doi: 10.1111/1365-2656.13009.

Gissi, E., E. Manea, A.D. Mazaris, S. Fraschetti, V. Almpanidou and S. Bevilacqua et al. 2020. A review of the combined effects of climate change and other local human stressors on the marine environment. Sci. Total Env. 755(Pt 1): 142564. doi: 10.1016/j.scitotenv.2020.142564.

Goyert, H.F., E.O. Garton and A.J. Poe. 2018. Effects of climate change and environmental variability on the carrying capacity of alaskan seabird populations. Auk 135: 975–991. doi: 10.1642/AUK-18-37.1.

Grecian, W.J., G.A. Taylor, G. Loh, R.A.R. McGill, C.M. Miskelly, R.A. Phillips et al. 2016. Contrasting migratory responses of two closely related seabirds to long-term climate change. Mar. Ecol. Prog. Ser. 559: 231–242. doi: 10.3354/meps11875.

Grémillet, D. and T. Boulinier. 2009. Spatial ecology and conservation of seabirds facing global climate change: A review. Mar. Ecol. Prog. Ser. 391: 121–137. doi: 10.3354/meps08212.

Grémillet, D. and A. Charmantier. 2010. Shifts in phenotypic plasticity constrain the value of seabirds as ecological indicators of marine ecosystems. Ecol. Appl. 20: 1498–1503. doi: 10.1890/09-1586.1.

Grémillet, D., J. Welcker, N.J. Karnovsky, W. Walkusz, M.E. Hall and J. Fort. 2012. Little Auks buffer the impact of current arctic climate change. Mar. Ecol. Prog. Ser. 454: 197–206. doi: 10.3354/meps09590.

Grémillet, D., J. Fort, F. Amélineau, E. Zakharova, T. Le Bot, E. Sala et al. 2015. Arctic warming: Nonlinear impacts of sea-ice and glacier melt on seabird foraging. Glob. Chang Biol. 21: 1116–1123. doi: 10.1111/gcb.12811.

Grémillet, D., A. Ponchon, M. Paleczny, M.-L.D. Palomares, V. Karpouzi and D. Pauly. 2018. Persisting worldwide seabird-fishery competition despite seabird community decline. Curr. Biol. 28: 4009–4013.e2. doi: 10.1016/j.cub.2018.10.051.

Hass, T., J. Hyman and B.X. Semmens. 2012. Climate change, heightened hurricane activity and extinction risk for an endangered tropical seabird, the black-capped petrel *Pterodroma hasitata*. Mar. Ecol. Prog. Ser. 454: 251–261. doi: 10.3354/meps09723.

Hatfield, J.S., M.H. Reynolds, N.E. Seavy and C.M. Krause. 2012. Population dynamics of hawaiian seabird colonies vulnerable to sea-level rise. Cons. Biol. 26: 667–678. doi: 10.1111/j.1523-1739.2012.01853.x.

Haug, F.D., V.H. Paiva, A.C. Werner and J.A. Ramos. 2015. Foraging by experienced and inexperienced Cory's shearwater along a 3-year period of ameliorating foraging conditions. Mar. Biol. 162: 649–660. doi: 10.1007/s00227-015-2612-1.

Hazen, E.L., B. Abrahms, S. Brodie, G. Carroll, M.G. Jacox, M.S. Savoca et al. 2019. Marine top predators as climate and ecosystem sentinels. Front. Ecol. Environ. 17: 565–574. doi: 10.1002/fee.2125.

Hazen, E.L., S. Jorgensen, R.R. Rykaczewski, S.J. Bograd, D.G. Foley, I.D. Jonsen et al. 2012. Predicted habitat shifts of pacific top predators in a changing climate. Nat. Clim. Change 3: 1–5. doi: 10.1038/nclimate1686.

Hipfner, J.M., M. Galbraith, D.F. Bertram and D.J. Green. 2020. Basin-Scale oceanographic processes, zooplankton community structure and diet and reproduction of a sentinel north pacific seabird over a 22-year period. Prog. Oceanogr. 182: 102290. doi: 10.1016/j.pocean.2020.102290.

Jakubas, D., K. Wojczulanis-Jakubas, L.M. Iliszko, H. Strøm and L. Stempniewicz. 2017. Habitat foraging niche of a high arctic zooplanktivorous seabird in a changing environment. Sci. Rep. 7: 16203. doi: 10.1038/s41598-017-16589-7.

Jenouvrier, S. 2013. Impacts of climate change on avian populations. Glob. Change Biol. 19: 2036–2057. doi: 10.1111/gcb.12195.

Jenouvrier, S., M. Desprez, R. Fay, C. Barbraud, H. Weimerskirch, K. Delord et al. 2018. Climate change and functional traits affect population dynamics of a long-lived seabird. J. Anim. Ecol. 87: 906–920. doi: 10.1111/1365-2656.12827.

Jones, T., J.K. Parrish, W.T. Peterson, E.P. Bjorkstedt, N.A. Bond, L.T. Balance et al. 2018. Massive mortality of a planktivorous seabird in response to a marine heatwave. Geophys. Res. Lett. 45: 3193–3202. doi: 10.1002/2017GL076164.

Keogan, K., F. Daunt, S. Wanless, R.A. Phillips, C.A. Walling, P. Agnew et al. 2018. Global phenological insensitivity to shifting ocean temperatures among seabirds. Nat. Clim. Change 8: 313–318. doi: 10.1038/s41558-018-0115-z.

Krüger, L., J.A. Ramos, J.C. Xavier, D. Grémillet, J. González-Solís, M.V. Petry et al. 2017. Projected distributions of southern ocean albatrosses, petrels and fisheries as a consequence of climatic change. Ecography. 41: 195–208. doi: 10.1111/ecog.02590.

Kuletz, K., D. Cushing and E. Labunski. 2020. Distributional shifts among seabird communities of the northern bering and chukchi seas in response to ocean warming during 2017–2019. Deep Sea Res. Part II.181-182: 104913. doi: 10.1016/j.dsr2.2020.104913.

Lauria, V., M.J. Attrill, A. Brown, M. Edwards and S.C. Votier. 2013. Regional variation in the impact of climate change: Evidence that bottom-up regulation from plankton to seabirds is weak in parts of the Northeast Atlantic. Mar. Ecol. Prog. Ser. 488: 11–22. doi: 10.3354/meps10401.

Lorenzo, E.D. and N. Mantua. 2016. Multi-year persistence of the 2014/15 North Pacific Marine Heatwave. Nat. Clim. Change 6: 1042–1047. doi: 10.1038/nclimate3082.

Louzao, M., O. Aumont, T. Hothorn, T. Wiegand and H. Weimerskirch. 2013. Foraging in a changing environment: Habitat shifts of an oceanic predator over the last half century. Ecography 36: 57–67. doi: 10.1111/j.1600-0587.2012.07587.x.

Louzao, M., R. Gallagher, I. García-Barón, G. Chust, I. Intxausti, J. Albisu et al. 2019. Threshold responses in bird mortality driven by extreme wind events. Ecol Ind. 99: 183–192. doi: 10.1016/j.ecolind.2018.12.030.

Lynam, C.P., M. Llope, C. Möllmann, P. Helaouët, G.A. Bayliss-Brown and N.C. Stenseth. 2017. Interaction between top-down and bottom-up control in marine food webs. Proc Nat. Acad. Sci. 114: 1952–1957. doi: 10.1073/pnas.1621037114.

Manno, K., G. Loh and Y. Van Heezik. 2014. Buffering against Food Availability? The physical environment has little influence on breeding performance of fairy prions (*Pachyptila turtur*). Austral Ecol. 39: 548–559. doi: 10.1111/aec.12115.

Matović, N., B. Cadiou, D. Oro and A. Sanz-Aguilar. 2017. Disentangling the effects of predation and oceanographic fluctuations in the mortality of two allopatric seabird populations. Pop. Ecol. 59: 225–238. doi: 10.1007/s10144-017-0590-5.

Mauck, R.A., D.C. Dearborn and C.E. Huntington. 2018. Annual global mean temperature explains reproductive success in a marine vertebrate from 1955 to 2010. Glob. Change Biol. 24: 1599–1613. doi:10.1111/gcb.13982.

Merkel, B., S. Descamps, N.G. Yoccoz, J. Danielsen, F. Daunt, K.E. Erikstad et al. 2019. Earlier colony arrival but no trend in hatching timing in two congeneric seabirds (*Uria* spp.) across the North Atlantic. Biol. Lett. 15: 20190634. doi: 10.1098/rsbl.2019.0634.

Merkel, B., S. Descamps, N.G. Yoccoz, D. Grémillet, F. Daunt, K.E. Erikstad et al. 2020. Individual migration strategy fidelity but no habitat specialization in two congeneric seabirds. J. Biogeogr. 48: 263–275. doi: 10.1111/jbi.13883.

Mesquita, M.D.S., K.E. Erikstad, H. Sandvik, R.T. Barrett, T.K. Reiertsen and T. Anker-Nilssen. 2015. There is more to climate than the north atlantic oscillation: A new perspective from climate dynamics to explain the variability in population growth rates of a long-lived seabird. Front. Ecol. Evol. 3: 43. doi: 10.3389/fevo.2015.00043.

Möllmann, C., B. Müller-Karulis, G. Kornilovs and M.A. St. John. 2008. Effects of climate and overfishing on zooplankton dynamics and ecosystem structure: Regime shifts, trophic cascade and feedback loops in a simple ecosystem. ICES J. Mar. Sci. 65: 302–310. doi: 10.1093/icesjms/fsm197.

Moore, J.K., W. Fu, F. Primeau, G.L. Britten, K. Lindsay, M. Long et al. 2018. Sustained climate warming drives declining marine biological productivity. Science 359: 1139–1143. doi: 10.1126/science.aao6379.

Morley, T., A. Fayet, H. Jessop, P. Veron, M. Veron, J. Clark et al. 2016. The seabird wreck in the bay of biscay and southwestern approaches in 2014: A review of reported mortality. Seabird 29: 22–38.

Nevoux, M., J. Forcada, C. Barbraud, J. Croxall and H. Weimerskirch. 2010. Bet-hedging response to environmental variability, an intraspecific comparison. Ecology 91: 2416–2427. doi: 10.1890/09-0143.1.

Nevoux, M., H. Weimerskirch and C. Barbraud. 2010. Long- and Short-Term influence of environment on recruitment in a species with highly delayed maturity. Oecologia 162: 383–392. doi: 10.1007/s00442-009-1482-y.

Nicoll, M.A.C., M. Nevoux, C.G. Jones, N. Ratcliffe, K. Ruhomaun, V. Tatayah et al. 2017. Contrasting effects of tropical cyclones on the annual survival of a pelagic seabird in the Indian Ocean. Glob. Change Biol. 23: 550–565. doi: 10.1111/gcb.13324.

Oro, D. 2014. Seabirds and climate: Knowledge, pitfalls and opportunities. Front. Ecol. Evol. 2: 79. doi: 10.3389/fevo.2014.00079.

Osborn, T.J. 2011. Winter 2009/2010 temperatures and a record-breaking North Atlantic oscillation index. Weather 66: 19–21. doi: 10.1002/wea.660.

Paiva, V.H, P. Geraldes, V. Marques, R. Rodriguez, S. Garthe and J.A. Ramos. 2013. Effects of Environmental Variability on Different Trophic Levels of the North Atlantic Food Web. Mar. Ecol. Prog. Ser. 477: 15–28. doi: 10.3354/meps10180.

Paiva, V.H., J. Pereira, F.R. Ceia and J.A. Ramos. 2017. Environmentally driven sexual segregation in a marine top predator. Sci. Rep. 7: 2590. doi: 10.1038/s41598-017-02854-2.

Pardo, D., C. Barbraud, M. Authier and H. Weimerskirch. 2013. Evidence for an age-dependent influence of environmental variations on a long-lived seabird's life-history traits. Ecology 94: 208–220. doi: 10.1890/12-0215.1.

Pardo, D., S. Jenouvrier, H. Weimerskirch and C. Barbraud. 2017. Effect of extreme sea surface temperature events on the demography of an age-structured albatross population. Phil. Trans. R. Soc. B. 372: 20160143. doi: 10.1098/rstb.2016.0143.

Parsons, M., I. Mitchell, A. Butler, N. Ratcliffe, M. Frederiksen, S. Foster et al. 2008. Seabirds as indicators of the marine environment. ICES J. Mar. Sci. 65: 1520–1526. doi: 10.1093/icesjms/fsn155.

Pereira, J.M., V.H. Paiva, F.R. Ceia and J.A. Ramos. 2020. Facing extremes: Cory's shearwaters adjust their foraging behaviour differently in response to contrasting phases of North Atlantic oscillation. Reg. Environ. Change 20: 77. doi: 10.1007/s10113-020-01662-1.

Péron, C., M. Authier, C. Barbraud and K. Delord. 2010. Interdecadal changes in at-sea distribution and abundance of subantarctic seabirds along a latitudinal gradient in the southern Indian Ocean. Glob. Change Biol. 16: 1895–1909. doi: 10.1111/j.1365-2486.2010.02169.x.

Péron, C., H. Weimerskirch and Charles-André Bost. 2012. Projected poleward shift of king penguins' (*Aptenodytes patagonicus*) foraging range at the crozet islands, Southern Indian Ocean. Proc. R. Soc. Lond. B: Biol. Sci. 279: 2515–2523. doi: 10.1098/rspb.2011.2705.

Piatt, J.F., A.M.A. Harding, M. Shultz, S.G. Speckman, T.I. van Pelt, G.S. Drew et al. 2007. Seabirds as indicators of marine food supplies: Cairns revisited. Mar. Ecol. Prog. Ser. 352: 221–234. doi: 10.3354/meps07078.

Poloczanska, E.S., C.J. Brown, W.J. Sydeman, W. Kiessling, D.S. Schoeman, P.J. Moore et al. 2013. Global imprint of climate change on marine life. Nat. Clim. Change 3: 919–925. doi: 10.1038/nclimate1958.

Poloczanska, E.S., M.T. Burrows, C.J. Brown, J.G. Molinos, B.S. Halpern, O. Hoegh-Guldberg et al. 2016. Responses of marine organisms to climate change across oceans. Front. Mar. Sci. 3: 62. doi: 10.3389/fmars.2016.00062.

Precheur, C., C. Barbraud, F. Martail, M. Mian, J.-C. Nicolas, R. Brithmer et al. 2016. Some like it hot: Effect of environment on population dynamics of a small tropical seabird in the caribbean region. Ecosphere 7: e01461. doi: 10.1002/ecs2.1461.

Quillfeldt, P., J.F. Masello, R.A.R. McGill, M. Adams and R.W. Furness. 2010. Moving polewards in winter: A recent change in the migratory strategy of a pelagic seabird? Front. Zool. 7: 15. doi: 10.1186/1742-9994-7-15.

Ramírez, F., A. Tarroux, J. Hovinen, J. Navarro, I. Afán, M.G. Forero et al. 2017. Sea ice phenology and primary productivity pulses shape breeding success in arctic seabirds. Sci. Rep. 7: 4500. doi: 10.1038/s41598-017-04775-6.

Ramos, J.A. and R.W. Furness. 2022. Seabirds as indicators of forage fish stocks. pp. 137–148. *In*: Ramos, J.A. and L. Pereira [eds.]. Seabird Biodiversity and Human Activities. CRC Press, Boca Raton, FL.

Renner, H.M., B.A. Drummond, A.-M. Benson and R. Paredes. 2014. Reproductive success of kittiwakes and murres in sequential stages of the nesting period: Relationships with diet and oceanography. Deep Sea Res. II: Top. Stud. Oceanogr. 109: 251–265. doi: 10.1016/j.dsr2.2014.03.006.

Reynolds, J., J. Hughes, C.P. Wearn, R.C. Dickey, J. Brown, N.L. Weber et al. 2019. Long-term dietary shift and population decline of a pelagic seabird—A health check on the tropical atlantic? Glob. Change Biol. 25: 1383–1394. doi: 10.1111/gcb.14560.

Reynolds, M.H., K.N. Courtot, P. Berkowitz, C.D. Storlazzi, J. Moore and E. Flint. 2015. Will the effects of sea-level rise create ecological traps for pacific island seabirds? Plos One 10: e0136773. doi: 10.1371/journal.pone.0136773.

Russell, D.J.F., S. Wanless, Y.C. Collingham, B. Huntley and K.C. Hamer. 2015a. Predicting future european breeding distributions of british seabird species under climate change and unlimited/no dispersal scenarios. Diversity 7: 342–359. doi: 10.3390/d7040342.

Russell, D.J.F., S. Wanless, Y.C. Collingham, B.J. Anderson, C. Beale, J.B. Reid et al. 2015b. Beyond climate envelopes: Bio-climate modelling accords with observed 25-year changes in seabird populations of the british isles. Divers. Distrib. 21: 211–222. doi: 10.1111/ddi.12272.

Sandvik, H., K.E. Erikstad, R.T. Barrett and N.G. Yoccoz. 2005. The effect of climate on adult survival in five species of north atlantic seabirds. J Anim. Ecol. 74: 817–831. doi: 10.1111/j.1365-2656.2005.00981.x.

Sandvik, H., K.E. Erikstad and B.E. Sæther. 2012. Climate affects seabird population dynamics both via reproduction and adult survival. Mar. Ecol. Prog. Ser. 454: 273–284. doi: 10.3354/meps09558.

Santora, J.A., L.B. Eisner, K.J. Kuletz, C. Ladd, M. Renner and G.L. Hunt. 2018. Biogeography of seabirds within a high-latitude ecosystem: Use of a data-assimilative ocean model to assess impacts of mesoscale oceanography. J Mar. Syst. 178: 38–51. doi: 10.1016/j.jmarsys.2017.10.006.

Santos, A.M.P., A. Chícharo, A. Dos Santos, T. Moita, P.B. Oliveira, A. Peliz et al. 2007. Physical-biological interactions in the life history of small pelagic fish in the western iberia upwelling ecosystem. Progr. Oceanogr. 74: 192–209. doi: 10.1016/j.pocean.2007.04.008.

Sauve, D., G. Divoky and V.L. Friesen. 2019. Phenotypic plasticity or evolutionary change? An examination of the phenological response of an arctic seabird to climate change. Func. Ecol. 33: 2180–2190. doi: 10.1111/1365-2435.13406.

Sherley, R.B., K. Ludynia, B.M. Dyer, T. Lamont, A.B. Makhado, J.-P. Roux et al. 2017. Metapopulation tracking juvenile penguins reveals an ecosystem-wide ecological trap. Curr. Biol. 27: 563–568. doi: 10.1016/j.cub.2016.12.054.

Shultz, M.T., J.F. Piatt, A.M.A. Harding, A.B. Kettle and T.I. Van Pelt. 2009. Timing of breeding and reproductive performance in murres and kittiwakes reflect mismatched seasonal prey dynamics. Mar. Ecol. Prog. Ser. 393: 247–258. doi: 10.3354/meps08136.

Soldatini, C., Y.V. Albores-Barajas, B. Massa and O. Gimenez. 2014. Climate driven life histories: The case of the mediterranean storm petrel. Edited by sébastien descamps. PLoS ONE 9: e94526. doi: 10.1371/journal.pone.0094526.

Stenseth, N., A. Mysterud, G. Ottersen, J.W. Hurrell, K.-S. Chan and M. Lima. 2002. Ecological effects of climate fluctuations. Science 297: 1292–1296. doi: 10.1126/science.1071281.

Stenseth, N., G. Ottersen, J.W. Hurrell, A. Mysterud, M. Lima, K.-S. Chan et al. 2003. Review article. Studying climate effects on ecology through the use of climate indices: The North Atlantic Oscillation, El Niño Southern Oscillation and Beyond. Proc. R. Soc. Lond. B: Biol. Sci. 270: 2087–2096. doi: 10.1098/rspb.2003.2415.

Studholme, K.R., J.M. Hipfner, A.D. Domalik, S.J. Iverson and G.T. Crossin. 2019. Year-round tracking reveals multiple migratory tactics in a sentinel north pacific seabird, Cassin's Auklet. Mar. Ecol. Prog. Ser. 619: 169–185. doi: 10.3354/meps12965.

Surman, C.A., L.W. Nicholson and J.A. Santora. 2012. Effects of climate variability on breeding phenology and performance of tropical seabirds in the eastern Indian Ocean. Mar. Ecol. Prog. Ser. 454: 147–157. doi: 10.3354/meps09729.

Sydeman, W.J., S.A. Thompson and A. Kitaysky. 2012. Seabirds and climate change: roadmap for the future. Mar. Ecol. Prog. Ser. 454: 107–117. doi: 10.3354/meps09806.

Sydeman, W.J., E. Poloczanska, T.E. Reed and S.A. Thompson. 2015. Climate change and marine vertebrates. Science 350: 772–777. doi: 10.1126/science.aac9874.

Tavares, D.C., J.F. de Moura and S. Siciliano. 2016. Environmental predictors of seabird wrecks in a tropical coastal area. PLoS ONE 11: e0168717. doi: 10.1371/journal.pone.0168717.

Tavares, D.C., J.F. Moura, A. Merico and S. Siciliano. 2020. Mortality of seabirds migrating across the tropical atlantic in relation to oceanographic processes. Anim. Cons. 23: 307–319. doi: 10.1111/acv.12539.

Thorne, L.H., E.L. Hazen, S.J. Bograd, D.G. Foley, M.G. Conners, M.A. Kappes et al. 2015. Foraging behavior links climate variability and reproduction in north Pacific Albatrosses. Mov. Ecol. 3: 27. doi: 10.1186/s40462-015-0050-9.

Timmermann, A., S.-I. An, J.-S. Kug, F.-F. Jin, W. Cai, A. Capotondi et al. 2018. El Niño-Southern Oscillation Complexity. Nature 559: 535–545. doi: 10.1038/s41586-018-0252-6.

Tremblay, Y., S. Bertrand, RW. Henry, M.A. Kappes, D.P. Costa and S.A. Shaffer. 2009. Analytical approaches to investigating seabird-environment interactions: A review. Mar. Ecol. Prog. Ser. 391: 153–163. doi: 10.3354/meps08146.

Veit, R.R. and L.L. Manne. 2015. Climate and changing winter distribution of alcids in the Northwest Atlantic. Front. Ecol. Evol. 3: 38. doi: 10.3389/fevo.2015.00038.

Watanuki, Y. and M. Ito. 2012. Climatic effects on breeding seabirds of the northern Japan Sea. Mar. Ecol. Prog. Ser. 454: 183–196. doi: 10.3354/meps09627.

Waugh, S.M., C. Barbraud, L. Adams, A.N.D. Freeman, K.-J. Wilson, G. Wood et al. 2015. Modeling the demography and population dynamics of a subtropical seabird and the influence of environmental factors. The Condor 117: 147–164. doi: 10.1650/CONDOR-14-141.1.

Weimerskirch, H., M. Louzao, S. de Grissac and K. Delord. 2012. Changes in wind pattern alter albatross distribution and life-history traits. Science 335: 211–214. doi: 10.1126/science.1210270.

Wolf, S.G., M.A. Snyder, W.J. Sydeman, D.F. Doak and D.A. Croll. 2010. Predicting population consequences of ocean climate change for an ecosystem sentinel, the seabird cassin's Auklet. Glob. Change. Biol. 16: 1923–1935. doi: 10.1111/j.1365-2486.2010.02194.x.

CHAPTER 3

Seabirds as Indicators of Metal and Plastic Pollution

Catarina S. Lopes,[1,*] *Maria I. Laranjeiro,*[1] *Jennifer L. Lavers,*[2] *Annett Finger*[3] and *Jennifer Provencher*[4]

Introduction

The light weight, robustness, versatility, durability and low cost of plastic materials make them suitable to create a wide range of useful products (Andrady and Neal 2009). Unfortunately, these properties are also the reason why plastics are a serious threat to natural ecosystems throughout the world (Galgani et al. 2015). Considering the large production, intense consumption and rapid disposal of plastics, they are widespread and ubiquitous in the environment, particularly in oceans. The absolute quantity of plastic that enter the marine environment is poorly described (Thompson et al. 2004, Barnes et al. 2009, Jambeck et al. 2015), but it is estimated that around 10% of produced plastics will end up in the oceans (Avio et al. 2017). Plastics are made by polymerizing monomers and other substances such as plastic additives, which are chemical compounds like plasticizers (polybrominated diphenyl ethers (PBDEs), polycyclic aromatic hydrocarbons (PAHs), bisphenol A (BPA), phthalates, among others), which provide the required properties of performance and appearance to a plastic polymer (e.g., Andrady 2011, Law 2017). As they do not break down into their chemical components, plastics will persist in the environment for very long time periods, typically hundreds or thousands of years, and in some cases might not degrade at all (Lambert et al. 2014). Instead, plastics are exposed to mechanical and chemical weathering processes (e.g., wind, waves, UV light) that break them into smaller pieces over time, leading to the formation of microplastics (Andrady 2017). In 2008, during the first international research workshop about

[1] University of Coimbra, MARE - Marine and Environmental Sciences Centre, Department of Life Sciences, Calçada Martim de Freitas, 3000-456 Coimbra, Portugal.
Email: mariaines.santos@outlook.pt
[2] Institute for Marine and Antarctic Studies, University of Tasmania, 20 Castray Esplanade, Battery Point, Tasmania, 7004, Australia.
Email: jennifer.lavers@utas.edu.au
[3] Institute for Sustainability & Innovation, Victoria University, PO Box 14428, Melbourne, VIC, 8001, Australia.
Email: annett.finger@vu.edu.au
[4] Environment and Climate Change Canada, National Wildlife Research Centre, C/O Carleton University, 1125 Colonel By Drive, Raven Road, Ottawa, ON, K1A 0H3, Canada.
Email: jennifer.provencher@canada.ca
* Corresponding author: catarina.s.lopes90@gmail.com

microplastics' occurrence, effects and fate (Arthur et al. 2009), microplastics were defined as plastic particles varying from a few μm to 5 mm (5000 μm) in diameter. Currently, microplastics' size classification, especially the upper size limit, is not unanimous (Andrady 2017), being a source of discussion (e.g., particles smaller than 1 mm, GESAMP 2015). Microplastics form a heterogeneous mix of polymers varying in shape (e.g., pellets, fibres, fragment and films), density, size and additives (Huang et al. 2020). There are also manufactured plastics of microscopic size used in some products such as beauty scrubbers and exfoliates (Fendall and Sewell 2009, Duis and Coors 2016), thus entering the ocean from both primary source products and as secondary sources derived from the breakup of larger plastic products (Cole et al. 2011, Hermabessiere et al. 2017). Consequently, plastics are rapidly increasing, ubiquitous marine pollutants, with significant ecological, cultural, aesthetic and economic impacts (Cole et al. 2011, Cózar et al. 2014, Wilcox et al. 2015), and largely ineffective prevention or collection technologies given the scale of the plastic pollution problem (Schmaltz et al. 2020).

Plastics are widely distributed in the marine environment, from coastal regions to the open ocean and throughout the water column, from the sea surface to the seafloor, with particularly worrying accumulation areas in convergence zones such as vortexes and central subtropical gyres (Barnes et al. 2009, van Sebille et al. 2015, van Franeker and Law 2015). Thus, plastic pollution represents a direct threat for a wide range of species including seabirds (Kühn et al. 2015, Kühn and van Franeker 2020). For instance, Kühn and van Franeker (2020) reported that 914 marine species were affected by marine debris, with high incidence on seabirds (at least 226 of the 409 known seabird species interacted with marine plastics either through ingestion and/or entanglement). Accordingly, Wilcox et al. (2015) predicted that by 2050, 99% of seabird species and 95% of the individuals will ingest at least some plastic.

Moreover, plastics can absorb contaminants from the surrounding sea water, including toxic trace elements like metals (Rochman et al. 2013), thus acting as vectors of contamination when ingested, by facilitating the leaching of contaminants to the animals' blood stream (Teuten et al. 2009, Lavers and Bond 2016b, Tanaka et al. 2020).

Metal pollution is also a major problem in marine ecosystems (Streets et al. 2009, Lamborg et al. 2014) which are consistently used as dumping grounds for pollutants from urban, industrial and agriculture run-off (Boye et al. 2012, Finger et al. 2015). Thus, seabirds exploiting and living in polluted regions are exposed to increasing and accumulating levels of such contaminants (Furness and Camphuysen 1997). Metals are very dense elements with high tendency to form positive ions and react with biomolecules (Koivula and Eeva 2010), influencing important biochemical reactions and causing damage in the metabolism of organisms (Scheuhammer 1987, Koivula and Eeva 2010). Some metals, such as zinc (Zn), iron (Fe) and manganese (Mn), occur ordinarily in nature and are important to organisms functioning, becoming problematic only when in atypical and excessive concentrations (Lemos et al. 2013, Mallory et al. 2017), in contrast with the non-essential metals like mercury (Hg), lead (Pb) and cadmium (Cd), which are toxic even at very low concentrations (Furness 1993, Theron et al. 2011, Mallory et al. 2017). Several studies indicate an increase in the concentration of hazardous contaminants in seabirds, particularly Hg, consistent with the increase of global atmospheric emissions (Thompson et al. 1992, Vo et al. 2011, Bond and Lavers 2020).

Given the widespread occurrence and increasing concentration levels of plastics and metals in the marine environment, one of the fundamental challenges is how to properly assess and monitor the increasing marine pollution by these two stressors (Streets et al. 2009, Ryan et al. 2009, Lamborg et al. 2014), and assess their effects on marine life.

Seabirds are typically long-lived top-predators, increasingly recognised as sensitive bioindicators of pollution and the health of marine ecosystems (Furness and Camphuysen 1997, Burger and Gochfeld 2004) due to their capacity to integrate and accumulate contaminants (e.g., plastics and metals) through their food webs (Bearhop et al. 2000, Bocher et al. 2003, Tanaka et al. 2020). Additionally, breeding seabirds normally gather in large and well-established breeding colonies every year (Coulson 2001), which allows monitoring their foraging areas over consecutive years (Furness and Camphuysen 1997, Piatt and Sydeman 2007, Ryan 2018). Most seabirds are migratory species, show high mobility and forage over wide areas, being more likely to interact with contaminants, and thus good sentinels of plastic and metal pollution sources in marine ecosystems (Ryan and Fraser 1988, Furness and Camphuysen 1997, van Franeker et al. 2011). Furthermore, they can be used as proxies of plastic and metal contamination hazards for humans which occupy similar trophic levels in the food chain (Furness 1993, Huang et al. 2020).

Plastic exposure in seabirds

Plastic pollution is known to cause either direct mortality or injury, and a range of sub-lethal effects to seabirds which may influence species' behaviour, physiology and survival, although subtle changes in individuals' health can be difficult to detect (Lavers et al. 2019). The main impacts of plastic pollution on seabirds arise from entanglement in large plastic items and ingestion of smaller particles (Laist 1987, Kühn et al. 2015, Gall and Thompson 2015).

Entanglement in marine debris, such as lost or discarded fishing gear, nets, ropes or plastic bags, is the most visible consequence of plastic pollution to seabirds (Laist 1997, Ryan 2018). Impacts of entanglement include injuries, impeded mobility that decreases seabirds' ability to obtain sufficient food or to effectively avoid predators, and drowning (Derraik 2002, Kühn et al. 2015, Ryan 2018). The most recent review by Kühn and van Franeker (2020) reported that 112 of the 409 known seabird species (27.4%) were entangled in marine debris, however this number is certainly higher as many entangled seabirds may die far from land and are not detected (Laist 1987).

Seabirds are also known to interact directly with marine plastics through collection of items which are then incorporated in their nests (reviewed by Battisti et al. 2019), and this is particularly evident in seabirds using urban environments such as gulls (Lopes et al. 2020). Incorporation of plastic strings, fishing nets and other threadlike plastics in nests (Fig. C3.1A) may lead to entanglement of chicks and adults, which can cause direct injury or death (e.g., over 80% of Northern Gannet *Morus bassanus* nests in Wales contained plastic resulting in the entanglement of 62 individuals on average each year; Votier et al. 2011). The presence of plastics in seabirds' nests may also lead to a higher exposure to some potentially harmful chemicals (Lithner et al. 2011, O'Hanlon et al. 2017). As suggested by Verlis et al. (2014), nesting on top of certain plastics could potentially lead to the absorption of contaminants through the skin of chicks and adults. This may interfere with birds' physiology, causing negative effects on reproduction, behaviour and survival (Herzke et al. 2016, O'Hanlon et al. 2017). Despite this postulation by Verlis et al. (2014), the authors did not directly assess this issue or provide any quantitative results, hence this dermal exposure route of contaminants requires further study.

Figure C3.1. Examples of seabirds' interactions with plastics. (A) A Silver Gull (*Chroicocephalus novaehollandiae*) nest from southern Tasmania with a plastic bag incorporated, pointed out with an arrow (photo: Jennifer Lavers). (B) Plastic fragments found in a single Yellow-legged Gull (*Larus michahellis*) regurgitated pellet (bolus) collected in an urban breeding location in northern Portugal (photo: Catarina Lopes).

The ingestion of marine plastics by seabirds is also concerning due to the harmful effects caused to birds' digestive system. These indigestible materials (Fig. C3.1B) can cause stomach lesions and perforations, gastrointestinal blockage with the obstruction of food passage, reduced appetite caused by a false sensation of satiety, decreased dietary efficiency and consequently reduced growth, general debilitation often leading to death (Ryan 1988, Kühn et al. 2015, Roman et al. 2019). Ingestion of plastics may potentially disturb the absorption and assimilation of nutrients (Gregory 2009). Seabirds can also feed their chicks with large quantities of plastics (parental transfer; Cadée 2002, Verlis et al. 2013). Alarmingly, studies for a number of seabird species at some specific locations and years had reported > 90% of fledglings contain plastics in their digestive tracts (e.g., Flesh-footed Shearwaters *Ardenna carneipes*, Lavers et al. 2021; Short-tailed Shearwaters *A. tenuirostris*, Cousin et al. 2015; Northern Fulmars *Fulmarus glacialis*, van Franeker et al. 2011; Laysan Albatrosses *Phoebastria immutabilis*, and Black-footed Albatrosses *P. nigripes*, Lavers and Bond 2016b).

Physical trophic transfer of plastics may also occur via prey that have ingested plastic debris (i.e., secondary ingestion; Provencher et al. 2019a). Small plastic particles are commonly ingested by many invertebrates (including zooplankton, reviewed by Lusher et al. 2017), entering in the base of marine food webs and transferred along food chains between prey and predator (Hermabessiere et al. 2017, Nelms et al. 2018, Tanaka et al. 2020). This trophic transfer of plastics has been observed directly in seabirds, as well as in other marine species in the wild (Farrell and Nelson 2013, Hammer et al. 2016, Santana et al. 2017).

In addition to physical damage (Wright et al. 2013), the ingestion of plastics can potentially interfere with birds' physiology (e.g., blood chemistry parameters, Lavers et al. 2019), body condition and fitness (Lavers et al. 2014, 2019). Plastics can serve as vectors of two main categories of contaminants: (1) plastic-associated contaminants originating from multiple sources that are adsorbed by plastics due to their hydrophobic features (e.g., polychlorinated biphenyls (PCBs), PBDEs, mercury, among others, Mato et al. 2001), and (2) plastic-derived contaminants that are chemical compounds added to plastics during their manufacture process (plastic additives, such as phthalates, UV stabilizers and others, Hermabessiere et al. 2017). Seabirds can be exposed to these compounds through several pathways (reviewed by Teuten et al. 2009): (1) direct plastic ingestion, with chemicals leached off those plastic particles entering seabirds' blood stream (Tanaka et al. 2013, Lavers and Bond 2016b, Padula et al. 2020); (2) indirect exposure through diet by the consumption of contaminated prey (Bakir et al. 2016, Herzke et al. 2016); and (3) through the environment itself as some compounds (e.g., phthalates) were detected in water and air and may be inhaled or absorbed by seabirds (Net et al. 2015). These contaminants (i.e., plastic-associated and plastic-derived) may have hazardous consequences for seabirds (e.g., some plastic additives commonly act as endocrine disruptors, Lithner et al. 2011). The transfer of plastic-associated contaminants to seabirds has received some attention lately (Rochman et al. 2013, Herzke et al. 2016, Lavers and Bond 2016b), however, the transfer of plastic-derived chemicals has been much less studied (Hermabessiere et al. 2017, Lu et al. 2019, Padula et al. 2020). In the future special attention should be paid to the transfer and the effects of potentially harmful plastic-derived chemicals to seabirds (Huang et al. 2020).

Plastic pollution monitoring tools

To monitor plastic pollution using seabirds, some parameters should be considered; these include the evaluation of the proportion of seabirds that (1) are found entangled in marine plastics, (2) incorporate plastic items on their nests and (3) ingest plastics (directly or indirectly). Quantifying entanglement rates and subsequent mortality can be difficult (Ryan 2018) but, despite not all seabird species incorporate plastics in their nests, monitoring plastics in nests can be a rapid, non-destructive, and simple method to quantify the magnitude of plastic pollution in the surrounding environment and the associated probability of entanglement (Tavares et al. 2016, Grant et al. 2018, Ryan 2018, O'Hanlon et al. 2019). Plastic ingestion by seabirds can be assessed through necropsies of intact birds and examination of food remains (Provencher et al. 2017). Necropsies may include beached birds, birds from human activities, such as fishery bycatch, and from rehabilitation centres. The major advantages of necropsies are the simplicity of collecting different tissues for complementary analysis (e.g., histopathological analysis) and the possibility of determining the entire plastic burden, as well as birds' basal information such as age, sex, possible cause of death and body condition. However, necropsies are opportunistic, depending on the availability of deceased birds, and pre-planned sampling can be difficult (Provencher et al. 2019b). Examination of food remains includes birds' regurgitations (e.g., pellet or bolus; Provencher et al. 2017). Although not all species can regurgitate and a complete sample is not guaranteed, induced regurgitations by water-offloading (lavage or flushing) or emetics can be repeated on the same populations throughout time allowing local comparisons of the plastic loads (Bond and Lavers 2013, Lavers et al. 2014). Pellet collection is the simplest method, allows repeatability and regularity in sampling, however, it is difficult to attribute pellets to specific individuals (Provencher et al. 2019b, Lopes et al. 2021). A portion of ingested plastics (typically microplastics) can also be excreted via seabirds' faeces, and therefore analysing seabirds' guano can elucidate the occurrence of microplastics in their system (Provencher et al. 2018, Bourdages et al. 2021). It is extremely important to standardise sample collection, processing, plastic quantification and reporting in order to ensure data on entanglements (Ryan 2018), nest plastic incorporation (O'Hanlon et al. 2017) and ingestion of plastics by seabirds (Provencher et al. 2017) can function as effective monitors of plastic pollution in the surrounding environment. Currently, there are many regional, national and international strategies to prevent and mitigate marine plastic pollution, although regional reviews indicate that they often do not link directly to seabird populations (Linnebjerg et al. 2021). Additionally, none has a level of commitment consistent with the global magnitude and rapid growth of the problem (Borrelle et al. 2017). With the exception of the OSPAR Convention that aim to achieve less than 10% of beached Northern Fulmars with less than 0.1 g of ingested plastics (van Franeker et al. 2011), international policies that acknowledge the globality of plastic pollution remain, most of the times, too ambitious and lack in defined and concise targets to be achieved (Borrelle et al. 2020). Only the standardisation of methods and the establishment of long-term monitoring programs will allow researchers to measure the impacts of plastic debris in marine vertebrates, including seabirds, and to reliably assess the progress and success in achieving the targets recommended by the international strategies (Provencher et al. 2020). This will provide data on the true magnitude of plastic pollution, enable meta-analysis and large-scale temporal, spatial and taxonomic comparisons of plastic accumulation.

Factors influencing seabird-plastic interactions

As marine plastic pollution is increasing, so is the probability of seabirds to become entangled (Ryan 2018). Kühn et al. (2015) concluded that all seabird species are at risk of entanglement, and the list of affected species is continuously increasing. Many seabird species use seaweed or vegetation to build their nests but given the resemblance between some plastics and natural nesting materials, birds actively collect plastics and incorporate them into their nests (Lavers et al. 2013, O'Hanlon et al. 2017, Battisti et al. 2019). As nesting material is presumably collected by seabirds locally (Hartwig et al. 2007, Bond et al. 2012, Thompson et al. 2020), the prevalence of plastic incorporated in seabird nests is thought to be related to both the availability of plastics and natural nesting materials in the surroundings of the breeding colony (Bond et al. 2012, Witteveen et al. 2017, O'Hanlon et al. 2017).

Ingestion of plastics by seabirds occur either accidentally while foraging on other prey items (i.e., indirectly by secondary ingestion) or directly by confounding plastics with prey (Laist 1997, Cadée 2002). Some seabirds also seem to exhibit a preference for ingesting different colours of plastic, which varies depending on the species and location (Lavers et al. 2014, Lavers and Bond 2016a). Foraging behaviour is thought to be an important predictor of the incidence of plastic ingestion

(Moser and Lee 1992, Avery-Gomm et al. 2013). Seabirds can be classified into three main foraging types: surface-feeders, plungers and pursuit divers (Ashmole 1971). Plastics are more concentrated at the water surface than in the subsequent layers of the water column due to buoyancy (Cózar et al. 2014, Reisser et al. 2015), therefore, surface-feeders are more susceptible to interacting with plastics than diving species that catch prey items beneath the water's surface. Thus, diving species are less likely (although not completely immune) to ingesting plastic (Provencher et al. 2015, Poon et al. 2017, O'Hanlon et al. 2017, Baak et al. 2020). Seabirds' foraging range may also be important in predicting the susceptibility to plastic ingestion due to the relative concentrations of plastic pollution in different regions (e.g., coastal, continental shelf, offshore waters). At-sea sampling and models indicate that large-scale currents transport plastic debris from the continents, where they are mostly but not exclusively (e.g., abandoned fishing nets) released, to central convergence zones in the open ocean, where marine plastics tend to accumulate in greater densities (Law et al. 2010, Lebreton et al. 2012). Therefore, seabirds that forage farther offshore may be more prone to interact with plastics than inshore, coastal seabird species (Avery-Gomm et al. 2013, Baak et al. 2020), however this is dependent on regional differences in plastic pollution. Variations in plastic accumulation in birds' upper digestive system may be attributed to species-specific regurgitation abilities (Ryan 1987). Some species (e.g., skuas, albatross, gulls) are known to regurgitate indigestible prey remains, which may limit the accumulation of plastics (Shealer 2002). Other species (e.g., fulmars, petrels, some auks) do not typically regurgitate indigestible prey due to a narrow passage between the proventriculus and the gizzard that prevent the returning of accumulated plastics to the proventriculus, retaining ingested plastics in the gizzard and acting as reservoirs (Ryan 1987, van Franeker and Law 2015).

Metal contamination in seabirds

Metals can affect seabirds indirectly by altering habitats, species ecological interactions and the availability and quality of resources through, for example, decreasing of essential antioxidants (e.g., carotenoids and vitamins) in their diet, thus increasing their oxidative stress levels (Koivula and Eeva 2010). On the other hand, impacts can be directly related to the level and duration of exposure to these compounds through diet, water, sediments, and air (Burger and Gochfeld 2002, Theron et al. 2011). In comparison to organic pollutants, ecotoxicological thresholds in birds have been reasonably well established for some metals (Scheuhammer 1987, Burger and Gochfeld 2002). However, some seabird species appear to have a natural tolerance to high metal concentrations (Hindell et al. 1999), potentially due to their higher uptake of these metals through marine food webs, for instance species feeding on cephalopods which naturally contain high levels of mercury (Neff 2002). Thus, high metal concentrations do not necessarily indicate pollution and sometimes can be related to natural accumulation or detoxification processes (Furness 1993).

Burger (1993) and Burger and Gochfeld (2000) established deleterious effect levels for non-essential metal concentrations in feathers of seabirds to be from 100 (shearwaters) to 2000 (terns) ppb for Cd, 4000 ppb for Pb and 5000 ppb for Hg. These detrimental effects can be related to reproduction, chick development, behavioural changes, parasitic infections and reduced immune function. For example, Pb levels can negatively affect the body condition of Common Eider Ducks (*Somateria mollissima*), suggesting a lower investment in reproduction (Provencher et al. 2016). Also, Hg is known to negatively affect the breeding probability of Southern Skua species (*Catharacta* spp.; Goutte et al. 2014b) and Wandering Albatrosses (*Diomedea exulans*) by endocrine disruption of the reproductive system (Goutte et al. 2014a). However, Hindell et al. (1999) suggest that Wandering Albatrosses can handle very high Hg concentrations with no negative effects. Another study found that Common Tern (*Sterna hirundo*) chicks from polluted areas had higher levels of oxidative stress and lower body mass, which can negatively affect the survival of juveniles (Oudi et al. 2019). Likewise, Burger and Gochfeld (2002, 2004) described the known effects of Cd and Pb in chick development and behaviour (e.g., locomotion, balance, recognition skills, feeding, etc.).

Moreover, metal contamination can be associated with increased parasitic infections and reduced immune function (Wayland et al. 2001, 2002). These toxic effects are in part due to metal-induced oxidative stress (Ercal et al. 2001), which is characterised by an imbalance between reactive oxygen species' production and antioxidant defences (Valko et al. 2005, Valavanidis et al. 2006). Thus, oxidative stress can induce morphological, neurological and physiological alterations like DNA damage, lipid peroxidation or enzyme damage, with serious implications for contaminated individuals (Valavanidis et al. 2006, Koivula and Eeva 2010). Metal-related oxidative stress has been recently described in free-living seabirds (Fenstad et al. 2016, Espín et al. 2016, Costantini et al. 2020, Soldatini et al. 2020), which is an important contribution to understand metal contamination hazards for seabirds.

Metal pollution monitoring tools

Some specific tissues tend to accumulate higher metal concentrations than others and therefore, are generally considered appropriate as monitoring tools (Monteiro and Furness 1995). Internal tissues (e.g., liver, kidneys, bones and muscles) have been used in studies of metal contamination assessment and detoxification mechanisms in seabirds. However, this is an invasive method often requiring the sacrifice of birds (Monteiro and Furness 1995), or utilising already dead individuals (e.g., disease or bycatch) which may be difficult, time consuming and biased (e.g., cause of death for beach-washed birds is unknown). Thus, for logical conservation and ethical reasons it became necessary to use non-invasive alternatives (Monteiro and Furness 1995). The eggs, feathers and blood can be collected with a minimum invasiveness and allow re-sampling from the same individuals (Monteiro and Furness 1995). Yet, there are some limitations, for instance, eggs do not allow collecting data from males, thus are not representative of the entire population (Monteiro and Furness 1995), and only reflect contamination during the pre- and egg-laying periods (Albert et al. 2019).

On the other hand, feathers are often appropriate for monitoring studies because they are the main excretion route of some metals like Hg (feathers may contain over 70–90% of the Hg body burden; Honda et al. 1986, Monteiro and Furness 2001). During moulting, concentration levels of metals are lowered in soft tissues as growing feathers are the main excretion routes for these contaminants (Burger 1993, Monteiro and Furness 1995). Thus, first-moult feathers tend to have higher metal concentrations, by combination of current dietary intake, long-term exposure and accumulation between moults, while later-moult feathers show a gradual decline in concentration levels of metals (Furness and Camphuysen 1997). This variation in metal concentrations among feathers has become useful when assessing contamination in migratory species, as it is possible to interpret metal levels in accordance with species moulting patterns and location during feather growth (Ramos et al. 2009, Albert et al. 2021). For an overall measure of exposure, it is recommended to sample from synchronous body feathers which should have the same chemical composition and thus little variation among feathers (Carravieri et al. 2014a).

Blood metal concentrations are representative of current dietary (medium- to short-term) exposure and transport of metals from other internal tissues to the bloodstream (Monteiro and Furness 2001, Albert et al. 2019). By using blood as a monitoring unit, it is possible to assess contamination when seabirds are accessible for sampling, often during the breeding season (from the pre-laying to the late chick-rearing period, see Lavoie et al. 2014). Additionally, metal levels in blood are also a good reflection of contamination burdens in internal tissues. For instance, strong correlations in Hg concentrations have been found between blood, the brain and muscle (Fort et al. 2015, Mallory et al. 2018). Moreover, blood is not as hampered by external contamination, as feathers often are (Bond and Diamond 2008) by atmospheric deposition of metals on their surface (Furness and Camphuysen 1997). More important, different tissues provide information on specific time periods and therefore areas and prey types exploited by these birds (e.g., feathers from different parts of the bird reflect exposure during the period of their growth; Finger et al. 2015, Albert et al.

2019), and such studies are needed for migratory birds which may be experiencing differing pollution levels throughout their annual cycle. Non-migratory seabirds (e.g., Little Penguin *Eudyptula minor*) that can be sampled throughout the entire annual cycle can give an insight into seasonal metal accumulation and detoxification of their local foraging grounds, as well as be an accurate indicator for bioavailable contamination (Finger et al. 2015, 2016, 2017).

Factors influencing metal contamination

Seabirds are relatively well-known organisms in terms of lifespan, ecology, foraging behaviour and distribution patterns (Furness and Monaghan 1986, Schreiber and Burger 2001), which makes them suitable to use as bioindicators of metal contamination.

Differences in feeding strategies are among the most important factors influencing metal concentrations in seabirds (Fig. C3.2). Birds feeding at higher trophic levels should show higher contamination levels due to bioaccumulation and biomagnification processes (Bearhop et al. 2000, Anderson et al. 2009). Therefore, metal concentrations should be corrected for the individual's trophic position (e.g., using stable isotope analyses) to determine the 'true' level of contamination in its' diet (Bond and Lavers 2014, 2020). Besides, specific prey preferences can also influence metal concentrations. For instance, seabird species like Audouin's Gull (*Ichthyaetus audouinii*), that feed mainly on fishery discards usually show higher metal concentrations than species feeding on smaller pelagic fish species and/or on other terrestrial/freshwater food resources such as the Yellow-legged Gull (*Larus michahellis*; Arcos et al. 2002, Laranjeiro et al. 2020). Demersal and benthic fish species from fishery discards are known to accumulate high burdens of non-essential metals like Hg, which are extra sources of contamination otherwise unavailable as natural prey to seabirds (Arcos et al. 2002, Bocher et al. 2003). Other specific prey like cephalopods are known to naturally carry high burdens of Cd in their system (Bustamante et al. 1998), and are the main vectors of bioaccumulation of this metal to some procellariform seabird species (e.g., *Calonectris* spp. shearwaters, Stewart et al.

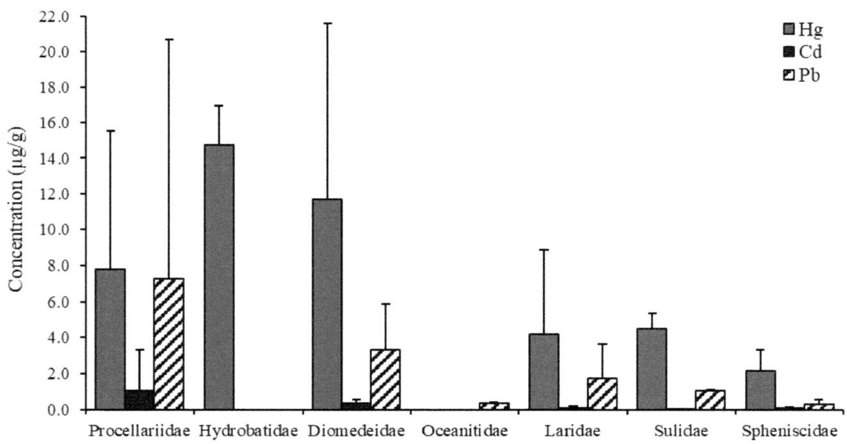

Figure C3.2. Mean concentrations of Hg, Cd and Pb (error bars indicate standard deviation), in feathers of seabirds from seven different families around the world: Procellariidae (15 species, 11 studies), Hydrobatidae (3 species, 2 studies), Diomedeidae (5 species, 5 studies), Oceanitidae (2 species, 1 study), Laridae (11 species, 11 studies), Sulidae (2 species, 2 studies) and Spheniscidae (7 species, 5 studies). For Hydrobatidae and Oceanitidae, only studies measuring Hg and Pb, respectively, were found. Study sources: Abdennadher et al. (2010), Anderson et al. (2010, 2009), Arcos et al. (2002), Barbieri et al. (2010), Bond and Lavers (2011), Burger et al. (2009, 2008), Burger and Gochfeld (2000), Carvalho et al. (2013), Cipro et al. (2014), Finger et al. (2015), Fromant et al. (2016), Jerez et al. (2011), Laranjeiro et al. (2020), Lavers et al. (2014), Lavers and Bond (2016b), Mendes et al. (2008), Monteiro et al. (1999, 1995), Pacyna et al. (2019), Pilcher et al. (2020), Ramos et al. (2013), Sanpera et al. (2007), Seco Pon et al. (2011), Squadrone et al. (2019, 2018).

1997; Grey-headed Albatrosses *Thalassarche chrysostoma* and White-chinned Petrels *Procellaria aequinoctialis*, Anderson et al. 2010).

Age is also very important when assessing contamination in seabirds. Overall, metal concentrations increase with age due to bioaccumulation during seabirds' long lifespan (Barbieri et al. 2010, Carravieri et al. 2014b). However, in chicks, these levels can decrease with age. Two studies suggested that some essential metals (Zn, Fe and Mn) for chick development (bone/tissue growth and feathering) might decrease as they are incorporated in tissues and feathers during their development (Kim et al. 2010, Lerma et al. 2020). In Little Terns (*Sternula albifrons*) and other seabird species, toxic metals like Hg can also decrease with chick age, as metal burdens from egg exposure can be diluted through increase of body mass/size and excretion into growing feathers (Paiva et al. 2008, Ackerman et al. 2011). In some species, like the Wandering Albatross, young pre-breeders might have higher metal concentrations than adults. While juveniles have lower frequency and efficiency in moulting, adults can moult regularly through their annual cycle and thus maintain more constant and lower levels of Hg in their system (Tavares et al. 2013, Bustamante et al. 2016).

Sex and reproductive processes could lead to differences in metal accumulation rates between males and females. During the breeding season, it is expected to find lower contamination levels in females than males, as egg production and laying presents an additional excretion route for females (Monteiro and Furness 1995, Ramos et al. 2009). However, metal excretion to eggs may be not so substantial due to differences in foraging strategies between males and females driven by sexual dimorphism or other factors such as intra-specific competition or individual specialisation. Some species, like Wandering Albatrosses, White-chinned and Giant petrels (*Macronectes* spp.), show sexual segregation in foraging areas and prey preferences which can lead to alterations in metal concentrations (González-Solís et al. 2002, Carvalho et al. 2013, Tavares et al. 2013, Bustamante et al. 2016).

Seabirds as vectors of plastic and metal pollution

Seabirds tend to carry high concentrations of contaminants, including metals and plastics, as a result of biomagnification and bioaccumulation processes through marine food webs (Blais et al. 2007, van Franeker et al. 2011, Kühn and van Franeker 2020). Thus, mobile and migratory species constitute a significant biovector of both microplastics and metals from marine (foraging areas) to terrestrial (breeding areas) habitats, through regurgitation, defecation, dropping food, moulting or mortality (Ellis et al. 2006, Bauer and Hoye 2014, Provencher et al. 2018). This process is particularly evident for species gathering in large colonies to reproduce, thus creating localized "hotspots" of contamination (Blais et al. 2005, Sun et al. 2013). Although this biological pathway of contamination is still poorly understood for plastics (even though work is expanding this knowledge—see Grant et al. 2021, Hamilton et al. 2021), it is well studied for metals, and it may be as significant as abiotic processes such as wind and water currents (Blais et al. 2007).

To date, plastics were thought to move mainly from the terrestrial environment to the marine ecosystem (Cole et al. 2011), however recent work has shown that plastics also move the other way around, from the marine environment to terrestrial landscapes (Allen et al. 2020). Microplastics were detected in guano of seabirds known to ingest plastic: the Northern Fulmar and the Thick-billed Murre (*Uria lomvia*; Provencher et al. 2018, Bourdages et al. 2021). Also, recently, Grant et al. 2021 estimated that a single seabird species (Flesh-footed Shearwater) is transporting and depositing a total of ~ 690 000 pieces of plastic to their colonies on a single island every year, suggesting that seabirds, especially colonial breeders, may contribute to the transport and concentration of plastics in the areas around their breeding grounds, possibly acting as a vector of plastic pollution (Provencher et al. 2018, Bourdages et al. 2021, Grant et al. 2021, Hamilton et al. 2021). This route of plastics through colonial breeding seabirds needs further studies and should be considered when monitoring for plastics in regions where these birds are known to nest in large numbers.

There are several studies showing that transportation of metals by seabirds to terrestrial environments influences the chemical properties of plants and soil in the proximity of breeding colonies (Liu et al. 2006, Mallory et al. 2015, Ziółek et al. 2017). Consequently, terrestrial consumers living in these areas are more prone to ingest contaminants of marine origin by feeding on resources from these contaminated habitats (Evenset et al. 2007, Choy et al. 2010). The enrichment of terrestrial habitats with metals by seabirds, close to breeding colonies, is well described in polar environments (Huang et al. 2014, Mallory et al. 2015, Outridge et al. 2016, Cipro et al. 2018), and throughout the globe (Liu et al. 2006, Xu et al. 2011, 2019, Otero et al. 2018) suggesting this issue is likely widespread, with impacts and especially mitigation measures poorly described.

How plastic exposure affects seabirds

Seabirds have long been regarded as sentinels of ocean health, and careful monitoring of their populations has provided information on changes in key marine features such as fish abundance (e.g., Montevecchi et al. 2009). More recently, chemical concentrations in seabird tissues, or plastic debris items incorporated in seabird nests, have been used as bioindicators of pollution in the marine environment (Finger et al. 2015, Grant et al. 2018, Bond and Lavers 2020). A handful of review papers have documented changes in the exposure of seabirds to chemical and physical pollution over time and suggest conditions are deteriorating. For example, over the past two decades, three comprehensive reviews of ingested debris by seabirds clearly indicate an increase in the number of species at risk from plastic ingestion. In the mid-1990s, 35.6% of seabird species had been documented to ingest plastic (111 out of 312 species; Laist 1997). By 2014, that number had increased to 39.1% (122 out of 312 seabirds; Gall and Thompson 2015) and in 2019, the proportion of seabirds known to ingest plastic was 44.0% (180 out of 409 species; Kühn and van Franeker 2020).

To date, the bulk of literature on ingested plastic in seabirds is focussed on reporting Frequency of Occurrence (FO), particle size, type, and colour of items (Provencher et al. 2017). While these studies provide valuable information on plastic pollution that can be used to track changes over time, the volume of literature might convey a false sense of the depth of our understanding of how plastic pollution affects seabirds. In reality, most of these studies focused on the litter as the central driving question, few were centred around seabirds, and specifically how individuals and populations may be affected by ingested plastic pollution. How much do we know about what life is like for seabirds living in or near an ocean that is rapidly filling up with plastics (Jambeck et al. 2015)? Quantitative data on the complex issues of how ingested plastic pollution may alter seabird physiology is extremely limited. For example, our understanding of how plastic exposure alters seabird blood chemistry is based on only one paper (Lavers et al. 2019), and while useful, this study documented the negative effects of ingested plastic in only one seabird species, Flesh-footed Shearwaters, at a single location, Lord Howe Island, New South Wales. Even when common metrics (i.e., changes in body condition) have been examined across multiple studies of the same species, no clear relationship between plastic pollution and bird health has emerged, and analyses are often based on small sample sizes (15–65 birds; Ryan 1987, Carey 2011, Lavers et al. 2014, Puskic et al. 2019).

In light of these and other limitations in the plastic literature, it is not surprising researchers have struggled to generate quantitative data directly linking marine plastics to population trends in seabirds and other marine wildlife (Lavers et al. 2014, Rochman et al. 2016). However, the data currently available does suggest exposure of seabirds to plastics is increasing (Kühn and van Franeker 2020). For example, the volume and number of items consumed per bird over time has increased in many species, such as the Northern Fulmar (van Franeker and Law 2015). Within seabirds, there are a handful of species generally considered to be the most at-risk from plastic based on high rates of ingestion (see Fig. 5.1 in Avery-Gomm 2020, Lavers et al. 2014): Short-tailed

Shearwater, Flesh-footed Shearwater, Buller's Shearwater (*Puffinus bulleri*), Laysan Albatross, and Fork-tailed Storm-petrel (*Oceanodroma fucata*). Current plastic ingestion rates in chicks/juveniles of these species ranges from 83–98% (Lavers et al. 2014, Avery-Gomm 2020) with individual birds containing 0–26 plastic items in Short-tailed Shearwaters (Puskic et al. 2020) to 18 ± 34 SD and 134 ± 104 items in Flesh-footed Shearwaters and Laysan Albatrosses, respectively (Lavers and Bond 2016a, 2021). Remarkably, despite > 30 studies investigating plastic ingestion in these five species over the past six decades, only three studies have examined the relationship between plastic load and trace element concentrations in the birds' tissues. One study found no relationship between ingested plastic and trace elements in Short-tailed Shearwater breast muscle (Puskic et al. 2020), while two studies found a relationship with chlorine, iron, lead, manganese and rubidium concentrations in Laysan Albatross feathers (Lavers and Bond 2016b), and chromium and silver in Flesh-footed Shearwater feathers (Lavers et al. 2014). No published data exists for species such as the Buller's Shearwater or Fork-tailed Storm-petrel, and overall, there does not appear to be a consistent pattern in which trace elements are derived from ingested plastic. While recent simulated digestion and *in vivo* feeding experimental studies indicate ingested plastics can introduce hazardous elements (e.g., arsenic, cadmium) into a bird's bloodstream (Turner 2018, Smith and Turner 2020), exposure to trace and essential elements is influenced by a range of factors, such as polymer type and seabird physiology (Smith and Turner 2020). Further studies, including those involving free-living birds, are needed to better understand this complex and critical issue.

There is currently a lack of evidence of the sub-lethal effects and physiological harm from ingested plastic pollution in seabirds. This is due, in part, to the metrics and approaches commonly used by scientists to track pollution in wildlife populations often not being designed to detect subtle changes in health. While there are some acute effects on seabirds from ingested plastic pollution (i.e., pierced gastrointestinal tract), most are likely to be subtle. In the case of the Flesh-footed Shearwater, detailed examination of free-living birds that appeared "normal" uncovered a range of impacts associated with ingested plastics that were invisible to the naked eye: these studies found altered blood chemistry, including elevated cholesterol, and increased concentrations of non-essential metals including chromium along with reduced body mass and wing length, both of which are critical to survival in young birds (Lavers et al. 2014, 2019). Generating this unique data for the Flesh-footed Shearwater required the application of a multi-disciplinary, physio-ecotoxicological approach. Application of these techniques to suite of seabird species would enhance our understanding of the subtle effects of plastic ingestion, and would require a critical review of priority issues, such as what elements are most likely to be associated with ingested plastics, and consequent physiological responses. This will help us elucidate the true cost of marine plastic pollution and envision future scenarios of what life will be like for seabirds living in oceans filling up with plastic.

Acknowledgments

We are very grateful for the opportunity to participate in this project. We thank the editorial team (Jaime Ramos and Leonel Pereira) for the invitation and for helping us along this challenging process. We also thank Jaime Ramos and Filipe Ceia for their constructive comments and suggestions, which contributed considerably to the improvement of the manuscript.

References cited

Abdennadher, A., F. Ramírez, M.S. Romdhane, X. Ruiz, L. Jover and C. Sanpera. 2010. Biomonitoring of coastal areas in Tunisia: Stable isotope and trace element analysis in the Yellow-legged Gull. Mar. Pollut. Bull. 60: 440–447. doi: 10.1016/j.marpolbul.2009.10.003.

Ackerman, J.T., C.A. Eagles-Smith and M.P. Herzog. 2011. Bird mercury concentrations change rapidly as chicks age: Toxicological risk is highest at hatching and fledging. Environ. Sci. Technol. 45: 5418–5425. doi: 10.1021/es200647g.

Albert, C., M. Renedo, P. Bustamante and J. Fort. 2019. Using blood and feathers to investigate large-scale Hg contamination in Arctic seabirds: A review. Environ. Res. 177: 108588. doi: 10.1016/j.envres.2019.108588.

Albert, C., H.H. Helgason, M. Brault-Favrou, G.J. Robertson, S. Descamps and F. Amélineau. 2021. Seasonal variation of mercury contamination in Arctic seabirds: A pan-Arctic assessment. Sci. Total Environ. 750: 142201. doi: 10.1016/j.scitotenv.2020.142201.

Allen, S., D. Allen, K. Moss, G. Le Roux, V.R. Phoenix and J.E. Sonke. 2020. Examination of the ocean as a source for atmospheric microplastics. PLoS One. 15: e0232746. doi: 10.1371/journal.pone.0232746.

Anderson, O.R.J., R.A. Phillips, R.A. McDonald, R.F. Shore, R.A.R. McGill and S. Bearhop. 2009. Influence of trophic position and foraging range on mercury levels within a seabird community. Mar. Ecol. Prog. Ser. 375: 277–288. doi: 10.3354/meps07784.

Anderson, O.R.J., R.A. Phillips, R.F. Shore, R.A.R. McGill, R.A. McDonald and S. Bearhop. 2010. Element patterns in albatrosses and petrels: Influence of trophic position, foraging range, and prey type. Environ. Pollut. 158: 98–107. doi: 10.1016/j.envpol.2009.07.040.

Andrady, A.L. and M.A. Neal. 2009. Applications and societal benefits of plastics. Philos. Trans. R. Soc. B364: 1977–1984. doi: 10.1098/rstb.2008.0304.

Andrady, A.L. 2011. Microplastics in the marine environment. Mar. Pollut. Bull. 62: 1596–1605. doi: 10.1016/j.marpolbul.2011.05.030.

Andrady, A.L. 2017. The plastic in microplastics: a review. Mar. Pollut. Bull. 119: 12–22. doi: 10.1016/j.marpolbul.2017.01.082.

Arcos, J.M., X. Ruiz, S. Bearhop and R.W. Furness. 2002. Mercury levels in seabirds and their fish prey at the Ebro Delta (NW Mediterranean): The role of trawler discards as a source of contamination. Mar. Ecol. Prog. Ser. 232: 281–290. doi: 10.3354/meps232281.

Arthur, C., J. Baker and H. Bamford. 2009. Proceedings of the international research workshop on the occurrence, effects and fate of microplastic marine debris. Sept 9–11, 2008. National Oceanic and Atmospheric Administration Technical Memorandum NOS-OR & R-30. University of Washington Tacoma, Tacoma, WA, USA.

Ashmole, P.N. 1971. Seabird ecology and the marine environment. pp. 223–286. *In:* Farner, D.S. and J.R. King [eds.]. Avian Biology. Academic Press, New York.

Avery-Gomm, S.M., J.F. Provencher, K.H. Morgan and D.F. Bertram. 2013. Plastic ingestion in marine-associated bird species from the eastern North Pacific. Mar. Pollut. Bull. 72: 257–259. doi: 10.1016/j.marpolbul.2013.04.021.

Avery-Gomm, S.M. 2020. Plastic pollution and conservation of imperilled seabird species. Ph.D. Thesis, The University of Queensland, Australia.

Avio, C.G., S. Gorbi and F. Regoli. 2017. Plastics and microplastics in the oceans: From emerging pollutants to emerged threat. Mar. Environ. Res. 128: 2–11. doi: 10.1016/j.marenvres.2016.05.012.

Baak, J.E., J.F. Provencher and M.L. Mallory. 2020. Plastic ingestion by four seabird species in the Canadian Arctic: Comparisons across species and time. Mar. Pollut. Bull. 158: 111386. doi: 10.1016/j.marpolbul.2020.111386.

Bakir, A., I.A. O'Connor, S.J. Rowland, A.J. Hendriks and R.C. Thompson. 2016. Relative importance of microplastics as a pathway for the transfer of hydrophobic organic chemicals to marine life. Environ. Pollut. 219: 56–65. doi: 10.1016/j.envpol.2016.09.046.

Barbieri, E., E.D.A. Passos, A. Filippini, I.S. dos Santos and C.A.B. Garcia. 2010. Assessment of trace metal concentration in feathers of seabird (*Larus dominicanus*) sampled in the Florianópolis, SC, Brazilian coast. Environ. Monit. Assess. 169: 631–638. doi: 10.1007/s10661-009-1202–4.

Barnes, D.K.A., F. Galgani, R.C. Thompson and M. Barlaz. 2009. Accumulation and fragmentation of plastic debris in global environments. Philos. Trans. R. Soc. B 364: 1985–1998. doi: 10.1098/rstb.2008.0205.

Battisti, C., E. Staffieri, G. Poeta, A. Sorace, L. Luiselli and G. Amori. 2019. Interactions between anthropogenic litter and birds: a global review with a 'black-list' of species. Mar. Pollut. Bull. 138: 93–114. doi: 10.1016/j.marpolbul.2018.11.017.

Bauer, S. and B.J. Hoye. 2014. Migratory animals couple biodiversity and ecosystem functioning worldwide. Science. 344: 1242552. doi: 10.1126/science.1242552.

Bearhop, S., S. Waldron, D. Thompson and R.W. Furness. 2000. Bioamplification of mercury in Great Skua *Catharacta skua* chicks: The influence of trophic status as determined by stable isotope signatures of blood and feathers. Mar. Pollut. Bull. 40: 181–185. doi: 10.1016/S0025-326X(99)00205-2.

Blais, J.M., L.E. Kimpe, D. McMahon, B.E. Keatley, M.L. Mallory, M.S.V. Douglas et al. 2005. Arctic seabirds transport marine-derived contaminants. Science. 309: 445. doi: 10.1126/science.1112658.

Blais, J.M., R.W. Macdonald, D. Mackay, E. Webster, C. Harvey and J.P. Smol. 2007. Biologically mediated transport of contaminants to aquatic systems. Environ. Sci. Technol. 41: 1075–1084. doi: 10.1021/es061314a.

Bocher, P., F. Caurant, P. Miramand, Y. Cherel and P. Bustamante. 2003. Influence of the diet on the bioaccumulation of heavy metals in zooplankton-eating petrels at Kerguelen archipelago, Southern Indian Ocean. Polar Biol. 26: 759–767. doi: 10.1007/s00300-003-0552-6.

Bond, A.L. and A.W. Diamond. 2008. High within-individual variation in total mercury concentration in seabird feathers. Environ. Toxicol. Chem. 27: 2375–2377. doi: 10.1897/08–163.1

Bond, A.L. and J.L. Lavers. 2011. Trace element concentrations in feathers of Flesh-footed Shearwaters (*Puffinus carneipes*) from across their breeding range. Arch. Environ. Contam. Toxicol. 61: 318–326. doi: 10.1007/s00244-010-9605–3.

Bond, A.L., W.A. Montevecchi, N. Guse, P.M. Regular, S. Garthe and J.-F. Rail. 2012. Prevalence and composition of fishing gear debris in the nests of northern gannets (*Morus bassanus*) are related to fishing effort. Mar. Pollut. Bull. 64: 907–911. doi: 10.1016/j.marpolbul.2012.03.011.

Bond, A.L. and J.L. Lavers. 2013. Effectiveness of emetics to study plastic ingestion by Leach's Storm-petrels (*Oceanodroma leucorhoa*). Mar. Pollut. Bull. 70: 171–175. doi: 10.1016/j.marpolbul.2013.02.030.

Bond, A.L. and J.L. Lavers. 2014. Climate change alters the trophic niche of a declining apex marine predator. Glob. Chang. Biol. 20: 2100–2107. doi: 10.1111/gcb.12554.

Bond, A.L. and J.L. Lavers. 2020. Biological archives reveal contrasting patterns in trace element concentrations in pelagic seabird feathers over more than a century. Environ. Pollut. 263: 114631. doi: 10.1016/j.envpol.2020.114631.

Borrelle, S.B., C.M. Rochman, M. Liboiron, A.L. Bond, A. Lusher, H. Bradshaw et al. 2017. Why we need an international agreement on marine plastic pollution. Proc. Natl. Acad. Sci. 114: 9994–9997. doi: 10.1073/pnas.1714450114.

Borrelle, S.B., J. Ringma, K.L. Law, C.C. Monnahan, L. Lebreton, A. McGivern et al. 2020. Predicted growth in plastic waste exceeds efforts to mitigate plastic pollution. Science. 369: 1515–1518. doi: 10.1126/science.aba3656.

Bourdages, M.P.T., J.F. Provencher, J.E. Baak, M.L. Mallory and J.C. Vermaire. 2021. Breeding seabirds as vectors of microplastics from sea to land: Evidence from colonies in Arctic Canada. Sci. Total Environ. 764: 142808.doi: 10.1016/j.scitotenv.2020.142808.

Boye, M., B.D. Wake, P. Lopez Garcia, J. Bown, A.R. Baker and E.P. Achterberg. 2012. Distributions of dissolved trace metals (Cd, Cu, Mn, Pb, Ag) in the southeastern Atlantic and the Southern Ocean. Biogeosciences. 9: 3231–3246. doi: 10.5194/bg-9-3231-2012.

Burger, J. 1993. Metals in avian feathers: Bioindicators of environmental pollution. Rev. Environ. Toxicol. 5: 203–311.

Burger, J. and M. Gochfeld. 2000. Metal levels in feathers of 12 species of seabirds from Midway Atoll in the northern Pacific Ocean. Sci. Total Environ. 257: 37–52. doi: 10.1016/S0048–9697(00)00496–4.

Burger, J. and M. Gochfeld. 2002. Effects of chemicals and pollution on seabirds. pp. 485–525. *In:* Schreiber, E.A. and J. Burger [eds.]. Biology of Marine Birds. CRC Press, Boca Raton, Fl., USA. doi: 10.1201/9781420036305.

Burger, J. and M. Gochfeld. 2004. Marine birds as sentinels of environmental pollution. EcoHealth. 1: 263–274. doi: 10.1007/s10393–004–0096–4.

Burger, J., M. Gochfeld, K. Sullivan, D. Irons and A. McKnight. 2008. Arsenic, cadmium, chromium, lead, manganese, mercury, and selenium in feathers of Black-legged Kittiwake (*Rissa tridactyla*) and Black Oystercatcher (*Haematopus bachmani*) from Prince William Sound, Alaska. Sci. Total Environ. 398: 20–25. doi: 10.1016/j.scitotenv.2008.02.051.

Burger, J., M. Gochfeld, C. Jeitner, S. Burke, C.D. Volz, R. Snigaroff et al. 2009. Mercury and other metals in eggs and feathers of glaucous-winged gulls (*Larus glaucescens*) in the Aleutians. Environ. Monit. Assess. 152: 179–194. doi: 10.1007/s10661-008-0306-6.

Bustamante, P., F. Caurant, S.W. Fowler and P. Miramand. 1998. Cephalopods as a vector for the transfer of cadmium to top marine predators in the north-east Atlantic Ocean. Sci. Total Environ. 220: 71–80. doi: 10.1016/S0048-9697(98)00250-2.

Bustamante, P., A. Carravieri, A. Goutte, C. Barbraud, K. Delord, O. Chastel et al. 2016. High feather mercury concentrations in the wandering albatross are related to sex, breeding status and trophic ecology with no demographic consequences. Environ. Res. 144: 1–10. doi: 10.1016/j.envres.2015.10.024.

Cadée, G.C. 2002. Seabirds and floating plastic debris. Mar. Pollut. Bull. 44: 1294–1295. doi: 10.1016/S0025-326X(02)00264-3.

Carey, M.J. 2011. Intergenerational transfer of plastic debris by Short-tailed Shearwaters (*Ardenna tenuirostris*). Emu 111: 229–234. doi: 10.1071/MU10085.

Carravieri, A., P. Bustamante, C. Churlaud, A. Fromant and Y. Cherel. 2014a. Moulting patterns drive within-individual variations of stable isotopes and mercury in seabird body feathers: Implications for monitoring of the marine environment. Mar. Biol. 161: 963–968. doi: 10.1007/s00227-014-2394–x.

Carravieri, A., Y. Cherel, P. Blévin, M. Brault-Favrou, O. Chastel and P. Bustamante. 2014b. Mercury exposure in a large subantarctic avian community. Environ. Pollut. 190: 51–57. doi: 10.1016/j.envpol.2014.03.017.

Carvalho, P.C., L. Bugoni, R.A.R. McGill and A. Bianchini. 2013. Metal and selenium concentrations in blood and feathers of petrels of the genus *Procellaria*. Environ. Toxicol. Chem. 32. 1641–1648. doi: 10.1002/etc.2204.

Choy, E.S., M. Gauthier, M.L. Mallory, J.P. Smol, M.S.V. Douglas, D. Lean et al. 2010. An isotopic investigation of mercury accumulation in terrestrial food webs adjacent to an Arctic seabird colony. Sci. Total Environ. 408: 1858–1867. doi: 10.1016/j.scitotenv.2010.01.014.

Cipro, C.V.Z., Y. Cherel, F. Caurant, P. Miramand, P. Méndez-Fernandez and P. Bustamante. 2014. Trace elements in tissues of white-chinned petrels (*Procellaria aequinoctialis*) from Kerguelen waters, Southern Indian Ocean. Polar Biol. 37: 763–771. doi: 10.1007/s00300-014-1476–z.

Cipro, C.V.Z., P. Bustamante, M.V. Petry and R.C. Montone. 2018. Seabird colonies as relevant sources of pollutants in Antarctic ecosystems: Part 1 – Trace elements. Chemosphere. 204: 535–547. doi: 10.1016/j.chemosphere.2018.02.048.

Cole, M., P. Lindeque, C. Halsband and T.S. Galloway. 2011. Microplastics as contaminants in the marine environment: A review. Mar. Pollut. Bull. 62: 2588–2597. doi: 10.1016/j.marpolbul.2011.09.025.

Costantini, D., P. Bustamante, M. Brault-Favrou and G. Dell'Omo. 2020. Patterns of mercury exposure and relationships with isotopes and markers of oxidative status in chicks of a Mediterranean seabird. Environ. Pollut. 260: 114095. doi: 10.1016/j.envpol.2020.114095.

Coulson, J.C. 2001.Colonial breeding in seabirds. pp. 87–113. *In:* Schreiber, E.A. and J. Burger [eds.]. Biology of Marine Birds. CRC Press, Boca Raton, Fl., USA. doi: 10.1201/9781420036305.

Cousin, H.R., H.J. Auman, R. Alderman and P. Virtue. 2015. The frequency of ingested plastic debris and its effects on body condition of Short-tailed Shearwater (*Puffinus tenuirostris*) pre-fledging chicks in Tasmania, Australia. Emu 115: 6–11. doi: 10.1071/MU13086.

Cózar, A., F. Echevarria, J.I. Gonzalez-Gordillo, X. Irigoien, B. Ubeda, S. Hernandez-Leon et al. 2014. Plastic debris in the open ocean. Proc. Natl. Acad. Sci. 111: 10239–10244. doi: 10.1073/pnas.1314705111.

Derraik, J.G.B. 2002. The pollution of the marine environment by plastic debris: a review. Mar. Pollut. Bull. 44: 842–852. doi: 10.1016/S0025-326X(02)00220-5.

Duis, K. and A. Coors. 2016. Microplastics in the aquatic and terrestrial environment: Sources (with a specific focus on personal care products), fate and effects. Environ. Sci. Eur. 28: 2. doi: 10.1186/s12302-015-0069-y.

Ellis, J.C., J.M. Fariña and J.D. Witman. 2006. Nutrient transfer from sea to land: The case of gulls and cormorants in the Gulf of Maine. J. Anim. Ecol. 75: 565–574. doi: 10.1111/j.1365–2656.2006.01077.x.

Ercal, N., H. Gurer-Orhan and N. Aykin-Burns. 2001. Toxic metals and oxidative stress part I: Mechanisms involved in metal induced oxidative damage. Curr. Top. Med. Chem. 1: 529–539. doi: 10.2174/1568026013394831.

Espín, S., E. Martínez-López, P. Jiménez, P. María-Mojica and A.J. García-Fernández. 2016. Interspecific differences in the antioxidant capacity of two Laridae species exposed to metals. Environ. Res. 147: 115–124. doi: 10.1016/j.envres.2016.01.029.

Evenset, A., J. Carroll, G.N. Christensen, R. Kallenborn, D. Gregor and G.W. Gabrielsen. 2007. Seabird guano is an efficient conveyer of persistent organic pollutants (POPs) to Arctic lake ecosystems. Environ. Sci. Technol. 41: 1173–1179. doi: 10.1021/es0621142.

Farrell, P. and K. Nelson. 2013. Trophic level transfer of microplastic: *Mytilus edulis* (L.) to *Carcinus maenas* (L.). Environ. Pollut. 177: 1–3. doi: 10.1016/j.envpol.2013.01.046.

Fendall, L.S. and M.A. Sewell. 2009. Contributing to marine pollution by washing your face: microplastics in facial cleansers. Mar. Pollut. Bull. 58: 1225–1228. doi: 10.1016/j.marpolbul.2009.04.025.

Fenstad, A.A., A.J. Moody, M. Öst, K. Jaatinen, J.O. Bustnes, B. Moe et al. 2016. Antioxidant responses in relation to persistent organic pollutants and metals in a low- and a high-exposure population of seabirds. Environ. Sci. Technol. 50: 4817–4825. doi: 10.1021/acs.est.6b00478.

Finger, A., J.L. Lavers, P. Dann, D. Nugegoda, J.D. Orbell, B. Robertson et al. 2015. The Little Penguin (*Eudyptula minor*) as an indicator of coastal trace metal pollution. Environ. Pollut. 205: 365–377. doi: 10.1016/j.envpol.2015.06.022.

Finger, A., J.L. Lavers, J.D. Orbell, P. Dann, D. Nugegoda and C. Scarpaci. 2016. Seasonal variation and annual trends of metals and metalloids in the blood of the Little Penguin (*Eudyptula minor*). Mar. Pollut. Bull. 110: 261–273.doi: 10.1016/j.marpolbul.2016.06.055.

Finger, A., J.L. Lavers, P. Dann, N.D. Kowalczyk, C. Scarpaci, D. Nugegoda et al. 2017. Metals and metalloids in Little Penguin (*Eudyptula minor*) prey, blood and faeces. Environ. Pollut. 223: 567–574. doi: 10.1016/j.envpol.2017.01.059.

Fort, J., T. Lacoue-Labarthe, H.L. Nguyen, A. Boué, J. Spitz and P. Bustamante. 2015. Mercury in wintering seabirds, an aggravating factor to winter wrecks? Sci. Total Environ. 527–528: 448–454. doi: 10.1016/j.scitotenv.2015.05.018.

Fromant, A., A. Carravieri, P. Bustamante, P. Labadie, H. Budzinski, L. Peluhet et al. 2016. Wide range of metallic and organic contaminants in various tissues of the Antarctic prion, a planktonophagous seabird from the Southern Ocean. Sci. Total Environ. 544: 754–764. doi: 10.1016/j.scitotenv.2015.11.114.

Furness, R.W. and P. Monaghan. 1986. Seabird ecology. *In:* Furness, R.W. and P. Monaghan [eds.]. Seabird Ecology. Springer US, Boston, MA. doi: 10.1007/978-1-4613-2093-7.

Furness, R.W. 1993. Birds as monitors of pollutants. pp. 86–143. *In:* Furness, R.W. and J.J.D. Greenwood [eds.]. Birds as Monitors of Environmental Change. Springer, Dordrecht, Netherlands. doi: 10.1007/978-94-015-1322-7_3.

Furness, R.W. and K.C.J. Camphuysen. 1997. Seabirds as monitors of the marine environment. ICES J. Mar. Sci. 54: 726–737. doi: 10.1006/jmsc.1997.0243.

Galgani, F., G. Hanke and T. Maes. 2015. Global distribution, composition and abundance of marine litter. pp. 29–56. *In:* Bergmann, M., L. Gutow and M. Klages [eds.]. Marine Anthropogenic Litter. Springer International Publishing, Cham, Switzerland. doi: 10.1007/978-3-319-16510-3_2.

Gall, S.C. and R.C. Thompson. 2015. The impact of debris on marine life. Mar. Pollut. Bull. 92: 170–179. doi: 10.1016/j.marpolbul.2014.12.041.

GESAMP. 2015. Sources, fate and effects of microplastics in the marine environment: A global assessment. *In*: Kershaw, P.J. [ed.]. IMO/FAO/UNESCO-IOC/UNIDO/WMO/ IAEA/UN/UNEP/UNDP Joint Group of Experts on the Scientific Aspects of Marine Environmental Protection. Rep. Stud. GESAMP No. 90. International Maritime Organization, London, UK.

González-Solís, J., C. Sanpera and X. Ruiz. 2002. Metals and selenium as bioindicators of geographic and trophic segregation in giant petrels *Macronectes* spp. Mar. Ecol. Prog. Ser. 244: 257–264. doi: 10.3354/meps244257.

Goutte, A., C. Barbraud, A. Meillère, A. Carravieri, P. Bustamante, P. Labadie et al. 2014a. Demographic consequences of heavy metals and persistent organic pollutants in a vulnerable long-lived bird, the wandering albatross. Proc. R. Soc. B Biol. Sci. 281: 20133313. doi: 10.1098/rspb.2013.3313.

Goutte, A., P. Bustamante, C. Barbraud, K. Delord, H. Weimerskirch and O. Chastel. 2014b. Demographic responses to mercury exposure in two closely related Antarctic top predators. Ecology 95: 1075–1086. doi: 10.1890/13-1229.1.

Grant, M.L., J.L. Lavers, S. Stuckenbrock, P.B. Sharp and A.L. Bond. 2018. The use of anthropogenic marine debris as a nesting material by brown boobies (*Sula leucogaster*). Mar. Pollut. Bull. 137: 96–103. doi: 10.1016/j.marpolbul.2018.10.016.

Grant, M.L., J.L. Lavers, I. Hutton and A.L. Bond. 2021. Seabird breeding islands as sinks for marine plastic debris. Environ. Pollut. 276: 116734. doi: 10.1016/j.envpol.2021.116734.

Gregory, M.R. 2009. Environmental implications of plastic debris in marine settings—entanglement, ingestion, smothering, hangers–on, hitch-hiking and alien invasions. Philos. Trans. R. Soc. B Biol. Sci. 364: 2013–2025. doi: 10.1098/rstb.2008.0265.

Hamilton, B.M., M.P.T. Bourdages, C. Geoffroy, J.C. Vermaire, M.L. Mallory, C.M. Rochman et al. 2021. Microplastics around an Arctic seabird colony: particle community composition varies across environmental matrices. Sci. Total Environ. 773: 145536. doi: 10.1016/j.scitotenv.2021.145536.

Hammer, S., R.G. Nager, P.C.D. Johnson, R.W. Furness and J.F. Provencher. 2016. Plastic debris in great skua (*Stercorarius skua*) pellets corresponds to seabird prey species. Mar. Pollut. Bull. 103: 206–210. doi: 10.1016/j.marpolbul.2015.12.018.

Hartwig, E., T. Clemens and M. Heckroth. 2007. Plastic debris as nesting material in a Kittiwake (*Rissa tridactyla*) colony at the Jammerbugt, Northwest Denmark. Mar. Pollut. Bull. 54: 595–597. doi: 10.1016/j.marpolbul.2007.01.027.

Hermabessiere, L., A. Dehaut, I. Paul-Pont, C. Lacroix, R. Jezequel, P. Soudant et al. 2017. Occurrence and effects of plastic additives on marine environments and organisms: A review. Chemosphere. 182: 781–793. doi: 10.1016/j.chemosphere.2017.05.096.

Herzke, D., T. Anker-Nilssen, T.H. Nøst, A. Götsch, S. Christensen-Dalsgaard, M. Langset et al. 2016. Negligible impact of ingested microplastics on tissue concentrations of persistent organic pollutants in Northern Fulmars off coastal Norway. Environ. Sci. Technol. 50: 1924–1933. doi: 10.1021/acs.est.5b04663.

Hindell, M.A., N. Brothers and R. Gales. 1999. Mercury and cadmium concentrations in the tissues of three species of southern albatrosses. Polar Biol. 22: 102–108. doi: 10.1007/s003000050396.

Honda, K., T. Nasu and R. Tatsukawa. 1986. Seasonal changes in mercury accumulation in the black-eared kite, *Milvus migrans lineatus*. Environ. Pollut. Ecol. Biol. 42: 325–334. doi: 10.1016/0143-1471(86)90016-4.

Huang, T., L. Sun, Y. Wang, Z. Chu, X. Qin and L. Yang. 2014. Transport of nutrients and contaminants from ocean to island by emperor penguins from Amanda Bay, East Antarctic. Sci. Total Environ. 468-469: 578–583. doi: 10.1016/j.scitotenv.2013.08.082.

Huang, W., B. Song, J. Liang, Q. Niu, G. Zeng, M. Shen et al. 2020. Microplastics and associated contaminants in the aquatic environment: a review on their ecotoxicological effects, trophic transfer, and potential impacts to human health. J. Hazard. Mater. 405: 124187. doi: 10.1016/j.jhazmat.2020.124187.

Jambeck, J.R., R. Geyer, C. Wilcox, T.R. Siegler, M. Perryman, A. Andrady et al. 2015. Plastic waste inputs from land into the ocean. Science. 347: 768–771. doi: 10.1126/science.1260352.

Jerez, S., M. Motas, M.J. Palacios, F. Valera, J.J. Cuervo and A. Barbosa. 2011. Concentration of trace elements in feathers of three Antarctic penguins: Geographical and interspecific differences. Environ. Pollut. 159: 2412–2419. doi: 10.1016/j.envpol.2011.06.036.

Kim, J., D.P. Lee and T.H. Koo. 2010. Effects of age on heavy metal concentrations of black-crowned night herons *Nycticorax nycticorax* from Korea. J. Environ. Monit. 12: 600–607. doi: 10.1039/b911598f.

Koivula, M.J. and T. Eeva. 2010. Metal-related oxidative stress in birds. Environ. Pollut. 158: 2359–2370. doi: 10.1016/j.envpol.2010.03.013.

Kühn, S., E.L. Bravo Rebolledo and J.A. van Franeker. 2015. Deleterious effects of litter on marine life. pp. 75–116. *In:* Bergmann, M., L. Gutow and M. Klages [eds.]. Marine Anthropogenic Litter. Springer International Publishing, Cham, Switzerland. doi: 10.1007/978-3-319-16510-3_4.

Kühn, S. and J.A. van Franeker. 2020. Quantitative overview of marine debris ingested by marine megafauna. Mar. Pollut. Bull. 151: 110858. doi: 10.1016/j.marpolbul.2019.110858.

Laist, D.W. 1987. Overview of the biological effects of lost and discarded plastic debris in the marine environment. Mar. Pollut. Bull. 18: 319–326. doi: 10.1016/S0025-326X(87)80019-X.

Laist, D.W. 1997. Impacts of marine debris: entanglement of marine life in marine debris including a comprehensive list of species with entanglement and ingestion records. pp. 99–139. *In:* Coe, J.M. and D.B. Rogers [eds.]. Marine Debris: Sources, Impacts and Solutions. Springer, New York, USA. doi: 10.1007/978-1-4613-8486-1_10.

Lambert, S., C. Sinclair and A. Boxall. 2014. Occurrence, degradation, and effect of polymer-based materials in the environment. pp. 1–53. *In:* Whitacre, D.M. [ed.]. Reviews of Environmental Contamination and Toxicology. Springer International Publishing, Cham, Switzerland. doi: 10.1007/978-3-319-01327-5_1.

Lamborg, C.H., C.R. Hammerschmidt, K.L. Bowman, G.J. Swarr, K.M. Munson, D.C. Ohnemus et al. 2014. A global ocean inventory of anthropogenic mercury based on water column measurements. Nature. 512: 65–68. doi: 10.1038/nature13563.

Laranjeiro, M.I., L.M.F. Alves, J.M. Silva, J.G. Calado, A.C. Norte, V.H. Paiva et al. 2020. Assessment of environmental health based on a complementary approach using metal quantification, oxidative stress and trophic ecology of two gull species (*Larus michahellis* & *Larus audouinii*) breeding in sympatry. Mar. Pollut. Bull. 159: 111439. doi: 10.1016/j.marpolbul.2020.111439.

Lavers, J.L., J.C. Hodgson and R.H. Clarke. 2013. Prevalence and composition of marine debris in Brown Booby (*Sula leucogaster*) nests at Ashmore Reef. Mar. Pollut. Bull. 77: 320–324. doi: 10.1016/j.marpolbul.2013.09.026.

Lavers, J.L., A.L. Bond and I. Hutton. 2014. Plastic ingestion by Flesh-footed Shearwaters (*Puffinus carneipes*): Implications for fledgling body condition and the accumulation of plastic-derived chemicals. Environ. Pollut. 187: 124–129. doi: 10.1016/j.envpol.2013.12.020.

Lavers, J.L. and A.L. Bond. 2016a. Selectivity of flesh-footed shearwaters for plastic colour: Evidence for differential provisioning in adults and fledglings. Mar. Environ. Res. 113: 1–6. doi: 10.1016/j.marenvres.2015.10.011.

Lavers, J.L. and A.L. Bond. 2016b. Ingested plastic as a route for trace metals in Laysan Albatross (*Phoebastria immutabilis*) and Bonin Petrel (*Pterodroma hypoleuca*) from Midway Atoll. Mar. Pollut. Bull. 110: 493–500. doi: 10.1016/j.marpolbul.2016.06.001.

Lavers, J.L., I. Hutton and A.L. Bond. 2019. Clinical pathology of plastic ingestion in marine birds and relationships with blood chemistry. Environ. Sci. Technol. 53: 9224–9231. doi: 10.1021/acs.est.9b02098.

Lavers, J.L., I. Hutton and A.L. Bond. 2021. Temporal trends and interannual variation in plastic ingestion by Flesh-footed Shearwaters (*Ardenna carneipes*) using different sampling strategies. Environ. Pollut. 291: 118086. doi: 10.1016/j.envpol.2021.118086.

Lavoie, R.A., C.J. Baird, L.E. King, T.K. Kyser, V.L. Friesen and L.M. Campbell. 2014. Contamination of mercury during the wintering period influences concentrations at breeding sites in two migratory piscivorous birds. Environ. Sci. Technol. 48: 13694–13702. doi: 10.1021/es502746z.

Law, K.L., S. Moret-Ferguson, N.A. Maximenko, G. Proskurowski, E.E. Peacock, J. Hafner et al. 2010. Plastic accumulation in the North Atlantic subtropical gyre. Science. 329: 11851188. doi: 10.1126/science.1192321.

Law, K.L. 2017. Plastics in the marine environment. Ann. Rev. Mar. Sci. 9: 205–229. doi: 10.1146/annurev-marine-010816-060409.

Lebreton, L.C.M., S.D. Greer and J.C. Borrero. 2012. Numerical modelling of floating debris in the world's oceans. Mar. Pollut. Bull. 64: 653–661. doi: 10.1016/j.marpolbul.2011.10.027.

Lemos, L.S., J.F. de Moura, R.A. Hauser-Davis, R.C. de Campos and S. Siciliano. 2013. Small cetaceans found stranded or accidentally captured in southeastern Brazil: Bioindicators of essential and non-essential trace elements in the environment. Ecotoxicol. Environ. Saf. 97: 166–175. doi: 10.1016/j.ecoenv.2013.07.025.

Lerma, M., J.A. Castillo-Guerrero, J. García-Hernández and G. Fernández. 2020. Zinc concentrations in Blue-footed booby (*Sula nebouxii*) eggs, nestlings, and adults. J. Sea Res. 165: 101952. doi: 10.1016/j.seares.2020.101952.

Linnebjerg, J.F., J.E. Baak, T. Barry, M.V. Gavrilo, M.L. Mallory, F.R. Merkel et al. 2021. Review of plastic pollution policies of Arctic countries in relation to seabirds. Facets 6: 1–25. doi: 10.1139/FACETS-2020-0052.

Lithner, D., Å. Larsson and G. Dave. 2011. Environmental and health hazard ranking and assessment of plastic polymers based on chemical composition. Sci. Total Environ. 409: 3309–3324. doi: 10.1016/j.scitotenv.2011.04.038.

Liu, X., S. Zhao, L. Sun, X. Yin, Z. Xie, L. Honghao et al. 2006. P and trace metal contents in biomaterials, soils, sediments and plants in colony of red-footed booby (*Sula sula*) in the Dongdao Island of South China Sea. Chemosphere. 65: 707–715. doi: 10.1016/j.chemosphere.2006.01.043.

Lopes, C.S., J.P. de Faria, V.H. Paiva and J.A. Ramos. 2020. Characterization of anthropogenic materials on yellow-legged gull (*Larus michahellis*) nests breeding in natural and urban sites along the coast of Portugal. Environ. Sci. Pollut. Res. 27: 36954–36969. doi: 10.1007/s11356–020-09651-x.

Lopes, C.S., V.H. Paiva, P.T. Vaz, J.P. de Faria, J.G. Calado, J.M. Pereira et al. 2021. Ingestion of anthropogenic materials by yellow-legged gulls (*Larus michahellis*) in natural, urban, and landfill sites along Portugal in relation to diet composition. Environ. Sci. Pollut. Res. 28: 19046–19063. doi: 10.1007/s11356-020-12161-5.

Lu, Z., A.O. De Silva, J.F. Provencher, M.L. Mallory, J.L. Kirk, M. Houde et al. 2019. Occurrence of substituted diphenylamine antioxidants and benzotriazole UV stabilizers in Arctic seabirds and seals. Sci. Total Environ. 663: 950–957. doi: 10.1016/j.scitotenv.2019.01.354.

Lusher, A.L., N.A. Welden, P. Sobral and M. Cole. 2017. Sampling, isolating and identifying microplastics ingested by fish and invertebrates. Anal. Methods. 9: 1346–1360. doi: 10.1039/c6ay02415g.

Mallory, M.L., L. Mahon, M.D. Tomlik, C. White, G.R. Milton and I. Spooner. 2015. Colonial marine birds influence island soil chemistry through biotransport of trace elements. Water, Air, Soil Pollut. 226: 31. doi: 10.1007/s11270-015-2314-9.

Mallory, C.D., H.G. Gilchrist, G.J. Robertson, J.F. Provencher, B.M. Braune, M.R. Forbes et al. 2017. Hepatic trace element concentrations of breeding female common eiders across a latitudinal gradient in the eastern Canadian Arctic. Mar. Pollut. Bull. 124: 252–257. doi: 10.1016/j.marpolbul.2017.07.050.

Mallory, M.L., J.F. Provencher, G.J. Robertson, B.M. Braune, E.R. Holland, S. Klapstein et al. 2018. Mercury concentrations in blood, brain and muscle tissues of coastal and pelagic birds from northeastern Canada. Ecotoxicol. Environ. Saf. 157: 424–430. doi: 10.1016/j.ecoenv.2018.04.004.

Mato, Y., T. Isobe, H. Takada, H. Kanehiro, C. Ohtake and T. Kaminuma. 2001. Plastic resin pellets as a transport medium for toxic chemicals in the marine environment. Environ. Sci. Technol. 35: 318–324. doi: 10.1021/es0010498.

Mendes, P., C. Eira, J. Torres, A.M.V.M. Soares, P. Melo and J. Vingada. 2008. Toxic element concentration in the Atlantic Gannet *Morus bassanus* (Pelecaniformes, Sulidae) in Portugal. Arch. Environ. Contam. Toxicol. 55: 503–509. doi: 10.1007/s00244-008-9134-5.

Monteiro, L.R. and R.W. Furness. 1995. Seabirds as monitors of mercury in the marine environment. Water, Air, Soil Pollut. 80: 851–870. doi: 10.1007/BF01189736.

Monteiro, L.R., R.W. Furness and A.J. del Nevo. 1995. Mercury levels in seabirds from the Azores, Mid-North Atlantic Ocean. Arch. Environ. Contam. Toxicol. 28: 304–309. doi: 10.1007/BF00213106.

Monteiro, L.R., J.P. Granadeiro, R.W. Furness and P. Oliveira. 1999. Contemporary patterns of mercury contamination in the Portuguese Atlantic inferred from mercury concentrations in seabird tissues. Mar. Environ. Res. 47: 137–156. doi: 10.1016/S0141-1136(98)00110-X.

Monteiro, L.R. and R.W. Furness. 2001. Kinetics, dose-response, and excretion of methylmercury in free-living adult Cory's Shearwaters. Environ. Sci. Technol. 35: 739–746. doi: 10.1021/es000114a.

Montevecchi, W.A., S. Benvenuti, S. Garthe, G.K. Davoren and D. Fifield. 2009. Flexible foraging tactics by a large opportunistic seabird preying on forage and large pelagic fishes. Mar. Ecol. Prog. Ser. 385: 295–306. doi: 10.3354/meps08006.

Moser, M.L. and D.S. Lee. 1992. A fourteen–year survey of plastic ingestion by western north Atlantic seabirds. Colon. Waterbirds. 15: 83–94. doi: 10.2307/1521357.

Neff, J.M. 2002. Bioaccumulation mechanisms. pp. 37–56. *In:* Neff, J.M. [ed.]. Bioaccumulation in Marine Organisms: Effect of Contaminants from Oil Well Produced Water. Elsevier, Oxford. doi: 10.1016/B978-008043716-3/50003-8.

Nelms, S.E., T.S. Galloway, B.J. Godley, D.S. Jarvis and P.K. Lindeque. 2018. Investigating microplastic trophic transfer in marine top predators. Environ. Pollut. 238: 999–1007. doi: 10.1016/j.envpol.2018.02.016.

Net, S., A. Delmont, R. Sempéré, A. Paluselli and B. Ouddane. 2015. Reliable quantification of phthalates in environmental matrices (air, water, sludge, sediment and soil): A review. Sci. Total Environ. 515–516: 162–180. doi: 10.1016/j.scitotenv.2015.02.013.

O'Hanlon, N.J., N.A. James, E.A. Masden and A.L. Bond. 2017. Seabirds and marine plastic debris in the northeastern Atlantic: A synthesis and recommendations for monitoring and research. Environ. Pollut. 231: 1291–1301. doi: 10.1016/j.envpol.2017.08.101.

O'Hanlon, N.J., A.L. Bond, J.L. Lavers, E.A. Masden and N.A. James. 2019. Monitoring nest incorporation of anthropogenic debris by Northern Gannets across their range. Environ. Pollut. 255: 113152. doi: 10.1016/j.envpol.2019.113152.

Otero, X.L., S. de la Peña-Lastra, D. Romero, G.N. Nobrega, T.O. Ferreira and A. Pérez-Alberti. 2018. Trace elements in biomaterials and soils from a Yellow-legged gull (*Larus michahellis*) colony in the Atlantic Islands of Galicia National Park (NW Spain). Mar. Pollut. Bull. 133: 144–149. doi: 10.1016/j.marpolbul.2018.05.027.

Oudi, A., M.A. Chokri, A. Hammouda, R. Chaabane, R. Badraoui, A. Besnard et al. 2019. Physiological impacts of pollution exposure in seabird's progeny nesting in a Mediterranean contaminated area. Mar. Pollut. Bull. 142: 196–205. doi: 10.1016/j.marpolbul.2019.02.056.

Outridge, P.M., M.E. Goodsite, O. Bennike, N. Rausch and W. Shotyk. 2016. Seabird transfer of nutrients and trace elements from the north water polynya to land during the mid-holocene warm period, Carey Islands, northwest Greenland. Arctic. 69: 253–265. doi: 10.14430/arctic4577.

Pacyna, A.D., D. Jakubas, A.N.M.A. Ausems, M. Frankowski, Ż. Polkowska and K. Wojczulanis-Jakubas. 2019. Storm petrels as indicators of pelagic seabird exposure to chemical elements in the Antarctic marine ecosystem. Sci. Total Environ. 692: 382–392. doi: 10.1016/j.scitotenv.2019.07.137.

Padula, V., A.H. Beaudreau, B. Hagedorn and D. Causey. 2020. Plastic-derived contaminants in Aleutian Archipelago seabirds with varied foraging strategies. Mar. Pollut. Bull. 158: 111435. doi: 10.1016/j.marpolbul.2020.111435.

Paiva, V.H., P.C. Tavares, J.A. Ramos, E. Pereira, S. Antunes and A.C. Duarte. 2008. The influence of diet on mercury intake by Little Tern chicks. Arch. Environ. Contam. Toxicol. 55: 317–328. doi: 10.1007/s00244-007-9118-x.

Piatt, J.F., W.J. Sydeman and F. Wiese. 2007. Introduction: seabirds as indicators of marine ecosystems. Mar. Ecol. Prog. Ser. 352: 199–204. doi: 10.3354/meps07070.

Pilcher, N., S. Gaw, R. Eisert, T.W. Horton, A.M. Gormley, T.L. Cole et al. 2020. Latitudinal, sex and inter-specific differences in mercury and other trace metal concentrations in Adélie and Emperor penguins in the Ross Sea, Antarctica. Mar. Pollut. Bull. 154: 111047. doi: 10.1016/j.marpolbul.2020.111047.

Poon, F.E., J.F. Provencher, M.L. Mallory, B.M. Braune and P.A. Smith. 2017. Levels of ingested debris vary across species in Canadian Arctic seabirds. Mar. Pollut. Bull. 116: 517–520. doi: 10.1016/j.marpolbul.2016.11.051.

Provencher, J.F., A.L. Bond and M.L. Mallory. 2015. Marine birds and plastic debris in Canada: A national synthesis and a way forward. Environ. Rev. 23: 1–13. doi: 10.1139/er-2014-0039.

Provencher, J.F., M.R. Forbes, H.L. Hennin, O.P. Love, B.M. Braune, M.L. Mallory et al. 2016. Implications of mercury and lead concentrations on breeding physiology and phenology in an Arctic bird. Environ. Pollut. 218: 1014–1022. doi: 10.1016/j.envpol.2016.08.052.

Provencher, J.F., A.L. Bond, S. Avery-Gomm, S.B. Borrelle, E.L. Bravo Rebolledo, S. Hammer et al. 2017. Quantifying ingested debris in marine megafauna: A review and recommendations for standardization. Anal. Methods. 9: 14541469. doi: 10.1039/C6AY02419J.

Provencher, J.F., J.C. Vermaire, S. Avery-Gomm, B.M. Braune and M.L. Mallory. 2018. Garbage in guano? Microplastic debris found in faecal precursors of seabirds known to ingest plastics. Sci. Total Environ. 644: 1477–1484. doi: 10.1016/j.scitotenv.2018.07.101.

Provencher, J.F., J. Ammendolia, C.M. Rochman and M.L. Mallory. 2019a. Assessing plastic debris in aquatic food webs: What we know and don't know about uptake and trophic transfer. Environ. Rev. 27: 304–317. doi: 10.1139/er-2018-0079.

Provencher, J.F., S.B. Borrelle, A.L. Bond, J.L. Lavers, J.A. van Franeker, S. Kühn et al. 2019b. Recommended best practices for plastic and litter ingestion studies in marine birds: collection, processing, and reporting. Facets. 4: 111–130. doi: 10.1139/facets-2018-0043.

Provencher, J.F., M. Liboiron, S.B. Borrelle, A.L. Bond, C. Rochman, J.L. Lavers et al. 2020. A Horizon Scan of research priorities to inform policies aimed at reducing the harm of plastic pollution to biota. Sci. Total. Environ. 733: 139381. doi: 10.1016/j.scitotenv.2020.139381.

Puskic, P.S., J.L. Lavers, L.R. Adams, M. Grünenwald, I. Hutton and A.L. Bond. 2019. Uncovering the sub-lethal impacts of plastic ingestion by shearwaters using fatty acid analysis. Conserv. Physiol. 7: coz017. doi: 10.1093/conphys/coz017.

Puskic, P.S., J.L. Lavers, L.R. Adams and A.L. Bond. 2020. Ingested plastic and trace element concentrations in Short-tailed Shearwaters (*Ardenna tenuirostris*). Mar. Pollut. Bull. 155: 111143. doi: 10.1016/j.marpolbul.2020.111143.

Ramos, R., J. González-Solís, M.G. Forero, R. Moreno, E. Gómez-Díaz, X. Ruiz et al. 2009. The influence of breeding colony and sex on mercury, selenium and lead levels and carbon and nitrogen stable isotope signatures in summer and winter feathers of *Calonectris* shearwaters. Oecologia 159: 345–354. doi: 10.1007/s00442–008–1215–7.

Ramos, R., F. Ramírez and L. Jover. 2013. Trophodynamics of inorganic pollutants in a wide-range feeder: The relevance of dietary inputs and biomagnification in the Yellow-legged gull (*Larus michahellis*). Environ. Pollut. 172: 235–242. doi: 10.1016/j.envpol.2012.09.014.

Reisser, J., B. Slat, K. Noble, K. du Plessis, M. Epp, M. Proietti et al. 2015. The vertical distribution of buoyant plastics at sea: an observational study in the North Atlantic Gyre. Biogeosciences. 12: 1249–1256. doi: 10.5194/bg-12-1249-2015.

Rochman, C.M., E. Hoh, B.T. Hentschel and S. Kaye. 2013. Long-term field measurement of sorption of organic contaminants to five types of plastic pellets: Implications for plastic marine debris. Environ. Sci. Technol. 47: 1646–1654. doi: 10.1021/es303700s.

Rochman, C.M., M.A. Browne, A.J. Underwood, J.A. van Franeker, R.C. Thompson and L.A. Amaral-Zettler. 2016. The ecological impacts of marine debris: Unraveling the demonstrated evidence from what is perceived. Ecology 97: 302–312. doi: 10.1890/14-2070.1.

Roman, L., B.D. Hardesty, M.A. Hindell and C. Wilcox. 2019. A quantitative analysis linking seabird mortality and marine debris ingestion. Sci. Rep. 9: 3202. doi: 10.1038/s41598-018-36585-9.

Ryan, P.G. 1987. The incidence and characteristics of plastic particles ingested by seabirds. Mar. Environ. Res. 23: 175–206. doi: 10.1016/0141-1136(87)90028-6.

Ryan, P.G. 1988. Intraspecific variation in plastic ingestion by seabirds and the flux of plastic through seabird populations. Condor 90: 446–452. doi: 10.2307/1368572.

Ryan, P.G. and M.W. Fraser. 1988. The use of Great Skua pellets as indicators of plastic pollution in seabirds. Emu 88: 16–19. doi: 10.1071/MU9880016.

Ryan, P.G., C.J. Moore, J.A. van Franeker and C.L. Moloney. 2009. Monitoring the abundance of plastic debris in the marine environment. Philos. Trans. R. Soc. B 364: 1999–2012. doi: 10.1098/rstb.2008.0207.

Ryan, P.G. 2018. Entanglement of birds in plastics and other synthetic materials. Mar. Pollut. Bull. 135: 159–164. doi: 10.1016/j.marpolbul.2018.06.057.

Sanpera, C., X. Ruiz, R. Moreno, L. Jover and S. Waldron. 2007. Mercury and stable isotopes in feathers of Audouin's Gulls as indicators of feeding habits and migratory connectivity. Condor 109: 268–275. doi: 10.1093/condor/109.2.268.

Santana, M.F.M., F.T. Moreira and A. Turra. 2017. Trophic transference of microplastics under a low exposure scenario: Insights on the likelihood of particle cascading along marine food-webs. Mar. Pollut. Bull. 121: 154–159. doi: 10.1016/j.marpolbul.2017.05.061.

Scheuhammer, A.M. 1987. The chronic toxicity of aluminium, cadmium, mercury, and lead in birds: a review. Environ. Pollut. 46: 263–295. doi: 10.1016/0269-7491(87)90173-4.

Schmaltz, E., E.C. Melvin, Z. Diana, E.F. Gunady, D. Rittschof, J.A. Somarelli et al. 2020. Plastic pollution solutions: Emerging technologies to prevent and collect marine plastic pollution. Environ. Int. 144: 106067. doi:10.1016/j.envint.2020.106067.

Schreiber, E.A. and J. Burger. 2001. Biology of Marine Birds. CRC Press, New York. doi: 10.1201/9781420036305.

Seco Pon, J.P., O. Beltrame, J. Marcovecchio, M. Favero and P. Gandini. 2011. Trace metals (Cd, Cr, Cu, Fe, Ni, Pb, and Zn) in feathers of Black-browed Albatross *Thalassarche melanophrys* attending the Patagonian Shelf. Mar. Environ. Res. 72: 40–45. doi: 10.1016/j.marenvres.2011.04.004.

Shealer, D.A. 2002. Foraging behavior and food of seabirds. pp. 137–177. *In*: Schreiber, E.A. and J. Burger [eds.]. Biology of Marine Birds. CRC Press, Boca Raton, Fl., USA. doi: 10.1201/9781420036305.

Smith, E.C. and A. Turner. 2020. Mobilisation kinetics of Br, Cd, Cr, Hg, Pb and Sb in microplastics exposed to simulated, dietary-adapted digestive conditions of seabirds. Sci. Total Environ. 733: 138802. doi: 10.1016/j.scitotenv.2020.138802.

Soldatini, C., M. Sebastiano, Y.V. Albores-Barajas, H. Abdelgawad, P. Bustamante and D. Costantini. 2020. Mercury exposure in relation to foraging ecology and its impact on the oxidative status of an endangered seabird. Sci. Total Environ. 724: 138131. doi: 10.1016/j.scitotenv.2020.138131.

Squadrone, S., M.C. Abete, P. Brizio, D. Pessani and L. Favaro. 2018. Metals in feathers of African Penguins (*Spheniscus demersus*): Considerations for the welfare and management of seabirds under human care. Bull. Environ. Contam. Toxicol. 100: 465–471. doi: 10.1007/s00128-018-2293-9.

Squadrone, S., P. Brizio, L. Favaro, G. Todino, D. Florio, C. Da Rugna et al. 2019. Humboldt penguins' feathers as bioindicators of metal exposure. Sci. Total Environ. 650: 1734–1739. doi: 10.1016/j.scitotenv.2018.09.326.

Stewart, F.M., L.R. Monteiro and R.W. Furness. 1997. Heavy metal concentrations in Cory's shearwater, *Calonectris diomedea*, fledglings from the Azores, Portugal. Bull. Environ. Contam. Toxicol. 58: 115–122. doi: 10.1007/s001289900308.

Streets, D.G., Q. Zhang and Y. Wu. 2009. Projections of global mercury emissions in 2050. Environ. Sci. Technol. 43: 2983–2988. doi: 10.1021/es802474j.

Sun, L.G., S.D. Emslie, T. Huang, J.M. Blais, Z.Q. Xie, X.D. Liu et al. 2013. Vertebrate records in polar sediments: Biological responses to past climate change and human activities. Earth-Science Rev. 126: 147–155. doi: 10.1016/j.earscirev.2013.08.004.

Tanaka, K., H. Takada, R. Yamashita, K. Mizukawa, M.- aki Fukuwaka and Y. Watanuki. 2013. Accumulation of plastic–derived chemicals in tissues of seabirds ingesting marine plastics. Mar. Pollut. Bull. 69: 219–222. doi: 10.1016/j.marpolbul.2012.12.010.

Tanaka, K., Y. Watanuki, H. Takada, M. Ishizuka, R. Yamashita, M. Kazama et al. 2020. *In vivo* accumulation of plastic–derived chemicals into seabird tissues. Curr. Biol. 30: 723–728. doi: 10.1016/j.cub.2019.12.037.

Tavares, D.C., L.L. Da Costa, D.F. Rangel, J.F. De Moura, I.R. Zalmon and S. Siciliano. 2016. Nests of the brown booby (*Sula leucogaster*) as a potential indicator of tropical ocean pollution by marine debris. Ecol. Indic. 70: 10–14. doi: 10.1016/j.ecolind.2016.06.005.

Tavares, S., J.C. Xavier, R.A. Phillips, M.E. Pereira and M.A. Pardal. 2013. Influence of age, sex and breeding status on mercury accumulation patterns in the wandering albatross *Diomedea exulans*. Environ. Pollut. 181: 315–320. doi: 10.1016/j.envpol.2013.06.032.

Teuten, E.L., J.M. Saquing, D.R.U. Knappe, M.A. Barlaz, S. Jonsson, A. Björn et al. 2009. Transport and release of chemicals from plastics to the environment and to wildlife. Philos. Trans. R. Soc. B 364: 2027–2045. doi: 10.1098/rstb.2008.0284.

Theron, A.J., G.R. Tintinger and R. Anderson. 2011. Harmful interactions of non-essential heavy metals with cells of the innate immune system. J. Clin. Toxicol. s3: 005. doi: 10.4172/2161-0495.s3-005.

Thompson, D.L., T.S. Ovenden, T. Pennycott and R.G. Nager. 2020. The prevalence and source of plastic incorporated into nests of five seabird species on a small offshore island. Mar. Pollut. Bull. 154: 111076. doi: 10.1016/j.marpolbul.2020.111076.

Thompson, D.R., R.W. Furness and P.M. Walsh. 1992. Historical changes in mercury concentrations in the marine ecosystem of the north and north-east Atlantic Ocean as indicated by seabird feathers. J. Appl. Ecol. 29: 79–84. doi: 10.2307/2404350.

Thompson, R.C., Y. Olson, R.P. Mitchell, A. Davis, S.J. Rowland, A.W.G. John et al. 2004. Lost at Sea: Where is all the plastic? Science 304: 838. doi: 10.1126/science.1094559.

Turner, A. 2018. Mobilisation kinetics of hazardous elements in marine plastics subject to an avian physiologically-based extraction test. Environ. Pollut. 236: 1020–1026. doi: 10.1016/j.envpol.2018.01.023.

Valavanidis, A., T. Vlahogianni, M. Dassenakis and M. Scoullos. 2006. Molecular biomarkers of oxidative stress in aquatic organisms in relation to toxic environmental pollutants. Ecotoxicol. Environ. Saf. 64: 178–189. doi: 10.1016/j.ecoenv.2005.03.013.

Valko, M., H. Morris and M.T.D. Cronin. 2005. Metals, toxicity and oxidative stress. Curr. Med. Chem. 12: 1161–1208. doi: 10.2174/0929867053764635.

van Franeker, J.A., C. Blaize, J. Danielsen, K. Fairclough, J. Gollan, N. Guse et al. 2011. Monitoring plastic ingestion by the northern fulmar *Fulmarus glacialis* in the North Sea. Environ. Pollut. 159: 2609–2615. doi: 10.1016/j.envpol.2011.06.008.

van Franeker, J.A. and K.L. Law. 2015. Seabirds, gyres and global trends in plastic pollution. Environ. Pollut. 203: 89–96. doi: 10.1016/j.envpol.2015.02.034.

van Sebille, E., C. Wilcox, L. Lebreton, N. Maximenko, B.D. Hardesty, J.A. van Franeker et al. 2015. A global inventory of small floating plastic debris. Environ. Res. Lett. 10: 124006. doi: 10.1088/1748-9326/10/12/124006.

Verlis, K.M., M.L. Campbell and S.P. Wilson. 2013. Ingestion of marine debris plastic by the wedge-tailed shearwater *Ardenna pacifica* in the Great Barrier Reef, Australia. Mar. Pollut. Bull. 72: 244–249. doi: 10.1016/j.marpolbul.2013.03.017.

Verlis, K.M., M.L. Campbell and S.P. Wilson. 2014. Marine debris is selected as nesting material by the brown booby (*Sula leucogaster*) within the Swain Reefs, Great Barrier Reef, Australia. Mar. Pollut. Bull. 87: 180–190. doi: 10.1016/j.marpolbul.2014.07.060.

Vo, A.-T.E., M.S. Bank, J.P. Shine and S.V. Edwards. 2011. Temporal increase in organic mercury in an endangered pelagic seabird assessed by century-old museum specimens. Proc. Natl. Acad. Sci. 108: 7466–7471. doi: 10.1073/pnas.1013865108.

Votier, S.C., K. Archibald, G. Morgan and L. Morgan. 2011. The use of plastic debris as nesting material by a colonial seabird and associated entanglement mortality. Mar. Pollut. Bull. 62: 168–172. doi: 10.1016/j.marpolbul.2010.11.009.

Wayland, M., H.G. Gilchrist, D.L. Dickson, T. Bollinger, C. James, R.A. Carreno et al. 2001. Trace elements in king eiders and common eiders in the Canadian arctic. Arch. Environ. Contam. Toxicol. 41: 491–500. doi: 10.1007/s002440010276.

Wayland, M., H.G. Gilchrist, T. Marchant, J. Keating and J.E. Smits. 2002. Immune function, stress response, and body condition in Arctic-breeding Common Eiders in relation to cadmium, mercury, and selenium concentrations. Environ. Res. 90: 47–60. doi: 10.1006/enrs.2002.4384.

Wilcox, C., E. van Sebille and B.D. Hardesty. 2015. Threat of plastic pollution to seabirds is global, pervasive, and increasing. Proc. Natl. Acad. Sci. 112: 11899–11904. doi: 10.1073/pnas.1502108112.

Witteveen, M., M. Brown and P.G. Ryan. 2017. Anthropogenic debris in the nests of kelp gulls in South Africa. Mar. Pollut. Bull. 114: 699–704. doi: 10.1016/j.marpolbul.2016.10.052.

Wright, S.L., R.C. Thompson and T.S. Galloway. 2013. The physical impacts of microplastics on marine organisms: A review. Environ. Pollut. 178: 483–492. doi: 10.1016/j.envpol.2013.02.031.

Xu, L.Q., X.-D. Liu, L.-G. Sun, H. Yan, Y. Liu, Y.-H. Luo et al. 2011. Geochemical evidence for the development of coral island ecosystem in the Xisha Archipelago of South China Sea from four ornithogenic sediment profiles. Chem. Geol. 286: 135–145. doi: 10.1016/j.chemgeo.2011.04.015.

Xu, L.Q., L.B. Wu, Y.H. Zhang and J.J. Zhao. 2019. Transport of cobalt and silver from the ocean to a reef island by seabirds in the South China Sea. J. Geophys. Res. Biogeosciences. 124: 3005–3014. doi: 10.1029/2019JG005264.

Ziółek, M., P. Bartmiński and A. Stach. 2017. The influence of seabirds on the concentration of selected heavy metals in organic soil on the Bellsund Coast, Western Spitsbergen. Arctic, Antarct. Alp. Res. 49: 507–520. doi: 10.1657/AAAR0016-024.

CHAPTER 4

Antarctic Seabirds as Indicators of Climate Change

José C. Xavier,[1,2,*] *Richard A. Phillips*[2] and *Akinori Takahashi*[3]

Introduction

Seabird populations have long been considered as ecosystem indicators (Montevecchi 1993, Furness and Camphuysen 1997, Piatt et al. 2007). In particular, they can integrate the effects of variability in physical or biological parameters across wide spatial and temporal scales (Fraser and Trivelpiece 1996, Hindell et al. 2020). As such, they are often included as key components in assessments of marine ecosystem health (Halpern et al. 2012, Bestley et al. 2020), and for demonstrating effects of climate change (Durant et al. 2009, Oro 2014, Sydeman et al. 2015). Climate change is defined here as a "statistically significant variation in either the mean state of the climate or in its variability, persisting for an extended period (typically decades or longer)" (Solomon et al. 2007, Jenouvrier 2013).

The Antarctic is a key region in terms of global anthropogenic heat and carbon uptake, absorbing 43% of the total anthropogenic CO_2 and 75% of the heat (Frölicher et al. 2015, Brown et al. 2019, Bronselaer et al. 2020). Moreover, the Antarctic is arguably one of the regions that is most affected in the world by climate change (Turner et al. 2014, IPCC 2018, Rintoul et al. 2018, Convey and Peck 2019, Rogers et al. 2020, Xavier and Convey 2020). Research on seabirds can help us understand the impacts of climate change, particularly as they are one of the most numerous and well-studied groups of top predators in the Southern Ocean (Croxall et al. 2002, Trathan et al. 2007, Barbraud et al. 2012), and can be monitored in detail, particularly during the breeding season (Reid et al. 2005, Reid 2018). Along with climate change, human activities had major impacts on Antarctic marine ecosystems over the past 200 years, from hunting of seals and penguins, to whaling, to fisheries for toothfish (*Dissostichus* spp.), icefish (Channichthyidae) and Antarctic krill (*Euphausia superba*) which likely changed the structure and functioning of Southern Ocean food webs (Croxall 1992, Ainley and Blight 2009, Ainley et al. 2010, Rogers et al. 2015, Hofman 2017). Today, several

[1] University of Coimbra, MARE - Marine and Environmental Sciences Centre, Department of Life Sciences, 3000-456 Coimbra, Portugal.
[2] British Antarctic Survey, Natural Environment Research Council, High Cross, Madingley Road, Cambridge CB3 0ET, United Kingdom.
Email: raphil@bas.ac.uk
[3] National Institute of Polar Research, Tachikawa, Tokyo 190-8518, Japan.
Email: atak@nipr.ac.jp
* Corresponding author: jxavier@zoo.uc.pt

albatross, petrel and penguin species in the Antarctic and sub-Antarctic have poor conservation status according to the International Union for Conservation of Nature (IUCN) (Phillips et al. 2016, Boersma et al. 2019). In such a context, this chapter aims to: (1) assess the scientific evidence for climate change in the Antarctic region including the Southern Ocean (for the purposes of this review considered to be the Convention Area of the Convention on the Conservation of Antarctic Marine Living Resources; CCAMLR); (2) review the effects of climate change on Antarctic seabird populations, and; (3) discuss potential future consequences of climate change for Antarctic seabirds and strategies for their conservation, addressing their needs in terms of policies under the Antarctic Treaty System (ATS), Agreement for the Conservation of Albatrosses and Petrels (ACAP) and other international frameworks.

Climate-related changes in the physical environment

The most well-studied climatic variables in relation to distribution, ecology and life history of biota are temperature and precipitation (Jenouvrier 2013); however, other variables can be important in an Antarctic context, such as wind (Weimerskirch et al. 2012, Christensen et al. 2013, Turner et al. 2014, Cavanagh et al. 2017). The scientific evidence of change in the Antarctic region (Fig. C4.1; Table C4.1) includes the following. (1) Warming: an increase in air temperatures around

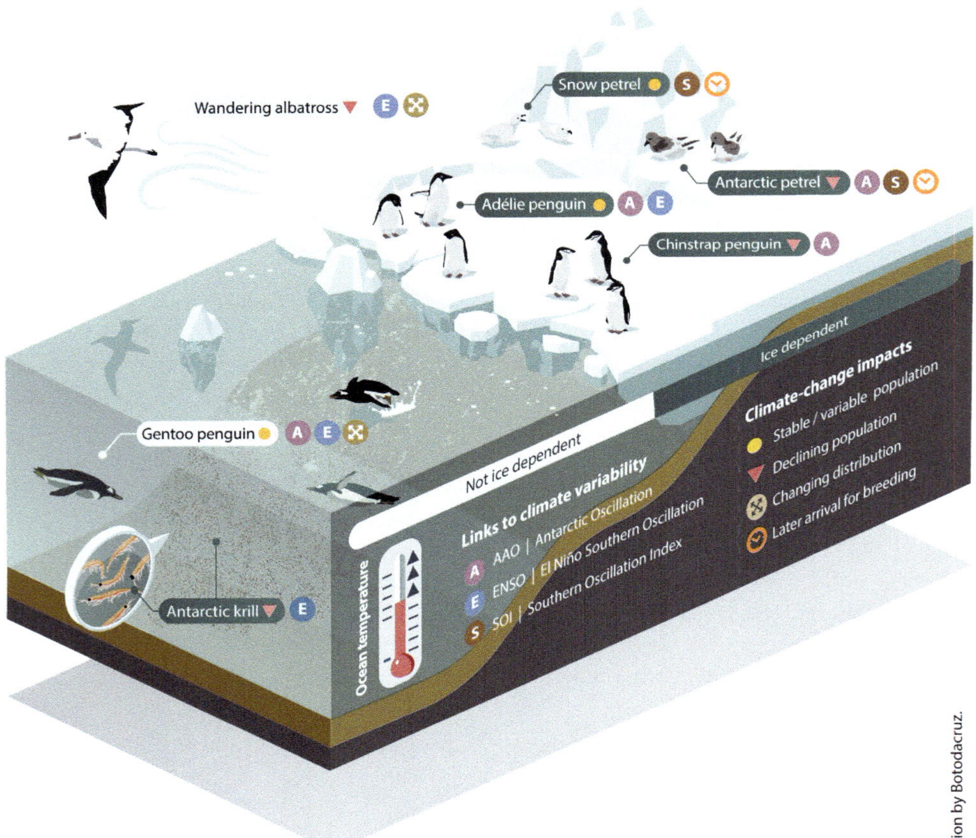

Illustration by Botodacruz.

Figure C4.1. Schematic diagram of the impacts of climate variability on Antarctic seabirds.

Table C4.1. Key environmental changes recorded in Antarctica and the Southern Ocean.

Parameter	Evidence	Region	Datasets (maximum range)	Reference
Air temperature	> 3°C	Parts of Antarctic Peninsula	1950–2000	(Marshall et al. 2006, Turner et al. 2009)
	> 2°C	Central West Antarctica	1958–2010	(Bromwich et al. 2013)
	At least > 1°C	From Antarctic continent to sub-Antarctic regions	1902–1999	(Budd 2000, Smith 2002, Weimerskirch et al. 2003, Meredith and King 2005, Turner et al. 2009, Turner et al. 2014)
Sea Temperature (sea surface temp./temp. at depth)	> 1°C	Antarctic Peninsula	1955–1994	(Meredith and King 2005, Turner et al. 2014)
	2.3°C	South Georgia	1925–2006	(Whitehouse et al. 2008)
	0.2°C	Southern Ocean	1930–2015	(Gille 2002, Swart et al. 2018)
Salinity	< 0.015‰/decade psu	Southern Ocean	1950–2015	(Böning et al. 2008, Swart et al. 2018)
pH	~ 0.1 units pH decrease	Southern Ocean (surface)	1880–1995	(Caldeira and Wickett 2003, Orr et al. 2005, McNeil and Matear 2007, Midorikawa et al. 2012, Bronselaer et al. 2020)
Carbonate ion concentrations	18 (10% reduction in concentration (u mol Kg^{-1}))	Southern Ocean (surface)	1800–1994	(Sabine et al. 2004, Orr et al. 2005, Hauri et al. 2016)
Dissolved oxygen	Decrease in concentration (u mol Kg^{-1})	Southern Ocean (Antarctic coast)	1995–2019	(Bronselaer et al. 2020)
Sea ice	Stable or growing in area (e.g., increase 1.7%/decade)	Southern Ocean	1979–2014	(Comiso and Nishio 2008, Turner and Comiso 2017)
	Declined in summer extent	Southern Ocean	2014–2017	(Parkinson 2019)

the Antarctic Peninsula and central West Antarctica as far north as the sub-Antarctic region (Meredith and King 2005, Barbraud et al. 2012, Bromwich et al. 2013, IPCC 2018), and in ocean temperatures, particularly the top 1500 m of the water column (Gille 2002, Gille 2008, Swart et al. 2018). However, there have been no clear trends in air temperature in most other regions of continental Antarctica (Christensen et al. 2013, Clem et al. 2020). (2) Freshening: salinity has decreased, particularly in surface waters south of 45°S since the 1950s (Böning et al. 2008, Durack et al. 2012, Swart et al. 2018); (3) Acidification: the pH of surface waters has decreased (Orr et al. 2005, McNeil and Matear 2007, Hauri et al. 2016, IPCC 2018); (4) Shrinkage in the extent of cold waters in the Southern Ocean: ocean fronts are shifting towards the Antarctic continent, affected by the change in wind patterns and in relation to the Antarctic Circumpolar Current (Cai 2006, Meijers 2014, Meijers et al. 2019, Bronselaer et al. 2020); (5) Variability in sea-ice extent and seasonality: overall sea-ice cover has been stable or increased for several decades (data from late 1970s), but the lowest yearly-average extent was registered recently (Comiso and Nishio 2008, Cavanagh et al. 2017, Turner and Comiso 2017, Parkinson 2019), (6) Change in the Southern Annular Mode (SAM; which drives much of the climate variability in Antarctica): the SAM has become more positive (associated with warmer conditions), which is also related to a southward shift in winds, particularly towards the Antarctic Peninsula (Cai 2006, Christensen et al. 2013).

Climate change and lower trophic levels in the Southern Ocean

These changes in physical parameters have demonstrable effects on marine organisms in the Southern Ocean (Constable et al. 2014, Rogers et al. 2020), which are well-adapted to cold and show high levels of endemism (Kock 1992, Knox 2007, De Broyer et al. 2014, Convey and Peck 2019). The abundance of algae (including some diatoms, flagellates and the haptophyte (*Phaeocystis antarctica*)) is positively affected by climate change, causing changes in the timing and magnitude of primary productivity, and these positive effects may continue in the future (Moreau et al. 2015, Schine et al. 2016, Deppeler and Davidson 2017, Kaufman et al. 2017, Schofield et al. 2017). The warming of the Southern Ocean has allowed a southward extension in the distribution of certain organisms, such as the coccolithophorid (*Emiliania huxleyi*) and the heterotrophic dinoflagellate (*Noctiluca scintillans*) (Cubillos et al. 2007, McLeod et al. 2012, Winter et al. 2014, Hallegraeff et al. 2020). In addition, marine productivity can be enhanced by an increase in the formation of giant icebergs through calving events (Duprat et al. 2016).

Further up in the food web, Antarctic animals that use calcium carbonate to form shells (e.g., the pteropod *Limacina helicina antarctica*) have been affected negatively by ocean acidification (Bednaršek et al. 2012a, Bednaršek et al. 2012b, Moy et al. 2009, Roberts et al. 2011). However, other organisms, such as sea urchins, seem equipped to cope with ocean acidification and most probably will not be affected by a slight decrease in pH of seawater (Collard et al. 2015, Cross et al. 2015, Suckling et al. 2015). The reduced sea-ice extent in certain Antarctic regions may result in lower mean phytoplankton size and benefit gelatinous organisms (e.g., the salp (*Salpa thompsoni*)), increasing their abundance (Moline et al. 2004). There has been a marked shift in benthic Antarctic communities related to increased sediment runoff triggered by glacier retreat because of climate change (Sahade et al. 2015), but exposure of new habitats by the break-up of ice shelves can also promote an increase in benthic fauna (Lagger et al. 2018).

At mid-trophic levels, Antarctic krill, an important species in Southern Ocean food webs (Everson 2000, Xavier and Peck 2015), has declined by > 50% in abundance since the 1970s in the Atlantic sector, coinciding with rapid sea-ice loss and a warming ocean (Loeb et al. 1997, Atkinson et al. 2004), with recognised links between El Niño Southern Oscillation index (ENSO) and Antarctic krill abundance (Murphy et al. 2007a, Murphy et al. 2007b). Furthermore, the distribution of Antarctic krill has contracted southward over the last 90 years following positive anomalies in the SAM associated with warm, windy and cloudy weather, and reduced sea ice (Atkinson et al. 2019); this effect may become even stronger with predicted future environmental change (Hückstädt et al. 2020). Other zooplankton have also been affected by climate change: there has been an increase in the abundance of another krill species *Thysanoessa macrura* in the northern part of the western Antarctic Peninsula and *Euphausia crystallorophias* in the south of this region (Steinberg et al. 2015, Gutt et al. 2021). Changes in the distribution of zooplankton in the Antarctic Peninsula and Drake Passage have also been described, when comparing historical records (from the 1920s) with recent data (Takahashi et al. 2010).

Various fish species have been affected by changes in sea ice and other aspects of oceanography in the Southern Ocean (Mintenbeck et al. 2012). Impacts seem to be greatest on fish that use sea ice (e.g., for foraging, spawning, nursery), including Antarctic silverfish (*Pleurogramma antarcticum*) and *Notothenia coriiceps* (Kellermann 1996, Vacchi et al. 2004, Moline et al. 2008). The distribution of *P. antarcticum* has reduced significantly in shelf waters of the west Antarctic Peninsula (Parker et al. 2015, Mintenbeck and Torres 2017, Gutt et al. 2021). Similarly, ocean acidification together with temperature change may affect the development of fish (e.g., *Gymnodraco acuticeps* (Flynn et al. 2015)). Notothenioid fish have limited ability to cope with the stress associated with warm temperatures (Bilyk and DeVries 2011, Enzor et al. 2017).

Cephalopods of the Southern Ocean are also likely to be affected by climatic changes, particularly in mesoscale oceanography, the position of oceanic fronts (through knock-on effects on marine productivity), sea-ice extent and ocean acidification (Rodhouse 2013, Xavier et al. 2016, Xavier et al. 2018). Cephalopods have certain biological traits (e.g., rapid growth, short lifespans, broad habitat preferences), which allow them to adapt well to climate change (Doubleday et al. 2016, Xavier et al. 2018, Arreguín-Sánchez 2019, Queirós et al. 2021). Evidence of a poleward shift in the distributions of cephalopods, including octopods and squid, have been recorded in association with warming conditions (Golikov et al. 2013, Ramos et al. 2014, Xavier et al. 2018). Based on recent analyses of long-term datasets, the distribution of *Taonius notalia*, *Gonatus antarcticus*, *Galiteuthis glacialis* and *Histioteuthis atlantica*, but not *Moroteuthopsis longimana*, changed over the last 50 years (1976 to 2016), yet the trophic levels of all five species remained constant (Abreu et al. 2020). Both the SAM and Southern Oscillation Index (SOI) had increased in strength and frequency over the same period but, of the five species, only in *Taonius notalia* did these indices correlate with trophic level and distribution (Abreu et al. 2020). Such changes in this mid-trophic levels may affect the trophic levels above where seabirds occupy.

Climate change and timing of breeding in Antarctic seabirds

Climate change can affect seabirds in various ways, not only through its effects on their food resources, distribution and population dynamics, but also because of concomitant changes in fishing activities, pollution, invasive species at breeding colonies and disease (Trathan et al. 2007, Trivelpiece et al. 2011, Barbraud et al. 2012, Constable et al. 2014, Waller et al. 2017, Bessa et al. 2019) (Fig. C4.1; Table C4.2). It is important to note that changes not only in magnitude but also in the timing of prey availability can be important for seabirds. However, a recent global analysis concluded that seabird populations have not shown a systematic adjustment in the timing of breeding in response to long-term changes in sea surface temperature, which suggests they are potentially vulnerable to a mismatch with peaks in prey availability in the future (Keogan et al. 2018).

There is some evidence of shifts in phenology in Antarctic seabirds in response to a changing environment. Overall, nine species in a seabird community in East Antarctica on average arrived 9.1 days later, and laid eggs 2.1 days later in the early 2000s than in the early 1950s, apparently linked to a decrease in sea-ice extent; however, timing did not change significantly in all species, and one species advanced laying, i.e., bred earlier (Barbraud and Weimerskirch 2006). Laying date in Wandering Albatrosses (*Diomedea exulans*) at South Georgia also advanced from the 1980s to the mid 2000s, but this was related to within-individual changes in the timing of breeding and not correlated with environmental indices (Lewis et al. 2012). Hence observed shifts in the timing of breeding are not all in the same direction, and the underlying drivers may not be discernible. Indeed, other behavioural or demographic responses to climate can lag due to climatic, ecological or demographic processes (see below).

Climate change and food availability for Antarctic seabirds

During the breeding season, seabirds are subject to central-place constraints associated with the need to return to the colony to incubate the egg or feed the chick. Therefore, seabirds shift their diet, distribution or foraging behavior in response to variation in the distribution and availability of prey in surrounding waters as a consequence of climate change (Schreiber and Burger 2001, Xavier et al. 2003, Ducklow et al. 2007, Sherley et al. 2013, Keogan et al. 2018, McMahon et al. 2019) (Fig. C4.1, Table C4.2). Declines in marine productivity have affected the composition of food webs in various regions of the Southern Ocean (Ducklow et al. 2007, Murphy et al. 2007b, Constable

Table C4.2. Studies linking climate change to long-term datasets on demography, distribution or diet of seabird populations in the Antarctic and sub-Antarctic islands.

	Timing of breeding	Diet	Breeding and at-sea distributions	Population dynamics	Reference
Adélie Penguin (*Pygoscelis adeliae*)	2.0 days later	Stable (West Antarctic Peninsula, Scotia Sea)	Southward shift (Antarctic Peninsula, East Antarctica)	Declining/Increasing/variable (Antarctic Peninsula, East Antarctica, Ross Sea)	(Barbraud and Weimerskirch 2006, Hinke et al. 2007, Carlini et al. 2009, Trivelpiece et al. 2011, LaRue et al. 2013, Lyver et al. 2014, Southwell et al. 2015, Dunn et al. 2016, Barbraud et al. 2020)
Chinstrap Penguin (*Pygoscelis antarcticus*)		Stable/no change in prey trophic level (Antarctic Peninsula, Scotia Sea)		Declining (Antarctic Peninsula, Scotia Sea)	(Hinke et al. 2007, Carlini et al. 2009, Trivelpiece et al. 2011, Barbosa et al. 2012, Dunn et al. 2016, Petry et al. 2018, McMahon et al. 2019)
Emperor Penguin (*Aptenodytes forsteri*)	0.7 days later		Southward shift (Across Antarctica)	Declining/variable (Antarctic Peninsula, East Antarctica, Ross Sea)	(Barbraud and Weimerskirch 2006, Barber-Meyer et al. 2008, Trathan et al. 2011, Robertson et al. 2014, Kooyman and Ponganis 2017, Barbraud et al. 2020)
Gentoo Penguin (*Pygoscelis papua*)		Stable/Higher trophic level in the last 40 years (South Georgia, Antarctic Peninsula)	Southward shift (Antarctic Peninsula, Scotia Sea)	Increasing/stable/variable (Antarctic Peninsula, Crozet, Kerguelen, Scotia Sea)	(Lynch et al. 2012, Dunn et al. 2016, Waluda et al. 2017, McMahon et al. 2019, Barbraud et al. 2020)
King Penguin (*Aptenodytes patagonicus*)		Stable (Crozet)		Increasing/Declining (Crozet)	(Péron et al. 2012, Weimerskirch et al. 2018, Barbraud et al. 2020)
Macaroni Penguin (*Eudyptes chrysolophus*)		Stable (South Georgia)		Declining/increasing/Stable (Crozet, Kerguelen, South Georgia)	(Trathan et al. 2012, Waluda et al. 2012, Barbraud et al. 2020)
Black-browed Albatross (*Thalassarche melanophris*)		Variable (South Geogia)		Declining (South Georgia)	(Xavier et al. 2003, Pardo et al. 2017, Poncet et al. 2017, Mills et al. 2020)
Grey-headed Albatross (*Thalassarche chrysostoma*)		Variable (South Geogia)		Declining (South Georgia)	(Xavier et al. 2003, Pardo et al. 2017, Poncet et al. 2017, Mills et al. 2020)
Wandering Albatross (*Diomedea exulans*)		Stable (Crozet, Kerguelen, South Georgia)	Southward shift (Crozet, Kerguelen)	Declining (Crozet, Kerguelen, South Georgia)	(Croxall et al. 1998, Weimerskirch et al. 2003, Xavier et al. 2003, Xavier et al. 2004, Weimerskirch et al. 2012, Cherel et al. 2017, Pardo et al. 2017, Poncet et al. 2017, IUCN 2021)
Antarctic Shag (*Leucocarbo bransfieldensis*)		Variable (Nelson Island)		Declining/increasing (Antarctic Peninsula)	(Casaux and Barrera-Oro 2016, Schrimpf et al. 2018, Phillips et al. 2019)

Southern Fulmar (*Fulmarus glacialoides*)	21.5 days later		Variable (East Antarctica)	(Weimerskirch et al. 2003, Barbraud and Weimerskirch 2006)
Antarctic Petrel (*Thalassoica antarctica*)	31.9 days later		Declining (Dronning Maud Land)	(Barbraud and Weimerskirch 2006, Descamps et al. 2016)
Cape Petrel (*Daption capense*)	10.0 days later			(Barbraud and Weimerskirch 2006)
Northern Giant Petrel (*Macronectes halli*)			Increasing (South Georgia)	(Poncet et al. 2020)
Snow Petrel (*Pagodrama nivea*)	2.2 days earlier		Stable (East Antarctica)	(Barbraud and Weimerskirch 2006, Barbraud et al. 2011)
Southern Giant Petrel (*Macronectes giganteus*)	8.4 days later		Increasing (South Georgia)	(Barbraud and Weimerskirch 2006, Poncet et al. 2020)
White-chinned Petrel (*Procellaria aequinoctialis*)		Northward shift (Crozet, Kerguelen)		(Péron et al. 2010)
Wilson´s Storm Petrel (*Oceanites oceanicus*)	4.9 days later			(Barbraud and Weimerskirch 2006)
Antarctic Prions (*Pachyptila desolata*)		Southward shift (Auckland Island, Adams Island)		(Grecian et al. 2016)
Brown Skua (*Stercorarius antarcticus*)			Declining (South Orkneys Islands)	(Carneiro et al. 2016)
South Polar Skua (*Stercorarius maccormicki*)	5.1 days later		Increasing (Antarctic Peninsula)	(Barbraud and Weimerskirch 2006, Phillips et al. 2019)

et al. 2014). A larger extent of sea ice during the winter has a positive effect on recruitment of Antarctic krill, resulting in higher abundance of prey for Antarctic seabirds breeding the next summer (Fraser and Hofmann 2003, Atkinson et al. 2004, Trivelpiece et al. 2011). Changes in sea surface temperature affect the abundance or availability of Antarctic krill, which has repercussions for the reproductive performance of dependent predators, including seabirds at South Georgia (Croxall et al. 1999, Xavier et al. 2003, Murphy et al. 2007a, Murphy et al. 2007b). In contrast, warming conditions around Kerguelen led seabirds to consume more crustaceans than myctophid fish, as the latter tend to dominate in colder conditions (Connan et al. 2008, Connan et al. 2019).

Penguins have been affected by the recent changes in prey abundance and accessibility associated with climate variability, but to differing extents depending on the species and region. King Penguins (*Aptenodytes patagonicus*) at the Crozet Islands forage intensively around the Antarctic Polar Front (APF) south of the breeding colony, and their travel distance to this productive zone varies according to the frontal position (300–500 km distant), due to climatic anomalies (Bost et al. 2015). Based on climate model projections from the Intergovernmental Panel on Climate Change (IPCC), Péron et al. (2012) suggested that the preferred foraging zones of King Penguins will shift even further southward, making this zone too distant to exploit, which will have repercussions for breeding success. However, a poleward shift of frontal zones may have positive effects on king penguin populations that breed south of the APF, including at South Georgia or Heard Island, as preferred foraging zones may become more accessible in the future (Cristofari et al. 2018). Accessibility of foraging areas is also key to successful reproduction in Emperor Penguins (*Aptenodytes forsteri*), which breed almost exclusively on land-fast sea-ice and travel beyond the ice edge to feed in open waters. Extensive sea ice during the breeding season is therefore detrimental to reproduction in emperor penguins (Massom et al. 2009), and also Adélie Penguins (*Pygoscelis adeliae*), by limiting access to foraging areas (Emmerson and Southwell 2008, Watanabe et al. 2020). However, high winter sea-ice extent was associated with high adult survival rates in both these species, possibly through the increased abundance of prey (Barbraud and Weimerskirch 2001). These results suggest that for ice-dependent penguin species, there are optimal sea-ice conditions which ensure that prey are both abundant and accessible (Barbraud et al. 2012, Le Guen et al. 2018).

The availability of Antarctic krill, associated with climate anomalies (Murphy et al. 2007b), appeared to drive annual variability in the diet of Gentoo (*Pygoscelis papua*) and Macaroni Penguins (*Eudyptes chrysolophus*) at South Georgia (Waluda et al. 2012, Waluda et al. 2017). The annual contribution of Antarctic krill to the diet of these two species was correlated in 1989–2010, with no evidence of a decline in occurrence of either Antarctic krill or fish (Waluda et al. 2017). A higher proportion of Antarctic krill was associated with higher fledging mass in macaroni penguins and higher breeding success in gentoo penguins. Recently, it has been suggested that harvesting of Antarctic krill by fisheries, as well as climate-associated variability in its abundance, has synergistic effects on breeding performance and population size of pygoscelid penguins in the northern Antarctic Peninsula, which warrants further investigation (Krüger et al. 2020, Watters et al. 2020).

Climate change in the distant past has been shown to affect the diet of Antarctic penguins. Adélie Penguins changed their diet in the Ross Sea, decreasing their consumption of Antarctic silverfish over the past 600 years (Polito et al. 2002, Lorenzini et al. 2009). At the Antarctic Peninsula, Adélie Penguins consumed more silverfish in cool periods and more of the squid *Psychroteuthis glacialis* in warm periods over the last 6000 years (Emslie and McDaniel 2002). Gentoo penguins in the Antarctic Peninsula changed their diet in the last century, from low-trophic level prey, such as Antarctic krill, 100 years ago (in the early 20th century, at the peak of Antarctic whaling), to more fish in the last 40 years; this was associated with a decline in abundance of krill because of higher competition with other predators (e.g., recovering marine mammal populations), declining sea-ice extent and commercial fishing (McMahon et al. 2019). Chinstrap penguins, living in the same region, did not change their trophic level during the same time period, possibly because they have a more specialised foraging niche (McMahon et al. 2019). This probably explains why their populations have declined in response to the decline in krill abundance (Trivelpiece et al. 2011).

Long-term analyses of the diets of breeding Black-browed (*Thalassarche melanophris*) and Grey-headed Albatrosses (*T. chrysostoma*) at South Georgia (1996–2017) showed high inter-annual variability and increased consumption of fish over time (Mills et al. 2020). There was a positive relationship between primary productivity (chlorophyll-a) and Antarctic krill in the diet of grey-headed albatrosses and a negative relationship with occurrence of the squid *Martialia hyadesi* in the diets of both albatross species (Xavier et al. 2003, Mills et al. 2020). The importance of *M. hyadesi* in the diet of Grey-headed Albatrosses was positively related to breeding success (Mills et al. 2020).

Climate change, breeding and at-sea distributions of Antarctic seabirds

There is some evidence that recent climate change is affecting the breeding (nesting) or foraging distributions of seabirds in the Southern Ocean (Barbraud et al. 2012) (Fig. C4.1, Table C4.2). Among penguins, there have been major shifts in the distributions of breeding Adélie Penguins since the Holocene at the Antarctic Peninsula, East Antarctica, and Ross Sea regions, associated with climate change (Trathan et al. 1996, Croxall et al. 2002, Emslie and McDaniel 2002, Emslie and Woehler 2005, Emslie et al. 2007, Hinke et al. 2007, Trivelpiece et al. 2011). During past decades, Adélie penguin colonies have decreased around the Antarctic Peninsula but increased in East Antarctica (see next subsection) (Trivelpiece et al. 2011, Southwell et al. 2015), effectively shifting the center of its species-level distribution southward. Similarly, a study on the demographic history and population structure of gentoo penguins at the west Antarctic Peninsula show that there was a population expansion around 13 000 years BP of birds that had survived in peri-Antarctic refugia (e.g., South Georgia, northern South Sandwich Islands), which recolonized the Antarctic Peninsula and South Shetland Islands after ice-sheet retreat (Clucas et al. 2014, Peña et al. 2014, Cole et al. 2019). Gentoo penguins at the Antarctic Peninsula are currently expanding southward as sea-ice cover has decreased around potential breeding colonies in recent decades (Lynch et al. 2012). There is also evidence of an increase of breeding attempts of king penguins in Antarctic Peninsula suggesting an expansion in their distribution (Petry et al. 2013, Juarés et al 2017, Borowicz et al. 2020). Finally, based on genomics, population sizes of emperor penguins also expanded in the warming period after the Last Glacial Maximum (Cole et al. 2019), but more recent declines at the Antarctic Peninsula and East Antarctica (see next subsection) are projected to shift their distribution pole ward during the 21st century (Jenouvrier et al. 2014).

In relation to albatrosses, the at-sea distribution of Wandering Albatrosses from the southern Indian Ocean has shifted southward in recent decades, due to increasing westerly wind intensity related to climate change, which allows increased flight speeds and has reduced trip duration (Weimerskirch et al. 2012). However, the location of oceanic fronts in relation to the few islands where Antarctic seabirds can breed is important. A southward shift of the APF away from Kerguelen Islands now favors seabirds that forage in warmer waters closer to the colony (Wandering Albatross and Sooty Albatross *Phoebetria fusca*) over the species that forage further away in Antarctic waters (Light-mantled Sooty Albatross *Phoebetria palpebrata*, Blue Petrel *Halobaena carulea*, Thin-billed Prion *Pachyptilla belcheri*, and White-headed Petrel *Pterodroma lessoni*) (Inchausti et al. 2003, Péron et al. 2010).

The current breeding distributions of Snow Petrels (*Pagodroma nivea*), Gentoo Penguins, Adélie Penguins and Chinstrap Penguins (*Pygoscelis antarcticus*) are apparently related to the emergence of new sites coincident with deglaciation and reductions in sea ice, which are within reach of productive foraging habitats (Younger et al. 2016). Antarctic Prions (*Pachyptila desolata*) have shifted their at-sea distribution during the non-breeding season southward in the last 100 years, which may be related to shifts in water temperature and prey distribution (Phillips et al. 2009, Quillfeldt et al. 2015, Grecian et al. 2016). In contrast, White-chinned Petrels (*Procellaria aequinoctialis*) have shifted their distribution northward; however, this may not relate to changing climate but to the exploitation of discards from fisheries (Péron et al. 2010).

Climate change and population dynamics in Antarctic seabirds

The full suite of demographic data (adult and immature survival, breeding frequency and success) is available from relatively few seabirds around Antarctica and on sub-Antarctic islands, and usually from only one site in a region; as such, emigration and immigration among populations may affect results and their interpretation (Clucas et al. 2016, Younger et al. 2016, Younger et al. 2017). Population trends in penguins and other seabirds vary greatly among and within species, and regionally (Gales 1998, Rogers et al. 2020) (Fig. C4.1, Table C4.2). The populations of ice-dependent Adélie Penguins decreased in Admiralty Bay, King George Island by > 50% in 1977–2009 (2.3%/year) (Trivelpiece et al. 2011), around US Palmer Station, west Antarctic Peninsula by 85% in 1975–2012 (2.2%/year) (Fountain et al. 2016), and also at Signy Island, South Orkney Islands by 42% in 1978–2015 (1.5%/year) (Forcada et al. 2006, Dunn et al. 2016), associated with atmospheric and ocean warming, and a decrease in winter sea-ice extent. Additionally, increased precipitation and snow accumulation related to the southward shift of a more maritime climate has reduced the survival of Adélie Penguin chicks (Ducklow et al. 2007, Carlini et al. 2009). Reduced population sizes of Adélie Penguins in the Antarctic Peninsula contrast with increases elsewhere, including in some regions of the western Ross Sea in 1981–2012 (Lyver et al. 2014), and across 99 study sites in five regions in East Antarctic (Croxall et al. 2002, Ainley et al. 2005, Kato and Ropert-Coudert 2006, LaRue et al. 2013), possibly associated with the decreasing trend in winter sea-ice extent (Southwell et al. 2015). Various (positive and negative) relationships between ENSO, the Antarctic oscillation index (AAO) and demography of Adélie Penguins (number of breeders, fledging mass, survival, population growth rate) have been found for colonies in the Antarctic Peninsula and East Antarctica (Ainley et al. 2005, Cimino et al. 2014, Jenouvrier et al. 2006, Wilson et al. 2001).

Similarly, Chinstrap Penguins declined by 36% at Deception Island in 1991–2008 (2.1%/year) (Barbosa et al. 2012), in Admiralty Bay by > 50% in 1977–2009 (2.5%/year) (Trivelpiece et al. 2011), in Stinker point (Elephant Island) by ~ 60% in 1980s–2010s (Petry et al. 2018) and in Signy Island by 68% in 1978–2015 (3.6%/year) (Dunn et al. 2016). A negative relationship between AAO and population sizes of chinstrap penguins was recorded at Signy Island (Forcada and Trathan 2009). Gentoo Penguin populations in the Antarctic Peninsula (1979–2009), Signy Island (1948–2015), and Crozet and Kerguelen Islands (1988–2018) are stable or increasing (Lynch et al. 2012, Dunn et al. 2016, Barbraud et al. 2020) as this is a less ice-dependent species, although there are some local population declines (e.g., Port Lockroy by up to 25% in 1996–2016) (Dunn et al. 2019), with (negative/positive) relationships to ENSO and AAO (Forcada and Trathan 2009).

King Penguins at the Crozet Islands which, unlike pygoscelids, feed primarily on myctophid fish, increased in the last 50 years (Guinet et al. 1996, Cherel et al. 2002, Weimerskirch et al. 2003, Barbraud et al. 2020). Negative and positive relationships between ENSO and breeding traits have been recorded at Crozet (Le Bohec et al. 2008, Saraux et al. 2011). In particular, there is evidence of negative impacts of large-scale climatic anomalies (subtropical dipole events) on their foraging behavior, reproduction and population dynamics (Bost et al. 2015), and a massive decline was recorded recently at the largest colony, in Ile aux Cochons, with no cause identified (Weimerskirch et al. 2018).

Emperor Penguins declined substantially between the mid-1970s and early 1980s, but then stabilised until 2016 at Pointe Géologie, East Antarctica (Barbraud et al. 2020), reflecting negative relationships with ENSO (Jenouvrier et al. 2005). The decline was associated with low adult survival rates, attributed to low winter sea-ice extent and reduced food availability. The gradual extirpation of the colony of emperor penguins at the Dion Islands, west Antarctic Peninsula was also attributed to decreasing sea-ice extent (Trathan et al. 2011).

Populations of wandering albatrosses have either shown a continuous decline in the last 50 years (South Georgia), or a decline until the 1980s followed by a slow recovery (Indian Ocean colonies) (Ryan et al. 2009, Poncet et al. 2017, Weimerskirch Delord et al. 2018). Although this can be attributed largely to anthropogenic factors, particularly incidental mortality (bycatch) in commercial fisheries, climate change, including ENSO also has an influence on population trends (Nel et al. 2003, Fay et al. 2017, Pardo et al. 2017, Phillips and Wood 2020). Changes in breeding success in other albatross species, including Black-browed Albatross, Sooty Albatross and Indian Yellow-nosed Albatross (*Thalassarche carteri*) are linked to climatic anomalies, particularly SOI (Rolland et al. 2010). Population trends and drivers vary regionally, and albatrosses in some areas of the Southern Ocean may be at higher risk, including wandering, black-browed and grey-headed albatrosses at South Georgia, which have all declined by > 40% in just 30–40 years (Poncet et al. 2017). Based on demographic modelling in relation to environmental variables, the main drivers appear to be extreme sea-surface temperature events in black-browed albatrosses, and the effects of changing climate in combination with fisheries bycatch on adult survival in Grey-headed Albatrosses (Pardo et al. 2017).

Population trends in most seabirds other than penguins and albatrosses are less well-known. Although there is considerable variation among sites, global populations of both Southern Giant Petrel (*Macronectes giganteus*) and Northern Giant Petrel (*M. halli*) are increasing at some sites associated with increased availability of carrion from recovering Antarctic fur seal populations (Poncet et al. 2020). These species show very high sexual size dimorphism, climatic factors including winds, ENSO, sea-ice concentration and SAM, as well as fisheries, had sex-specific effects on adult survival at South Georgia (Gianuca et al. 2019). Although data are limited for recent years and there is high variability, populations of Brown Skuas (*Stercorarius antarctica*) are generally decreasing, and South Polar Skuas (*S. maccormicki*) are generally stable or increasing, with trends in the latter probably related to sea-ice extent and duration, late snow cover at nesting sites or extreme weather (Carneiro et al. 2016, Phillips et al. 2019). The few robust trend data for Antarctic Shags (*Leucocarbo bransfieldensis*) also suggest an overall decreasing trend and a shift in distribution related to a gradient in environmental changes at the western Antarctic Peninsula (Schrimpf et al. 2018, Phillips et al. 2019). Populations of the Southern Fulmar (*Fulmarus glacialoides*) and Snow Petrel (*Pagodroma nivea*) have fluctuated periodically due to the SOI and sea-ice extent (Jenouvrier et al. 2005), as have those of the Antarctic Petrel (*Thalassoica antarctica*) (related to SOI and AAO) (Descamps et al. 2016).

Overall, population trends of Antarctic seabirds in recent decades have been more negative than positive, suggesting deleterious impacts of climate change in general. However, responses are often complex and vary among species in the same region (e.g., Chinstrap and Gentoo Penguins in the Antarctic Peninsula) and among regions in the same species (e.g., Adélie Penguins). Climate change may act in opposite directions on different aspects of life history of seabirds (Descamps et al. 2016) even within a single population (e.g., emperor penguins in East Antarctica; better breeding success associated with less fast ice, but better adult survival with more winter sea ice), underlining the need for a thorough understanding of population processes in order to make predictions about the future. Besides warming, other climate variables such as wind directly affect seabirds, particularly albatrosses, and anthropogenic factors (e.g., fisheries harvest and bycatch, invasive species) and disease have additive or potentially synergetic effects (Angel et al. 2009, Rolland et al. 2009, Phillips et al. 2016, Pardo et al. 2017). Long-term monitoring of seabirds across different regions of the Antarctic and sub-Antarctic is therefore paramount to fully assessing the effects of ongoing climate change and other threats.

Policy mechanisms for mitigating impacts of climate change on Antarctic seabirds

The Antarctic Treaty lists in its principles the need to ensure the preservation and protection of the Antarctic environment, using mechanisms provided by the Protocol on Environmental Protection and CCAMLR; these include the designation of Antarctic Specially Protected Areas (ASPAs), Antarctic Specially Managed Areas (ASMAs) or marine protected areas (MPAs) relevant to the conservation of seabirds (Phillips et al. 2016, SAT 2014, Southwell et al. 2017, Hughes et al. 2018). As the Southern Ocean is one of the regions experiencing the most rapid warming on the planet (Turner et al. 2014, IPCC 2018, IPCC 2019), Antarctic environmental management and regulation is being challenged to keep pace with such a change (Chown and Brooks 2019). Priorities for the governance of the Antarctic may change in the future as lack of progress in recent years hints at a gradual erosion of the mechanisms set up to ensure sustainable resource use, international cooperation and balancing of the human presence, science and conservation (Rintoul et al. 2018). This is despite what is predicted to be the increasing demand for ecosystem services in the Southern Ocean (Rogers et al. 2020).

Predictions of the impacts of future climate change on seabird communities of the Southern Ocean include some positive outcomes (e.g., shifting of oceanic fronts closer to breeding islands may facilitate access to prey) but are generally negative (e.g., impacts of decreasing sea ice on food web dynamics, long-term reduction in prey abundance, increase infrequency and magnitude of disease outbreaks, climate amelioration in the sub-Antarctic increasing over-winter survival of invasive alien species) (Bost et al. 2015, Descamps et al. 2016, Phillips et al. 2016, Cavanagh et al. 2017, Cristofari et al. 2018, Krüger et al. 2018, Cole et al. 2019). Some of the underlying drivers of population change across the Southern Ocean predators closer to the continent include glaciation, sea ice extent and duration, availability of ice-free ground for breeding, access to productive foraging grounds (Younger et al. 2016) and global climate environmental conditions (i.e., ENSO, AAO) (Descamps et al. 2016). Unfortunately, natural adaptive responses of animals, including seabirds, to climate change tend to occur slowly (Radchuk et al. 2019), and are unlikely to keep pace with global patterns of climatic variability which are generally detrimental to populations of Antarctic seabirds (Descamps et al. 2016). As temperatures continue to increase, the more likely scenarios are that marine species that exploit warmer conditions will expand their distribution southwards, and those that prefer colder conditions will decline, potentially to extinction (Fraser et al. 2012, Boersma et al. 2019, Cole et al. 2019, Ropert-Coudert et al. 2019). For example, under business-as-usual greenhouse emissions, 80% of the colonies of emperor penguins are projected to be quasi-extinct by 2100 (i.e., decline by 81%); however, if the targets of the Paris Agreement are met, only up to 31% of colonies are projected to be quasi-extinct (Jenouvrier et al. 2020).

Areas of ecological significance (i.e., regions preferred by multiple predator species as indicators of high levels of biomass at lower trophic levels, and biodiversity), based on > 4 000 tracks of 17 Antarctic mammal and bird species (including albatrosses, penguins and petrels), were recently identified around sub-Antarctic islands in the Atlantic and Indian Oceans and over the Antarctic continental shelf (Hindell et al. 2020). Habitat models in that study based on IPCC climate projections showed that climate change over the next century will likely reduce the size of those areas and impose more pressures. Indeed, range-restricted communities of sub-Antarctic islands will likely suffer some of the greatest negative consequences of climate change (Rogers et al. 2020). The areas of ecological importance in 2100 will overlap less with the current and proposed MPAs for the Antarctic (Hindell et al. 2020), suggesting that adaptive management (e.g., regular monitoring of MPAs, flexible MPA boundaries) will be important to deal with future change and ensure the best outcomes for conservation. Moreover, poleward shifts in the distribution of seabirds in 2100 will increase the overlap with fisheries, increasing bycatch risk (Krüger et al. 2018).

Effective assessment of the status of Antarctic seabirds through monitoring of their populations (e.g., through the CCAMLR Ecosystem Monitoring Programme; CEMP) is essential for developing conservation and management policies for the Southern Ocean, as is the engagement of the various stakeholders, including ACAP, ATS, CCAMLR, Scientific Committee on Antarctic Research (SCAR), the fishing industry and NGOs (Reid 2018, Constable 2002). Major gaps remain in knowledge of ecology, diet, distribution, population sizes and demography of seabirds of the Southern Ocean (Phillips et al. 2016, Ropert-Coudert et al. 2019), especially during the austral winter (Takahashi et al. 2018). Moreover, there is a need to develop methods for assessing current and future impacts of climate change on the structure and function of the Southern Ocean in which seabirds play important roles (Constable et al. 2014, McMahon et al. 2019). As an example, Southwell et al. (2017) outlined how improvements in the assessment of distribution, abundance and prey consumption of Adélie Penguins in East Antarctica would aid decisions on future conservation management within the context of ATS and CCAMLR.

In future, the development of new policies for the Antarctic region should strive to take into account the effects of climate change in Antarctic seabirds in relation to other threats (competition with fisheries, fisheries bycatch, invasive alien species, pollution, disease, habitat degradation and human disturbance) that may exacerbate the impacts (Croxall et al. 2012, Trathan et al. 2015, Phillips et al. 2016, Ropert-Coudert et al. 2019, Weimerskirch et al. 2020). Success will depend on more engagement between seabird ecologists and policy makers to facilitate and accelerate policy development (Southwell et al. 2017, Hughes et al. 2018). Although good progress has been made in the last decade, conservation efforts need to be improved urgently to safeguard threatened species in particular (Phillips et al. 2016, Boersma et al. 2019). As the Antarctic represents a challenging management scenario due to the rising pressure to exploit its resources (Rogers et al. 2020), it is clear that a deeper understanding of the role, and close monitoring of the population status of seabirds is essential for its effective conservation and management.

Acknowledgments

The study benefitted from the strategic program of MARE, financed by FCT (UIDB/04292/2020), Project Research KP309 of National Institute of Polar Research, Japan and the Ecosystems component of the British Antarctic Survey Polar Science for Planet Earth Programme, funded by the UK Natural Environment Research Council. We also acknowledge Christophe Barbraud for his detailed data on seabirds breeding ecology and the designer Bruno Cruz for producing the diagram.

References cited

Abreu, J., R.A. Phillips, F.R. Ceia, L. Ireland, V.H. Paiva and J.C. Xavier. 2020. Long-term changes in habitat and trophic level of Southern Ocean squid in relation to environmental conditions. Sci. Rep. 10: 15215. doi: 10.1038/s41598-020-72103-6.

Ainley, D., G. Ballard, L.K. Blight, S. Ackley, S.D. Emslie, A. Lescroel et al. 2010. Impacts of cetaceans on the structure of Southern Ocean food webs. Mar. Mammal Sci. 26: 482–498. doi: 10.1111/j.1748-7692.2009.00337.x.

Ainley, D.G., E.D. Clarke, K. Arrigo, W.R. Fraser, A. Kato, K.J. Barton et al. 2005. Decadal-scale changes in the climate and biota of the Pacific sector of the Southern Ocean, 1950s to the 1990s. Antarctic Sci. 17: 171. doi: 10.3354/meps10928.

Ainley, D.G. and L.K. Blight. 2009. Ecological repercussions of historical fish extraction from the Southern Ocean. Fish and Fish. 10: 13–38. doi: 10.1111/j.1467-2979.2008.00293.x.

Angel, A., R.M. Wanless and J. Cooper. 2009. Review of impacts of the introduced house mouse on islands in the Southern Ocean: are mice equivalent to rats? Biol. Invasions 11: 1743–1754. doi: 10.1007/s10530-008-9401-4.

Arreguín-Sánchez, F. 2019. Climate change and the rise of the octopus fishery in the Campeche Bank, México. Regional Studies in Mar. Sci. 32: 100852.

Atkinson, A., V. Siegel, E.A. Pakhomov and P. Rothery. 2004. Long-term decline in krill stock and increase in salps within the Southern Ocean. Nature 432: 100–103. doi: 10.1038/nature02996.

Atkinson, A., S.L. Hill, E.A. Pakhomov, V. Siegel, C.S. Reiss, V.J. Loeb et al. 2019. Krill (*Euphausia superba*) distribution contracts southward during rapid regional warming. Nat. Clim. Chang. 9: 142. doi: 10.1038/s41558-018-0370-z.

Barber-Meyer, S., G. Kooyman and P. Ponganis. 2008. Trends in western Ross Sea emperor penguin chick abundances and their relationships to climate. Antarctic Sci. 20: 3–11. doi: 10.1017/S0954102007000673.

Barbosa, A., J. Benzal, A. de Leon and J. Moreno. 2012. Population decline of chinstrap penguin (*Pygoscelis antarctica*) on Deception Island, South Shetlands, Antarctica. Polar Biol. 35: 1453–1457. doi: 10.1007/s00300-012-1196-1.

Barbraud, C. and H. Weimerskirch. 2001. Emperor penguins and climate change. Nature 411: 183–186. doi: 10.1038/35075554.

Barbraud, C. and H. Weimerskirch. 2006. Antarctic birds breed later in response to climate change. P. Natl. Acad. Sci. USA 103: 6248–6251. doi: 10.1073/pnas.0510397103.

Barbraud, C., P. Rivalan, P. Inchausti, M. Nevoux, V. Rolland and H. Weimerskirch. 2011. Contrasted demographic responses facing future climate change in Southern Ocean seabirds. J. Animal Ecol. 80: 89–100. doi: 10.1111/j.1365–2656.2010.01752.x.

Barbraud, C., V. Rolland, S. Jenouvrier, M. Nevoux, K. Delord and H. Weimerskirch. 2012. Effects of climate change and fisheries bycatch on Southern Ocean seabirds: A review. Mar. Ecol. Progr. Ser. 454: 285–307. doi: 10.3354/meps09616.

Barbraud, C., K. Delord, C. Bost, A. Chaigne, C. Marteau and H. Weimerskirch. 2020. Population trends of penguins in the French Southern Territories. Polar Biol. 43: 835–850. doi: 0.1007/s00300-020-02691-6.

Bednaršek, N., G.A. Tarling, D.C. Bakker, S. Fielding, A. Cohen, A. Kuzirian et al. 2012a. Description and quantification of pteropod shell dissolution: A sensitive bioindicator of ocean acidification. Glob. Change Biol. 18: 2378–2388. doi: 10.1111/j.1365-2486.2012.02668.x.

Bednaršek, N., G.A. Tarling, D.C.E. Bakker, S. Fielding, E.M. Jones, H.J. Venables et al. 2012b. Extensive dissolution of live pteropods in the Southern Ocean. Nature Geosci. 5: 881–885. doi: 10.1038/ngeo1635.

Bessa, F., N. Ratcliffe, V. Otero, P. Sobral, J.C. Marques and J.C. Xavier. 2019. Microplastics in gentoo penguins from the Antarctic region. Sci. Rep. 9: 14191. doi: 10.1038/s41598-019-50621-2.

Bestley, S., Y. Ropert-Coudert, S. Bengtson Nash, C.M. Brooks, C. Cotté, M. Dewar et al. 2020. Marine ecosystem assessment for the southern ocean: Birds and marine mammals in a changing climate. Front. Ecol. Evol. 8: 566936. doi: 10.3389/fevo.2020.566936.

Bilyk, K.T. and A.L. DeVries. 2011. Heat tolerance and its plasticity in Antarctic fishes. Comp. Biochem. Physiol. A158: 382–390. doi: 10.1016/j.cbpa.2010.12.010.

Boersma, P.D., P.G. Borboroglu, N.J. Gownaris, C.A. Bost, A. Chiaradia, S. Ellis et al. 2019. Applying science to pressing conservation needs for penguins. Conserv. Biol. 34: 103–112. doi: 10.1111/cobi.13378.

Böning, C.W., A. Dispert, M. Visbeck, S.R. Rintoul and F.U. Schwarzkopf. 2008. The response of the Antarctic circumpolar current to recent climate change. Nature Geosci. 1: 864–869. doi: 10.1038/ngeo362.

Borowicz, A., S. Forrest, M. Wethington, N. Strycker and H.J. Lynch. 2020. Presence of King penguins (*Aptenodytes patagonicus*) on Elephant Island provides further evidence of range expansion. Polar Biology 43: 2117–2120. doi: 10.1007/s00300-020-02760–w.

Bost, C.A., C. Cotté, P. Terray, C. Barbraud, C. Bon, K. Delord, O. Gimenez et al. 2015. Large-scale climatic anomalies affect marine predator foraging behaviour and demography. Nature Commun. 6: 8220. doi: 10.1038/ncomms9220.

Bromwich, D.H., J.P. Nicolas, A.J. Monaghan, M.A. Lazzara, L.M. Keller, G.A. Weidner et al. 2013. Central West Antarctica among the most rapidly warming regions on Earth. Nature Geosci. 6: 139–145. doi: 10.1038/ngeo1671.

Bronselaer, B., J.L. Russell, M. Winton, N.L. Williams, R.M. Key, J.P. Dunne et al. 2020. Importance of wind and meltwater for observed chemical and physical changes in the Southern Ocean. Nature Geosci. 13: 35–42. doi: 10.1038/s41561-019-0502-8.

Brown, M.S., D.R. Munro, C.J. Feehan, C. Sweeney, H.W. Ducklow and O.M. Schofield. 2019. Enhanced oceanic CO2 uptake along the rapidly changing West Antarctic Peninsula. Nat. Clim. Change 9: 678–683. doi: 10.1038/s41558–019–0552–3.

Budd, G. 2000. Changes in Heard Island glaciers, king penguins and fur seals since 1947. Paper presented at the Papers and Proceedings of the Royal Society of Tasmania 133: 47–60.

Cai, W. 2006. Antarctic ozone depletion causes an intensification of the Southern Ocean super-gyre circulation. Geophys. Res. Lett. 33: L03712. doi: 10.1029/2005GL024911.

Caldeira, K. and M.E. Wickett. 2003. Anthropogenic carbon and ocean pH. Nature 425: 365–365. doi: 10.1038/425365[a].

Carlini, A.R., N.R. Coria, M.M. Santos, J. Negrete, M.A. Juares and G.A. Daneri. 2009. Responses of *Pygoscelis adeliae* and *P. papua* populations to environmental changes at Isla 25 de Mayo (King George Island). Polar Biol. 32: 1427–1433. doi: 10.1007/s00300-009-0637-y.

Carneiro, A.P., A. Manica and R.A. Phillips. 2016. Long-term changes in population size, distribution and productivity of skuas (*Stercorarius* spp.) at Signy Island, south Orkney Islands. Polar Biol. 39: 617–625. doi: 10.1007/s00300-015-1817-6.

Casaux, R. and E. Barrera-Oro. 2016. Linking population trends of Antarctic shag (*Phalacrocorax bransfieldensis*) and fish at Nelson Island, South Shetland Islands (Antarctica). Polar Biol. 39: 1491–1497. doi: 10.1007/s00300-015-1850-5.

Cavanagh, R.D., E.J. Murphy, T.J. Bracegirdle, J. Turner, C.A. Knowland, S.P. Corney et al. 2017. A synergistic approach for evaluating climate model output for ecological applications. Frontiers in Marine Science 4: 308. doi: 10.3389/fmars.2017.00308.

Cherel, Y., K. Pütz and K.A. Hobson. 2002. Summer diet of king penguins (*Aptenodytes patagonicus*) at the Fakland Islands, southern Atlantic ocean. Polar Biol. 25: 898–906. doi: 10.1007/s00300-002-0419-2.

Cherel, Y., J.C. Xavier, S. de Grissac, C. Trouvé and H. Weimerskirch. 2017. Feeding ecology, isotopic niche, and ingestion of fishery-related items of the wandering albatross *Diomedea exulans* at Kerguelen and Crozet Islands. Mar. Ecol. Prog. Ser. 565: 197–215. doi: 10.3354/meps11994.

Chown, S.L. and C.M. Brooks. 2019. The state and future of Antarctic environments in a global context. Annu. Rev. Env. Resour. 44: 1–30. doi: 10.1146/annurev-environ-101718-033236.

Christensen, J.H., K.K. Kanikicharla, E. Aldrian, S.I. An, I.F.A. Cavalcanti, M. de Castro et al. 2013. Climate phenomena and their relevance for future regional climate change. pp. 1217–1308. *In*: Stocker, T.F., D. Qin, G.-K. Plattner, M. Tignor, S.K. Allen, J. Boschung et al. [eds.]. Climate Change 2013 the Physical Science Basis: Working Group I Contribution to the Fifth Assessment Report of the Intergovernmental Panel on Climate Change. Cambridge University Press, Cambridge, UK.

Cimino, M.A., W.R. Fraser, D.L. Patterson-Fraser, V.S. Saba and M.J. Oliver. 2014. Large-scale climate and local weather drive interannual variability in Adélie penguin chick fledging mass. Mar. Ecol. Prog. Ser. 513: 253–268. doi: 10.3354/meps10928.

Clem, K.R., R.L. Fogt, J. Turner, B.R. Lintner, G.J. Marshall, J.R. Miller et al. 2020. Record warming at the South Pole during the past three decades. Nat. Clim. Change 10: 762–770. doi: 10.1038/s41558-020-0815-z.

Clucas, G.V., M.J. Dunn, G. Dyke, S.D. Emslie, H. Levy, R. Naveen et al. 2014. A reversal of fortunes: Climate change 'winners' and 'losers' in Antarctic Peninsula penguins. Sci. Rep. 4: 5024. doi: 10.1038/srep05024.

Clucas, G.V., J.L. Younger, D. Kao, A.D. Rogers, J. Handley, G.D. Miller, P. Jouventin et al. 2016. Dispersal in the sub-Antarctic: King penguins show remarkably little population genetic differentiation across their range. BMC Evol. Biol. 16: 211. doi: 10.1186/s12862-016-0784-z.

Cole, T.L., L. Dutoit, N. Dussex, T. Hart, A. Alexander, J.L. Younger et al. 2019. Receding ice drove parallel expansions in Southern Ocean penguins. P. Natl. Acad. Sci. USA 116: 26690–26696. doi: 10.1073/pnas.1904048116.

Collard, M., C. De Ridder, B. David, F. Dehairs and P. Dubois. 2015. Could the acid-base status of Antarctic sea urchins indicate a better-than-expected resilience to near-future ocean acidification? Glob. Change Biol. 21: 605–617. doi: 10.1111/gcb.12735.

Comiso, J.C. and F. Nishio. 2008. Trends in the sea ice cover using enhanced and compatible AMSR-E, SSM/I, and SMMR data. J. Geophys. Res. 113: C2. doi: 10.1029/2007JC004257.

Connan, M., P. Mayzaud, C. Trouvé, C. Barbraud and Y. Cherel. 2008. Interannual dietary changes and demographic consequences in breeding blue petrels from Kerguelen Islands. Mar. Ecol. Prog. Ser. 373: 123–135. doi: 10.3354/meps07723.

Connan, M., B.J. Dilley, T.O. Whitehead, D. Davies, C.D. McQuaid and P.G. Ryan. 2019. Multidimensional stable isotope analysis illuminates resource partitioning in a sub-Antarctic island bird community. Ecography 42: 1948–1959. doi: 10.1111/ecog.04560.

Constable, A.J. 2002. CCAMLR ecosystem monitoring and management: Future work. CCAMLR Science 9: 233–253.

Constable, A.J., J. Melbourne-Thomas, S.P. Corney, K. Arrigo, C. Barbraud, D. Barnes et al. 2014. Change in Southern Ocean ecosystems I: How changes in physical habitats directly affect marine biota. Glob. Change Biol. 20: 3004–3025. doi: 10.1111/gcb.12623.

Convey, P. and L.S. Peck. 2019. Antarctic environmental change and biological responses. Sci. Advances 5: eaaz0888. doi: 10.1126/sciadv.aaz0888.

Cristofari, R., X. Liu, F. Bonadonna, Y. Cherel, P. Pistorius, Y. Le Maho et al. 2018. Climate-driven range shifts of the king penguin in a fragmented ecosystem. Nat. Clim. Change 8: 245–251. doi: 10.1038/s41558-018-0084-2.

Cross, E.L., L.S. Peck and E.M. Harper. 2015. Ocean acidification does not impact shell growth or repair of the Antarctic brachiopod *Liothyrella uva* (Broderip, 1833). J. Exp. Mar. Biol. Ecol. 462: 29–35. doi: 10.1016/j.jembe.2014.10.013.

Croxall, J.P. 1992. Southern Ocean environmental changes—Effects on Seabird, Seal and Whale Populations. Philos. Trans. R. Soc. Lond. Ser. B–Biol. Sci. 338: 319–328. doi: 10.1098/rstb.1992.0152.

Croxall, J.P., P.A. Prince, P. Rothery and A.G. Wood. 1998. Population changes in albatrosses at South Georgia. pp. 69–83. *In*: Robertson, G. and R. Gales [eds.]. Albatross Biology and Conservation. Surrey Beatty and Sons, Chipping Norton, Australia.

Croxall, J.P., K. Reid and P.A. Prince. 1999. Diet, provisioning and productivity responses of marine predators to differences in availability of Antarctic krill. Mar. Ecol. Prog. Ser. 177: 115–131. doi: 10.3354/meps177115.

Croxall, J.P., P.N. Trathan and E.J. Murphy. 2002. Environmental change and Antarctic Seabird populations. Science 297: 1510–1514. doi: 10.1126/science.1071987.

Croxall, J.P., S.H.M. Butchart, B. Lascelles, A.J. Stattersfield, B. Sullivan, A. Symes et al. 2012. Seabird conservation status, threats and priority actions: A global assessment. Birdlife Conserv. Int. 22: 1–34. doi: 10.1017/S0959270912000020.

Cubillos, J., S. Wright, G. Nash, M. De Salas, B. Griffiths, B. Tilbrook et al. 2007. Calcification morphotypes of the coccolithophorid *Emiliania huxleyi* in the Southern Ocean: changes in 2001 to 2006 compared to historical data. Mar. Ecol. Prog. Ser. 348: 47–54. doi: 10.3354/meps07058.

De Broyer, C., P. Koubbi, H.J. Griffiths, B. Raymond, C.d. Udekem d'Acoz, A.P. Van de Putte et al. 2014. Biogeographic Atlas of the Southern Ocean: Scientific Committee on Antarctic Research, Cambridge, UK.

Deppeler, S.L. and A.T. Davidson. 2017. Southern ocean phytoplankton in a changing climate. Front. Mar. Sci. 4: 40. doi: 10.3389/fmars.2017.00040.

Descamps, S., A. Tarroux, S.H. Lorentsen, O.P. Love, Ø. Varpe and N.G. Yoccoz. 2016. Large-scale oceanographic fluctuations drive Antarctic petrel survival and reproduction. Ecography 39: 496–505. doi: 10.5061/dryad.ds63c.

Doubleday, Z.A., T.A.A. Prowse, A. Arkhipkin, G.J. Pierce, J. Semmens, M. Steer et al. 2016. Global proliferation of cephalopods. Current Biol. 26: R406–R407. doi: 10.1016/j.cub.2016.04.002.

Ducklow, H.W., K. Baker, D.G. Martinson, L.B. Quetin, R.M. Ross, R.C. Smith et al. 2007. Marine pelagic ecosystems: The West Antarctic Peninsula. Philos. Trans. R. Soc. Lond. Ser. B–Biol. Sci. 362: 67–94. doi: 10.1098/rstb.2006.1955.

Dunn, M.J., J.A. Jackson, S. Adlard, A.S. Lynnes, D.R. Briggs, D. Fox et al. 2016. Population size and decadal trends of three penguin species nesting at Signy Island, South Orkney Islands. PLoS One 11: e0164025. doi: 10.1371/journal.pone.0164025.

Dunn, M.J., J. Forcada, J.A. Jackson, C.M. Waluda, C. Nichol and P.N. Trathan. 2019. A long-term study of gentoo penguin (*Pygoscelis papua*) population trends at a major Antarctic tourist site, Goudier Island, Port Lockroy. Biodivers. Conserv. 28: 37–53. doi: 10.1007/s10531-018-1635-6.

Duprat, L.P., G.R. Bigg and D.J. Wilton. 2016. Enhanced Southern Ocean marine productivity due to fertilization by giant icebergs. Nat. Geosci. 9: 219–221. doi: 10.1038/ngeo2633.

Durack, P.J., S.E. Wijffels and R.J. Matear. 2012. Ocean salinities reveal strong global water cycle intensification during 1950 to 2000. Science 336: 455–458. doi: 10.1126/science.1212222.

Durant, J.M., D. Hjermann, M. Frederiksen, J.B. Charrassin, Y. Le Maho, P.S. Sabarros et al. 2009. Pros and cons of using seabirds as ecological indicators. Climate Res. 39: 115–129. doi: 10.3354/cr00798.

Emmerson, L. and C. Southwell. 2008. Sea ice cover and its influence on Adélie penguin reproductive performance. Ecology 89: 2096–2102. doi: 10.3354/cr00798.

Emslie, S.D. and J.D. McDaniel. 2002. Adélie penguin diet and climate change during the middle to late Holocene in northern Marguerite Bay, Antarctic Peninsula. Polar Biol. 25: 222–229. doi: 10.1007/s00300-001-0334-y.

Emslie, S.D. and E.J. Woehler. 2005. A 9000–year record of Adélie penguin occupation and diet in the Windmill Islands, East Antarctica. Antarctic Sci. 17: 57–66. doi: 10.1017/S0954102005002427.

Emslie, S.D., L. Coats and K. Licht. 2007. A 45,000 yr record of Adélie penguins and climate change in the Ross Sea, Antarctica. Geology 35: 61–64. doi: 10.1130/G23011A.1.

Enzor, L.A., E.M. Hunter and S.P. Place. 2017. The effects of elevated temperature and ocean acidification on the metabolic pathways of notothenioid fish. Conserv. Physiol. 5: cox019. doi: 10.1093/conphys/cox019.

Everson, I. 2000. Krill: Biology, ecology and fisheries. Blackwell Science Ltd., Oxford, UK.

Fay, R., C. Barbraud, K. Delord and H. Weimerskirch. 2017. Contrasting effects of climate and population density over time and life stages in a long-lived seabird. Funct. Ecol. 31: 1275–1284. doi: 10.1111/1365-2435.12831.

Flynn, E.E., B.E. Bjelde, N.A. Miller and A.E. Todgham. 2015. Ocean acidification exerts negative effects during warming conditions in a developing Antarctic fish. Conserv. Physiol. 3: cov033. doi: 10.1093/conphys/cov033.

Forcada, J., P.N. Trathan, K. Reid, E.J. Murphy and J.P. Croxall. 2006. Contrasting population changes in sympatric penguin species in association with climate warming. Glob. Change Biol. 12: 411–423. doi: 10.1111/j.1365–2486.2006.01108.x.

Forcada, J. and P.N. Trathan. 2009. Penguin responses to climate change in the Southern Ocean. Glob. Change Biol. 15: 1618–1630. doi: 10.1111/j.1365-2486.2009.01909.x.

Fountain, A.G., G. Saba, B. Adams, P. Doran, W. Fraser, M. Gooseff et al. 2016. The impact of a large-scale climate event on Antarctic ecosystem processes. BioScience 66: 848–863. doi: 10.1093/biosci/biw110.

Fraser, C.I., R. Nikula, D.E. Ruzzante and J.M. Waters. 2012. Poleward bound: Biological impacts of Southern Hemisphere glaciation. Trends Ecol. Evol. 27: 462–471. doi: 10.1016/j.tree.2012.04.011.

Fraser, W.R. and W.Z. Trivelpiece. 1996. Factors controlling the distribution of seabirds: Winter-summer heterogeneity in the distribution of Adélie penguin populations. pp. 257–272. *In*: Ross, R.M., E.E. Hofmann and L.B. Quetin [eds.]. Foundations for ecological research west of the Antarctic Peninsula. American Geophysical Union, vol. 70., Washingotn D.C., USA.

Fraser, W.R. and E.E. Hofmann. 2003. A predator's perspective on causal links between climate change, physical forcing and ecosystem response. Mar. Ecol. Prog. Ser. 265: 1–15. doi: 10.3354/meps265001.

Frölicher, T.L., J.L. Sarmiento, D.J. Paynter, J.P. Dunne, J.P. Krasting and M. Winton. 2015. Dominance of the Southern Ocean in anthropogenic carbon and heat uptake in CMIP5 models. J. Climate 28: 862–886. doi: 10.1175/JCLI-D-14-00117.1.

Furness, R.W. and K. Camphuysen. 1997. Seabirds as monitors of the marine environment. ICES J. Mar. Sci. 54: 726–737. doi: 10.1006/jmsc.1997.0243.

Gales, R. 1998. Albatross populations: status and threats. pp 20–45. *In*: Robertson, G. and R. Gales [eds.]. Albatross Biology and Conservation: Surrey Beatty and Sons, Chipping Norton, Australia.

Gianuca, D., S.C. Votier, D. Pardo, A.G. Wood, R.B. Sherley, L. Ireland et al. 2019. Sex-specific effects of fisheries and climate on the demography of sexually dimorphic seabirds. J. Animal Ecol. 88: 1366–1378. doi: 10.1111/1365-2656.13009.

Gille, S.T. 2002. Warming of the Southern Ocean since the 1950s. Science 295: 1275–1277. doi: 10.1126/science.1065863.

Gille, S.T. 2008. Decadal-scale temperature trends in the Southern Hemisphere ocean. J. Climate 21: 4749–4765. doi: 10.1175/2008JCLI2131.1.

Golikov, A.V., R.M. Sabirov, P.A. Lubin and L.L. Jørgensen. 2013. Changes in distribution and range structure of Arctic cephalopods due to climatic changes of the last decades. Biodiversity 14: 28–35. doi: 10.1080/14888386.2012.702301.

Grecian, W.J., G.A. Taylor, G. Loh, R.A. McGill, C.M. Miskelly, R.A. Phillips et al. 2016. Contrasting migratory responses of two closely related seabirds to long-term climate change. Mar. Ecol. Prog. Ser. 559: 231–242. doi: 10.3354/meps11875.

Guinet, C., Y. Cherel, V. Ridoux and P. Jouventin. 1996. Consumption of marine resources by seabirds and seals in Crozet and Kerguelen waters: Changes in relation to consumer biomass 1962–65. Antarctic Sci. 8: 23–30. doi: 10.1017/S0954102096000053.

Gutt, J., E. Isla, J.C. Xavier, B. Adams, I.-Y. Ahn and C.-H.C. Cheng et al. 2021. Antarctic ecosystems in transition—life between stresses and opportunities. Biol. Rev. 96: 798–821. doi: 10.1111/brv.12679.

Hallegraeff, G., C. Davies and W. Rochester. 2020. Range expansion of the red tide dinoflagellate *Noctiluca scintillans*. pp. 3.6.1–3.6.5. *In*: Richardson, A.J., R. Eriksen, T. Moltmann, I. Hodgson-Johnston and J.R. Wallis [eds.]. State and Trends of Australia's Ocean Report. Australia. doi: 10.26198/5e16ad2e49e85.

Halpern, B.S., C. Longo, D. Hardy, K.L. McLeod, J.F. Samhouri, S.K. Katona et al. 2012. An index to assess the health and benefits of the global ocean. Nature 488: 615–620. doi: 10.1038/nature11397.

Hauri, C., T. Friedrich and A. Timmermann. 2016. Abrupt onset and prolongation of aragonite undersaturation events in the Southern Ocean. Nat. Clim. Change 6: 172–176. doi: 10.1038/nclimate2844.

Hindell, M.A., R.R. Reisinger, Y. Yan Ropert-Coudert, L.A. Hückstädt, P.N. Trathan, H. Bornemann et al. 2020. Tracking of marine predators to protect Southern Ocean ecosystems. Nature 580: 87–92. doi: 10.1038/s41586-020-2126-y.

Hinke, J.T., K. Salwicka, S.G. Trivelpiece, G.M. Watters and W.Z. Trivelpiece. 2007. Divergent responses of *Pygoscelis* penguins reveal a common environmental driver. Oecologia 153: 845. doi: 10.1007/s00442-007-0781-4.

Hofman, R. 2017. Sealing, whaling and krill fishing in the Southern Ocean: Past and possible future effects on catch regulations. Polar Record 53: 88–99. doi: 10.1017/S0032247416000644.

Hückstädt, L.A., A. Piñones, D.M. Palacios, B.I. McDonald, M.S. Dinniman, E.E. Hofmann et al. 2020. Projected shifts in the foraging habitat of crabeater seals along the Antarctic Peninsula. Nat. Clim. Change 10: 472–477. doi: 10.1038/s41558-020-0745-9.

Hughes, K.A., A. Constable, Y. Frenot, J. López-Martínez, E. McIvor, B. Njåstad et al. 2018. Antarctic environmental protection: Strengthening the links between science and governance. Environ. Sci. Policy 83: 86–95. doi: 10.1016/j.envsci.2018.02.006.

Inchausti, P., C. Guinet, M. Koudil, J.P. Durbec, C. Barbraud, H. Weimerskirch et al. 2003. Inter-annual variability in the breeding performance of seabirds in relation to oceanographic anomalies that affect the Crozet and the Kerguelen sectors of the Southern Ocean. J. Avian Biol. 34: 170–176. doi: 10.1034/j.1600-48X.2003.03031.x.

IPCC. 2018. Global warming of 1.5°C. An IPCC Special Report on the impacts of global warming of 1.5°C above pre-industrial levels and related global greenhouse gas emission pathways, in the context of strengthening the global response to the threat of climate change, sustainable development, and efforts to eradicate poverty. V. Masson–Delmotte, P. Zhai, H. O. Pörtner, D. Roberts, J. Skea and P. R. Shukla et al. [eds.]. World Meteorological Organization, Geneva, Switzerland.

IPCC. 2019. IPCC Special Report on the Ocean and Cryosphere in a Changing Climate. H.O. Pörtner, D. C. Roberts, V. Masson–Delmotte, P. Zhai, M. Tignor, E. Poloczanska et al. [eds.]. Intergovernmental Panel on Climate Change (IPCC), Geneva, Switzerland.

IUCN. 2021. IUCN Red List of Threatened Species. Version 2020–3 (www.iucnredlist.org).

Juáres, M.A., F. Ferrer, N.R. Coria and M.M. Santos. 2017. Breeding events of king penguin at the South Shetland Islands: Has it come to stay? Polar Biology 40: 457–461. doi: 10.1007/s00300-016-1947-5.

Jenouvrier, S., H. Weimerskirch, C. Barbraud, Y.-H. Park and B. Cazelles. 2005. Evidence of a shift in the cyclicity of Antarctic seabird dynamics linked to climate. P. Roy. Soc. B272(1566): 887–895. doi: 10.1098/rspb.2004.2978.

Jenouvrier, S., C. Barbraud and H. Weimerskirch. 2006. Sea ice affects the population dynamics of Adélie penguins in Terre Adélie. Polar Biol. 29: 413–423. doi: 10.1007/s00300-005-0073-6.

Jenouvrier, S. 2013. Impacts of climate change on avian populations. Glob. Change Biol. 19: 2036–2057. doi: 10.1111/gcb.12195.

Jenouvrier, S., M. Holland, J. Stroeve, M. Serreze, C. Barbraud, H. Weimerskirch et al. 2014. Projected continent-wide declines of the emperor penguin under climate change. Nat. Clim. Change 4: 715–718. doi: 10.1038/nclimate2280.

Jenouvrier, S., M. Holland, D. Iles, S. Labrousse, L. Landrum, J. Garnier et al. 2020. The Paris Agreement objectives will likely halt future declines of emperor penguins. Glob. Change Biol. 26: 1170–1184. doi: 10.1111/gcb.14864.

Kato, A. and Y. Ropert-Coudert. 2006. Rapid increase in Adelie penguin populations in the Lutzow–Holm Bay area since the mid 1990s. Polar Biosci. 20: 55–62.

Kaufman, D.E., M.A. Friedrichs, W.O. Smith, Jr., E.E. Hofmann, M.S. Dinniman et al. 2017. Climate change impacts on southern Ross Sea phytoplankton composition, productivity, and export. J. Geophys. Res. 122: 2339–2359. doi: 10.1002/2016JC012514.

Kellermann, A.K. 1996. Midwater fish ecology. Ipp. 231–256. *In:* Ross, R.M., E.E. Hofmann and L.B. Quetin [eds.]. Foundations for Ecological Research West of the Antarctic Peninsula, vol 70. American Geophysical Union, Washington, DC, USA.

Keogan, K., F. Daunt, S. Wanless, R.A. Phillips, C.A. Walling, P. Agnew et al. 2018. Global phenological insensitivity to shifting ocean temperatures among seabirds. Nat. Clim. Change 8: 313–318. doi: 10.1038/s41558–018–0115–z.

Knox, G.A. 2007. The biology of the Southern Ocean. CRC Press/Taylor and Francis, Boca Raton, FL,USA.

Kock, K.-H. 1992. Antarctic fish and fisheries. Cambridge University Press, Cambridge, UK.

Kooyman, G. and P. Ponganis. 2017. Rise and fall of Ross Sea emperor penguin colony populations: 2000 to 2012. Antarctic Sci. 29: 201. doi: 10.1017/S0954102016000559.

Krüger, L., J. Ramos, J. Xavier, D. Grémillet, J. González-Solís, M. Petry et al. 2018. Projected distributions of Southern Ocean albatrosses, petrels and fisheries as a consequence of climatic change. Ecography 41: 195–208. doi: 10.1111/ecog.02590.

Krüger, L., M.F. Huerta, F. Santa Cruz and C.A. Cárdenas. 2020. Antarctic krill fishery effects over penguin populations under adverse climate conditions: Implications for the management of fishing practices. Ambio. doi: 10.1007/s13280-020-01386-w.

Lagger, C., M. Nime, L. Torre, N. Servetto, M. Tatián and R. Sahade. 2018. Climate change, glacier retreat and a new ice-free island offer new insights on Antarctic benthic responses. Ecography 41: 579–591. doi: 10.1111/ecog.03018.

LaRue, M.A., D.G. Ainley, M. Swanson, K.M. Dugger, O. Phil, B. Lyver et al. 2013. Climate change winners: Receding ice fields facilitate colony expansion and altered dynamics in an Adélie penguin metapopulation. PLoS One 8: e60568. doi: 10.1371/journal.pone.0060568.

Le Bohec, C., J.M. Durant, M. Gauthier-Clerc, N.C. Stenseth, Y.-H. Park, R. Pradel et al. 2008. King penguin population threatened by Southern Ocean warming. P. Natl Acad. Sci. 105: 2493–2497. doi: 10.1073/pnas.0712031105.

Le Guen, C., A. Kato, B. Raymond, C. Barbraud, M. Beaulieu, C.A. Bost et al. 2018. Reproductive performance and diving behaviour share a common sea-ice concentration optimum in Adélie penguins (*Pygoscelis adeliae*). Glob. Change Biol. 24: 5304–5317. doi: 10.1111/gcb.14377.

Lewis, S., D.H. Nussey, A.G. Wood, J.P. Croxall and R.A. Phillips. 2012. Intrinsic determinants of a population trend in timing of breeding in the wandering albatross. Oikos 121: 2061–2071. doi: 10.1111/j.1600-0706.2012.20293.x.

Loeb, V., V. Siegel, O. Holm-Hansen, R. Hewitt, W. Fraser, W. Trivelpiece et al. 1997. Effects of sea-ice extent and krill or salp dominance on the Antarctic food web. Nature 387: 897–900. doi: 10.1038/43174.

Lorenzini, S., S. Olmastroni, F. Pezzo, M.C. Salvatore and C. Baroni. 2009. Holocene Adélie penguin diet in Victoria Land, Antarctica. Polar Biol. 32: 1077–1086. doi: 10.1007/s00300-009-0607-4.

Lynch, H.J., R. Naveen, P.N. Trathan and W.F. Fagan. 2012. Spatially integrated assessment reveals widespread changes in penguin populations on the Antarctic Peninsula. Ecology 93: 1367–1377. doi: 10.1890/11-1588.1.

Lyver, P.O.B., M. Barron, K.J. Barton, D.G. Ainley, A. Pollard, S. Gordon et al. 2014. Trends in the breeding population of Adélie penguins in the Ross Sea, 1981–2012: A coincidence of climate and resource extraction effects. PLoS One 9: e91188. doi: 10.1371/journal.pone.0091188.

Marshall, G.J., A. Orr, N.P. Van Lipzig and J.C. King. 2006. The impact of a changing Southern Hemisphere Annular Mode on Antarctic Peninsula summer temperatures. J. Climate 19: 5388–5404. doi: 10.1175/JCLI3844.1.

Massom, R.A., K. Hill, C. Barbraud, N. Adams, A. Ancel, L. Emmerson et al. 2009. Fast ice distribution in Adélie Land, East Antarctica: Interannual variability and implications for emperor penguins *Aptenodytes forsteri*. Mar. Ecol. Prog. Ser. 374: 243–257. doi: 10.3354/meps07734.

McLeod, D.J., G.M. Hallegraeff, G.W. Hosie and A.J. Richardson. 2012. Climate-driven range expansion of the red-tide dinoflagellate *Noctiluca scintillans* into the Southern Ocean. J. Plankton Res. 34: 332–337. doi: 10.1093/plankt/fbr112.

McMahon, K.W., C.I. Michelson, T. Hart, M.D. McCarthy, W.P. Patterson and M.J. Polito. 2019. Divergent trophic responses of sympatric penguin species to historic anthropogenic exploitation and recent climate change. P. Natl. Acad. Sci. 116: 25721–25727. doi: 10.1073/pnas.1913093116.

McNeil, B.I. and R.J. Matear. 2007. Climate change feedbacks on future oceanic acidification. Tellus B59: 191–198. doi: 10.1111/j.1600–0889.2006.00241.

Meijers, A.J., M.P. Meredith, E.J. Murphy, D. Chambers, M. Belchier and E.F. Young. 2019. The role of ocean dynamics in king penguin range estimation. Nat. Clim. Chang. 9: 120–121. doi: 10.1038/s41558-018-0084-2.

Meijers, A.J.S. 2014. The Southern Ocean in the coupled model intercomparison project phase 5. Philos. T. Roy. Soc. A372: 20130296. doi: 10.1098/rsta.2013.0296.

Meredith, M.P. and J.C. King. 2005. Rapid climate change in the ocean west of the Antarctic Peninsula during the second half of the 20th century. Geophys. Res. Lett. 32: L19604. doi: 10.1029/2005GL024042.

Midorikawa, T., H.Y. Inoue, M. Ishii, D. Sasano, N. Kosugi, G. Hashida et al. 2012. Decreasing pH trend estimated from 35-year time series of carbonate parameters in the Pacific sector of the Southern Ocean in summer. Deep-Sea Res. Pt I 61: 131–139. doi: 10.1016/j.dsr.2011.12.003.

Mills, W.F., J.C. Xavier, S. Bearhop, Y. Cherel, S.C. Votier, C.M. Waluda et al. 2020. Long-term trends in albatross diets in relation to prey availability and breeding success. Marine Biol. 167: 29. doi: 10.1007/s00227-019-3630-1.

Mintenbeck, K., E.R. Barrera-Oro, T. Brey, U. Jacob, R. Knust, F.C. Mark et al. 2012. Impact of climate change on fishes in complex Antarctic ecosystems. pp. 351–426. *In:* Jacob, U. and G. Woodward [eds.]. Advances in Ecological Research, vol. 46. Academic Press, Burlington. doi: 10.1016/B978-0-12-396992-7.00006-X.

Mintenbeck, K. and J.J. Torres. 2017. Impact of climate change on the Antarctic silverfish and its consequences for the Antarctic ecosystem. pp. 253–286. *In*: Vacchi, M., E. Pisando and L. Ghigliotti [eds.]. The Antarctic Silverfish: A Keystone Species in a Changing Ecosystem. Advances in Polar Ecology 3. Springer, Cham, Switzerland. doi: 10.1007/978-3-319-55893-6_12.

Moline, M.A., H. Claustre, T.K. Frazer, O. Schofield and M. Vernet. 2004. Alteration of the food web along the Antarctic Peninsula in response to a regional warming trend. Glob. Change Biol. 10: 1973–1980. doi: 10.1111/j.1365–2486.2004.00825.x.

Moline, M.A., N.J. Karnovsky, Z. Brown, G.J. Divoky, T.K. Frazer, C.A. Jacoby et al. 2008. High latitude changes in ice dynamics and their impact on polar marine ecosystems. Ann. NY Acad. Sci. 1134: 267. doi: 10.1196/annals.1439.010.

Montevecchi, W.A. 1993. Birds as indicators of change in marine prey stocks. pp. 217–266. *In:* Furness, R.W. and J.J.D. Greenwood [eds.]. Birds as monitors of environmental change. Springer, Dordrecht. doi: 10.1007/978-94-015-1322-7.

Moreau, S., B. Mostajir, S. Bélanger, I.R. Schloss, M. Vancoppenolle and S. Demers et al. 2015. Climate change enhances primary production in the western Antarctic Peninsula. Glob. Change Biol. 21: 2191–2205. doi: 10.1111/gcb.12878.

Moy, A.D., W.R. Howard, S.G. Bray and T.W. Trull. 2009. Reduced calcification in modern Southern Ocean planktonic foraminifera. Nature Geosci. 2: 276–280. doi: 10.1038/ngeo460.

Murphy, E.J., P.N. Trathan, J.L. Watkins, K. Reid, M.P. Meredith, J. Forcada et al. 2007a. Climatically driven fluctuations in Southern Ocean ecosystems. P. Roy. Soc. B274: 3057–3067. doi: 10.1098/rspb.2007.1180.

Murphy, E.J., J.L. Watkins, P.N. Trathan, K. Reid, M.P. Meredith, S.E. Thorpe et al. 2007b. Spatial and temporal operation of the Scotia Sea ecosystem: A review of large-scale links in a krill centred food web. Philos. Trans. R. Soc. Lond. Ser. B–Biol. Sci. 362: 113–148. doi: 10.1098/rstb.2006.1957.

Nel, D., F. Taylor, P. Ryan and J. Cooper. 2003. Population dynamics of the wandering albatross *Diomedea exulans* at Marion Island: Longline fishing and environmental influences. African J. Mar. Sci. 25: 503–517. doi: 10.2989/18142320309504039.

Oro, D. 2014. Seabirds and climate: Knowledge, pitfalls, and opportunities. Front. Ecol. Evol. 2. doi: 10.3389/fevo.2014.00079.

Orr, J.C., V.J. Fabry, O. Aumont, L. Bopp, S.C. Doney, R.A. Feely et al. 2005. Anthropogenic ocean acidification over the twenty-first century and its impact on calcifying organisms. Nature 437: 681–686. doi: 10.1038/nature04095.

Pardo, D., J. Forcada, A.G. Wood, G.N. Tuck, L. Ireland, R. Pradel et al. 2017. Additive effects of climate and fisheries drive ongoing declines in multiple albatross species. P. Natl. Acad. Sci. 114: E10829–E10837. doi: 10.1073/pnas.1618819114.

Parker, M.L., W.R. Fraser, J. Ashford, T. Patarnello, L. Zane and J.J. Torres. 2015. Assemblages of micronektonic fishes and invertebrates in a gradient of regional warming along the Western Antarctic Peninsula. J. Mar. Syst. 152: 18–41. doi: 10.1016/j.jmarsys.2015.07.005.

Parkinson, C.L. 2019. A 40-y record reveals gradual Antarctic sea ice increases followed by decreases at rates far exceeding the rates seen in the Arctic. P. Natl. Acad. Sci. 116: 14414–14423. doi: 10.1073/pnas.1906556116.

Peña, M.F., E. Poulin, G.P.M. Dantas, D. González-Acuña, M.V. Petry and J.A. Vianna. 2014. Have historical climate changes affected Gentoo Penguin (*Pygoscelis papua*) populations in Antarctica? PLoS ONE9: e95375. doi: 10.1371/journal.pone.0095375.

Péron, C., M. Authier, C. Barbraud, K. Delord, D. Besson and H. Weimerskirch. 2010. Interdecadal changes in at-sea distribution and abundance of subantarctic seabirds along a latitudinal gradient in the Southern Indian Ocean. Glob. Change Biol. 16: 1895–1909. doi: 10.1111/j.1365–2486.2010.02169.x.

Péron, C., H. Weimerskirch and C.-A. Bost. 2012. Projected poleward shift of king penguins (*Aptenodytes patagonicus*) foraging range at the Crozet Islands, southern Indian Ocean. P. Roy. Soc. B279: 2515–2523. doi: 10.1098/rspb.2011.2705.

Petry, M.V., A.B. Basler, F.C.L. Valls and L. Krüger. 2013. New southerly breeding location of king penguins (*Aptenodytes patagonicus*) on Elephant Island (Maritime Antarctic). Polar Biology 36: 603–606. doi: 10.1007/s00300-012-1277-1.

Petry, M.V., F.C. Valls, E.S. Petersen, J.V. Finger and L. Krüger. 2018. Population trends of seabirds at Stinker Point, Elephant Island, Maritime Antarctica. Antarctic Science 30: 220–226. doi: 10.1017/S0954102018000135.

Phillips, R.A., S. Bearhop, R.A.R. McGill and D.A. Dawson. 2009. Stable isotopes reveal individual variation in migration strategies and habitat preferences in a suite of seabirds during the nonbreeding period. Oecologia 160: 795–806. doi: 10.1007/s00442-009-1342-9.

Phillips, R.A., R. Gales, G.B. Baker, M.C. Double, M. Favero, F. Quintana et al. 2016. The conservation status and priorities for albatrosses and large petrels. Biol. Conserv. 201: 169–183. doi: 10.1016/j.biocon.2016.06.017.

Phillips, R.A., J.R.D. Silk, A. Massey and K.A. Hughes. 2019. Surveys reveal increasing and globally important populations of south polar skuas and Antarctic shags in Ryder Bay (Antarctic Peninsula). Polar Biol. 42: 423–432. doi: 10.1007/s00300-018-2432-0.

Phillips, R.A. and A.G. Wood. 2020. Variation in live-capture rates of albatrosses and petrels in fisheries, post-release survival and implications for management. Biol. Conserv. 247: 108641. doi: 10.1016/j.biocon.2020.108641.

Piatt, J.F., W.J. Sydeman and F. Wiese. 2007. Introduction: Seabirds as indicators of marine ecosystems. Mar. Ecol. Prog. Ser. 352: 199–204. doi: 10.3354/meps07070.

Polito, M., S.D. Emslie and W. Walker. 2002. A 1000-year record of Adélie penguin diets in the southern Ross Sea. Antarctic Sci. 14: 327. doi: 10.1017/S0954102002000184.

Poncet, S., A.C. Wolfaardt, A. Black, S. Browning, K. Lawton, J. Lee et al. 2017. Recent trends in numbers of wandering (*Diomedea exulans*), black-browed (*Thalassarche melanophris*) and grey-headed (*T. chrysostoma*) albatrosses breeding at South Georgia. Polar Biol. 40: 1347–1358. doi: 10.1007/s00300-016-2057-0.

Poncet, S., A.C. Wolfaardt, C. Barbraud, R. Reyes-Arriagada, A. Black, R.B. Powell et al. 2020. The distribution, abundance, status and global importance of giant petrels (*Macronectes giganteus* and *M. halli*) breeding at South Georgia. Polar Biol. 43: 17–34. doi: 10.1007/s00300-019-02608-y.

Queirós, J.P., A. Hilário, D. Thompson, F.R. Ceia, G. Elliott, K. Walker et al. 2021. From hot to cold waters: New insights in the habitat and trophic ecology of Southern Ocean squids during their life cycle. Mar. Ecol. Prog. Ser. 659: 113–126. doi: 10.3354/meps13551.

Quillfeldt, P., Y. Cherel, K. Delord and H. Weimerkirch. 2015. Cool, cold or colder? Spatial segregation of prions and blue petrels is explained by differences in preferred sea surface temperatures. Biol. Lett. 11: 20141090. doi: 10.1098/rsbl.2014.1090.

Radchuk, V., T. Reed, C. Teplitsky, M. Van De Pol, A. Charmantier, C. Hassall et al. 2019. Adaptive responses of animals to climate change are most likely insufficient. Nature Commun. 10: 1–14. doi: 10.1038/s41467-019-10924-4.

Ramos, J.E., G.T. Pecl, N.A. Moltschaniwskyj, J.M. Strugnell, R.I. León and J.M. Semmens. 2014. Body size, growth and life span: Implications for the polewards range shift of *Octopus tetricus* in south-eastern Australia. PLoS One 9: e103480. doi: 10.1371/journal.pone.0103480.

Reid, K., J.P. Croxall, D.R. Briggs and E.J. Murphy. 2005. Antarctic ecosystem monitoring: Quantifying the response of ecosystem indicators to variability in Antarctic krill. ICES J. Mar. Sci. 62: 366–373. doi: 10.1016/j.icesjms.2004.11.003.

Reid, K. 2018. Climate change impacts, vulnerabilities and adaptations: Southern Ocean marine fisheries. pp. 363–374. *In:* Barange, M., T. Bahri, M.C.M. Beveridge, K.L. Cochrane, S. Funge-Smith and F. Poulain [eds.]. Impacts of Climate change on fisheries and aquaculture: Synthesis of current knowledge adaptation and mitigation options, Vol. 627. FAO Fisheries and Aquaculture Tecnhical Paper, Rome.

Rintoul, S.R., S.L. Chown, R.M. DeConto, M.H. England, H.A. Fricker, V. Masson-Delmotte et al. 2018. Choosing the future of Antarctica. Nature 558: 233–241. doi: 10.1038/s41586-018-0173-4.

Roberts, D., W.R. Howard, A.D. Moy, J.L. Roberts, T.W. Trull, S.G. Bray et al. 2011. Interannual pteropod variability in sediment traps deployed above and below the aragonite saturation horizon in the Sub-Antarctic Southern Ocean. Polar Biol. 34: 1739–1750. doi: 10.1007/s00300-011-1024-z.

Robertson, G., B. Wienecke, L. Emmerson and A.D. Fraser. 2014. Long-term trends in the population size and breeding success of Emperor Penguins at the Taylor Glacier colony, Antarctica. Polar Biol. 37: 251–259. doi: 10.1007/s00300-013-1428-z.

Rodhouse, P.G. 2013. Role of squid in the Southern Ocean pelagic ecosystem and the possible consequences of climate change. Deep-Sea Res. Pt II95: 129–138. doi: 10.1016/j.dsr2.2012.07.001.

Rogers, A., B. Frinault, D. Barnes, N. Bindoff, R. Downie, H. Ducklow et al. 2020. Antarctic futures: An assessment of climate-driven changes in ecosystem structure, function, and service provisioning in the Southern Ocean. Annu. Rev. Mar. Sci. 12: 87–120. doi: 10.1146/annurev-marine-010419-011028.

Rogers, A.D., C. Yesson and P. Gravestock. 2015. A Biophysical and Economic Profile of South Georgia and the South Sandwich Islands as Potential Large-Scale Antarctic Protected Areas. Adv. Mar. Biol. 70: 1–286. doi: 10.1016/bs.amb.2015.06.001.

Rolland, V., C. Barbraud and H. Weimerskirch. 2009. Assessing the impact of fisheries, climate and disease on the dynamics of the Indian yellow-nosed Albatross. Biol. Conserv. 142: 1084–1095. doi: 10.1016/j.biocon.2008.12.030.

Rolland, V., H. Weimerskirch and C. Barbraud. 2010. Relative influence of fisheries and climate on the demography of four albatross species. Glob. Change Biol. 16: 1910–1922. doi: 10.1111/j.1365–2486.2009.02070.x.

Ropert–Coudert, Y., A. Chiaradia, D. Ainley, A. Barbosa, P.D. Boersma, R. Brasso et al. 2019. Happy Feet in a Hostile World? The Future of Penguins Depends on Proactive Management of Current and Expected Threats. Front. Mar. Sci. 6. doi: 10.3389/fmars.2019.00248.

Ryan, P., M.G. Jones, B. Dyer, L. Upfold and R.J. Crawford. 2009. Recent population estimates and trends in numbers of albatrosses and giant petrels breeding at the sub-Antarctic Prince Edward Islands. African J. Mar. Sci. 31: 409–417. doi: 10.2989/AJMS.2009.31.3.13.1001.

Sabine, C.L., R.A. Feely, N. Gruber, R.M. Key, K. Lee, J.L. Bullister et al. 2004. The oceanic sink for anthropogenic CO2. Science 305: 367–371. doi: 10.1126/science.1097403.

Sahade, R., C. Lagger, L. Torre, F. Momo, P. Monien, I. Schloss et al. 2015. Climate change and glacier retreat drive shifts in an Antarctic benthic ecosystem. Sci. Advances 1: e1500050. doi: 10.1126/sciadv.1500050.

Saraux, C., C. Le Bohec, J.M. Durant, V.A. Viblanc, M. Gauthier-Clerc and D. Beaune et al. 2011. Reliability of flipper-banded penguins as indicators of climate change. Nature 469: 203–206. doi: 10.1038/nature09630.

SAT. 2014. Compilation of key documents of the Antarctic Treaty System (second edition). Secretariat of the Antarctic Treaty, Buenos Aires, Argentina.

Schine, C.M., G. van Dijken and K.R. Arrigo. 2016. Spatial analysis of trends in primary production and relationship with large-scale climate variability in the Ross Sea, Antarctica (1997–2013). J. Geophys. Res. 121: 368–386. doi: 10.1002/2015JC011014.

Schofield, O., G. Saba, K. Coleman, F. Carvalho, N. Couto, H. Ducklow et al. 2017. Decadal variability in coastal phytoplankton community composition in a changing West Antarctic Peninsula. Deep-Sea Res. Pt I124: 42–54. doi: 10.1016/j.dsr.2017.04.014.

Schreiber, E.A. and J. Burger [eds.]. 2001. Biology of marine birds. CRC press, Boca Raton, FL, USA.

Schrimpf, M., R. Naveen and H.J. Lynch. 2018. Population status of the Antarctic shag *Phalacrocorax (atriceps) bransfieldensis*. Antarctic Sci. 30: 151–159. doi: 10.1017/S0954102017000530.

Sherley, R.B., L.G. Underhill, B.J. Barham, P.J. Barham, J.C. Coetzee, R.J. Crawford et al. 2013. Influence of local and regional prey availability on breeding performance of African penguins Spheniscus demersus. Mar. Ecol. Prog. Ser. 473: 291–301. doi: 10.3354/meps10070.

Smith, V. 2002. Climate change in the sub-Antarctic: An illustration from Marion Island. Climatic Change 52: 345–357. doi: 10.1023/A:1013718617277.

Solomon, S., M. Manning, M. Marquis and D. Qin. 2007. Climate change 2007—the physical science basis: Working group I contribution to the fourth assessment report of the IPCC (Vol. 4). Cambridge University Press, Cambridge, UK.

Southwell, C., L. Emmerson, J. McKinlay, K. Newbery, A. Takahashi, A. Kato et al. 2015. Spatially extensive standardized surveys reveal widespread, multi-decadal increase in East Antarctic Adélie penguin populations. PLoS ONE 10: e0139877. doi: 10.1371/journal.pone.0139877.

Southwell, C., L. Emmerson, A. Takahashi, C. Barbraud, K. Delord and H. Weimerskirch. 2017. Large-scale population assessment informs conservation management for seabirds in Antarctica and the Southern Ocean: A case study of Adélie penguins. Glob. Ecol. Conserv. 9: 104–115. doi: 10.1016/j.gecco.2016.12.004.

Steinberg, D.K., K.E. Ruck, M.R. Gleiber, L.M. Garzio, J.S. Cope, K.S. Bernard et al. 2015. Long-term (1993–2013) changes in macrozooplankton off the Western Antarctic Peninsula. Deep-Sea Res. Pt I 101: 54–70. doi: 10.1016/j.dsr.2015.02.009.

Suckling, C.C., M.S. Clark, J. Richard, S.A. Morley, M.A. Thorne, E.M. Harper et al. 2015. Adult acclimation to combined temperature and pH stressors significantly enhances reproductive outcomes compared to short-term exposures. J. Animal Ecol. 84: 773–784. doi: 10.1111/1365-2656.12316.

Swart, N.C., S.T. Gille, J.C. Fyfe and N.P. Gillett. 2018. Recent Southern Ocean warming and freshening driven by greenhouse gas emissions and ozone depletion. Nature Geosci. 11: 836–841. doi: 10.1038/s41561-018-0226-1.

Sydeman, W.J., E. Poloczanska, T.E. Reed and S.A. Thompson. 2015. Climate change and marine vertebrates. Science 350: 772–777. doi: 10.1126/science.aac9874.

Takahashi, A., M. Ito, K. Nagai, J.B. Thiebot, H. Mitamura, T. Noda et al. 2018. Migratory movements and winter diving activity of Adélie penguins in East Antarctica. Mar. Ecol. Prog. Ser. 589: 227–239. doi: 10.3354/meps12438.

Takahashi, K.T., S. Kawaguchi, G.W. Hosie, T. Toda, M. Naganobu and M. Fukuchi. 2010. Surface zooplankton distribution in the Drake Passage recorded by Continuous Plankton Recorder (CPR) in late austral summer of 2000. Polar Sci. 3: 235–245. doi: 10.1016/j.polar.2009.08.004.

Trathan, P., J. Forcada and E. Murphy. 2007. Environmental forcing and Southern Ocean marine predator populations: Effects of climate change and variability. Philos. T. Roy. Soc. B 362: 2351–2365. doi: 10.1098/rstb.2006.1953.

Trathan, P., N. Ratcliffe and E. Masden. 2012. Ecological drivers of change at South Georgia: The krill surplus, or climate variability. Ecography 35: 983–993. doi: 10.1111/j.1600-0587.2012.07330.x.

Trathan, P.N., J.P. Croxall and E.J. Murphy. 1996. Dynamics of Antarctic penguin populations in relation to the annual variability in sea ice distribution. Polar Biol. 16: 321–330. doi: 10.1007/BF02342178.

Trathan, P.N., P.T. Fretwell and B. Stonehouse. 2011. First recorded loss of an emperor penguin colony in the recent period of Antarctic regional warming: Implications for other colonies. PLoS One 6: e14738. doi: 10.1371/journal.pone.0014738.

Trathan, P.N., P. García-Borboroglu, D. Boersma, C.-A. Bost, R.J.M. Crawford, G.T. Crossin et al. 2015. Pollution, habitat loss, fishing and climate change as critical threats to penguins. Conserv. Biol. 29: 31–41. doi: 10.1111/cobi.12349.

Trivelpiece, W.Z., J.T. Hinke, A.K. Miller, C.S. Reiss, S.G. Trivelpiece and G.M. Watters. 2011. Variability in krill biomass links harvesting and climate warming to penguin population changes in Antarctica. Proc. Natl. Acad. Sci. USA 108: 7625–7628. doi: 10.1073/pnas.1016560108.

Turner, J., R. Bindschadler, P. Convey, G. di Prisco, E. Fahrbach, J. Gutt et al. 2009. Antarctic climate change and the environment. Scientific Committee for Antarctic Research, Cambridge, UK.

Turner, J., N.E. Barrand, T.J. Bracegirdle, P. Convey, D. Hodgson, M. Jarvis et al. 2014. Antarctic climate change and the environment: an update. Polar Record 50: 237–259. doi: 10.1017/S0032247413000296.

Turner, J. and J. Comiso. 2017. Solve Antarctica's sea-ice puzzle. Nature 547: 275. doi: 10.1038/547275a.

Vacchi, M., M. La Mesa, M. Dalu and J. MacDonald. 2004. Early life stages in the life cycle of Antarctic silverfish, *Pleuragramma antarcticum* in Terra Nova Bay, Ross Sea. Antarctic Sci. 16: 299. doi: 10.1017/S0954102004002135.

Waller, C.L., H.J. Griffiths, C.M. Waluda, S.E. Thorpe, I. Loaiza, B. Moreno et al. 2017. Microplastics in the Antarctic marine system: an emerging area of research. Sci. Total Environ. 598: 220–227. doi: 10.1016/j.scitotenv.2017.03.283.

Waluda, C.M., S.L. Hill, H.J. Peat and P.N. Trathan. 2012. Diet variability and reproductive performance of macaroni penguins (*Eudyptes chrysolophus*) at Bird Island, South Georgia. Mar. Ecol. Prog. Ser. 466: 261–274. doi: 10.3354/meps09930.

Waluda, C.M., S.L. Hill, H.J. Peat and P.N. Trathan. 2017. Long-term variability in the diet and reproductive performance of penguins at Bird Island, South Georgia. Marine Biol. 164: 39. doi: 10.1007/s00227-016-3067-8.

Watanabe, Y.Y., K. Ito, N. Kokubun and A. Takahashi. 2020. Foraging behavior links sea ice to breeding success in Antarctic penguins. Sci. Advances 6: eaba4828. doi: 10.1126/sciadv.aba4828.

Watters, G.M., J.T. Hinke and C.S. Reiss. 2020. Long-term observations from Antarctica demonstrate that mismatched scales of fisheries management and predator-prey interaction lead to erroneous conclusions about precaution. Sci. Rep. 10: 1–9. doi: 10.1038/s41598-020-59223-9.

Weimerskirch, H., P. Inchausti, C. Guinet and C. Barbraud. 2003. Trends in bird and seal populations as indicators of a system shift in the Southern Ocean. Antarctic Sci. 15: 249. doi: 10.1017/S0954102003001202.

Weimerskirch, H., M. Louzao, S. De Grissac and K. Delord. 2012. Changes in wind pattern alter albatross distribution and life-history traits. Science 335: 211–214. doi: 10.1126/science.1210270.

Weimerskirch, H., K. Delord, C. Barbraud, F. Le Bouard, P.G. Ryan, P. Fretwell et al. 2018. Status and trends of albatrosses in the french southern territories, western Indian Ocean. Polar Biol. 41: 1963–1972. doi: 10.1007/s00300-018-2335-0.

Weimerskirch, H., F. Le Bouard, P.G. Ryan and C. Bost. 2018. Massive decline of the world's largest king penguin colony at Ile aux Cochons, Crozet. Antarctic Sci. 30: 236–242. doi: 10.1017/S0954102018000226.

Weimerskirch, H., J. Collet, A. Corbeau, A. Pajot, F. Hoarau, C. Marteau et al. 2020. Ocean sentinel albatrosses locate illegal vessels and provide the first estimate of the extent of nondeclared fishing. P. Natl. Acad. Sci. USA 117: 3006–3014. doi: 10.1073/pnas.1915499117.

Whitehouse, M.J., M.P. Meredith, P. Rothery, A. Atkinson, P. Ward and R. Korb. 2008. Rapid warming of the ocean around South Georgia, Southern Ocean, during the 20th Century: Forcings, characteristics and implications for lower trophic levels. Deep-Sea Res. Pt I 55: 1218–1228. doi: 10.1016/j.dsr.2008.06.002.

Wilson, P., D. Ainley, N. Nur, S. Jacobs, K. Barton and G. Ballard et al. 2001. Adélie penguin population change in the pacific sector of Antarctica: Relation to sea-ice extent and the Antarctic Circumpolar Current. Mar. Ecol. Prog. Ser. 213: 301–309. doi: 10.3354/meps213301.

Winter, A., J. Henderiks, L. Beaufort, R.E. Rickaby and C.W. Brown. 2014. Poleward expansion of the coccolithophore *Emiliania huxleyi*. J. Plankton Res. 36: 316–325. doi: 10.1093/plankt/fbt110.

Xavier, J.C., J.P. Croxall and K. Reid. 2003. Inter-annual variation in the diet of two albatross species breeding at South Georgia: Implications for breeding performance. Ibis 145: 593–610. doi: 10.1046/j.14749-19X.2003.00196.x.

Xavier, J.C., J.P. Croxall, P.N. Trathan and P.G. Rodhouse. 2003. Inter-annual variation in the cephalopod component of the diet of wandering albatrosses *Diomedea exulans* breeding at Bird Island, South Georgia. Marine Biol. 142: 611–622. doi: 10.1007/s00227-002-0962-y.

Xavier, J.C., P.N. Trathan, J.P. Croxall, A.G. Wood, G.P. Podestá and P.G. Rodhouse. 2004. Foraging ecology and interactions with fisheries of wandering albatrosses at South Georgia. Fish. Oceanogr. 13: 324–344. doi: 10.1111/j.1365-2419.2004.00298.x.

Xavier, J.C. and L.S. Peck. 2015. Life beyond the ice. pp. 229–252. *In*: Liggett, D., B. Storey, Y. Cook and V. Meduna [eds.]. Exploring the Last Continent. Springer International Publishing, Cham, Switzerland.

Xavier, J.C., B. Raymond, D.C. Jones and H. Griffiths. 2016. Biogeography of cephalopods in the Southern Ocean using habitat suitability prediction models. Ecosystems 19: 220–247. doi: 10.1007/s10021-015-9926-1.

Xavier, J.C., Y. Cherel, L. Allcock, R. Rosa, R.M. Sabirov, M.E. Blicher and A.V. Golikov. 2018. A review on the biodiversity, distribution and trophic role of cephalopods in the Arctic and Antarctic marine ecosystems under a changing ocean. Marine Biol. 165: 93. doi: 10.1007/s00227-018-3352-9.

Xavier, J.C. and P. Convey. 2020. Antarctic: Climate change, fisheries, and governance. pp. 1–12. *In*: Leal Filho, W., A.M. Azul, L. Brandli, A. Lange Salvia and T. Wall [eds.]. Life Below Water. Springer International Publishing, Cham, Switzerland.

Younger, J.L., L.M. Emmerson and K.J. Miller. 2016. The influence of historical climate changes on Southern Ocean marine predator populations: A comparative analysis. Glob. Change Biol. 22: 474–493. doi: 10.1111/gcb.13104.

Younger, J.L., G.V. Clucas, D. Kao, A.D. Rogers, K. Gharbi, T. Hart et al. 2017. The challenges of detecting subtle population structure and its importance for the conservation of emperor penguins. Molecular Ecol. 26: 3883–3897. doi: 10.1111/mec.14172.

Part D
Actions for Seabird Conservation

CHAPTER 1

Light Pollution as a Seabirds' Conservation Threat

Cátia S.A. Gouveia

Introduction

Turning on the lights has become a very common habit for all of us, and lighting has made humans feel safe and protected since prehistoric times. Throughout the years, more and improved urban night-time lighting was needed and seen as a symbol of development and progress (Stone 2017). Light pollution is the widely accepted term for the negative or adverse effects of artificial night-time illumination (Hölker et al. 2010, Stone 2017, IDA 2020). Even though electric lighting has increased, its impacts have received little attention when compared to other environmental problems (Hölker et al. 2010). The extent of light pollution is global (Elvidge et al. 2009), and in the last century its effects have gained research interest and became a major concern for conservationists. Excessive lighting has proven to have negative consequences on biodiversity and the ecological consequences of Artificial Light At Night (ALAN) are only beginning to be understood (Rich and Longcore 2006).

Ecological systems are structured in daily and seasonal light, and darkness cycles have been profoundly, and, in some cases, permanently altered by humans (Bradshaw and Holzapfel 2010). Considering that natural light cycles had been stable for a very long period in the past, the introduction and development of artificial lighting with human settlements, transport networks and industry constitute a particular environmental disruptive pressure (Gaston et al. 2014). Understanding the damaging effects of light pollution on nocturnal animals is a long-standing goal in conservation. In particular, light pollution erodes circadian cycles of organisms, with negative consequences on reproductive cycles, feeding behaviour, sleeping patterns and exposure to predators (Gaston et al. 2017). ALAN is now recognised as a major threat to biodiversity (Hölker et al. 2010), and the inappropriate use of artificial lighting, particularly outdoors and at night, has negative consequences for organisms and ecosystems. Both direct illumination and skyglow is of significant concern, and their effects have been reported on a wide variety of organisms, from microbial communities to amphibians, birds and humans (Gaston et al. 2013, 2017). ALAN can attract individuals from long distances, whilst repelling others, affecting many of their activities (foraging, interspecific

SPEA – Portuguese Society for the Study of Birds, SPEA Madeira, Rua da Mouraria nº9, 4ºB, 9000-047 Funchal, Portugal.
Email: catia.gouveia@spea.pt

interactions, reproductive behaviour and daily, monthly or annual movements), thus altering community structure and causing cascading negative effects on ecosystem functioning (Longcore and Rich 2004, Gaston et al. 2014, 2015).

Natural habitats are not only impacted by direct light but also by reflections/glare of light from faraway sources. Skyglow can be detected at places located tens (or even hundreds) of kms away from the source of light pollution. For this reason, it is essential to control light pollution in urban areas in order to reduce its effects on protected natural and semi-natural areas. Although ALAN was identified as an important driver of environmental change in the 21st century (Davies and Smyth 2018), increasing at an annual rate of 2% (Kyba et al. 2017), it is still an underestimated threat for species and ecosystems and may exacerbate already precarious conditions for endangered species. For birds, light pollution affects both nocturnal and diurnal species because changes in day length often cause deviations in the reproductive cycle and breeding rituals, making it difficult to mate, to look for food or to nest (Da Silva 2014, 2015). For migratory species, it can disrupt the direction, form and timing of night-time migratory behaviour (Gaston et al. 2017). ALAN is also an important environmental stressor for pelagic seabirds, which use a stellar-orientation system for visual spatial orientation at night, to locate colonies or to carry out long migrations (Telfer et al. 1987). In addition to stars, they have a well-developed olfactory system that is used not only for orientation, but also to locate nests and partners (Bonadonna and Bretagnolle 2002, Bonadonna 2003), and in the recognition of foraging areas in the ocean (Nevitt 2008). These sensory features allow them to use nocturnal and pelagic environments over vast areas without any visual guidance marks (Hart 2004, Mitkus et al. 2016). Some studies indicate that vision, contrary to smell, is developed over time. As a result, vision is poorly developed in fledglings (Mitkus et al. 2018), making them more sensitive to artificial night-lights and vulnerable to the increase of light pollution (Troy et al. 2011, Rodríguez et al. 2017, Atchoi et al. 2020).

This chapter provides an overview of the impacts of light pollution on seabirds, and of the mitigation measures to reduce this problem. Firstly, the reasons why artificial lights have a particular impact on seabirds are summarized. Secondly, a hierarchical strategy to minimize and mitigate the impact of light pollution on seabirds based on four main objectives: increasing knowledge, reducing light pollution sources, involving population on rescue campaign and raising community awareness in the adoption of light pollution mitigation measures are addressed. Throughout, most of the examples that are provided are from the Macaronesia region, Azores, Madeira and Canary Islands, where the effects of light pollution on seabirds and associated mitigation measures have been addressed in an integrative way.

Why light pollution affects seabirds?

The extensive and continuous growth of artificial light at night is an increasing threat to seabird species of the order Procellariiformes, one of the most endangered groups of birds (Croxall et al. 2012, Dias et al. 2019, Rodríguez et al. 2019). Pelagic seabirds are particularly vulnerable to light pollution as these species navigate in a vast and uniform ocean, and only visit land during the breeding season, attending the nest at night. Although the reasons for their attraction to ALAN and consequent disorientation are not well understood, Rodríguez et al. (2017) suggested three main hypotheses: (a) perception of light as a source of food, (b) association of light with the food source, and (c) light as a navigation pattern. During earlier stages, Procellariiform chicks are provisioned by their parents with regurgitated meals, and after fledging they do not receive any parental care. When leaving the nest, they must learn to find their food at sea on their own. Due to their foraging inexperience, and because they consume bioluminescent prey (squid and fish), inexperienced fledglings may misunderstand artificial lights as a source of food (Imber 1975). Juvenile seabirds may associate the light with food provisioning. These seabirds breed in burrows and, for the first months in the dark inside the nest, the only food source is provided by the parents, via the bright burrow entrance; thus, food might be associated with light and newly fledged birds might confuse artificial lights with food delivery (Rodríguez et al. 2017). However, these two hypotheses do not explain why adults are also affected by light pollution. The third hypothesis is that light pollution may interfere with their navigation capabilities, either by surpassing natural stellar clues or by obstructing vision-based magnetic navigation control systems (Telfer et al. 1987, Guilford et al. 2018).

Burrow-nesting petrels, and in lower numbers of other seabirds such as auklets and puffins (Alcidae) and sea ducks such as eiders (Anatidae) (Dick and Donalson 1977, Whitworth et al. 1997, Merkel and Johansen 2011, Wilhelm et al. 2013), are attracted to artificial lights at night. They get disorientated and are forced to land (Imber 1975), in a phenomenon called *fallout* (Reed et al. 1985). Studies indicate that birds flying over or near light-polluted areas are susceptible to disorientation by lights, which shows a local and direct effect of light pollution. In Tenerife Island, Cory's Shearwaters (*Calonectris borealis*) were rescued at locations less than 16 km from their colonies, with half of the birds being rescued within a radius of 3 km of their colonies (Rodríguez et al. 2019). However, in Hawaii, Newell's Shearwater (*Puffinus newelli*) fledglings may be attracted to lights from a long distance: according to several models containing basic assumptions related to fledgling movement and attraction to light, a substantial part of the fallout may correspond to birds attracted back to land after successfully reaching the sea (Troy et al. 2013). Therefore, the geographical distribution (coastal or inland colonies), the altitude of seabird colonies, and their distance from the light sources are variables that influence the attraction of seabirds to light (Rodríguez et al. 2019). Thus, ALAN impact on seabirds can diverge greatly according to intensity and distribution of light pollution along the coastline.

The *fallout* of seabirds may cause mass-mortality events (Telfer et al. 1987, Ainley et al. 2001, Le Corre et al. 2002, Rodríguez and Rodríguez 2009, Rodríguez et al. 2014, 2017). Most of the grounded birds do not crash directly into lights as they would do if they were feeding. More commonly, they fly over the illuminated areas in circles and get disoriented, causing them to collide into structures (buildings, walls, electric wires and light poles) or even hitting the ground. The reduction in *fallout* numbers during full moon, when artificial lights are less noticeable due to moonlight, corroborates this theory. Even if they are not injured, grounded birds encounter barriers to taking off again, and are susceptible to predation by introduced predators, starvation, dehydration and even poaching (reviewed in Rodríguez et al. 2017). Such fatality events are also intensified by environmental drivers such as adverse weather (rain, fog, strong winds) (Rodríguez et al. 2014) and dark moonless nights (Telfer et al. 1987), when breeding nocturnal seabirds exhibit increased activity in the colonies (Warham 1990). It is important to note that breeding colonies near the coast regularly experience foggy and misty conditions, which scatters light and amplifies its local effects. Sometimes, even in nearby unilluminated islets, light travels unimpeded across open water and its reach is amplified by reflection from high clouds.

Light pollution can also influence the behaviour of seabird predators. Burrow-nesting petrels visit their colonies at night, at least partly to avoid predators (Miles et al. 2013); therefore, an increase in light levels as a consequence of light pollution near the colonies can increase the predation by diurnal predators (Oro et al. 2005, Gouveia, pers. obs.). The excessive lighting at night might also reduce adult attendance to the colonies, which is more frequent during darker nights (Warham 1990). Since seabirds are long-lived species, changes in colony attendance and mortality of adults (Rodríguez et al. 2015a) can result in major fluctuations of their sensitive population dynamics.

Petrels are also affected by light pollution at sea. Mortality, attraction and disorientation at sea have been reported worldwide by lights used by fisheries (Ryan 1991, Black 2005, Merkel and Johansen 2011, Glass and Ryan 2013). For example, and according to Glass and Ryan (2013), during lobster fisheries at the Tristan archipelago and Gough Island in the South Atlantic Ocean, 5.4% of the petrels died after colliding with ships, and species such as the White-faced Storm-Petrel (*Pelagodroma marina*), the White-bellied Storm-Petrel (*Fregetta grallaria*) and the Black-bellied Storm-Petrel (*Fregetta tropica*), were highly affected. Besides petrels, other seabirds, such as the endangered Audouin's Gulls (*Ichthyaetus audouinii*), are attracted to fishing areas, where they became more vulnerable to bycatch (Arcos and Oro 2002). Light-induced nocturnal fisheries and baitfish harvest commonly occur near islands where seabirds' breed, and these high light levels can facilitate the predation of night-hunting gulls (Gouveia, pers. obs.), reduce attendance rates from adults for incubation shifts and chick provisioning (Keitt 1998), and attract juveniles during their first flights. Offshore oil and gas exploration platforms are expanding and turning into artificial reefs that harbour marine communities, such us plants, invertebrates, and fishes (De Groot 1996). Seabirds congregate around these areas and are suffering from the negative impacts of artificial light (Day et al. 2015, Ronconi et al. 2015). Collision with the structures, oiling and incineration by the flare are episodes that often result on seabirds' mortality (Wiese et al. 2001). Overall, light-induced mortality at sea is difficult to assess (Montevecchi 2006), but light attraction seems to affect both adults and juveniles, with events occurring outside fledging periods.

Is turning off the lights enough to save seabirds?

The rapid expansion in the global extent of light pollution exacerbates the challenge for 21st century policymakers in providing appropriate outdoor light, while reducing costs and climate change impact, enhancing visibility and minimizing any adverse effects on plants, animals and humans caused by exposure to unnatural levels of nocturnal light (Kyba et al. 2014, Gaston 2017). These decisions have implications that include economic issues, social acceptance and environmental considerations that can even be contradictory. For example, technological solutions for increasing energy efficiency and reducing climate change impact (i.e., LED—Light-Emitting Diode) may increase light pollution if this issue is not considered.

For these reasons, and in order to mitigate the negative effects of light pollution on seabirds and biodiversity in general, it is important to articulate action plans around four main objectives: (a) increase knowledge about light pollution and its effects on seabirds, (b) reduce light pollution, (c) promote rescue campaigns for seabirds and (d) raise awareness on light pollution impacts.

Increasing knowledge about light pollution and its effects on seabirds

Promoting multidisciplinary research of light pollution and its effects should be the first step in the definition of a mitigation strategy. Updating, mapping and characterising light pollution sources at a sufficient planning scale is crucial to evaluate the potential association between light intensity and the distance over which birds are attracted, allowing the quantification of the exposure of dark areas and colonies to light pollution. With clear limitations, light intensity has been mapped by satellite imagery to better understand spatial patterns of seabirds' *fallout* occurrences (Rodrigues et al. 2011, Troy et al. 2011, 2013, Rodríguez et al. 2015a, 2015b). Artificial lights have been charted in terms of light intensity with respect to human vision, and, given the differences in light sensitivity between seabirds and humans (Reed 1986), no attention has been paid to the spectra of lights responsible for *fallouts*. Finding the appropriate light spectra that minimise the stimulation of seabird photoreceptors, and knowing the taxon-specific responses to light spectra and colours, may reduce the number of birds grounded by light pollution (Telfer et al. 1987, Wiese et al. 2001, Poot et al. 2008, Doppler et al. 2015), and should provide important clues for urban light planning.

Combining satellite imagery (currently limited in sensitivity, resolution and time scale) with photometers will allow to identify highly polluted areas close to colonies, creating better light pollution maps in terms of sensitivity, and involving both direct light impact and skyglow. This information combined with adequate monitoring and registry of *fallout* occurrences can provide valuable information on the impact of environmental variables (e.g., prevailing wind, normal cloud cover during fledging) on seabirds' attraction to light.

Reducing light pollution

The best way to mitigate light attraction is by removing all the sources of light pollution. However, as one knows, this solution is unfeasible in most areas, and measures that minimises the duration, intensity and extent of the effects of artificial lights on seabirds are necessary. One of the easiest measures to undertake is the reduction of lights during fledging periods. There have been positive results in several islands: Barau's Petrel (*Pterodroma baraui*) in Reunion Island (Rodríguez et al. 2017), Newell's Shearwater in Hawaii (The Associated Press 2010), Cory's Shearwaters in Corvo Island, Azores (Luminaves 2020), Leach's Storm-petrel (*Hydrobates leucorhous*) in Saint Kilda in Scotland (Miles et al. 2010) and Short-tailed Shearwater (*Ardenna tenuirostris*) in Phillip Island in Australia (Rodríguez et al. 2014). Changing the spectral composition of light to a lower colour temperature (< 2200º K) can also reduce the number of grounded seabirds (Longcore et al. 2018). Replacing light infrastructures with more efficient lighting systems, shielding, adjusting and reorienting light downwards pursue to accomplish the necessary illumination for humans while reducing the visible light for seabirds is also beneficial. The use of timers to reduce duration of light emission during the night is not always effective, because the timing of fledging is not well known for most seabird species. Thus, the blackout time may not match the timing of the fledging period. For example, most Cory's Shearwaters fledglings initiate their flights toward the ocean during the first hours of darkness (Rodríguez et al. 2015b), which overlaps with the hours of peak light pollution (until midnight).

Financial concerns and the unwilling of humans to practical actions that prevent or minimise light pollution are, most of the time, the major limitation in urban light changes. It is important that authorities and citizens understand that more effective light-solutions can save both money and animal lives, while providing security for populations at night. Therefore, raising awareness of the community, which will be discussed later, is essential to ensure the implementation of these measures. It is also important to enhance the support from local electrical companies, and the agreement of stakeholders by showing economic savings, positive differences in the number of grounding cases, the definition and dissemination of guidelines and the establishment of a legal framework to promote the reduction of light pollution to gradually reduce seabird mortality. Whenever possible, the awareness and encouragement of municipalities, administrations, private entities and citizens, in order to implement periods of blackouts in critical areas and periods (usually the first hours of the night), will also contribute to a reduction of fledgling attraction to lit areas. Despite the vast worldwide knowledge as well as the rescue campaigns directed to seabirds, there is still a lack of standardisation between different areas, and no legal framework on national and regional levels to minimise the impact of artificial night light on biodiversity.

Although new developments in LED provides longer life lights and energy efficiency, this technology might have negative impacts on animals because of their spectral composition peaks. A joint path between biodiversity conservation and energy efficiency must be undertaken to promote energy sustainability and financial savings, thereby halting the loss of biodiversity. The selective adjustment of spectral composition of light emission in LED technology, local legislation to reduce light pollution and definition of guidelines for lighting near protected areas will have a decisive impact on nature conservation in general, and on seabirds in particular. Through the support from the local electric company and the agreement of local municipalities, Madeira Island has been on the frontline in implementing those guidelines and best practices to reduce coastal light pollution and its impacts on biodiversity. The municipalities of Santa Cruz and Machico have already adopted a Municipal Plan for Public Lighting and are willing to implement solutions that contribute to energy efficiency and nature conservation. Financial and community security concerns limited their actions. This example reinforces the importance of advocacy and engagement of local authorities, stakeholders (e.g., architects, engineers) and the population to adopt appropriate biodiversity friendly solutions for lighting.

Rescue campaigns to mitigate the impact of light pollution on seabirds

Throughout the world, rescue campaigns have played an important role in recovering grounded birds, particularly juveniles, before they died of starvation, dehydration or predation, and in releasing them near the ocean. Rescue campaigns involve volunteers who collect fallen birds and release them back to the sea. Thousands of volunteers have saved more than 100,000 seabirds globally and have collected important data (Rodríguez et al. 2017), such as information about the most affected species, age vulnerability and body condition of fledglings, thus providing basic knowledge for the monitoring of populations where the information coming from other sources is scarce. Local governments, environmental nongovernmental organizations and volunteers contributed to rescue campaigns. If those campaigns were not in place, the negative impact of light pollution on seabird populations would be higher, as suggested by models of population dynamics (Ainley et al. 2001, Fontaine et al. 2011, Griesemer and Holmes 2011).

Since the 1990's, rescue campaigns have been helpful in saving thousands of endangered seabirds in Macaronesia, and therefore in mitigating light-induced seabird mortality. In Madeira, more than 200 light-induced grounded birds from eight seabird species have been collected annually by volunteers, since 2009. Seven of these species are listed in Annex I of the European Union Birds Directive, two of them are of priority concern (Zino's Petrel (*Pterodroma madeira*) and Desertas Petrel (*Pterodroma deserta*)), and three other species are already showing decreasing population trends. Most of the grounded seabirds occurred near the South coast, in highly lighted areas (southern municipalities of Funchal, Santa Cruz and Machico, Fig. D1.1). Artificial light at night has been identified as a main threat for seabirds in Macaronesia under the Marine Strategy

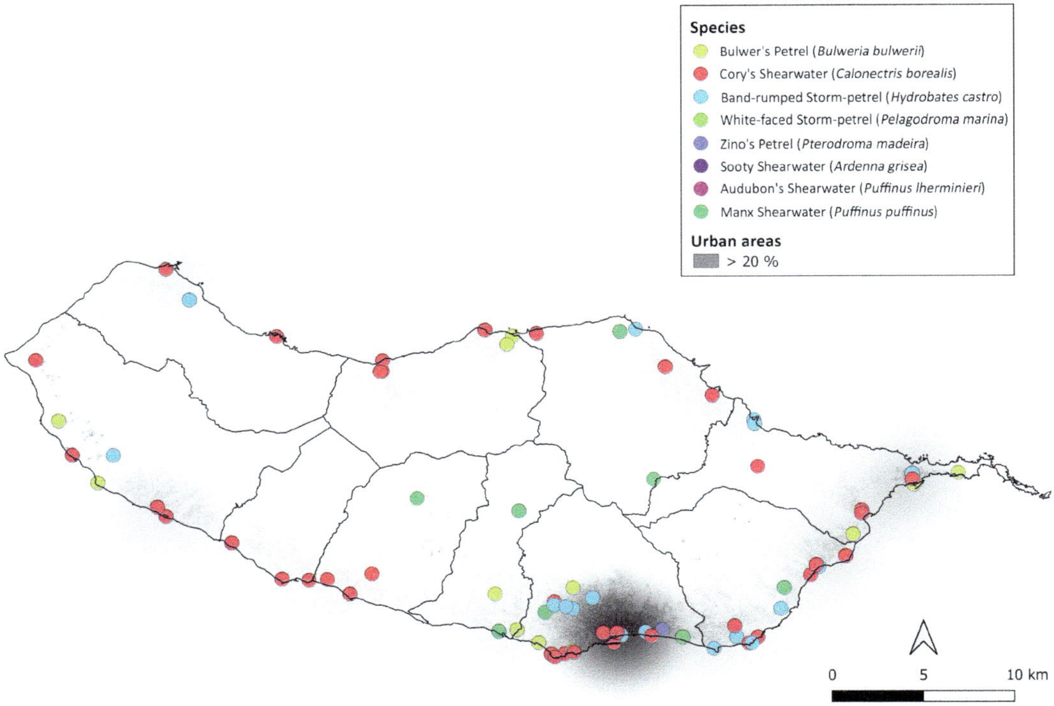

Figure D1.1. Fall out map of seabird species from Madeira Island in relation to Artificial Light At Night (ALAN). The location where each seabird was rescued was marked with a GPS and overlaid with urban areas (in grey). Data compiled and analysed during INTERREG Luminaves (non-systematic records dating since 2006) and collected during Save a Seabird Campaign led by the Portuguese Society for the Study of Birds (SPEA) in collaboration with the Natural Park of Madeira Island.

Framework Directive of the European Union (Saavedra et al. 2018), to the extent that seabirds are now used as a bioindicator of light pollution impacts in the Macaronesia region (Luminaves 2020).

In the Azores, a well established citizen science campaign since 1995 collects around 3,000 grounded birds per year of six seabird species (four listed in Annex I of the European Union Birds Directive, Bulwer's Petrel (*Bulweria bulwerii*), Band-rumped Storm-petrel (*Hydrobates castro*), Audubon's Shearwater (*Puffinus lherminieri*) and Cory's Shearwater) and one species of priority conservation, Monteiro's Storm-petrel (*Hydrobates monteiroi*). In the Canaries, 5,000 grounded birds of six species (four listed in Annex I of Birds Directive, Bulwer's Petrel, Audubon's Shearwater, Cory's Shearwater and European Storm-petrel (*Hydrobates pelagicus*)) are collected every breeding season, since 1998 (Luminaves 2020). Studies indicate that 50% of Procellariform species are affected by artificial light at night and up to 14% of the grounded seabirds die (Rodríguez et al. 2017), which would represent 1,100 birds/year in the Macaronesian region (Luminaves 2020). Nevertheless, these numbers are most likely an underestimation, because it is not possible to find all the grounded birds during rescuing campaigns.

Rescue programs also represent opportunities to gather information about the biology and ecology of several breeding seabird species, especially those that are poorly known or rare, such as the Manx Shearwater (*Puffinus puffinus*) in Tenerife Island (Rodríguez et al. 2008, 2020). Also in Tenerife, the number of rescued Audubon's Shearwaters has decreased since rescue programs started, suggesting a population decline (Rodríguez et al. 2012). In Hawaii, records of the number of fledglings in rescue campaigns have been used as a population index because they reflect overall population size and breeding success (Day et al. 2003). Monitoring programs can use rescued birds as indicators for the Good Environmental Status of the Ocean (e.g., European Union Marine Strategy Framework Directive) as well as for studies on epidemiology, pollution impacts (e.g., marine litter presence), marine productivity and foraging ecology of these species (Rodríguez et al. 2017). However, it is important to be aware that the increase of light-pollution levels and public awareness can influence the reported numbers of grounded seabirds, making these data difficult to interpret in terms of population size.

Rescue campaigns provide important clues about the most "problematic places" where most grounded birds fall out (Ainley et al. 2001, Rodrigues et al. 2011, 2012, Troy et al. 2011, 2013), which is the first step to prioritise interventions in the adaptation and removal of artificial light sources. Maintaining rescue programs and combining gathered information with rigorous scientific methods, such as ringing and tracking of fledglings (Rodríguez et al. 2019), use of radar to quantify light attraction (Day et al. 2015), and use of remote sensing to better map light pollution (Kyba et al. 2014) are crucial to understand the relationship between artificial light and seabirds. Rehabilitation of grounded seabirds is a crucial part of the rescue campaigns, and maximising the survival of rescued birds is imperative. The establishment of rescue protocols based of species' needs, the number of resources allocated to campaigns (personnel, facilities, and materials), and the creation of recovery centres should be a reality in every region where seabird *fallout* occurs.

The lack of systematic monitoring of all the grounded birds, and the accuracy of the numbers recorded by volunteers is dependent on the rescue effort (e.g., spatial planning of the area searched and probability of detecting birds), the survival of the individual on the ground (where they are highly vulnerable to predation), and the success of the long-term awareness programs. Rescue campaigns are normally centred on the peak fledging period, during moonless and windy nights but, most of the times, they face serious challenges. The real number of grounded birds caused by light pollution is difficult to assess and is often biased for four main reasons: (1) the lack of information regarding the distribution of breeding colonies, (2) the thresholds for the distance between the light source and colonies, (3) the light intensity that can affect each seabird species, and (4) the difficulties in maintaining a constant sampling effort by voluntary based rescue programs (Rodríguez et al. 2019). Finally, it is important to note that rescue protocols cannot completely eliminate fatalities. A systematic survey and correct quantification of grounded birds are crucial to understand the magnitude of the fallout events and light-induced mortality, and in planning the efforts to minimise the effects of light pollution.

Community awareness in the adoption of light pollution mitigation measures

Light pollution is a serious threat to both biodiversity and human health. To reduce the impact of light pollution, it is crucial that local communities and stakeholders are supportive and are aware of the constraints, as well as the conservation benefits of mitigation measures. The shared knowledge and insights of local communities should have a major role in the implementation of new eco-friendly light solutions. In this sense, it is important to modify the idea that the increased artificial lighting at night is a symbol of economic development and well-being. It is essential to reduce the anxiety related to security concerns associated with the reduction of public lighting by involving the local population in public discussions of Municipal Plans for Public Lighting, and promoting their involvement in rescue campaigns that simultaneously help seabirds and raise awareness on the problem and impacts of light pollution. The challenge of undergoing a mind-set change of the civil society regarding lighting at night faces different risks. It will be necessary to show local success stories providing examples of how development and environment can go together by being more efficient in the use of energy, using biodiversity friendly technologies, and adjusting our normal routines in relation to the use of lighting. Linking these behavioural changes with international events, such as Earth Hour or celebrations, like the Lusco-Fusco festival conducted in Corvo Island, Azores, during a full blackout for Cory's Shearwater fledgling's protection, may increase the willingness of the local people to adhere to this kind of initiatives (Saavedra et al. 2018, Luminaves 2020).

An effective awareness raising campaign on light pollution impact on seabirds will necessarily require to: (a) gather the support of the population by linking rescue programs with light pollution awareness activities, (b) promote environmental education on light pollution within the school community, (c) endorse public sessions with mayors and residents on light pollution, and its mitigation in order to promote the definition of municipal regulations on outdoor lighting, and (d) encourage economic activities linked to the use of the dark sky and reduced light pollution (such as astrotourism or nocturnal birdwatching).

Many challenges arise from the expansion of artificial lighting, and its ecological consequences on seabirds and other *taxa* are broad. A joint path between biodiversity conservation and energy efficiency should be undertaken to promote energy sustainability, financial savings and halting the loss of biodiversity. Therefore, it is very important to engage decision makers in the solution by taking advantage of the current political agenda linked to governmental strategies such as those of the European Union policies in energy efficiency and climate change; disclose best practices for public lighting developments and changes of lighting infrastructures and ensure population support for these changes. Financial concerns and population anxiety about light changes are real, and volunteer-based seabirds' rescue programs effects are limited both in operational terms as well as in scientific knowledge gathering (Luminaves 2020).

Overall, turning off the light is simply insufficient. A multilevel approach and building solid alliances between local authorities, I&D community and the civil society must be constructed to undertake measurable changes in light pollution levels, and to achieve long-term sustainability across the areas where light pollution affects seabirds. This is also an exciting new area for interdisciplinary research, involving biological and social sciences, and seabirds can act as flagship species for raising awareness and promoting the implementation of mitigation measures for light pollution issues. In research terms, studying the impacts of light pollution in other animal groups, such as bats and nocturnal insects, is important to guarantee an ecosystem-based conservation approach and define measures to mitigate the impacts of artificial light in endangered fauna. In addition, increasing knowledge of light pollution impacts, as well as improving modelling for sentinel seabird species such as Cory's Shearwater in the Macaronesian region, and the creation of sensibility maps will allow a better resource allocation during rescue campaigns (Rodrigues et al. 2011).

While the European Union strategies for 2030 target the increase of energy efficiency and climate change reduction, it fails to balance it with the protection of biodiversity. In this sense, including existent knowledge on light pollution impacts on biodiversity and health will contribute to the development of a more appropriate and sustainable local, regional, national and international approach to ecosystem-friendly lighting systems that benefit biodiversity and local communities.

Acknowledgments

Thanks to Fabrício Nunes and Paul Dangel for providing constructive comments on a previous version of this manuscript, and to Azucena de la Cruz Martín and Airam Rodríguez for the review of a final version.

References cited

Ainley, D.G., R. Podolsky, N. Nur, L. Deforest and G.A. Spencer. 2001. Status and population trends of the Newell's shearwater on Kauai: Insights from modeling. Stud. Avian Biol. 22: 108–123.

Arcos, J. and D. Oro. 2002. Significance of nocturnal purse seine fisheries for seabirds: A case study off the Ebro Delta (NW Mediterranean). Mar. Biol. 141: 277–286. doi: 10.1007/s00227-002-0828-3.

Atchoi, E., M. Mitkus and A. Rodríguez. 2020. Is seabird light-induced mortality explained by the visual system development? Conserv. Sci. Pract. 2: 94. doi: 10.1111/csp2.195.

BirdLife International. 2021. Data Zone: Species fact sheet. Available from http://www.birdlife.org (accessed on 28th of April of 2021).

Black, A. 2005. Light induced seabird mortality on vessels operating in the Southern Ocean: Incidents and mitigation measures. Antarct. Sci. 17: 67–68. doi: 10.1017/S0954102005002439.

Bonadonna, F. and V. Bretagnolle. 2002. Smelling home: a good solution for burrow-finding in nocturnal petrels? J. Exp. Biol. 205: 2519–2523. doi: 10.1242/jeb.205.16.2519.

Bonadonna, F. 2003. Evidence for nest-odour recognition in two species of diving petrel. J. Exp. Biol. 206: 3719–3722. doi: 10.1242/jeb.00610.

Bradshaw, W. and C. Holzapfel. 2010. What season is it anyway? Circadian Tracking vs. Photoperiodic Anticipation in Insects. J. Biol. Rhythms 25: 155–65. doi: 10.1177/0748730410365656.

Cabral, M.J., J. Almeida, P.R. Almeida, T. Delliger, N.F. Almeida, M.E. Oliveira et al. 2005. Livro Vermelho dos Vertebrados de Portugal. Instituto da Conservação da Natureza, Lisboa, Portugal.

Croxall, J.P., S.H.M. Butchart, B. Lascelles, A.J. Stattersfield, B. Sullivan, A. Symes et al. 2012 Seabird conservation status, threats and priority actions: A global assessment. Bird Conserv. Int. 22: 1–34. doi: 10.1017/S0959270912000020.

Da Silva, A., J.M. Samplonius, E. Schlicht, M. Valcu and B. Kempenaers. 2014. Artificial night lighting rather than traffic noise affects the daily timing of dawn and dusk singing in common European songbirds. Behav. Ecol. 25: 1037–1047. doi: 10.1093/beheco/aru103.

Da Silva, A., M. Valcu and B. Kempenaers. 2015. Light pollution alters the phenology of dawn and dusk singing in common European songbirds. Philos. Trans. R. Soc. B Biol. Sci. 370: 20140126. doi: 10.1098/rstb.2014.0126.

Davies, T.W. and T. Smyth. 2018. Why artificial light at night should be a focus for global change research in the 21st century. Glob. Chang. Biol. 24: 872–882. doi: 10.1111/gcb.13927.

Day, R.H., B.A. Cooper and T.C. Telfer. 2003. Decline of Townsend's (Newell's) shearwaters (*Puffinus auricularis newelli*) on Kauai, Hawaii. Auk 120: 669–679. doi: 10.1093/auk/120.3.669.

Day, J., J. Baker, H. Schofield, F. Mathews and K.J. Gaston. 2015. Part-night lighting: Implications for bat conservation. Anim. Conserv. 18: 512–516. doi: 10.1111/acv.12200.

De Groot, S.J. 1996. Quantitative assessment of the development of the offshore oil and gas industry in the North Sea. ICES J. Mar. Sci. 53: 1045–1050. doi: 10.1006/jmsc.1996.0130.

Dias, M.P., R. Martin, E. Pearmain, I. Burfield, C. Small, R. Phillips et al. 2019. Threats to seabirds: A global assessment. Biol. Conserv. 237: 525–537. doi: 10.1016/j.biocon.2019.06.033.

Dick, M.H. and W. Donalson. 1977. Fishing vessel endangered by crested auklet landings. Condor 80: 235–236. doi: 10.2307/1367924.

Doppler, M.S., B.F. Blackwell, T.L. DeVault and E. Fernandez-Juricic. 2015. Cowbird responses to aircraft with lights tuned to their eyes: Implications for bird–aircraft collisions. Condor 117: 165–177. doi: 10.1650/CONDOR-14-157.1.

Elvidge, C., P. Sutton, B. Tuttle, T. Ghosh and K. Baugh. 2009. Global urban mapping based on nighttime lights. pp. 129–145. *In:*Gamba, P. and M. Herold [eds.]. Global Mapping of Human Settlement: Experiences, Datasets and Prospects. CRC Press, New York, USA. doi: 10.1201/9781420083408-c6.

Fontaine, R., O. Gimenez and J. Bried. 2011. The impact of introduced predators, light-induced mortality of fledglings and poaching on the dynamics of the Cory's shearwater (*Calonectris diomedea*) population from the Azores, northeastern subtropical Atlantic. Biol. Conserv. 144: 1998–2011. doi: 10.1016/j.biocon.2011.04.022.

Gaston, K. J., J. Bennie, T.W. Davies and J. Hopkins. 2013. The ecological impacts of nighttime light pollution: A mechanistic appraisal. Biol. Rev. 88: 912–927.doi: 10.1111/brv.12036.

Gaston, K.J., J.P. Duffy, S. Gaston, J. Bennie and T.W. Davies. 2014. Human alteration of natural light cycles: Causes and ecological consequences. Oecologia 176: 917–931. doi: 10.1007/s00442-014-3088-2.

Gaston K.J., M.E. Visser and F. Holker. 2015. The biological impacts of artificial light at night: The research challenge. Philos. Trans. R. Soc. B Biol. Sci. 370: 20140133.doi: 10.1098/rstb.2014.0133.

Gaston, K.J., T.W. Davies, S.L. Nedelec and L.A. Holt. 2017. Impacts of Artificial Light at Night on Biological Timings. Annu. Rev. Ecol. Evol. Syst. 48: 49–68. doi: 10.1146/annurev-ecolsys-110316-022745.

Glass, J.P. and P.G. Ryan. 2013. Reduced seabird night strikes and mortality in the Tristan rock lobster fishery. Afr. J. Mar. Sci. 35: 589–592. doi: 10.2989/1814232X.2013.860049.

Griesemer, A.M. and N.D. Holmes. 2011. Newell's shearwater population modelling for HCP and recovery planning. Technical report, Pacific Cooperative Studies Unit, University of Hawaii, Manoa, USA.

Guilford, T., O. Padget, S. Bond and M. Syposz. 2018. Light pollution causes object collisions during local nocturnal manoeuvring flight by adult Manx Shearwaters *Puffinus puffinus*. Seabird 31: 48–55.

Hart, S. 2004. Microspectrophotometry of visual pigments and oil droplets in a marine bird, the wedge-tailed shearwater *Puffinus pacificus*: topographic variations in photoreceptor spectral characteristics. J. Exp. Biol. 207: 1229–1240. doi: 10.1242/jeb.00857.

Hölker, F., T. Moss, B. Griefahn, W. Kloas, C. Voigt, D. Henckel et al. 2010. The dark side of light: A transdisciplinary research agenda for light pollution policy. Ecol. Soc. 15: 13. doi: 10.5751/ES-03685-150413.

IDA. 2020. International Dark-Sky Association. Available from https://darksky.org/(accessed on the 30th of December of 2020).

Imber, M.J. 1975. Behaviour of petrels in relation to the moon and artificial lights. Notornis 22: 302–306.

IUCN. 2020. International Union for Conservation of Nature. The IUCN red list of threatened species. Version 2016-2. Cambridge, UK. Available from http://www.iucnredlist.org (accessed on 27th of October of 2020).

Keitt, B.S. 1998. Ecology and conservation biology of the Black-vented Shearwater (*Puffinus opisthomelas*) on Natividad Island, Vizcaino Biosphere Reserve, Baja California Sur, Mexico. MSc thesis, University of California, Santa Cruz, USA.

Kyba, C.C.M., S. Garz, H. Kuechly, A.S. de Miguel, J. Zamorano, J. Fischer et al. 2014. High-resolution imagery of earth at night: new sources, opportunities and challenges. Remote Sens. 7: 1–23. doi: 10.3390/rs70100001.

Kyba, C.C.M., T. Kuester, A.S. Miguel, K. Baugh, A. Jechow, F. Hölker et al. 2017. Artificially lit surface of Earth at night increasing in radiance and extent. Sci. Adv. 3: e1701528. doi: 10.1126/sciadv.1701528.

Le Corre, M., A. Ollivier, S. Ribes and P. Jouventin. 2002. Light-induced mortality of petrels: A 4-year study from Reunion Island (Indian Ocean). Biol. Conserv. 105: 93–102. doi: 10.1016/S0006-3207(01)00207-5.

Longcore, T. and C. Rich. 2004. Ecological light pollution. Front. Ecol. Environ. 2: 191–198. doi: 10.1890/1540-9295(2004)002[0191:ELP]2.0.CO;2.

Longcore, T., A. Rodríguez, B. Witherington, J.F. Penniman, L. Herf and M. Herf. 2018. Rapid assessment of lamp spectrum to quantify ecological effects of light at night. J. Exp. Zool. Part A, 329: 511–521. doi: 10.1002/jez.2184.

Luminaves. 2020. Light pollution and conservation in Macaronesian archipelagos: reducing light pollution effects in seabirds' populations: 2017–2020. Technical report, INTERREG MAC, Canary Island, Spain.

Merkel, F.R. and K.L. Johansen. 2011. Light-induced bird strikes on vessels in Southwest Greenland. Mar. Pollut. Bull. 62: 2330–2336. doi: 10.1016/j.marpolbul.2011.08.040.

Miles, W., S. Money, R. Luxmoore and R.W. Furness. 2010. Effects of artificial lights and moonlight on petrels at St Kilda. Bird Study 57: 244–251. doi: 10.1080/00063651003605064.

Miles, W.T.S., M. Parsons, A.J. Close, R. Luxmoore and R.W. Furness. 2013. Predator-avoidance behaviour in a nocturnal petrel exposed to a novel predator. Ibis 155: 16–31 doi: 10.1111/ibi.12007.

Mitkus, M., G.A. Nevitt, J. Danielsen and A. Kelber. 2016. Vision on the high seas: Spatial resolution and optical sensitivity in two procellariiform seabirds with different foraging strategies. J. Exp. Biol. 219: 3329–3338. doi: 10.1242/jeb.140905.

Mitkus, M., G.A. Nevitt and A. Kelber. 2018. Development of the visual system in a burrow-nesting seabird: Leach's storm petrel. Brain Behav. Evol. 91: 4–16. doi: 10.1159/000484080.

Montevecchi, W.A. 2006. Influences of artificial light on marine birds. pp. 94–113. *In*: Rich, C. and T. Longcore [eds.]. Ecological Consequences of Artificial night Lighting. Island Press, Washington, D.C., USA.

Nevitt, G.A. 2008. Sensory ecology on the high seas: the odor world of the procellariform seabirds. J. Exp. Biol. 211: 1706–1713. doi: 10.1242/jeb.015412.

Oro, D., A. De Leon, E. Minguez and R.W. Furness. 2005. Estimating predation on breeding European storm-petrels (*Hydrobates pelagicus*) by yellow-legged gulls (*Larus michahellis*). J. Zool. Lond. 265: 421–429. doi: 10.1017/S0952836905006515.

Poot, H., B.J. Ens, H. de Vries, M.A.H. Donners, M.R. Wernand and J.M. Marquenie. 2008. Green light for nocturnally migrating birds. Ecol. Soc. 13: 47. doi: 10.5751/ES-02720-130247.

Reed, J.R., J.L. Sincock and J.P. Hailman. 1985. Light attraction in endangered Procellariiform birds: Reduction by shielding upward radiation. Auk 102: 377–383. doi: 10.2307/4086782.

Reed, J.R. 1986. Seabird vision: Spectral sensitivity and light attraction behavior. PhD dissertation. University of Wisconsin, Madison, USA.

Rich, C. and T. Longcore. 2006. Ecological consequences of artificial night lighting. Island Press, Washington, DC, USA.

Rodrigues, P., C. Aubrecht, A. Gil, T. Longcore and C. Elvidge. 2011. Remote sensing to map influence of light pollution on Cory's shearwater in São Miguel Island, Azores Archipelago. Eur. J. Wildl. Res. 58: 147–155. doi: 10.1007/s10344-011-0555-5.

Rodríguez, A., B. Rodríguez, R. Barone, B. Perez and A. Hernandez. 2008. Status and conservation requirements of Manx Shearwaters *Puffinus puffinus* on Tenerife (Canary Islands). Alauda 76: 72–74.

Rodríguez, A. and B. Rodríguez. 2009. Attraction of petrels to artificial lights in the Canary Islands: Effects of the moon phase and age class. Ibis 151: 299–310. doi: 10.1111/j.1474-919X.2009.00925.x.

Rodríguez, A., B. Rodríguez and M.P. Lucas. 2012. Trends in numbers of petrels attracted to artificial lights suggest population declines in Tenerife, Canary Islands. Ibis 154: 167–172. doi: 10.1111/j.1474-919X.2011.01175.x.

Rodríguez, A., G. Burgan, P.M. Dann, R. Jessop, J.J. Negro and A. Chiaradia. 2014. Fatal attraction of short-tailed shearwaters to artificial lights. PLoS ONE 9(10): e110114. doi: 10.1371/journal.pone.0110114.

Rodríguez, A., B. Rodríguez and J.J. Negro. 2015a. GPS tracking for mapping seabird mortality induced by light pollution. Sci. Rep. 5: 10670. doi: 10.1038/srep10670.

Rodríguez, A., D. García, B. Rodríguez, E. Cardona, L. Parpal and P. Pons. 2015b. Artificial lights and seabirds: Is light pollution a threat for the threatened Balearic petrels? J. Ornithol.156: 893–902. doi: 10.1007/s10336-015-1232-3.

Rodríguez, A., N.D. Holmes, P.G. Ryan, K.J. Wilson, L. Faulquier, Y. Murillo et al. 2017. Seabird mortality induced by land-based artificial lights. Conserv. Biol. 31: 986–1001. doi: 10.1111/cobi.12900.

Rodríguez, A., J.M. Arcos, V. Bretagnolle, M.P. Dias, N.D. Holmes, M. Louzão et al. 2019. Future directions in conservation research on petrels and shearwaters. Front. Mar. Sci. 6: 94. doi: 10.3389/fmars.2019.00094.

Rodríguez, A., B. Rodríguez. T. Montelongo, J. Garcia-Porta, T. Pipa, M. Carty et al. 2020. Cryptic differentiation in the Manx Shearwater hinders the identification of a new endemic subspecies. J. Avian Biol. 51: 11. doi: 10.1111/jav.02633.

Ronconi, R.A., K. Allard and P.D. Taylor. 2015. Bird interactions with offshore oil and gas platforms: Review of impacts and monitoring techniques. J. Environ. Manage. 147: 34–45. doi: 10.1016/j.jenvman.2014.07.031.

Ryan, P.G. 1991. The impact of the commercial lobster fishery on seabirds at the Tristan da Cunha Islands, South Atlantic Ocean. Biol. Conserv. 57: 339–350. doi: 10.1016/0006-3207(91)90076-L.

Saavedra, C., M.B. Santos, P. Valcarce, L. Freitas, M. Silva, T. Pipa et al. 2018. Macaronesian Roof Report. Technical report, MISTIC SEAS Project, Canary Island, Spain.

Stone, T. 2017. Light pollution: A case study in framing an environmental problem. Ethics Policy Environ. 20: 279–293 doi: 10.1080/21550085.2017.1374010.

Telfer, T.C., J.L. Sincock, G.V. Byrd and J.R. Reed. 1987. Attraction of Hawaiian seabirds to lights, conservation efforts and effects of moon phase. Wildl. Soc. Bull.15: 406–413.

The Associated Press. 2010. In Hawaii, birds' Friday night flights turn out the lights on prep games. New York Times 23 October 2010. Available from http://www.nytimes.com/2010/10/24/sports/24birds.html (accessed on 25 May 2021).

Troy, J.R., N.D. Holmes and M.C. Green. 2011. Modeling artificial light viewed by fledgling seabirds. Ecosphere 2: 109. doi: 10.1890/ES11-00094.1.

Troy, J.R., N.D. Holmes, J.A Veech and M.C. Green. 2013. Using observed seabird fallout records to infer patterns of attraction to artificial light. Endanger. Species Res. 22: 225–234. doi: 10.3354/esr00547.

Warham, J. 1990. The Petrels: Their Ecology and Breeding Systems. Academic Press, London, UK.

Whitworth, D.L., J.Y. Takewaka, H.R. Carter and W.R. McIver. 1997. A night lighting technique for at-sea capture of Xantus' murrelets. Colon. Waterbirds 20: 525–531.

Wiese, F.K., W.A. Montevecchi, G.K. Davoren, F. Huettmann and A.W. Diamond. 2001. Seabirds at risk around offshore oil platforms in the Northwest Atlantic. Mar. Pollut. Bull. 42: 1285–1290. doi: 10.1016/S0025-326X(01)00096-0.

Wilhelm, S.I., J.J. Schau, E. Schau, M. Suzanne, D.L. Wiseman, H.A. Hogan et al. 2013. Atlantic Puffins are attracted to coastal communities in Eastern Newfoundland. Northeast Nat. 20: 624–630. doi: 10.1656/045.020.0409.

CHAPTER 2

Eradication and Control of Invasive Mammal Species as a Seabird Conservation Tool

Paulo Oliveira,[1,*] *Félix M. Medina,*[2] *Manuel Nogales*[3] and *Pedro Luís Geraldes*[4]

Introduction

Invasive Alien Species (IAS) are amongst the top three threats to seabirds in terms of the number of species affected and the level of impact. Presently, IAS directly affect 165 species across all the most threatened groups. It is historically documented that IAS are implicated in the extinction of more than 70 seabird species. This problem is widespread but especially critical on islands, where breeding seabirds lack appropriate defences against introduced predators such as rats and cats.

Eradication and control of invasive mammal populations to benefit native species are essential tools to overcome these threats. Such measures contribute dramatically to restoration of the breeding habitats of these intrinsically vulnerable species. The first intentional invasive mammal (rodent) eradication programme on islands took place in New Zealand in 1980, and since then, many others followed. In Europe, the first initiative was in the late 1980s, and the earliest relatively large European island to undergo an intervention was Deserta Grande in the Madeira archipelago (central Macaronesia, North East Atlantic). Since then, many other projects have been carried out throughout this archipelago. We review in this chapter the general state of the art applied worldwide in habitat restoration projects involving the eradication and control of invasive mammal species, and zoom in to a case study of the Madeira and Selvagens archipelagos. The techniques, benefits and challenges of these projects, the constraints deriving from the currently enforced legal framework, and novel

[1] Instituto das Florestas e Conservação da Natureza, Rua João de Deus 12 E, RC C, 9050-027 Funchal, Madeira, Portugal.
[2] Consejería de Medio Ambiente, Cabildo Insular de La Palma, Avenida Los Indianos, 20, 2°, Santa Cruz de La Palma, Canary Islands 38700, Spain.
Email: felix.medina@cablapalma.es
[3] Instituto de ProductosNaturales y Agrobiología (IPNA-CSIC), Island Ecology and Evolution Research Group, AstrofísicoFrancisco Sánchez, 3, La Laguna, Tenerife, Canary Islands, 38206, Spain.
Email: mnogales@ipna.csic.es
[4] Sociedade Portuguesa para o Estudo das Aves, Avenida Columbano Bordalo Pinheiro, 87, 3°, 1070-062 Lisboa, Portugal.
Email: pedro.geraldes@spea.pt
* Corresponding author: paulo.oliveira@madeira.gov.pt

social trends are specifically discussed. Most of the data and examples that are described come from European jurisdiction islands, but the discussion that they raise also apply to other areas of the world.

Direct and indirect effects of introduced mammals on seabirds

Seabirds are globally threatened at sea mostly due to by-catch, overfishing, climate change, hunting, disturbance and pollution (Rodríguez et al. 2019). In their breeding colonies seabirds are threatened by habitat modification, diseases and the introduction of alien invasive species (Croxall et al. 2012, Dias et al. 2019, Rodríguez et al. 2019). As a result of these impacts, at least 31% of all seabird species are threatened worldwide (Dias et al. 2019). The occurrence of threats from invasive mammal species is greater on islands than in mainland areas (Hulme 2007, Bellard et al. 2016, Doherty et al. 2016). Moreover, oceanic islands are particularly vulnerable because native species evolved in the absence of predation (Courchamp et al. 2003). In fact, 92% of all globally threatened vertebrates exclusively breed on islands (Spatz et al. 2017). Historically, 615 of total known extinctions and 37% of the critically endangered species have also occurred there (Tershy et al. 2015).

A list of introduced predators affecting seabirds was recorded by Towns et al. (2011). All of them are mammals, except the barn owl (*Tyto alba*); this list includes cats (*Felis catus*), dogs (*Canis familiaris*), European hedgehogs (*Erinaceus europaeus*), mice (*Mus musculus*), Norway rats (*Rattus norvegicus*), Pacific rats (*R. exulans*), Australian possums (*Trichosurus vulpecula*), pigs (*Sus scrofa*), ship rats (*R. rattus*), stoat (*Mustela erminea*), Argentine grey foxes (*Pseudalopex griseus*), red foxes (*Vulpes vulpes*), mongooses (*Herpestes auropunctatus*), Algerian hedgehog (*Atelerixalgirus*), weasels (*Mustela nivaris*), pine martens (*Martes martes*), Arctic foxes (*V. lagopus*), Arctic ground squirrels (*Spermophilus parryii*), voles (*Microtus* spp.), deer mice (*Peromyscus* sp.), Grivet monkeys (*Clorocepbus* sp.), pig-tailed macaques (*Macaca nemestrina*), long-tailed macaques (*M. fascicularis*) and stump-tailed macaques (*M. arctoides*). For this reason, mammals have been globally considered the most pernicious species, especially on islands (Lever 1994, Courchamp et al. 2003, Long 2003, Doherty et al. 2016). Practically all of them have been introduced by humans as livestock, companion animals or as commensals; species belonging to Artiodactyla, Perissodactyla, Lagomorpha, Rodentia and Carnivora were the most successful invaders (Clout and Russell 2007). Of these, six species: ship rats, cats, dogs, pigs, goats (*Capra aegagrus hircus*) and cattle (*Bos taurus*), have been included as the seven most pernicious species worldwide (Bellard et al. 2016). Furthermore, mustelids and mongooses have also been considered as key groups in the decline of native species (McCreless et al. 2016). Hulme (2007) revised the impact of alien vertebrates on specific animal species in European island territories, where 39.4% of the total native species affected were seabirds, and rats were the most recorded pernicious invasive mammals (76.9% of the total invasive predators). Besides these, other invasive mammals are also affecting threatened species at different ecological levels in Europe, such as rabbits (*Oryctolagus cuniculus*), Barbary sheep (*Ammotragus lervia*), sheep (*Ovis aries*), Corsican mouflons (*O. a. musimon*), reindeer (*Rangifer tarandus*) and American minks (*Neovison vison*) (Chapuis et al. 1994, Nogales et al. 2006, Genovesi et al. 2015a).

Introduced mammals commonly exert their impact on native biodiversity through predation, competition for resources, habitat modification (herbivory: grazing, browsing and trampling), disease transmission and hybridisation (Clout 2002, Genovesi et al. 2015a). In particular, impacts on seabirds have mainly been summarised as direct impacts of predation and competition for resources, with indirect implications due to habitat modification by herbivores (McCreless et al. 2016).

Direct effects of seabird predators

Direct effects on seabirds are caused through predation by introduced mammals. Of all seabird species threatened, 46% are affected by invasive alien species, including mammalian predators and herbivores, plants, birds, insects and even fishes (Dias et al. 2019). Of these, 23.7% of seabird species are affected by predators such as rats, cats and mice. Although a single impact of only one of these predators is usually described, multiple effects caused by the presence of several predator species upon a single seabird species in the same location have also been reported (see Hervías et al. 2013a).

The effects of rats on seabirds were globally reviewed by Jones et al. (2008). Three rat species (*Rattus exulans*, *R. norvegicus* and *R. rattus*) are equally responsible for seabird extirpation and population decline of at least 75 species on 61 islands worldwide, due to predation on eggs, chicks and adults. Hydrobatids are the most affected seabird family, while albatrosses, frigatebirds and larids suffer the lowest predation impact (Jones et al. 2008). Towns et al. (2006) also list the effects of rats on several seabird species for islands of European jurisdiction, mainly affecting shearwaters, puffins, fulmars, prions and petrels. For example, Caut et al. (2008) detected impacts of rats on particular species such as Wedge-tailed Shearwater (*Puffinus pacificus*), Masked Booby (*Sula dactylatra*) or Brown Booby (*S. leucogaster*) on Surprise Island (New Caledonia); Bell et al. (2019) reported impacts on Manx Shearwater (*Puffinus puffinus*), European Storm Petrel (*Hydrobates pelagicus*) and Lesser Black-backed Gull (*Larus fuscus*) on the Isles of Scilly (UK), and Ruffino et al. (2009) described the presence of rats as a limiting factor in the abundance of European Storm Petrel on Mediterranean Islands.

Cats are top predators and co-responsible for the extinction of 14% of the total vertebrate species on islands, affecting 8% of the total of critically endangered mammals, birds and reptiles on islands worldwide (Medina et al. 2011). Among these globally threatened species, 65 (39%) out of 167 cases are seabirds. Cats predate upon large birds, including albatrosses, burrowing petrels, cormorants, frigatebirds, murrelets, penguins and shearwaters, with many cases occurring on islands under European jurisdiction. They are now present on islands in all the oceans, such as Amsterdam, Crozet, Réunion, Henderson, Madeira, Menorca, Ascension or Grande Terre (more details in Medina et al. 2011).

Mice were initially considered an invasive species due to their effects on invertebrates and plants, and on habitat modification (see review by Courchamp et al. 2003). Their impacts were described for smaller ground-nesting birds such as Spotted Sandpiper (*Actitis macularius*) (Maxon and Oring 1978). However, predation on chicks and eggs of large seabird species such as Tristan Albatross (*Diomedea dabbenena*), Atlantic Petrel (*Pterodroma incerta*), Salvin's Prion (*Pachyptila macgillivrayi*) and Great Shearwater (*Puffinus gravis*) has been recorded on Gough Island (Cuthbert and Hilton 2004, Wanless et al. 2007, 2012, Caravaggi et al. 2019). Predation by mice has drastically affected the populations of Grey-backed Storm Petrel (*Garrodia nereis*) and Wilson's Storm Petrel (*Oceanites oceanicus*) in the Falkland Islands (Bolton et al. 2014). Examples have also been documented for other islands and seabird species worldwide (see, for instance, Duhr et al. 2019, Preston et al. 2019).

Effects of other introduced predators have been scarcely documented. Specific cases were reported for the Western Isles of Scotland where mink (*Mustela vison*) affected colonies of Arctic Tern (*Sterna paradisaea*) (Moore et al. 2003), and for the Azores archipelago, where two other mustelid species (weasels *M. nivaris* and ferrets *M. furo*) threatened breeding colonies of Common Tern (*S. hirundo*) (Monteiro et al. 1996). Hilton and Cuthbert (2010) reported on the general impact of the Small Indian Mongoose (*Herpestes scaurus punctatus*) on seabird populations in the British Virgin Islands.

Indirect effects of invasive species

Considering the effect of only three introduced predator species on seabirds (superpredator: cats, mesopredator: rats and herbivore: goats), Russell (2011) described five different predator-related indirect effects: (1) interspecific competition, (2) apparent competition, (3) trophic cascade, (4) symmetrical intraguild predation and (5) asymmetrical intraguild predation. One of the most studied indirect effects is the hyperpredation process, which is related to the apparent competition effect. It occurs when introduced prey species like rabbits, rats or mice can increase the numbers of an introduced predator (cats, dogs, mustelids), thus increasing the pressure on native species (Courchamp et al. 2000).

However, predators can also have other underlying indirect effects on species and at the ecosystem level. Rats predating on burrowing seabirds can influence soil nutrients and litter decomposition (Fukami et al. 2006, Wardle et al. 2009), indirectly affecting native invertebrate communities through reduced nutrient subsidies and other consequent trophic cascades (Towns et al. 2009, Thoresen et al. 2017). Furthermore, decreasing seabird populations due to rat predation may be an important determinant of ecosystem carbon sequestration (Wardle et al. 2007). Rats can also affect seabirds by disrupting adults attempting to brood or provision for young chicks (Major et al. 2006).

Feral cats compete with native species for food and habitat resources, and also alter ecological processes, given their interference in bird migration, breeding behaviour or seed dispersal systems (Medina et al. 2014, Nogales et al. 2015). One of the significant indirect impacts of cats on seabirds is the transmission of diseases such as toxoplasmosis. This parasite has been transmitted to several species such as Red-footed Booby (*Sula sula*), Galápagos Penguin (*Spheniscus mendiculus*) and Flightless Cormorant (*Phalacrocorax harrisi*) (Medina et al. 2014). Cats also provoke changes in seabird breeding behaviour. Red-footed Booby breeds in taller trees on islands with cats, rather than in shorter bushes when cats are absent (Medina et al. 2014). This same behavioural change has been observed in European Shags (*Phalacrocorax aristotelis*), which moved their natural nest-sites to a more protected place when harassed by American minks, leading to a high cost in their reproductive success (Barros et al. 2016).

Mice can predate on and kill large seabird species, in some cases reducing nesting success by abandonment or mortality of one or both parents, due to secondary bacterial infection of wounds (Work et al. 2021). High predator activity around the nests can induce stress in breeding seabirds. Such predation pressure makes seabirds more susceptible to parasitism and population ill-health, which in turn reduce breeding success. This is the case of Cory´s Shearwaters (*Calonectris borealis*) on Corvo, Azores archipelago (Hervías et al. 2013b).

Introduced herbivores (rabbits, goats, sheep, Barbary sheep, mouflons and cattle) can also cause critical negative impacts on the nesting habitat of seabird species. The most significant damage is seen in their vegetation (Clout 2002, Courchamp et al. 2003, Campbell and Donlan 2005, Nogales et al. 2006, McCreless et al. 2016). Direct impact on vegetation alters plant community structure and composition, causing species extinction, faster soil erosion and finally entire habitat degradation (Courchamp et al. 2003, Campbell and Donlan 2005). These effects can also indirectly degrade and limit the available nesting habitat for burrowing and other seabirds (Chapuis et al. 1994, Monteiro et al. 1996, McChesney and Tershy 1998, Carlile et al. 2003, Swinnerton et al. 2018).

Most adverse effects of exotic herbivores have been assigned to rabbits and goats. As a result of such grazing and browsing on vegetation, the breeding habitat of tree-nesting seabirds such as frigatebirds and boobies can be drastically reduced (Russell and Le Corre 2009). Furthermore, removal of natural vegetation by these introduced herbivores also increases the visibility and accessibility of nests, leading to higher predation on nest contents (Priddel et al. 2000, Carlile et al. 2003, Chan 2013). Another disturbance, described for goats, is nest abandonment by breeding seabird adults when goats visit breeding colony areas, as recorded in the case of Red-billed Tropicbird (*Phaeton aethereus*) in the Dutch Caribbean (Chan 2013). Indirect impacts on seabirds are mainly described for exotic herbivores and other non-predators such as pigs. However, some direct effects have also been mentioned, such as trampling and consumption of eggs (Pierce et al. 2020), and occupancy of nesting burrows by rabbits replacing burrowing seabirds (Imber et al. 2000, Courchamp et al. 2003).

Control and eradication of invasive species: methods and techniques

When the impact of invasive species is detected in sensitive areas, they are often already past control (Lodge et al. 2006, Pitt and Witmer 2014). Until now, the decision to eradicate or control these species has been based on several factors, ranging from availability of human and technical resources and financial capacity, to even the personal interests of managers or local conservation officers. Over the years, several guidelines to orient these efforts have been compiled, and nowadays national and local managers have a variety of tools to help them define priorities and objective achievable targets (Carboneras et al. 2018, Holmes et al. 2019). Several techniques are available when facing the need to decide whether to control or eradicate an exotic invasive species. Their choice, expected results and constraints can help determine the final goal and best approach. This choice will have to consider possible secondary target species, the human resources available, ideal and possible schedules and time-frames for implementation and monitoring, the local weather, the necessary equipment and training, financial resources, long-term commitment possibilities, likelihood of reinvasions and so on. Overall, eradication proves to be more cost-effective in the long run, more humane and with better results in terms of habitat recovery (Gherardi and Angiolini 2004, Jones et al. 2016). Programmes for IAS eradication resort to the same methods as for control, and when feasible they are recommended by the IUCN as the best approach (IUCN 2000). Eradication and control methods can broadly be divided into Biological, Mechanical and Chemical (Orueta and Ramos 2001).

Biological methods rely on the introduction of a particular species to control or eradicate the invasive species. Most attempts to introduce predators to control or eradicate vertebrates have failed and some even intensified the impacts on the ecosystem, becoming invasive themselves (Orueta and Ramos 2001). Specific pathogens can also be released into the wild to control or eradicate IAS, and if successful, they can complement other methods and lead to successful eradication or control (Trout et al. 1992, Huntley 1996, Pech 1996, Van Rensburg et al. 1987, Cooke 1998, Sheppard et al. 2019). However, biological methods are by far the least used since they require a great deal of prior specific research; they are difficult to isolate, transport and release and their effects are hard to monitor. Thus, they cannot be used alone (Bell 1995).

Mechanical methods rely on trapping and hunting the target species. This can be done in a myriad of ways; indeed, techniques are under constant development and refinement. These are the most selective, but also very demanding in terms of human effort and therefore expensive. They can be adapted to various species, habitats and communities, but can also be the most prone to error. Traps can be expensive, difficult to transport and subject to maintenance issues, since they must be moved, relocated and finally removed at the end of the programme. Mechanical methods are the most frequently used for large and medium-sized mammal species, either as single methods or as complements. They prevent the release of large quantities of toxic bait into the environment, are highly selective in limiting accidents with non-target species, and are generally more humane.

Live traps require regular short-spaced monitoring, while kill traps require high selectivity to protect non-target species. Both have to be deployed by expert trappers for full effectiveness in terms of location, baits, etc. Monitoring and re-setting them is a critical factor, especially in remote or difficult-access locations, as also is the number of traps in the grid. Shooting and hunting can be used as stand-alone or complementary techniques. In several cases, use of hunting techniques can determine the success of eradication, being applied to eliminate the last recalcitrant animals. Such techniques include direct shooting, use of hunting dogs to locate or reveal the target species' refuges or of observation hides. Some specialised aids can be recurred to, such as ferrets, nocturnal shooting with night vision scopes or silencers. The main disadvantages of shooting are the disturbance caused to non-target species, the expertise necessary for personnel using the guns and in most cases the permits and licenses that can be very hard to obtain. A good example is the case of Cape Verde Islands, where there are no game species, and as such there is no hunting law and access to guns is reserved for police or military personnel. Some programmes resort to PCP (Pre-Charged Pneumatic) rifles (air guns) to circumvent these constraints, since they are legally easier to obtain. This type of gun can be effective for use on medium-sized animals, although with diminished range and effectiveness.

The latest developments in mechanical methods include several new kinds of traps (Warburton and Orchard 1996, Orueta and Ramos 2001, Forsyth et al. 2005): from self-resetting traps that allow for longer intervals between trap checking, long-lasting baits, to bait-release devices and even GSM-based alarms that warn the operator whenever the trap is triggered. All these advances reduce human effort and especially operational costs, allowing larger areas to be covered.

Chemical methods rely on the use of chemicals to control and kill the target species. The majority of successful eradications on islands have used chemical methods (Howald et al. 2007, Fisher et al. 2019). These methods were first used on a large scale in New Zealand (Fisher et al. 2019). Since then, toxic baits have been used in hundreds of eradications throughout the world. They are the method of choice for rats and mice, but are also applied on a large scale to carnivores such as stoats or feral cats (Orueta and Ramos 2001). Their disadvantages include toxic active ingredients that can remain in the environment for some time with the possibility of secondary poisoning, and bait degradation after setting in place. Special release methods may be required when no aerial dispersal means are available. In large-scale operations, sometimes animals of non-target species have to be removed from the island to guarantee a viable genetic pool to repopulate the area if they are affected (Ward et al. 2019).

Baits have different compositions and presentations. Anticoagulants like warfarin were first developed in the 1950s and widely used to control rodents (Pelz et al. 2005, Lim 2017). At first, they were presented in liquid form or powders, and had to be mixed with food items and other attractants. Since then, new research and development of resistance by some animals has led to the development of novel active ingredients and presentations to avoid non-target species or humans, while being more effective (Berny et al. 2018). They are widely used as an effective tool to eliminate rodents, rabbits or ungulates. Second-generation coagulants have antidotes that can be used in case of accidental poisoning, or even to prevent non-target species and domestic animals from being affected by the bait (Berny et al. 2018). This requires access to the antidote to be species-specific. Most operations on larger islands require the use of aerial means, but baits can also be laid in bait-boxes, by hand spraying or using a combination of these methods. Whilst for small rodents the application grids usually require large amounts of bait spread over the whole area they occupy, for larger mammals hand placement in suitable locations can be a better approach, allowing the release of lesser amounts of bait and increasing its consumption by the target species. For carnivores, 1080 (the manufactured version of sodium fluoroacetate) was the usual bait of choice for most reported eradications, being already used in various types of baits for pest control operations. It is particularly highly toxic to mammals, and is still in widespread use in Australia and New Zealand (Eason et al. 2011). More humane baits have been developed for carnivores, and recently PAPP (para-aminopropiophenone) was introduced as a very useful Vertebrate Toxic Agent for the control and eradication of larger carnivores. Although expensive, it has several advantages: (a) an antidote, (b) similar biodegradability in the environment, (c) faster action, and (d) more effectiveness in mammals than in birds (Eason et al. 2014). In Europe it is still not commercially available and requires special permits, but it was already used in a cat eradication project in Cape Verde Islands, with very promising results (authors' unpublished data). Special new traps are also being developed to enable the use of toxic bait in areas where domestic animals are present. Some designs recognise the chip on those animals, thus not spraying the toxic bait meant to eliminate the targeted feral or wild species (Read et al. 2019). Today we can safely say that there are baits available that are suitable for almost every eradication operation necessary. They present themselves in several different formulations, with different attractants, colours, smells and resistance to humidity or weather, depending on what is required for the site in question or for the target species (Witmer and Moulton 2014, Suárez and Cueto 2018).

Monitoring is an essential part of any control or eradication programme. It permits adaptive responses to the effectiveness of the methods chosen. Presence or absence is usually the minimum information required to determine a change of methods or trap locations or to (dis)continue the programme. Precise estimates of abundance of the target species are seldom sought during eradications, usually not being effective enough to compensate their logistic requirements and cost. Monitoring methods and tools have also evolved. Today, IAS programme managers have sophisticated camera-traps available that can be remotely controlled or send videos or photos over GSM networks. Drones can be used to prospect for animal shelters and refuges, even with thermal imaging cameras to detect animals at night (Morley et al. 2017, Hyun et al. 2020, Witt et al. 2020). DNA sampling of droppings can be used to detect the animals present and discriminate males from females individually.

New Zealand led the way in rodent eradication and over the years they have turned island rat/mouse eradications into a very precise operation. Research is today focussed on rodent invasion patterns in order to strengthen biosecurity effectiveness, and on the type and techniques of bait spreading as well as on the calendar of campaigns and monitoring. Nevertheless, these programmes have to be highly adaptable to local fauna and habitats; indeed, the same protocol can rarely be applied elsewhere without modifications. The cases in the Madeira archipelagos described in the next section illustrate such adaptations.

Difficulties in earlier eradication projects were mainly logistic, attributable to a lack of technical knowledge and poorly trained ecologists. Today most of those problems have been overcome but public awareness has increased, and with it other constraints have arisen. Animal welfare is sovereign for several interest groups and social networks, furthermore the media arbitrarily amplify critics' voices, often without scientific support. The generally very limited understanding of this threat by the public and decision-makers leads to inadequate participation and political commitment, which is particularly acute in Europe (Genovesi 2001, Genovesi et al. 2015b). New Zealand still leads in this conservation field by making a national commitment to a Predator Free country by 2050, and allocating resources and funding for this goal of Zero Invasive Predators (Russell et al. 2015).

European conservationists and ecologists face constraints of two different types. On the one hand, the legal frameworks are inexistent, outdated or inadequate. On the other hand, as misinformed public can today generate major negative campaigns using the new social networks and pressure decision-makers,whom are often susceptible to negative public viewpoints despite sound scientific data showing the contrary. Eradications are often a controversial management practice, and in several cases opposition by the public or specific interest groups has resulted in failure of the project (Genovesi and Bertolino 2001). To summarise, the limited number of eradications carried out in Europe so far is probably due to the limited awareness of the public and decision-makers, inadequacy of legal frameworks and scarce resources (Genovesi 2005). Recently, adoption by the European Union of Regulation 1143/2014 on invasive alien species (Carboneras et al. 2018) is the most important European environmental policy measure to date, aimed at meeting the Convention on Biological Diversity Aichi, target 9. Despite this, several methods widely used outside Europe are not available to European conservationists. Although most methods are merely improvements or refinements of control techniques already used for public health reasons, population density corrections (culls), leisure activities (such as hunting) or agricultural crop protection, and practical legal issues limit or prevent the use of the same techniques for IAS control or eradication.

Today, animal welfare and rights movements are well funded and supported by thousands of members, and may even rely on international support. Examples of their more extreme actions are abundant: (1) Petitions to decision-makers, such as those to avoid rat culling in Paris or Portugal (https://www.mesopinions.com/petition/animaux/stoppez-genocide-rats/26805 or https://peticaopublica.com/pview.aspx?pi=salvar-rato-berlenga), (2) death threats to rat eradication programme managers, (3) legal action such as questions in the European or national parliaments, legal prosecution, injunctions and criminal processes, and (4) strong media campaigns (https://www.publico.pt/2015/07/27/ciencia/noticia/o-ratopreto-da-berlenga-e-inocente-1702138). Even a more moderate reaction of this calibre can severely undermine conservationist efforts and erode project budgets and operational capacity. It leads to loss of support from funders and public entities, and is directly taking its toll on conservation. Although the scientific data is unmistakable and shows tremendous positive effects for ecosystems after the removal of IAS (Howald et al. 2007, McGeoch et al. 2010, Jones et al. 2016), conservationists still have to justify every action to the public, sometimes not based on their merits or using scientific techniques and data, but on a better sound-bite or stronger media campaign. They often have to dedicate most of their time to legal action, advertising and awareness campaigns, and end up debating degrees of "cuteness" or the lack of effectiveness of tried and tested TNR (Trap, Neuter, Release) methods to reduce IAS populations (McCarthy et al. 2013, Hostetler et al. 2020). Today, this kind of movements are one of the major problems facing the species threatened by IAS, and may even contribute to the threats. Conservationists have to find new ways to justify their work and methods, in order to earn public funding and support for their actions.

Research on new, more effective and humane methods is always ongoing and new techniques may be the response to this societal problem. Gene editing tools such as CRISPR may in the near future open up the possibility to eradicate IAS from larger inhabited islands (Leitschuh et al. 2018, Grunwald et al. 2019). Environmental DNA is currently a new prospect tool that can be very powerful for detecting the presence of IAS (Herder et al. 2014). Although still not commonly used, it can make a difference in assuring successful programmes or determining the continuance of eradication efforts.

The eradication and control of invasive mammals: a case study of Macaronesia Region and the Madeira archipelago

Efforts to protect seabirds and their breeding grounds from invasive vertebrate species have been carried out throughout the world. We illustrate this by presenting (1) a review of projects carried out in the Macaronesia Biogeographic Region, (2) the long-term programme in the Madeira archipelago, and (3) insights into the recovery and habitat restoration of the North Atlantic Gadfly Petrels (*Pterodroma* spp.) breeding in the Madeira archipelago.

The Macaronesian Biogeographic Region is situated in the North Atlantic between 39° 45' and 14° 49' N and 31° 17' and 13° 20 W. From north to south, it comprises the archipelagos of the Azores, Madeira, Selvagens, Canaries and CapeVerde (Fig. D2.1). These island systems share a wide range of biological affinities (Báez and Sanchez-Pinto 1983) and are characterised by high endemism levels (e.g., Borges et al. 2008, 2010). Although the Selvagens are geographically a different archipelago, they are administratively part of the Madeira archipelago. Hereinafter, when not otherwise mentioned, we will consider them together. Pelagic seabirds (belonging to the Order Procelariformes, Suliformes and Charadriformes) are consistently present on all islands, small islands and islets in these archipelagos, with a total of 15 breeding species, most of them having a global conservation status of some concern (IUCN 2020). Therefore, due to their characteristics and location, the Macaronesian lands are vital breeding areas for globally important seabird species.

Invasive species are among the most severe threats to seabirds in Macaronesia. These archipelagos were no exception in the introduction of a remarkable array of plant and animal species, with extremely disruptive characteristics for the pre-human ecosystems in general, and

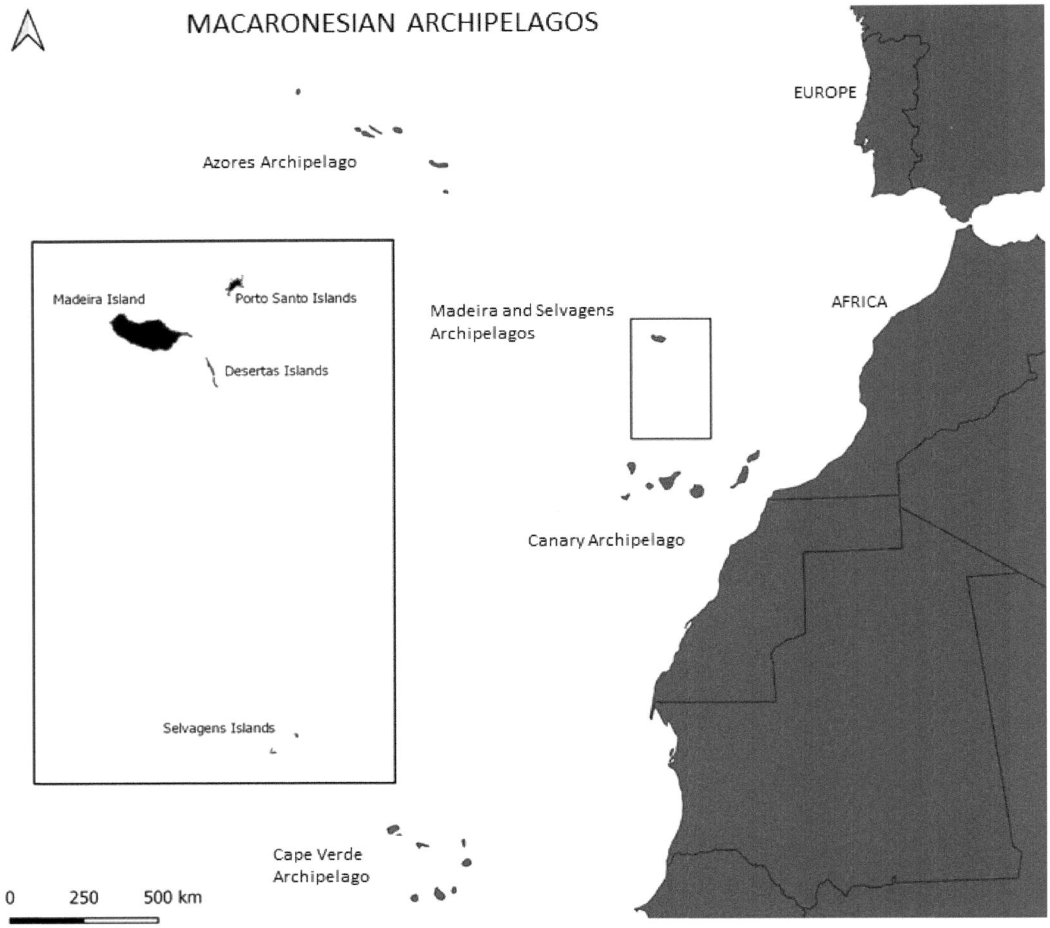

Figure D2.1. Map of the Macaronesian archipelagos, with a close-up to the Madeira and Selvagens Archipelagos.

seabirds inparticular. Historically, invasive vertebrate species have caused the greatest impacts. Non-vertebrate invasive species are also of concern nowadays, such as the Argentine ant (*Linepithema humile*). This threat will not be dealt with here, but it is of major concern for the modern management of seabird breeding colonies.

A total of 14 invasive mammal species can be found in these archipelagos (Direção Geral do Ambiente 2004, Nogales et al. 2006, Borges et al. 2010, Masseti 2010, Oliveira and Menezes 2012), with the number of species usually related to island size and human presence (e.g., Oliveira and Menezes 2012). All these species are considered amongst the 100 worst invasive alien species worldwide (Silva et al. 2008), with direct and/or indirect impact on seabird populations. Consequently, seabird conservation and protection, together with the necessary eradication and control of introduced species, are among the leading nature conservation priorities in these archipelagos (Borges et al. 2008). For the last 35 years, many projects to eradicate/control invasive vertebrate species have been carried out throughout all the Macaronesian archipelagos (Table D2.1).

However, in the Madeira archipelago (Fig. D2.1), a more systematic long-term programme was implemented. This long-term approach was made possible owing to the European Union funding through several life projects in the field between 1995 and 2019. It is essential to point out that these funded projects also largely contributed to building up an effective capacity, mainly through training a highly specialised and skilled team. As a result, invasive species were eradicated from many difficult small islands and controlled locally in areas with high conservation value, where eradication was not feasible. This programme has become a powerful tool to prevent extinctions and restore ecosystems in the Madeira archipelago, directly or indirectly benefitting large and important seabird colonies. The Deserta Grande eradication project was a landmark in the history of these projects in Europe. This was the first remote difficult access small island in Europe where such a project was successfully carried out. It is still the largest North Atlantic Island from where rabbits were eradicated. This was also an important capacity-building opportunity for the local administration staff, which allowed the implementation of other ambitious projects (e.g., the Selvagens and Bugio rabbit and mice eradication projects).

In most projects for the control and eradication of invasive mammals, the adequate application of bait is the key to their success (Oliveira and Menezes 2012). Rodenticide in a cereal-based bait (pellets) or as wax blocks were used in all but one campaign (Cima Islet) (see Table D2.1). Wax blocks were used when it was important to have good palatable bait available during winter (Selvagem Grande, Cal Islet and Desembarcadouros Islet) or in wet conditions (Madeira); elsewhere pellets were used. Brodifacoum (0.02 and 0.05%) was consistently used because it is a very efficient second-generation anti-coagulant rodenticide (e.g., Howald et al. 2007). Bait was hand broadcasted in most campaigns following a 25 × 25 m grid for rabbits and 12.5 × 12.5 m for mice. When non-target threatened species were present or the bait was exposed to severe weather conditions, bait stations were used. During the eradication campaign on Cima Islet, bait was not used, and most of the rabbits were caught in traps. This is an important technique, especially to avoid killing non-target species. However, due to the high probability of 'by-catch', it is a technique that might not work in areas where the density of breeding seabirds is high.

Goat eradication on mainland Madeira was a long process due to social and political constraints. This process was made possible by compensating the owners for each animal caught and removed from the wild. This was a critical project strongly contributing to the recovery of the breeding habitat of the Madeira or Zino's Petrel (*Pterodroma madeira*), which will be dealt with in more detail below. Additionally, goats are nowadays the major conservation problem on Deserta Grande, and, because eradication so far was not feasible, population density control (culling) is carried out regularly. Goats were secondarily also removed from Bugio Islet during the challenging campaign to eradicate mice and rabbits. This project aimed to recover the habitat of the unique and globally threatened Desertas Petrel (*Pterodroma deserta*). Further details about this campaign are also given below.

In Madeira, as frequently happens all over the world, the direct positive impacts of these efforts on seabirds, and on the wider ecosystems, were poorly quantified. However, Oliveira et al. (2010) monitored the response of different groups of fauna and flora following rabbit and mice eradication on Selvagem Grande, and showed a positive response for plants, invertebrates and terrestrial birds to the eradication program. Zino et al. (2008) also showed an increase in the numbers of fledged Cory's Shearwaters on Selvagem Grande, after five consecutive years of rabbit and mouse eradication. Consistent systematic monitoring of the impacts of these eradication/control projects should be higher on the priority list of project and area managers. Directly or indirectly, all these efforts contributed to enhancing the conservation status of most seabirds and their habitats. The efforts targeting the endemic Gadfly Petrels of Madeira Island and Bugio Islet are good case studies to illustrate two different approaches: long-term control and a quick eradication event.

Table D2.1. List of the main eradication/control campaigns in the Macaronesian Archipelagos, showing target species, outcome (target species controlled/eradicated) and the seabird species with enhanced conservation status (* or ** applies to * and ** in the following column).

Archipelago/Island	*Size (ha)/ Altitude (m)*	*Year*	*Type of Intervention*	*Target species*	*Outcome*	*Seabird pop. enhanced (target species of the project in bold)*
Madeira/Madeira	74175/1862	1986–ongoing	Control of an area of ca. 300 ha in the Central Mountainous Massif	*Felis catus* *Mustela putorius* *Mus musculus* *Rattus rattus*	Success	*Pterodroma madeira*
Madeira/Madeira	74175/1862	1994–2003	Eradication	*Capra aegagrushircus*	Success	*Pterodroma madeira*
Madeira/Deserta Grande	1028/479	1995–ongoing	Eradication*/Population control**	*Oryctolagus cuniculus** *Capra aegagrus hircus***	Success	*Bulweria bulwerii* *Calonectris borealis, Hydrobates pelagicus* *Puffinus lherminieri*
Madeira/Selvagem Grande	241/163		Eradication	*Mus musculus* *Oryctolagus cuniculus*	Success	*Bulweria bulwerii* *Calonectris diomedea* *Hydrobates pelagicus* *Pelagodroma marina* *Puffinus lherminieri*
Madeira/Bugio Islet	321/388	2007–2010	Eradication	*Oryctolagus cuniculus* *Capra aegagrus hircus*	Success	*Bulweria bulwerii* *Calonectris diomedea* *Hydrobates pelagicus* *Pterodroma deserta*
Madeira/Cima Islet	31/124	2010–2011	Eradication	*Oryctolagus cuniculus*	Success	*Bulweria bulwerii* *Calonectris diomedea* *Hydrobates pelagicus* *Puffinus lherminieri*
Madeira/Cal Islet	140/178	2012–2013	Eradication	*Mus musculus* *Oryctolagus cuniculus*	Success	*Bulweria bulwerii* *Calonectris diomedea* *Hydrobates pelagicus* *Puffinus lherminieri*
Madeira/ Desembarcadouros Islet	40/101	2014-2015	Eradication	*Mus musculus* *Rattus rattus* *Felis catus*	Success	*Bulweria bulwerii* *Calonectris diomedea* *Hydrobates pelagicus*
Madeira/Madeira	74175/1862	2014–ongoing	Control of an area of ca. 20 ha at the eastern end of the island	*Mus musculus* *Rattus rattus* *Felis catus*	Success	*Bulweria bulwerii* *Calonectris diomedea* *Hydrobates pelagicus*

Madeira/Chão Islet	100/112	2015–2016	Eradication	*Mus musculus*	Success	*Bulweria bulwerii* *Calonectris diomedea* *Hydrobates pelagicus*
Canaries/Lobos	440/127	1992, 1997, 2002	Eradication	*Felis catus* *Mus musculus* *Oryctolagus cuniculus*	Success	*Bulweria bulwerii* *Calonectris diomedea* *Hydrobates pelagicus* *Pelagodroma marina* *Puffinus lherminieri*
Canaries/Alegranza	1019/289	1998	Eradication	*Felis catus** *Oryctolagus cuniculus***	Success*/No success**	*Bulweria bulwerii* *Calonectris diomedea* *Hydrobates pelagicus* *Pelagodroma marina* *Puffinus lherminieri*
Canaries/Montaña Clara	1,3/256	2002	Eradication	*Oryctolagus cuniculus*	Success	*Pelagodroma marina*
Azores/Praia Islet	10/52	1997	Eradication	*Oryctolagus cuniculus*	Success	*Calonectris diomedea* *Hydrobates monteiroi* *Hydrobates castro* *Puffinus lherminieri*
Azores/FenoIslet (Terceira)	1,6/14	2006–2007	Eradication	*Oryctolagus cuniculus*	Success	*Calonectris diomedea* *Hydrobates monteiroi*
Cabo Verde/St. Luzia	3420/395	2014–ongoing	Eradication	*Felis catus*	Ongoing	*Hydrobates jabejabe* *Alauda razae* (terrestrial bird species)*

North Atlantic Gadfly Petrels (*Pterodroma* spp.) case study

In Macaronesia, three species of the *Pterodroma* genus can be found: Zino's petrel (*P. madeira*) on Madeira Island, Desertas Petrel (*P. deserta*) on Bugio Islet and Gon-gon (*P. feae*) in Cape Verde Islands (Fogo, São Nicolau, Santo Antão and Santiago) (Ratcliffe et al. 2000). The actions to conserve these three species are very different. On Madeira Island and Bugio Islet, the birds and their habitat are well protected, and many conservation activities are under way. On the other hand, in Cape Verde Islands only a small part of the population benefits from the National Park established at Chão das Caldeiras on Fogo Island, and concrete conservation efforts are scarce.

Zino's Petrel is endemic to the island of Madeira. It is listed as Endangered in the IUCN Red Data Book because of its extremely small population, about 80 pairs breeding in an area with less than 300 ha in the central mountain massif of Madeira (BirdLife International 2021). Subfossil evidence indicates that Zino's petrel was once abundant and more widely distributed (Illera et al. 2012). In 1903, when first recorded, it was already limited to Madeira's high central massif. After being considered extinct for many decades, in 1969 a relict population was discovered. By 1985, all known breeding attempts were disrupted by introduced rats to the extent that breeding success was nil (Zino et al. 1996). In 1987 a rat exclusion zone was set up around the breeding ledges with bait boxes containing rodenticide (Brodifacoum 0.05%). This exclusion zone was progressively enlarged, and nowadays, there are 110 baiting stations in the area, including on the breeding ledges themselves. Additionally, in more remote areas where it is difficult to monitor bait boxes regularly, strategic bait is placed at the beginning of the breeding season and just before hatching. This is the longest reported ongoing rodent control at world level. Forty years of regular monitoring of bait consumption provides an array of information that, when analysed, should be of great importance to evaluate this type of effort elsewhere.

This exclusion zone was extended to cats in 1991, following a predation event that killed 10 adult birds. As it happened with rat control, this cat exclusion zone was improved with time and enlarged throughout the area. Nowadays, there are 25 traps strategically placed in the field. When cat signs are identified, extra traps (ca. 3 or 4) are temporarily placed in the area. Since 1991, cats have remained mostly under control, with an average of 14 catches per year for the period 2014–2020. However, in 2019 another dramatic predation event occurred, and eight adult birds were killed over three weeks. It was probably an experienced cat that, during this period, managed to avoid all the extra traps set up in the area. Nowadays, cat traps are visited daily between March and November. In an area with extreme relief and weather, this is a very demanding and time-consuming activity (ca. 5 hours each visit). At the end of 2020, trap activation sensors (GSM signal based) were successfully tested. It is expected that this technology will enable the exclusion zone to be enlarged, which will dramatically reduce the necessary labour and improve exclusion zone efficiency. The results of these efforts have been positive, and the Zino's petrel population is making a small but steady recovery. The number of fledged birds has increased from 0 in 1986 to an average of 28 fledgings for the last 4 years (2016–2020), with an average breeding success above 75% (non-published data, Menezes and Zino in litt. 2021). These figures could be even more positive, but a devastating fire in 2010 destroyed or degraded 80% (n = 43) of the active nests. Only one chick of a total of 38 fledged in that year. Three adults were found dead, but the number of casualties was probably higher.

The Desertas Petrel is endemic to the Desertas group of islands and is presently restricted to the islet of Bugio (ca 300 ha). There is fossil evidence that this petrel was once widely abundant within the archipelago (Pieper 1985). The majority of the present breeding population, estimated to be about 180 pairs (BirdLife International 2021), uses a restricted plateau on Bugio Island (less than 15 ha). For these reasons, this recently split species (Jesus et al. 2009) is listed as Vulnerable in the IUCN Red Data Book (BirdLife International 2021). Habitat degradation due to overgrazing by goats and rabbits was identified as the main threat and limiting factor for this burrow nesting species

(Zino et al. 1996, Menezes et al. 2010). Rabbits were also the cause of major disturbance to breeding birds, invading and modifying burrows, digging new entrances and inter-connections and promoting their collapse. As a consequence, in 1993, a feasibility study on the removal of these introduced species was carried out. The steep terrain and lack of facilities on Bugio led to the conclusion that such an eradication project would be impractical at that time (Bell 2001). About 20 years later, this was made possible in 2006–2007 due to the capacity built up during the Deserta Grande and Selvagem Grande projects carried out by the local nature conservation authorities.

During these projects, bait was hand laid by a combination of field operators and aerial broadcast. The inaccessibility of Bugio's cliffs was overcome by hand broadcast from a helicopter. A specific new technique was used to accommodate the bait inside paper bags to ensure that part of it would stay near the impact area for mice, and another part would be spread for rabbits. It is also worth mentioning that goats were indirectly eradicated because they also consumed the bait, due to scarcity of their usual food at the time (Oliveira and Menezes 2012). As a result of these efforts, recovery of the breeding habitat of Bugio's Desertas petrel is now evident. Accurate monitoring of the population is challenging on Bugio, but the existing data shows that it is stable (D. Menezes, pers. comm.), and that soil erosion is now controlled and overall habitat conditions have improved greatly.

As a final note, it is evident that worldwide, many seabird species and their habitats have strongly benefitted from eradicating and controlling invasive vertebrate species.

Acknowledgments

We thank Sandra Hervías for conducting a critical revision of the last version. This manuscript has been edited by Guido Jones, currently funded by the Cabildo de Tenerife, under the TFinnova Programme supported by MEDI and FDCAN funds.

References cited

Báez, M. and L. Sánchez-Pinto. 1983. Islas de Fuego y Agua. Canarias, Azores, Madeira, Salvajes, Cabo Verde. La Macaronesia. Editorial Regional Canaria, Las Palmas de Gran Canaria, Spain.

Barros, A., R. Romero, I. Munilla, C. Pérez and A. Velando. 2016. Behavioural plasticity in nest-site selection of a colonial seabird in response to an invasive carnivore. Biol. Invasions 18: 3149–3161. doi: 10.1007/s10530-016-1205-3.

Bell, B.D. 1995. The effects of goats and rabbits on breeding seabirds: Methods of eradication and control. Bol. Mus. Mun. Funchal Sup. 4: 83–95.

Bell, B.D. 2001. Removal of rabbits from Deserta Grande Island, Madeira Archipelago. Arquipélago. Life and Marine Sciences. Supplement 2 (Part B): 115–117. Ponta Delgada, Azores.

Bell, E., K. Floyd, D. Boyle, J. Pearson, P. St. Pierre, L. Lock et al. 2019. The Isles of Scilly seabird restoration project: The eradication of brown rats (*Rattus norvegicus*) from the inhabited islands of St Agnes and Gugh, Isles of Scilly. pp. 88–94. *In*: Veitch, C.R., M.N. Clout, A.R. Martin, J.C. Russell and C.J. West [eds.]. Island Invasives: Scaling up to Meet the Challenge, Occasional Paper SSC no. 62. IUCN, Gland, Switzerland

Bellard, C., P. Genovesi and J.M. Jeschke. 2016. Global patterns in threats to vertebrates by biological invasions. Proc. R. Soc. B 283: 20152454. doi: 10.1098/rspb.2015.2454.

Berny P., A. Esther, J. Jacob and C. Prescott. 2018. Development of resistance to anticoagulant rodenticides in rodents. pp. 259–286. *In*: van den Brink, N., J. Elliott, R. Shore and B. Rattner [eds.]. Anticoagulant Rodenticides and Wildlife. Emerging Topics in Ecotoxicology (Principles, Approaches and Perspectives) vol. 5. Springer, Cham, Switzerland. doi: 10.1007/978-3-319-64377-9_10.

BirdLife International. 2021. Species factsheet: *Pterodroma deserta*. Downloaded from http://www.birdlife.org on 07/03/2021.

Bolton, M., A. Stanbury, A.M.M. Baylis and R. Cuthbert. 2014. Impact of introduced house mice (*Mus musculus*) on burrowing seabirds on Steeple Jason and Grand Jason Islands, Falklands, South Atlantic. Polar Biol. 37: 1659–1668. doi: 10.1007/s00300-014-1554-2.

Borges, P., C. Abreu, A. Aguiar, P. Carvalho, R. Jardim, I. Melo et al. [eds.]. 2008. A list of the Terrestrial Fungi, Flora and Fauna of Madeira and Selvagens Archipelagos. Direcção Regional do Ambiente da Madeira and Universidade dos Açores, Funchal and Angra do Heroísmo, Portugal.

Borges, P., A. Costa, R. Cunha, R. Gabriel, V. Gonçalves, A. Martins et al. [eds.]. 2010. A List of the Terrestrial and Marine Biota from the Azores. Princípia, Cascais, Portugal.

Campbell, K. and C.J. Donlan. 2005. Feral goat eradications on islands. Conserv. Biol. 19: 1362–1374. doi: 10.1111/j.1523-1739.2005.00228.x.

Carlile, N., D. Priddel, F. Zino, C. Natividad and D.B. Wingate. 2003. A review of four successful recovery programmes for threatened, sub-tropical petrels. Marine Ornithol. 31: 185–192.

Caravaggi, A., R.J. Cuthbert, P.G. Ryan, J. Cooper and A.L. Bond. 2019. The impact of introduced house mice on the breeding success of nesting seabirds on Gough Island. Ibis 161: 648–661. doi: 10.1111/ibi.12664.

Carboneras, C., P. Genovesi, M. Vilà, T.M. Blackburn, M. Carrete, M. Clavero et al. 2018. A prioritised list of invasive alien species to assist the effective implementation of EU legislation. J. Appl. Ecol. 55: 539–547. doi: 10.1111/1365-2664.12997.

Caut, S., E. Angulo and F. Courchamp. 2008. Dietary shift of and invasive predator: Rats, seabirds and sea turtles. J. Appl. Ecol. 45: 428–437. doi: 10.1111/j.1365-2664.2007.01438.x.

Chan, C.J. 2013. Interactions between invasive mammals and their effects on red-billed tropicbird (*Phaeton aethereus*) nesting productivity. Ms. Thesis, Imperial College London, London, UK

Chapuis, J.L., P. Boussès and G. Barnaud. 1994. Alien mammals, impact and management in the French Subantarctic Islands. Biol. Conserv. 67: 97–104. doi: 10.1016/0006-3207(94)90353-0.

Clout, M.N. 2002. Biodiversity loss caused by invasive alien vertebrates. Z. Jagdwiss. 48, Supp: 51–58. doi: 10.1007/BF02192392.

Clout, M.N. and J.C. Russell. 2007. The invasion ecology of mammals: A global perspective. Wildl. Res. 35: 180–184. doi: 10.1071/WR07091.

Cooke, B. 1998. Rabbit haemorragic disease: Advances in field epidemiology. pp. 86. *In:* Reig, S. [ed.]. Abstracts Euro-American Mammal Congress, Santiago de Compostela, Spain.

Courchamp, F., M. Langlais and G. Sugihara. 2000. Rabbits killing birds: Modelling the hyperpredation process. J. Ani. Ecol. 69: 154–164. doi: 10.1046/j.1365-2656.2000.00383.x.

Courchamp, F., J.-L. Chapuis and M. Pascal. 2003. Mammal invaders on islands: impact, control and control impact. Biol. Rev. 78: 347–383. doi: 10.1017/S1464793102006061.

Croxall, J.P., S.H.M. Butchart, B. Lascelles, A.J. Stattersfield, B. Sullivan, A. Symes et al. 2012. Seabird conservation status, threats and priority actions: A global assessment. Bird Conserv. Int. 22: 1–34. doi: 10.1017/S0959270912000020.

Cuthbert, R. and G. Hilton. 2004. Introduced house mice *Mus musculus*: A significant predator of threatened and endemic birds on Gough Island, South Atlantic Ocean? Biol. Conserv. 117: 483–489. doi: 10.1016/j.biocon.2003.08.007.

Dias, M.P., R. Martin, E.J. Pearmain, I.J. Burfield, C. Small, R.A. Phillips et al. 2019. Threats to seabirds: A global assessment. Biol. Conserv. 237: 525–537. doi: 10.1016/j.biocon.2019.06.033.

Direcção Geral do Ambiente. 2004. Livro Branco Sobre o Estado do Ambiente em Cabo Verde. Ministério do Ambiente Agricultura e Pesca, Praia, Santiago, Cabo Verde.

Doherty, T.S., A.S. Glen, D.G. Nimmo, E.G. Ritchie and C.R. Dickman. 2016. Invasive predators and global biodiversity loss. PNAS 113: 11261–11265. doi: 10.1073/pnas.1602480113.

Duhr, M., E.N. Flint, S.A. Hunter, R.V. Taylor, B. Flanders, G. Howald et al. 2019. Control of house mice preying on adult albatrosses at Midway Atoll National Wildlife Refuge. pp. 21–25. *In*: Veitch, C.R., M.N. Clout, A.R. Martin, J.C. Russell and C.J. West [eds.]. Island Invasives: Scaling up to Meet the Challenge, Occasional Paper SSC no. 62. IUCN, Gland, Switzerland

Eason, C.T., A. Miller, S. Ogilvie and A. Fairweather. 2011. An updated review of the toxicology and ecotoxicology of sodium fluoroacetate (1080) in relation to its use as a pest control tool in New Zealand. N. Z. J. Ecol. 35: 1–20. https://www.jstor.org/stable/24060627.

Eason, C.T., A. Miller, D.B. MacMorran and E.C. Murphy. 2014. Toxicology and ecotoxicology of para-aminopropiophenone (PAPP)—a new predator control tool for stoats and feral cats in New Zealand. N. Z. J. Ecol. 38: 177–188. https://www.jstor.org/stable/24060796.

Fisher, P., K.J. Campbell, G.R. Howaldand and B. Warburton. 2019. Anticoagulant rodenticides, islands and animal welfare accountancy. Animals 9: 919. doi: 10.3390/ani9110919.

Forsyth, D.M., A.J. Robley and B. Reddiex. 2005. Review of Methods Used to Estimate the Abundance of Feral Cats. Arthur Rylah Institute for Environmental Research, Department of Sustainability and Environment, Melbourne, Australia.

Fukami, T., D.A. Wardle, P.J. Bellingham, C.P.H. Mulder, D.R. Towns, G.W. Yeates et al. 2006. Above- and below-ground impacts of introduced predators in seabird-dominated island ecosystems. Ecol. Lett. 9: 1299–1307. doi: 10.1111/j.1461-0248.2006.00983.x.

Genovesi, P. 2001. Guidelines for the Eradication of Terrestrial Vertebrates: a European Contribution to the Invasive Alien Species Issue. Council of Europe, Strasbourg, France.

Genovesi, P. and S. Bertolino. 2001. Human dimension aspects in invasive alien species issues: The case of the failure of the grey squirrel eradication project in Italy. pp. 113–120. *In*: McNeely, J.A. [ed.]. The Great Reshuffling: Human Dimensions of Invasive Alien Species. IUCN, Gland, Switzerland and Cambridge, UK.

Genovesi, P. 2005. Eradications of invasive alien species in Europe: A review. Biol. Invasions 7: 127–133. doi: 10.1007/s10530-004-9642-9.

Genovesi, P., L. Carnevali and R. Scalera. 2015a. The Impact of Invasive Alien Species on Native Threatened Species in Europe. Technical report for the European Commission, ISPRA – ISSG, Rome, Italy.

Genovesi, P., C. Carboneras, M. Vilà and P. Walton. 2015b. EU adopts innovative legislation on invasive species: A step towards a global response to biological invasions? Biol. Invasions 17: 1307–1311. doi: 10.1007/s10530-014-0817-8.

Gherardi, F. and C. Angiolini. 2004. Eradication and control of invasive species. pp. 271–299. *In*: Gherardi, F., M. Gualtieri and C. Corti [eds.]. Biodiversity Conservation and Habitat Management. Eolss Publishers, Oxford, UK.

Grunwald, H.A., V.M. Gantz, G. Poplawski, X. Xu, E. Bier and K.L. Cooper. 2019. Super-mendelian inheritance mediated by CRISPR–Cas9 in the female mouse germline. Nature 566: 105–109. doi: 10.1038/s41586-019-0875-2.

Herder, J., A. Valentini, E. Bellemain, T. Dejean, J. Delft, P. Thomsen and P. Taberlet. 2014. Environmental DNA—a review of the possible applications for the detection of (invasive) species. Report 2013–104. Stichting RAVON, Nijmegen.

Hervías, S., A. Henriques, N. Oliveira, T. Pipa, H. Cowen, J.A. Ramos et al. 2013a. Studying the effects of multiple invasive mammals on Cory's shearwaters nest survival. Biol. Invasions 15: 143–155. doi: 10.1007/s10530-012-0274-1.

Hervías, S., J.A. Ramos, M. Nogales and R. Ruiz de Ybáñez. 2013b. Effect of exotic mammalian predators on parasites of Cory´s shearwater: Ecological effect on population health and breeding success. Parasitol. Res. 112: 2721–2730. doi: 10.1007/s00436-013-3443-y.

Hilton, G.M. and R.J. Cuthbert. 2010. The catastrophic impact of invasive mammalian predators on birds of the UK Overseas Territories: A review and synthesis. Ibis 152: 443–458. doi: 10.1111/j.1474-919X.2010.01031.x.

Holmes, N.D., D.R. Spatz, S. Oppel, B. Tershy, D.A. Croll, B. Keitt et al. 2019. Globally important islands where eradicating invasive mammals will benefit highly threatened vertebrates. PLoS ONE 14: e0212128. doi: 10.1371/journal.pone.0212128.

Hostetler, M., S. Wisely, S. Johnson, E. Pienaar and M. Main. 2020. How effective and humane is Trap-Neuter-Release (TNR) for feral cats? EDIS 2: 8. doi: 10.13140/RG.2.2.35571.27682.

Howald, G., C. Donlan, J.P. Galvan, J. Russell, J. Parkes, A. Samaniego-Herrera et al. 2007. Invasive rodent eradication on Islands. Conserv. Biol. 21: 1258–1268. doi: 10.1111/j.1523-1739.2007.00755.x.

Hulme, P.E. 2007. Biological invasions in Europe: Drivers, pressures, states, impacts and responses. pp. 56–80. *In*: Hester, R.E. and R.M. Harrison [eds.]. Biodiversity UnderThreat. Issues in Environmental Science and Technology, 2007, 25. The Royal Society of Chemistry, Cambridge, UK.

Huntley, B.J. 1996. South Africa's experience regarding alien species: impact and controls. pp. 182–188. *In:* Sandlund, O.T., P.J. Schei and A. Viken [eds.]. Proceedings of the Norway/UN Conference on Alien Species 1996. DN and NINA, Trondheim, Norway.

Hyun, C.U., M. Park and W.Y. Lee. 2020. Remotely piloted aircraft system (RPAS)-based wildlife detection: A review and case studies in maritime Antarctica. Animals 10: 2387. doi: 10.3390/ani10122387.

Illera, J.C., J.C. Rando, D.S. Richardson and B.C. Emerson. 2012. Age, origins and extinctions of the avifauna of Macaronesia: A synthesis of phylogenetic and fossil information. Quat. Sci. Rev. 50: 14–22. doi: 10.1016/j.quascirev.2012.07.013.

Imber, M., M. Harrison and J. Harrison. 2000. Interactions between petrels, rats and rabbits on Whale Island, and effects of rat and rabbit eradication. N. Z. J. Ecol. 24: 153–160. https://www.jstor.org/stable/24054669.

IUCN. 2000. Guidelines for the prevention of biodiversity loss caused by alien invasive species. Prepared by the SSC Invasive Species Specialist Group. Approved by the 51st Meeting of the IUCN Council, Gland Switzerland.

IUCN. 2020. The IUCN Red List of Threatened Species. Version 2020–3. https://www.iucnredlist.org. Downloaded on [3/2/2021].

Jesus, J., D. Menezes, S. Gomes, P. Oliveira, M. Nogales and A. Brehm. 2009. Phylogenetic relationships of gadfly petrels *Pterodroma* spp. From the Northeastern Atlantic Ocean: Molecular evidence for specific status of Bugio and Cape Verde petrels and implications for conservation. Bird Conserv. Int. 19: 199–214. doi: 10.1017/S0959270909008296.

Jones, H.P., B.R. Tershy, E.S. Zavaleta, D.A. Croll, B.S. Keitt, M.E. Funkelstein et al. 2008. Severity of the effects of invasive rats on seabirds: A global review. Conserv. Biol. 22: 16–26. doi: 10.1111/j.1523-1739.2007.00859.x.

Jones, H.P., N.D. Holmes, S.H.M. Butchart, B.R. Tershy, P.J. Kappes, I. Corkery et al. 2016. Invasive mammal eradication on islands results in substantial conservation gains. Proc. Natl. Acad. Sci. USA 113: 4033–4038. doi: 10.1073/pnas.1521179113.

Leitschuh, C.M., D. Kanavy, G.A. Backus, R.X. Valdez, M. Serr, E.A. Pitts et al. 2018. Developing gene drive technologies to eradicate invasive rodents from islands. J. Responsible Innov. 5, sup. 1: S121–S138. doi: 10.1080/23299460.2017.1365232.

Lever, C. 1994. Naturalised Animals. Poyser, London, UK.

Lim, G. 2017. Warfarin: from rat poison to clinical use. Nat. Rev. Cardiol. 172. doi: 10.1038/nrcardio.2017.172.

Lodge, D.M., S.L. Williams, H.J. MacIsaac, K.R. Hayes, B. Leung, S. Reichard et al. 2006. Biological invasions: Recommendations for U.S. policy and management. Ecol. Appl. 16: 2035–2054. doi: 10.1890/1051-0761(2006)016[2035:BIRFUP]2.0.CO;2.

Long, J.L. 2003. Introduced Mammals of the World: Their History, Distribution and Influence. CSIRO Publishing, Melbourne, Australia.

Major, H.L., I.L. Jones, G.V. Byrd and J.C. Williams. 2006. Assessing the effects of introduced Norway rats (*Rattus norvegicus*) on survival and productivity of least auklets (*Aethia pusilla*). The Auk 123: 681–694. doi: 10.1093/auk/123.3.681.

Masseti, M. 2010. Mammals of the Macaronesian islands (the Azores, Madeira, the Canary and Cape Verde islands): Redefinition of the ecological equilibrium. Mammalia 74: 3–34. doi: 10.1515/MAMM.2010.011.

Maxon, S.J. and L.W. Oring. 1978. Mice as a source of egg loss among ground-nesting birds. The Auk 95: 582–584.

McCarthy, R.J., S.H. Levine and J.M. Reed. 2013. Estimation of effectiveness of three methods of feral cat population control by use of a simulation model. J. Am. Vet. Med. Assoc. 243: 502–511. doi: 10.2460/javma.243.4.502.

McChesney, G.J. and B.R. Tershy. 1998. History and status of introduced mammals and impacts to breeding seabirds on the California Channel and Northwestern Baja California Islands. Col. Waterbirds 21: 335–347. doi: 10.2307/1521646.

McCreless, E.E., D.D. Huff, D.A. Croll, B.R. Tershy, D.R. Spatz, N.D. Holmes et al. 2016. Past and estimated future impact of invasive alien mammals on insular threatened vertebrate populations. Nat. Commun. 7: 12488. doi: 10.1038/ncomms12488.

McGeoch, M.A., S.H. Butchart, D. Spear, E. Marais, E.J. Kleynhans and A. Symes. 2010. Global indicators of biological invasion: species numbers, biodiversity impact and policy responses. Divers. Distrib. 16: 95–108. doi: 10.1111/j.1472-4642.2009.00633.x.

Medina, F.M., E. Bonnaud, E. Vidal, B.R. Tershy, E.S. Zavaleta, C.J. Donlan et al. 2011. A global review of the impacts of invasive cats on islands endangered vertebrates. Glob. Chang. Biol. 17: 3503–3510. doi: 10.1111/j.1365-2486.2011.02464.x.

Medina, F.M., E. Bonnaud, E. Vidal and M. Nogales. 2014. Underlying impacts of invasive cats on islands: Not only a question of predation. Biodivers. Conserv. 23: 327–342. doi: 10.1007/s10531-013-0603-4.

Menezes, D., P. Oliveira and I. Ramírez. 2010. Pterodromas do Arquipélago da Madeira. Duas Espécies em Recuperação. Serviço do Parque Natural da Madeira, Funchal, Madeira, Portugal.

Monteiro, L.R., J.A. Ramos and R.W. Furness. 1996. Past and present status and conservation of the seabirds breeding in the Azores archipelago. Biol. Conserv. 78: 319–328. doi: 10.1016/S0006-3207(96)00037-7.

Moore, N.P., S.S. Roy and A. Helyar. 2003. Mink (*Mustela vison*) eradication to protect ground-nesting birds in the Western Isles, Scotland, United Kingdom. N. Z. J. Zool. 30: 443–452. doi. 10.1080/03014223.2003.9518351.

Morley C.G., J. Broadley, R. Hartley, D. Herries, D. McMorran and I.G. McLean. 2017. The potential of using Unmanned Aerial Vehicles (UAVs) for precision pest control of possums (*Trichosurus vulpecula*). Rethinking Ecol. 2: 27–39. doi: 10.3897/rethinkingecology.2.14821.

Nogales, M., J.L. Rodríguez-Luengo and P. Marrero. 2006. Ecological effects and distribution of invasive non-native mammals on the Canary Islands. Mammal Rev. 36: 49–65. doi: 10.1111/j.1365-2907.2006.00077.x.

Nogales, M., I. Castañeda, M. López-Darias, F.M. Medina and E. Bonnaud. 2015. The unnoticed effect of a top predator on complex mutualistic ecological interactions. Biol. Invasions 17: 1655–1665. doi: 10.1007/s10530-014-0823-x.

Oliveira, P., D. Menezes, R. Trout, A. Buckle, P. Geraldes and J. Jesus. 2010. Successful eradication of the European rabbit (*Oryctolagus cuniculus*) and house mouse (*Mus musculus*) from the island of Selvagem Grande (Macaronesian archipelago), in the Eastern Atlantic. Integr. Zool. 5: 70–83. doi: 10.1111/j.1749-4877.2010.00186.x.

Oliveira, P. and D. Menezes. 2012. Eradication and control of vertebrate invasive species in Madeira and the Selvagens Archipelago: Short review. Airo 22: 43–48.

Orueta, J.F. and Y.A. Ramos. 2001. Methods to control and eradicate non-native terrestrial vertebrate species. Nature and Environment, No. 118. Council of Europe Publishing, Strasbourg, France.

Pech, R.P. 1996. Managing alien species: The Australian experience. pp. 198–203. *In*: Sandlund, O.T., P.J. Schei and A. Viken [eds.]. Proceedings of the Norway/UN conference on alien species 1996. DN and NINA, Trondheim, Norway.

Pelz, H.J., S. Rost, M. Hünerberg, A. Fregin, A.C. Heiberg, K. Baert et al. 2005. The genetic basis of resistance to anticoagulants in rodents. Genetics 170: 1839–1847. doi: 10.1534/genetics.104.040360.

Pieper, H. 1985. The fossil land birds of Madeira and Porto Santo. Bocagiana 88: 1–6.

Pierce, R., E. VanderWerf, S. Cranwell, K. Taabu, T. Ghestemme and T. Withers. 2020. A Conservation Action Plan for Two Endangered Seabirds – Phoenix Petrel (*Pterodromaalba*) and Polynesian Storm-petrel (*Nesofregetta fuliginossa*), 2020-2025. Eco Oceania Pty Ltd., Queensland, Australia.

Pitt, W. and G. Witmer. 2014. Invasive vertebrate species and the challenges of management. Proceedings of the Vertebrate Pest Conference 26. Doi: 10.5070/V426110455.

Preston, G.R., B.J. Dilley, J. Cooper, J. Beaumont, L.F. Chauke, S.L. Chown et al. 2019. South Africa works towards eradicating introduced house mice from sub-Antarctic Marion Island: the largest island yet attempted for mice. pp. 40–46. *In*: Veitch, C.R., M.N. Clout, A.R. Martin, J.C. Russell and C.J. West [eds.]. Island Invasives: Scaling up to Meet the Challenge, Occasional Paper SSC no. 62. IUCN, Gland, Switzerland.

Priddel, D., N. Carlile and R. Wheeler. 2000. Eradication of european rabbits (*Oryctolagus cuniculus*) from Cabbage Tree Island, NSW, Australia, to protect the breeding habitat of Gould's petrel (*Pterodroma leucoptera leucoptera*). Biol. Conserv. 94: 115–125. doi: 10.1016/S0006-3207(99)00155-X.

Ratcliffe, N., F.J. Zino, P. Oliveira, A, Vasconcelos, C.J. Hazevoet, H.C. Neves et al. 2000. The status and distribution of Fea's petrel *Pterodroma feae* in the Cape Verde Islands. Atl. Seab. 2: 73–86.

Read, J.L., T. Bowden, P. Hodgens, M. Hess, H. McGregor and K Moseby. 2019. Target specificity of the felixer grooming "trap". Wildl. Soc. Bull. 43: 112–120. doi: 10.1002/wsb.942.

Rodríguez, A., J.M. Arcos, V. Bretagnolle, M.P. Dias, N.D. Holmes, M. Louzao et al. 2019. Future directions in conservation research on petrels and shearwaters. Front. Mar. Sci. 6: 94. doi: 10.3389/fmars.2019.00094.

Ruffino, L., K. Bourgeois, E. Vidal, C. Duhem, M. Paracuellos, F. Escribano et al. 2009. Invasive rats and seabirds after 2,000 years of an unwanted coexistence on Mediterranean islands. Biol. Invasions 11: 1631–1651. doi: 10.1007/s10530-008-9394-z.

Russell, J.C. 2011. Indirect effects of introduced predators on seabird islands. pp. 261–279. *In*: Mulder, C.P.H., W.B. Anderson, D.R. Towns and P.J. Bellingham [eds.]. Seabird Islands: Ecology, Invasion, and Restoration. Oxford University Press, New York, USA.

Russell, J.C. and M. Le Corre. 2009. Introduced mammal impacts on seabirds in the ÎlesÉparses, Western Indian Ocean. Marine Ornithol. 37: 121–129.

Russell J.C., J.G. Innes, P.H. Brown and A.E. Byrom. 2015. Predator-free New Zealand: Conservation country. BioScience 65: 520–525. doi: 10.1093/biosci/biv012.

Sheppard, A.W., Q. Paynter, P. Mason, S. Murphy, P. Stoett, P. Cowan et al. 2019. The application of biological control for the management of established invasive alien species causing environmental impacts. IUCN SSC Invasive Species Specialist Group. The Secretariat of the Convention on Biological Diversity. Technical Series No. 91. Montreal, Canada.

Silva, L., E. Ojeda Land and J.L. Rodríguez Luengo [eds.]. 2008. Invasive Terrestrial Flora and Fauna of Macaronesia. TOP 100 in Azores, Madeira and Canarias. ARENA, Ponta Delgada, Portugal.

Spatz, D.R., K.M. Zilliacus, N.D. Holmes, S.H.M. Butchart, P. Genovesi, G. Ceballos et al. 2017. Globally threatened vertebrates on islands with invasive species. Sci. Adv. 3: e1603080. doi: 10.1126/sciadv.1603080.

Suárez, O.V. and G.R. Cueto. 2018. Comparison of efficacy of second-generation anticoagulant rodenticides: Effect of active ingredients, type of formulation and commercial suppliers. Cogent Food and Agriculture 4: 1525147. doi: 10.1080/23311932.2018.1525147.

Swinnerton, K.J., H.G. Young and P. Sangan. 2018. Options for the recovery of nesting seabirds on Jersey, Channel Islands. Report to Durrell Wildlife Conservation Trust, Jersey, France.

Tershy, B.R., K.-W. Shen, K.M. Newton, N.D. Holmes and D.A. Croll. 2015. The importance of islands for the protection of biological and linguistic diversity. Bioscience 65: 592–597. doi: 10.1093/biosci/biv031.

Thoresen, J., D. Towns, S. Leuzinger, M. Durrett, C.P.H. Mulder and D.A. Wardle. 2017. Invasive rodents have multiple indirect effects on seabird island invertebrate food web structure. Ecol. Appl. 27: 1190–1198. doi: 10.1002/eap.1513.

Towns, D.R., I.A.E. Atkinson and C.H. Daugherty. 2006. Have the harmful effects of introduced rats on islands been exaggerated? Biol. Invasions 8: 863–891. doi: 10.1007/s10530-005-0421-z.

Towns, D.R., D.A. Wardle, C.P.H. Mulder, G.W. Yeates, B.M. Fitzgerald, D.R. Parrish et al. 2009. Predation of seabirds by invasive rats: Multiple indirect consequences for invertebrate communities. Oikos 118: 420–430. doi: 10.1111/j.1600-0706.2008.17186.x.

Towns, D.R., G.V. Byrd, H.P. Jones, M.J. Rauzon, J.C. Russell and C. Wilcox. 2011. Impacts of introduced predators on seabirds. pp. 56–90. *In*: Mulder, C.P.H., W.P. Anderson, D.R. Towns and P.J. Bellingham [eds.]. Seabird Island: Ecology, Invasion, and Restoration. Oxford University Press, New York, USA.

Trout, R.C., J. Ross, A.M. Tittensor and A.P. Fox. 1992. The effect on a British wild rabbit population (*Oryctolagus cuniculus*) of manipulating myxomatosis. J. Appl. Ecol. 29: 679–686. doi: 10.2307/2404476.

Van Rensburg, P.J.J., J.D. Skinner and R.J. van Aarde. 1987. Effects of feline panleucopaenia on the population characteristics of feral cats on Marion Island. J. Appl. Ecol. 24: 63–73. doi: 10.2307/2403787.

Wanless, R.N., A. Angel, R.J. Cuthbert, G.M. Hilton and P.G. Ryan. 2007. Can predation by invasive mice drive seabird extinctions? Biol. Lett. 3: 241–244. doi: 10.1098/rsbl.2007.0120.

Wanless, R.N., N. Ratcliffe, A. Angel, B.C. Bowie, K. Cita, G.M. Hilton et al. 2012. Predation of Atlantic petrel chicks by house mouse on Gough Island. Anim. Conserv. 15: 472–479. doi: 10.1111/j.1469-1795.2012.00534.x.

Warburton, B. and I. Orchard. 1996. Evaluation of five kill traps for effective capture and killing of Australian brushtail possums (*Trichosurus vulpecula*). New Zealand Journal of Zoology 23: 4, 307–314. doi: 10.1080/03014223.1996.9518089.

Ward, S., A.M.V. Fournier and A.L. Bond. 2019. Assessing gaps in reporting non-target mortality in island rodent eradication operations. Biol. Invasions 21: 3101–3108. doi: 10.1007/s10530-019-02032-7.

Wardle, D.A., P.J. Bellingham, T. Fukami and C.P.H. Mulder. 2007. Promotion of ecosystem carbon sequestration by invasive predators. Biol. Lett. 3: 479–482. doi: 10.1098/rsbl.2007.0163.

Wardle, D.A., P.J. Bellingham, K.I. Bonner and C.P.H. Mulder. 2009. Indirect effects of invasive predators on litter decomposition and nutrient resorption on seabird-dominated islands. Ecology 90: 452–464. doi: 10.1890/08-0097.1.

Witmer, G.W. and R.S. Moulton. 2014. Improving invasive house mice control and eradication strategies via more effective rodenticides. pp. 67–72. *In*: Timm, R.M. and J.M. O´Brien [eds.]. Proceedings of the 26th Vertebrate Pest Conference. University of California, Davis, California, USA.

Witt, R.R., C.T. Beranek, L.G. Howell, S.A. Ryan, J. Clulow, N.R. Jordan et al. 2020. Real-time drone derived thermal imagery outperforms traditional survey methods for an arboreal forest mammal. PLoS One 15: e0242204. doi: 10.1371/journal.pone.0242204.

Work, T.M., M. Duhr and B. Flint. 2021. Pathology of house mouse (*Mus musculus*) predation on Laysan albatross (*Phoebastria immutabilis*) on Midway Atoll National Wildlife Refuge. J. Wildl. Dis. 57: 125–131. doi: 10.7589/JWD-D-20-00065.

Zino, F., B. Heredia and M.J. Biscoito. 1996. Action plan for Fea's Petrel (*Pterodroma feae*). pp. 25–31. *In*: Heredia, B., L. Rose and M. Painter [eds.]. Globally Threatened Birds in Europe: Action Plans. Council of Europe and BirdLife International, Strasbourg, France.

Zino, F., M.V. Hounsome, A.P. Buckle and M. Biscoito. 2008. Was the removal of rabbits and house mice from Selvagem Grande beneficial to the breeding of Cory's Shearwater *Calonectris diomedea borealis*? Oryx 42: 151–154. doi: 10.1017/S0030605308000665.

CHAPTER 3

Identifying and Establishing Marine Protected Areas Worldwide

The Contribution of Seabird Data

Lucas Krüger

Introduction

The need for Marine Protected Areas

The influence of humans over planetary ecosystems has been growing substantially to the point that human-induced changes now can be recorded even at the remotest corners on earth. Halpern et al. (2015, 2019) estimated that 97.7% of global ocean surface is affected by multiple stressors: more than 40% of the ocean experiences medium to very high impacts, and over the last decade 66% of the ocean has experienced increasing cumulative impacts. Such impacts have triggered abrupt shifts in marine ecosystems (Beaugrand et al. 2019) risking the balance of fundamental ecosystem services (i.e., climate regulation, carbon absorption, oxygen generation) to human populations (Lenton et al. 2019, Myers 2017, Steffen et al. 2015). Most widespread stressors are those related to climate change (Halpern et al. 2015), industrial fisheries—75 to 95% of the global ocean is exploited by fisheries (Kroodsma et al. 2018, Tickler et al. 2018, Watson and Tidd 2018), and plastic pollution (van Sebille et al. 2015). Other factors likely with significant impact, include noise pollution (Markus and Sánchez 2018), oil pollution (Zhang et al. 2019) and invasive species (Geburzi and McCarthy 2018).

In order to prevent these environmental changes to jeopardize marine ecosystem services, management of marine exploitation is of utmost importance (Palumbi et al. 2009). Marine Protected Areas (MPAs) are now recognized as one of the most important tools available to manage marine ecosystems. MPAs are zones where certain human activities are controlled, limited or completely excluded with the objective of reducing impacts over important species, communities, ecosystems, ecological processes or geological structures such as seamounts (Laffoley et al. 2018). MPAs can allow ecosystems to recover from impacts/exploitation (Anderson et al. 2014, Cabral et al. 2020, Di Lorenzo et al. 2016), and by reducing pressures over marine organisms on key sites, MPAs can

---

Instituto Antártico Chileno, Programa Áreas Marinas Protegidas, Plaza Muñoz Gamero, 1055, Punta Arenas, Chile.
Email: lkruger@inach.cl

increase the resilience and adaptiveness of ecosystems to changing climate conditions (Roberts et al. 2017, Wilson et al. 2020). For a better view on MPAs benefits see (Cabral et al. 2020, Laffoley et al. 2018, Sala and Giakoumi 2018, Toropova et al. 2010, Wing and Jack 2013).

While the number of MPAs has substantially increased over time, it still represents a small amount of the global ocean (Fig. D3.1). Currently, around 7% of global oceans are protected (Marine Conservation Institute 2021, UNEP-WCMC and IUCN 2021). However, only 2.7% of ocean areas are under fully or highly protected status (Marine Conservation Institute 2021), which is necessary to guarantee optimal ecosystem protection (Costello and Ballantine 2015, Edgar et al. 2014, Sala and Giakoumi 2018). Most increases in MPAs coverage occurred within national waters (Laffoley et al. 2018), but areas beyond national jurisdiction (ABNJs), which represent more than 60% of global oceans, are still poorly protected—some of the oceanic least protected provinces are in ABNJs (UNEP-WCMC and IUCN 2021).

Several countries targeted protecting 10% of global oceans by 2012 (Boonzaier and Pauly 2016, Sala et al. 2018), but that goal has not been achieved even after eight years. However, it is likely that a minimum of 30% of national and 80% of international oceans should be under high protection to avoid human activities jeopardizing the balance of marine ecosystems (Sala and Rechberger 2018); therefore 10% was a low aim. The goal of 30% of global ocean protection was recognized and a resolution was approved by participants of the IUCN World Conservation Congress in 2016 (WCC 2016 Res 050). Recently, this aim was also adopted by 50 countries participating on the High Ambition Coalition for Nature and People for 2030 (Waldron et al. 2020). With less than 3% of the ocean currently fully protected, we are way behind schedule to reach the target of 30%.

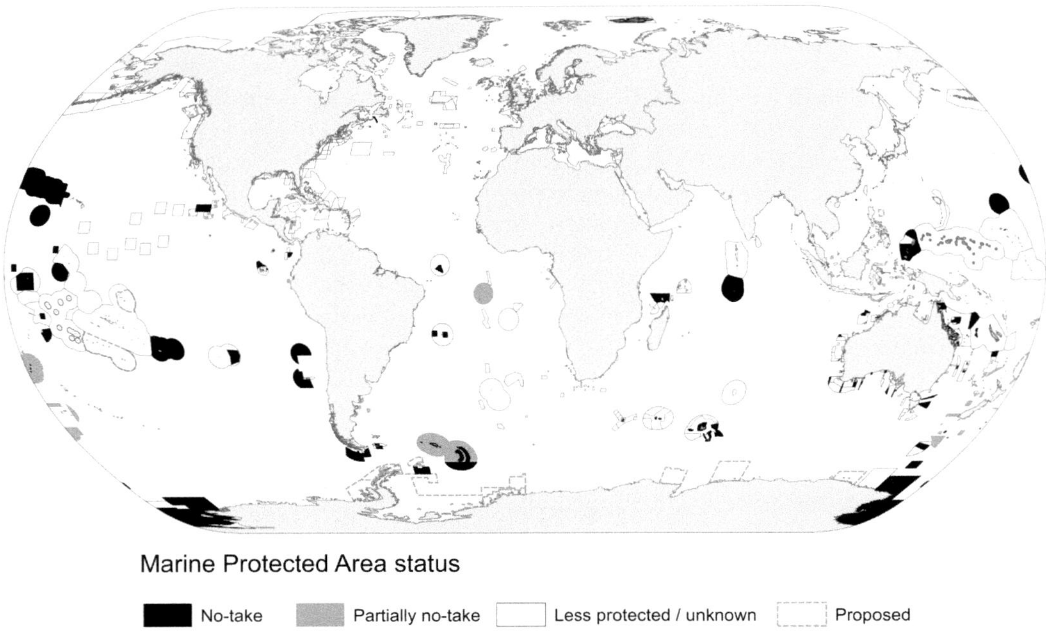

Figure D3.1. Current global network of Marine Protected Areas (MPAs) based on their classification, fully no-take (fishing not allowed), partially no-take, less protected or unknown, and PROPOSED but not implemented MPAs. Data from Marine Conservation Institute (2021) and UNEP-WCMC and IUCN (2021).

What makes an area prone to protection? How to select areas to protect?

Priority areas for conservation are usually identified based on the presence of a representative amount of one or many conservation targets, such as particular ecosystems, significant oceanographic features and processes, species or biological communities of significant importance (Dunn et al. 2014, Dunstan et al. 2016, Sundblad et al. 2011). Targeting umbrella/charismatic species have been historically important for determining areas to protect, and can be a useful method when such species depend on large habitats to fulfil their needs (Cabeza et al. 2008, Sergio et al. 2008, 2006). Therefore, protecting such areas can benefit a large number of other species. Top-predators usually fit this description and have been extensively used as indicators of key marine habitats to protect (Hazen et al. 2019, Hindell et al. 2020).

Implemented Marine Protected Areas aiming at seabird conservation

Several MPAs worldwide have been designed taking into consideration seabirds. Special Protection Areas (Birds Directive) form a network of protected areas in Europe including more than 200,000 km^2 of marine protection that effectively cover the range of many seabird species of at least seven orders (Ramirez et al. 2017). The Pacific Remote Islands Marine National and the Papahānaumokuākea Marine National Monuments are two large MPAs on a significant network on the Central Pacific Ocean that were designed to protect large seabird breeding colonies and foraging areas, covering around 18 million nesting seabirds of 22 species (Young et al. 2015). Young et al. (2015) point out that these two MPAs facilitate the protection of large vagile fish species, although this was not their primary conservation target. One of the largest single MPAs in the world, the South Georgia and South Sandwich Islands (SGSS), was designed taking into consideration several conservation targets (GSGSSI 2013, Handley et al. 2020) that include significant populations of several threatened seabird species, such as the Wandering Albatross (*Diomedea exulans*), Grey-headed Albatross (*Thalassarche chrysostoma*), White-chinned Petrel (*Procellaria aequinoctialis*) and Macaroni Penguin (*Eudyptes chrysolophus*). This MPA has a zonation system for fishing activities, including no-take areas that optimize protection for those species (GSGSSI 2013) from their main impacting factor (Dias et al. 2019). By doing so, the MPA also protects significant proportion of important marine habitat for key zooplankton species, pelagic fish and squids and underwater forests of giant kelps that holds a high diversity of coastal benthic communities (Barnes et al. 2006, GSGSSI 2013).

Why using seabirds to identify Marine Protected Areas?

Useful indicators

Seabirds represent good indicators of ecosystems' state (Hazen et al. 2019, Heerah et al. 2019, Piatt et al. 2007), and their distribution overlap with a wide diversity of other marine organisms (Kuletz et al. 2015, Raymond et al. 2015, Waggitt et al. 2020), including species exploited by humans (Clua and Grosvalet 2001, Miller et al. 2018). For instance, seabirds can exploit similar habitats of several species of marine mammals (Hindell et al. 2020, Patterson et al. 2016), and forage in close association with whales, dolphins, seals (Kuletz et al. 2015, Waggitt et al. 2020) and large fishes such as tunas (Clua and Grosvalet 2001, Miller et al. 2018, Morato et al. 2008). Seabirds rely on certain spatially and temporally predictable oceanic features that are indicators of food availability, such as upwellings (Crawford 2007, Grecian et al. 2016), shelf breakfronts (Cox et al. 2016, Serratosa et al. 2020, Yamamoto et al. 2016), eddies (Assali et al. 2017, van der Boog et al. 2019), gyre edges (Clay et al. 2017, Hátún et al. 2017), freshwater discharge plumes (Daudt et al. 2019, Urbanski et al. 2017) and seamounts (Rogers 2018, Rogers et al. 2017), that usually are patchy zones of higher primary productivity and biodiversity (Condie and Condie 2016, Cox et al. 2018, Weidberg et al. 2020). Those features are important not only from a biodiversity point of view, but they also sustain ecosystem processes that are fundamental for the balance of the biosphere (i.e., Levin and Lubchenco 2008). Therefore, by protecting the most important foraging habitats for seabirds, those processes are consequently protected (Lascelles et al. 2012, Ronconi et al. 2012, Thaxter et al. 2012).

Highly studied

Compared to other pelagic top-predators (for instance cetaceans, sharks or other highly mobile fishes), seabirds are easier to study. As they breed on land, it is easier to access seabirds and study them. As they depend on sea, it is possible for researchers to monitor their responses to oceanic changes (Lascelles et al. 2012). Onboard censuses of seabirds are also easier when compared to other marine predators, as their presence above the water is an obvious advantage for observation and identification.

Populations of some seabirds have been counted over decades and there are some interesting initiatives compiling long-term data (for instance, the Mapping Application for Penguin Populations and Projected Dynamics MAPPPD is perhaps one of the finest examples; Humphries et al. 2017). Similarly, mark-recapture studies have been used for a long time to assess population status (i.e., Gianuca et al. 2019, Rolland et al. 2010, Weimerskirch et al. 2012), providing important long-term information and allowing prioritization of protection based on population trends and vulnerability to change. More contemporary studies now incorporate spatial information of population movements and habitat preferences to make the portrait of population state, trends and vulnerabilities more accurate (Gianuca et al. 2019, Tuck et al. 2015), and providing better background for conservation actions (Carneiro et al. 2020, Oppel et al. 2018).

The use of tracking technology on an increasing number of species has been growing exponentially following technological advances over the last decade (Pereira et al. 2022). As an example, the Birdlife Seabird Tracking database (http://seabirdtracking.org/) has over 14 million points of 132 species tracked with different technologies. Some ocean areas have received periodically at-sea counting of seabirds, providing useful information on distribution and habitat preferences (Oppel et al. 2012, Pereira et al. 2018, Péron et al. 2010). Open data banks and citizen science have also started to become a reliable source of information for data on seabirds (Gorta et al. 2019, Martín et al. 2020, Ward et al. 2015), supporting conservation actions as well (Haywood et al. 2016, McKinley et al. 2017).

Highly threatened

Seabirds are highly threatened organisms (Croxall et al. 2012, Dias et al. 2019): the increasing pressures on oceans makes some groups of marine animals as the most endangered on earth (Butchart et al. 2010, Ripple et al. 2017, Webb and Mindel 2015), and that applies to seabirds (see Días et al. 2019 for a complete review of threats to seabirds). Several seabird species face pressures in both their foraging (at-sea) and nesting (breeding colonies) habitats, which is also the reason for their poor conservation status (Días et al. 2019). When analyzing the IUCN redlist data (IUCN 2021) for species from several bird orders it is possible to see that some seabird families have higher percentage of species under risk of extinction—tubenoses (albatrosses and petrels, Procelariiformes) and penguins (Sphenisciformes) are at the top—and at least two other groups have the highest percentage of species experiencing global decrease (Fig. D3.2). It is important to emphasize that the introduction of invasive species on the nesting sites is an important threat for many seabird species (Dias et al. 2019, Oliveira et al. 2022).

The high vulnerability of some seabird species can be used as a rule of thumb: the more threatened a species is, the highest is the priority that is given to protect its habitat. As a matter of fact, several studies have used such an approach to identify key areas and propose MPAs (Dias et al. 2017, Krüger et al. 2017a, Lascelles et al. 2016).

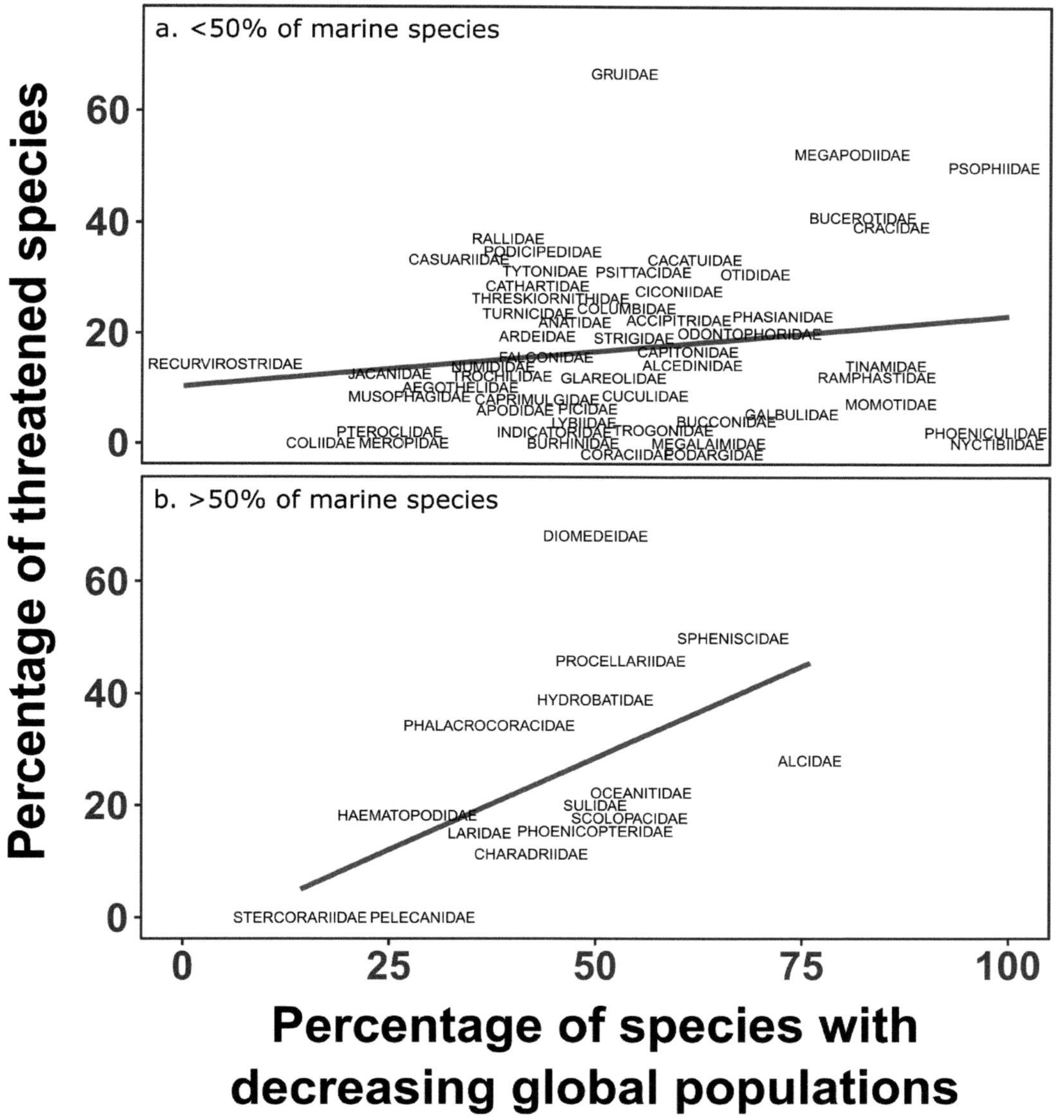

Figure D3.2. Percentage of threatened non-passerine bird species (classified as Vulnerable, Endangered, Critically Endangered or Extinct in Nature) and percentage of species with decreasing global population trends. The figure includes all non-passerine bird families with more than five recognized species. Families with less than 50% of species using any marine habitat (a), and only selected families with a minimum of 50% of species using marine habitats (b). Data from IUCN (2021).

Practical examples of the use of seabirds as MPAs tools

Important Bird and Biodiversity Areas

One of the largest programs to propose key areas for protection using seabirds is the Important Bird and Biodiversity Areas (IBAs) program (Donald et al. 2019, Waliczky et al. 2019). IBAs form a "network of sites that are significant for the long-term viability of naturally occurring bird populations" (Birdlife International 2021). The marine counterpart of the IBAs, the marine IBAs (https://maps.birdlife.org/marineIBAs/), incorporates both population sizes and known population distribution to identify highly important marine areas for seabird species throughout the globe (Arcos et al. 2012, BirdLife International 2009, Lascelles et al. 2016), which, with proper evaluation, are useful layers on spatial planning analysis for prioritization (McGowan et al. 2018). Marine IBAs are based on standardized criteria which facilitates its application worldwide. IBAs criteria are (Donald et al. 2019): (1) sites that regularly hold important numbers of threatened species; (2) sites that hold significant numbers of breeding distribution of endemic species; (3) sites that hold a significant number of biome-restricted species, and; (4) sites known to hold more than 1% of the global population of one or more species. Marine IBAs are mostly based on criteria '1' and '4'. Studies have built upon the marine IBAs criteria incorporating a range of technical approaches and standardization procedures to control for sampling bias, differences in population size, differences in population/species phenology and age of the sampled birds (Dias et al. 2017, Handley et al. 2021, Heerah et al. 2019, Lascelles et al. 2016). Recently, Handley et al. (2020) proposed the concept of Key Biodiversity Areas as an extension of the marine IBA also incorporating data from marine mammals.

Niche Models

Ecological Niche Models ENMs (or suitability models) are used to identify species likely habitat and distribution based on species association with a set of environmental conditions (Guisan et al. 2013, Melo-Merino et al. 2020). The ENMs provide robust insights on the potential distribution of species' habitats (Guisan et al. 2013), which may also be a proxy for abundance or density at certain spatial scales (Weber et al. 2017). In this context, ENMs outputs can be of use for conservation by: (1) identifying potential areas to protect (Cleasby et al. 2020, Engler et al. 2017, Pereira et al. 2018), (2) identifying areas of vulnerability to stressors where management actions are required (Krüger et al. 2017a, Krüger et al. 2018a, Pereira et al. 2018, Welch et al. 2018), (3) identifying optimal habitat for re-introduction of species (including suitable marine habitat surrounding a breeding colony) or nest habitat recovery (Melo-Merino et al. 2020), (4) estimating population sizes (Catry et al. 2003, Rayner et al. 2007, Krüger et al. 2017b), and (5) measuring coverage and/or efficiency of MPAs (Lambert et al. 2017). ENMs are robust and powerful tools whose applicability towards conservation of seabirds are enormous, with the design of MPAs constituting only a single of their multiple applications.

Distribution range maps

The well-known distribution range of several species allows for the use of relatively coarse-scale spatial data to identify diversity hotspots and conservation priorities (Fourcade 2016, Maréchaux et al. 2017, Moudrý and Devillers 2020). At global scales, two studies deserve mention, the studies of Jenkins and Van Houtan (2016) and O'Hara et al. (2019), which identified zones of conservation priority based on range maps of hundreds of species (not only seabirds) weighted by extinction risk and rarity rankings. The highly known distribution range of several seabird species makes them ideal species for such applications. By applying a ranking of seabird biodiversity out of

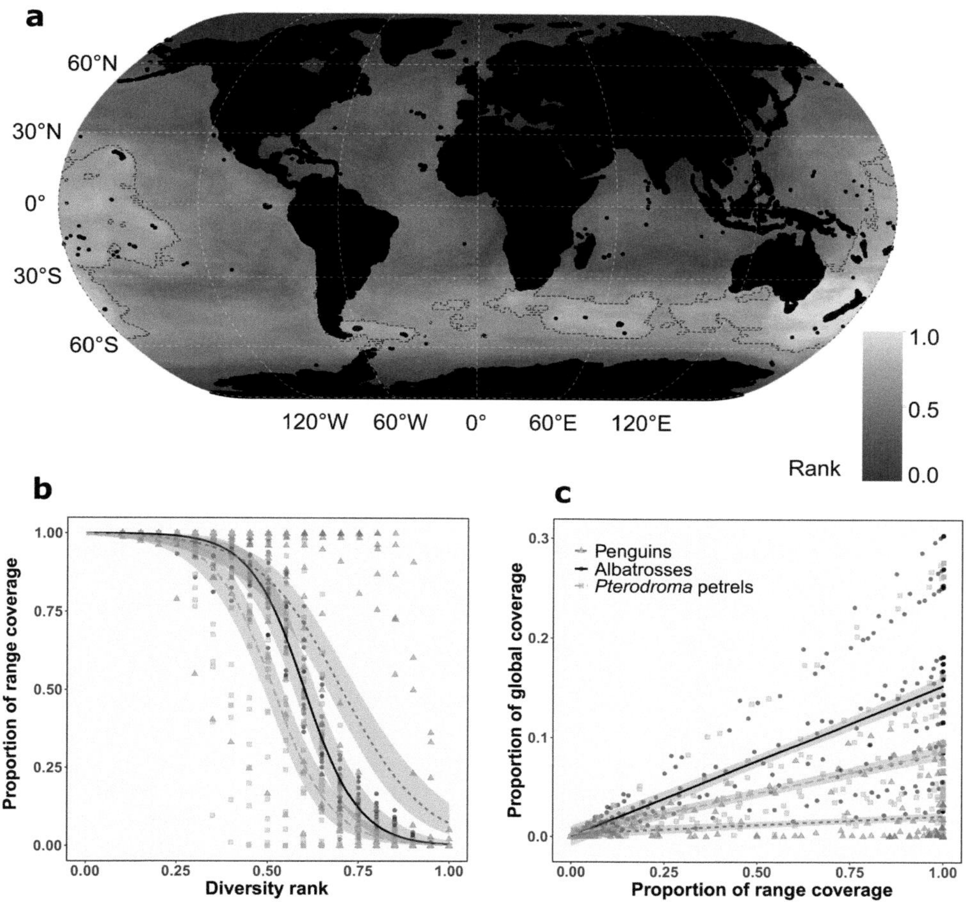

Figure D3.3. Global diversity of seabirds (a) using data from range maps (Birdlife International 2021) of 332 species of seabirds from the orders Procelariiformes (tubenoses), Sphenisciformes (penguins), Suliformes (gannets), Families Pelecanidae and Fregatidae (Pelecaniformes), Laridae (gulls), Stercorariidae (skuas) and Alcidae (auks) and ranked to the maximum value of simultaneous species occurring in any grid cell (values of 1 would mean 57 species occurring simultaneously); proportion of species range covered by the diversity rank in 'a' for groups of highly threatened seabirds (b), and the amount of global area corresponding to increases in coverage of species range (c). Currently, 15 out of 22 Albatross (68.2%), 23 out of 35 *Pterodroma* petrels (65.7%), and 9 out of 18 Penguin (50.0%) species are classified as Critically Endangered, Endangered or Vulnerable, meaning that they are the likely candidates for the most threatened seabirds.

332 species of seabirds in a similar fashion to that of those authors (Fig. D3.3a), it was possible to identify continuous large areas of high seabird diversity as dominant in the southern oceans (south of 40ºS) and central tropical Pacific Ocean waters. Smaller areas of intermediate to high diversity were found in the central tropical Atlantic Oceans, the East coast of South America and Galapagos Islands. The use of such an approach of ranking seabird diversity from range maps to identify conservation priorities (Fig. D3.3b,c) can substantially increase coverage for target species of seabirds (such as species from highly threatened groups). Fully protecting areas with more than 70% of the higher seabird diversity would mean adding more than 40 million km^2 to the existing network of global protection, taking it close to reaching 10% of fully global ocean protection. While the use of range maps is useful to identify large scale areas of potential high diversity, this approach does not consider abundance or density, therefore its use has the disadvantage of limiting coverage of key foraging areas, particularly for highly mobile species (Moudrý and Devillers 2020). For instance, seabirds such as *Pterodroma* petrels and *Diomedea* albatrosses, whose foraging strategy is

to maximize prey encounter by covering long distances instead of focusing on hotspots (i.e., Oppel et al. 2018, Ventura et al. 2020), such an approach would probably have less benefit than to species with more limited distribution or mobility.

Seabirds and marine reserves: challenges

The total area of MPAs coverage is low and likely insufficient to guarantee enough protection and a substantial representation of marine habitats (Sala and Rechberger 2018, Waldron et al. 2020). The existing network is composed mostly of lower protection MPAs which do not provide a real buffer against impacts (Handley et al. 2020, O'Hara et al. 2019). As a consequence, several seabird species still remain largely unprotected, or existing MPA networks fail to cover important parts of their usually large ranges (Critchley et al. 2018, Dias et al. 2017, Mason et al. 2018). Furthermore, while most of the oceans' biodiversity is within national and coastal waters of EEZs (where most of MPAs are located), there is a substantial part of seabirds' range in ABNJs, particularly for some of the most threatened seabirds such as albatrosses (Beal et al. 2021) and gadfly petrels (i.e., Ventura et al. 2021). Proposition/designation of protected areas under such international regime is possible only under the management of few international commissions, and progress usually follows a very slow pace under multinational negotiations (Brooks et al. 2020, Hislop and Jabour 2015, Sylvester and Brooks 2019). Several seabird species also require management actions that moves beyond the scope of the MPAs. Most pelagic seabird species depend on millions of squared kilometres of marine areas, and the implementation of fully protected reserves on a substantial proportion of their ranges is not realistic. For those species, MPAs should be followed by rigorous management actions and implementation of international regulations, for instance, preventing bycatch and depletion of prey (Gaines et al. 2010, Oppel et al. 2018). Similarly, many species would benefit from rigorous conservation actions on their breeding grounds. Invasive predators on seabird colonies are responsible for steep declines in breeding success (Borrelle et al. 2018, Spatz et al. 2014, 2017). Seabirds breeding in isolated islands largely benefit from eradication of the invasive species (i.e., Ratcliffe et al. 2010, Borrelle et al. 2018, Oliveira et al. 2022). Efforts of conservation more likely have been focused on protecting highly important marine areas under low impacts (O'Hara et al. 2019). However, the fact that seabirds target oceanic features which are prone to human interference—for instance, fisheries target highly productive habitats also used by seabirds—provides a basis for shifting the view towards protecting and/or managing highly important habitats under high impacts, which is conceptualized as reactive protection (O'Hara et al. 2019). In that context, the threat posed by industrial fisheries is of concern. Studies showing that high conservation valued areas are targeted by industrial fisheries have started to accumulate in literature (Krüger et al. 2017a, 2021, Pereira et al. 2018, Watters et al. 2020). Seabird populations have declined globally: competition with industrial fisheries (Grémillet et al. 2018), higher rates of mortality due to incidental bycatch or injuries due to fishing lines, nets or cables (Anderson et al. 2011, Gianuca et al. 2017) are among the most common threats to seabirds (but invasive predators on breeding colonies also is an issue, see Oliveira et al. 2022, and Dias et al. 2019 for a full evaluation of threats to seabirds). Not surprisingly, several studies pointed out that the global MPA network seems to fail covering some of the most unique marine ecosystems and/or habitats (Brooks 2013, Brooks et al. 2020, Jantke et al. 2018, Klein et al. 2015), which may have been a result of countries prioritizing fishing production instead of sustainability (Brooks 2013, O'Hara et al. 2019), and targeting protection only on areas of lower fishing yields with smaller conservation value.

Changes in climatic conditions also poses an important issue for MPAs efficiency. On the mid to long-term, shifting climate might force seabirds to change their distribution (Krüger et al. 2018a, b, Scales et al. 2017, but see Somveille et al. 2020), which is already evident in some studies (Borowicz et al. 2020, Péron et al. 2010, Poloczanska et al. 2016). On one hand, shifting conditions may misplace seabirds off MPAs designed based on current conditions (Bruno et al. 2018, Critchley et al. 2018). On the other hand, a well-planned network of large MPAs can account for changing distributions of seabirds (Davies et al. 2017, Hindell et al. 2020).

Conclusions

The need to increase oceans protection is unequivocal. The number of threats experienced by marine organisms is unprecedent. Seabirds, given their general characteristics, can be used as selecting criteria to identify key areas to protect. As indicators of ocean ecosystems productivity and diversity, their distribution and habitat requirements can enter as layers on spatial planning and prioritization analysis for designing MPAs. The fact that seabirds are highly threatened organisms also provides a logical rule of thumb for prioritization, and, in fact, many studies have used such approaches to identify hotspots of marine protection using seabird data worldwide. This criterion alone can help world governments to identify large areas of marine habitat to protect in the open oceans, helping to increase substantially the world's oceans MPA coverage.

Acknowledgments

Thanks to the Marine Conservation Institute, with special thanks to Jessica Coulter, for providing the Marine Protected Areas shapefiles from the MPAtlas database. Thanks to Jaime A. Ramos and Maria P. Dias for constructive comments on the manuscript. Thanks to the Marine Protected Areas program of the Instituto Antártico Chileno INACH.

References cited

Anderson, A., R. Bonaldo, D. Barneche, C. Hackradt, F. Félix-Hackradt, J. García-Charton and S. Floeter. 2014. Recovery of grouper assemblages indicates effectiveness of a marine protected area in Southern Brazil. Mar. Ecol. Prog. Ser. 514: 207–215. doi: 10.3354/meps11032.

Anderson, O., C. Small, J. Croxall, E. Dunn, B. Sullivan, O. Yates et al. 2011. Global seabird bycatch in longline fisheries. Endanger. Species Res. 14: 91–106. doi: 10.3354/esr00347.

Arcos, J.M., J. Bécares, D. Villero, L. Brotons, B. Rodríguez and A. Ruiz. 2012. Assessing the location and stability of foraging hotspots for pelagic seabirds: An approach to identify marine Important Bird Areas (IBAs) in Spain. Biol. Conserv. 156: 30–42. doi: 10.1016/j.biocon.2011.12.011.

Assali, C., N. Bez and Y. Tremblay. 2017. Seabird distribution patterns observed with fishing vessel's radar reveal previously undescribed sub-meso-scale clusters. Sci. Rep. 7: 1–10. doi: 10.1038/s41598-017-07480-6.

Barnes, D.K.A., K. Linse, C. Waller, S. Morely, P. Enderlein, K.P.P. Fraser et al. 2006. Shallow benthic fauna communities of South Georgia Island. Polar Biol. 29: 223–228. doi: 10.1007/s00300-005-0042-0.

Beal, M., M. Dias, R.A. Phillips, S. Oppel, C. Hazin, E.J. Pearmain et al. 2021. Global political responsibility for the conservation of albatrosses and large petrels. Sci. Adv. 7: eabd7225. doi: 10.1126/sciadv.abd7225.

Beaugrand, G., A. Conversi, A. Atkinson, J. Cloern, S. Chiba, S. Fonda-Umani et al. 2019. Prediction of unprecedented biological shifts in the global ocean. Nat. Clim. Chang. 9: 237–243. doi: 10.1038/s41558-019-0420-1.

BirdLife International. 2009. Designing networks of marine protected areas : Exploring the linkages between Important Bird Areas and ecologically or biologically significant marine areas. Cambridge, UK, BirdLife International. https://www.cbd.int/doc/meetings/mar/ewbcsima-01/other/ewbcsima-01-birdlife-02-en.pdf.

Birdlife International. 2021. Birdlife International: partnership for nature and people. Cambridge, UK, BirdLife International. Retrieved November 2020 from http://www.birdlife.org/.

Boonzaier, L. and D. Pauly. 2016. Marine protection targets: An updated assessment of global progress. Oryx 50: 27–35. doi: 10.1017/S0030605315000848.

Borowicz, A., S. Forrest, M. Wethington, N. Strycker and H.J. Lynch. 2020. Presence of King penguins (*Aptenodytes patagonicus*) on Elephant Island provides further evidence of range expansion. Polar Biol. 43: 2117–2120. doi: 10.1007/s00300-020-02760-w.

Borrelle, S.B., P.H. Boersch-Supan, C.P. Gaskin and D.R. Towns. 2018. Influences on recovery of seabirds on islands where invasive predators have been eradicated, with a focus on Procellariiformes. Oryx 52: 346–358. doi: 10.1017/S0030605316000880.

Brooks, C.M. 2013. Competing values on the Antarctic high seas: CCAMLR and the challenge of marine-protected areas. Polar J. 3: 277–300. doi: 10.1080/2154896X.2013.854597.

Brooks, C.M., S.L. Chown, L.L. Douglass, B.P. Raymond, J.D. Shaw, Z.T. Sylvester and C.L. Torrens. 2020. Progress towards a representative network of Southern Ocean protected areas. PLoS One 15: e0231361. doi: 10.1371/journal.pone.0231361.

Bruno, J.F., A.E. Bates, C. Cacciapaglia, E.P. Pike, S.C. Amstrup, R. Van Hooidonk et al. 2018. Climate change threatens the world's marine protected areas. Nat. Clim. Chang. 8: 499–503. doi: 10.1038/s41558-018-0149-2.

Butchart, S.H.M., M. Walpole, B. Collen, A. van Strien, J.P.W. Scharlemann, R.E.A. Almond et al. 2010. Global Biodiversity: Indicators of Recent Declines. Science 328: 1164–1168. doi: 10.1126/science.1187512.

Cabeza, M., A. Arponenand and A. Van Teeffelen. 2008. Top predators: Hot or not? A call for systematic assessment of biodiversity surrogates. J. Appl. Ecol. 45: 976–980. doi: 10.1111/j.1365-2664.2007.01364.x.

Cabral, R.B., D. Bradley, J. Mayorga, W. Goodell, A.M. Friedlander, E. Sala et al. 2020. A global network of marine protected areas for food. Proc. Natl. Acad. Sci. 117: 28134–28139. doi: 10.1073/pnas.2000174117.

Carneiro, A.P.B., E.J. Pearmain, S. Oppel, T.A. Clay, R.A. Phillips, A.S. Bonnet-Lebrun et al. 2020. A framework for mapping the distribution of seabirds by integrating tracking, demography and phenology. J. Appl. Ecol. 57: 514–525. doi: 10.1111/1365-2664.13568.

Catry, P., A. Campos, P. Segurado, M. Silvaand and I. Strange. 2003. Population census and nesting habitat selection of thin-billed prion *Pachyptila belcheri* on New Island, Falkland Islands. Polar Biol. 26: 202–207. doi: 10.1007/s00300-002-0465-9.

Clay, T.A., R.A. Phillips, A. Manica, H.A. Jackson and M.D.L. Brooke. 2017. Escaping the oligotrophic gyre? The year-round movements, foraging behaviour and habitat preferences of Murphy's petrels. Mar. Ecol. Prog. Ser. 579: 139–155. doi: 10.3354/meps12244.

Cleasby, I.R., E. Owen, L. Wilson, E.D. Wakefield, P. O'Connell and M. Bolton. 2020. Identifying important at-sea areas for seabirds using species distribution models and hotspot mapping. Biol. Conserv. 241: 108375. doi: 10.1016/j.biocon.2019.108375.

Clua, É. and F. Grosvalet. 2001. Mixed-species feeding aggregation of dolphins, large tunas and seabirds in the Azores. Aquat. Living Resour. 14: 11–18. doi: 10.1016/S0990-7440(00)01097-4.

Condie, S. and R. Condie. 2016. Retention of plankton within ocean eddies. Glob. Ecol. Biogeogr. 25, 1264–1277. doi: 10.1111/geb.12485.

Costello, M.J. and B. Ballantine. 2015. Biodiversity conservation should focus on no-take Marine Reserves: 94% of Marine Protected Areas allow fishing. Trends Ecol. Evol. 30: 507–509. doi: 10.1016/j.tree.2015.06.011.

Cox, S.L., P.I. Miller, C.B. Embling, K.L. Scales, A.W.J. Bicknell, P.J. Hosegood et al. 2016. Seabird diving behaviour reveals the functional significance of shelf-sea fronts as foraging hotspots. R. Soc. Open Sci. 3. doi: 10.1098/rsos.160317.

Cox, S.L., C.B. Embling, P.J. Hosegood, S.C. Votier and S.N. Ingram. 2018. Oceanographic drivers of marine mammal and seabird habitat-use across shelf-seas: A guide to key features and recommendations for future research and conservation management. Estuar. Coast. Shelf Sci. 212: 294–310. doi: 10.1016/j.ecss.2018.06.022.

Crawford, R.J.M. 2007. Food, fishing and seabirds in the Benguela upwelling system. J. Ornithol. 148: 253–260. doi: 10.1007/s10336-007-0228-z.

Critchley, E.J., W.J. Grecian, A. Kane, M.J. Jessopp and J.L. Quinn. 2018. Marine protected areas show low overlap with projected distributions of seabird populations in Britain and Ireland. Biol. Conserv. 224: 309–317. doi: 10.1016/j.biocon.2018.06.007.

Croxall, J.P., S.H.M. Butchart, B. Lascelles, A.J. Stattersfield, B. Sullivan, A. Symes and P. Taylor. 2012. Seabird conservation status, threats and priority actions: a global assessment. Bird Conserv. Int. 22: 1–34. doi: 10.1017/S0959270912000020.

Daudt, N.W., S.P. Martins, E.P. Kirinus and L. Bugoni. 2019. Seabird assemblage at the mouth of the Amazon River and its relationship with environmental characteristics. J. Sea Res. 155: 101826. doi: 10.1016/j.seares.2019.101826.

Davies, T.E., S.M. Maxwell, K. Kaschner, C. Garilao and N.C. Ban. 2017. Large marine protected areas represent biodiversity now and under climate change. Sci. Rep. 7: 1–7. doi: 10.1038/s41598-017-08758-5.

Di Lorenzo, M., J. Claudet and P. Guidetti. 2016. Spillover from marine protected areas to adjacent fisheries has an ecological and a fishery component. J. Nat. Conserv. 32: 62–66. doi: 10.1016/j.jnc.2016.04.004.

Dias, M.P., S. Oppel, A.L. Bond, A.P.B. Carneiro, R.J. Cuthbert, J. González-Solís et al. 2017. Using globally threatened pelagic birds to identify priority sites for marine conservation in the South Atlantic Ocean. Biol. Conserv. 211: 76–84. doi: 10.1016/j.biocon.2017.05.009.

Dias, M.P., R. Martin, E.J. Pearmain, I.J. Burfield, C. Small, R.A. Phillips et al. 2019. Threats to seabirds: A global assessment. Biol. Conserv. 237: 525–537. doi: 10.1016/j.biocon.2019.06.033.

Donald, P.F., L.D.C. Fishpool, A. Ajagbe, L.A. Bennun, G. Bunting, I.J. Burfield et al. 2019. Important Bird and Biodiversity Areas (IBAs): The development and characteristics of a global inventory of key sites for biodiversity. Bird Conserv. Int. 29: 177–198. doi: 10.1017/S0959270918000102.

Dunn, D.C., J. Ardron, N. Bax, P. Bernal, J. Cleary, I. Cresswell et al. 2014. The convention on biological diversity's ecologically or biologically significant areas: Origins, development, and current status. Mar. Policy 49: 137–145. doi: 10.1016/j.marpol.2013.12.002.

Dunstan, P.K., N.J. Bax, J.M. Dambacher, K.R. Hayes, P.T. Hedge, D.C. Smith et al. 2016. Using ecologically or biologically significant marine areas (EBSAs) to implement marine spatial planning. Ocean Coast. Manag. 121: 116–127. doi: 10.1016/j.ocecoaman.2015.11.021.

Edgar, G.J., R.D. Stuart-Smith, T.J. Willis, S. Kininmonth, S.C. Baker, S. Banks et al. 2014. Global conservation outcomes depend on marine protected areas with five key features. Nature 506: 216–220. doi: 10.1038/nature13022.

Engler, J.O., D. Stiels, K. Schidelko, D. Strubbe, P. Quillfeldt and M. Brambilla. 2017. Avian SDMs: Current state, challenges, and opportunities. J. Avian Biol. 48: 1483–1504. doi: 10.1111/jav.01248.

Fourcade, Y. 2016. Comparing species distributions modelled from occurrence data and from expert-based range maps. Implication for predicting range shifts with climate change. Ecol. Inform. 36: 8–14. doi: 10.1016/j.ecoinf.2016.09.002.

Gaines S.D., C. White, M.H. Carr and S.P. Palumbi. 2010. Designing marine reserves networks for both conservation and fisheries management. PNAS 107: 18286–18293. doi: 10.1073/pnas.0906473107.

Geburzi, J.C. and M.L. McCarthy. 2018. How Do They Do It?—Understanding the Success of Marine Invasive Species, in: YOUMARES 8—Oceans Across Boundaries: Learning from Each Other. pp. 109–124. doi: 10.1007/978-3-319-93284-2_8.

Gianuca, D., R.A. Phillips, S. Townley and S.C. Votier. 2017. Global patterns of sex- and age-specific variation in seabird bycatch. Biol. Conserv. 205: 60–76. doi: 10.1016/j.biocon.2016.11.028.

Gianuca, D., S.C. Votier, D. Pardo, A.G. Wood, R.B. Sherley, L. Ireland et al. 2019. Sex-specific effects of fisheries and climate on the demography of sexually dimorphic seabirds. J. Anim. Ecol. 88: 1366–1378. doi: 10.1111/1365-2656.13009.

Gorta, S.B.Z., J.A. Smith, J.D. Everett, R.T. Kingsford, W.K. Cornwell, I.M. Suthers et al. 2019. Pelagic citizen science data reveal declines of seabirds off south-eastern Australia. Biol. Conserv. 235: 226–235. doi: 10.1016/j.biocon.2019.05.007.

Grecian, W.J., M.J. Witt, M.J. Attrill, S. Bearhop, P.H. Becker, C. Egevang et al. 2016. Seabird diversity hotspot linked to ocean productivity in the Canary Current Large Marine Ecosystem. Biol. Lett. 12: 86–90. doi: 10.1098/rsbl.2016.0024.

Grémillet, D., A. Ponchon, M. Paleczny, M.L.D Palomares, V. Karpouzi and D. Pauly. 2018. Persisting Worldwide Seabird-Fishery Competition Despite Seabird Community Decline. Curr. Biol. 28: 4009–4013.e2. doi: 10.1016/j.cub.2018.10.051.

GSGSSI. 2013. South Georgia and the South Sandwich Islands Marine Protected Area Management Plan v 2.0 [online]. Government of South Georgia and the South Sandwich Islands. Stanley, Falkland Islands <https://www.gov.gs/docsarchive/environment/#tab-2>.

Guisan, A., R. Tingley, J.B. Baumgartner, I. Naujokaitis-Lewis, P.R. Sutcliff, A.I.T. Tulloch et al. 2013. Predicting species distributions for conservation decisions. Ecol. Lett. 16: 1424–1435. doi: 10.1111/ele.12189.

Halpern, B.S., M. Frazier, J. Potapenko, K.S. Casey, K. Koenig, C. Longo et al. 2015. Spatial and temporal changes in cumulative human impacts on the world's ocean. Nat. Commun. 6: 7615. doi: 10.1038/ncomms8615.

Halpern, B.S., M. Frazier, J. Afflerbach, J.S. Lowndes, F. Micheli C. O'Hara, C. Scarborough and K.A. Selkoe. 2019. Recent pace of change in human impact on the world's ocean. Sci. Rep. 9: 11609. doi: 10.1038/s41598-019-47201-9.

Handley, J.M., E.J. Pearmain, S. Oppel, A.P.B. Carneiro, C. Hazin, R.A. Phillips et al. 2020. Evaluating the effectiveness of a large multi-use MPA in protecting Key Biodiversity Areas for marine predators. Divers. Distrib. 26: 715–729. doi: 10.1111/ddi.13041.

Handley, J.M., M.M. Rouyer, E.J. Pearmain, V. Warwick-Evans, K. Teschke, J.T. Hinke et al. 2021. Marine Important Bird and Biodiversity Areas for penguins in Antarctica, targets for conservation targets. Front. Mar. Sci. 7: 602972. doi: 10.3389/fmars.2020.602972.

Hátún, H., B. Olsen and S. Pacariz. 2017. The dynamics of the North Atlantic subpolar gyre introduces predictability to the breeding success of kittiwakes. Front. Mar. Sci. 4: 1–7. doi: 10.3389/fmars.2017.00123.

Haywood, B.K., J.K. Parrish and J. Dolliver. 2016. Place-based and data-rich citizen science as a precursor for conservation action. Conserv. Biol. 30: 476–486. doi: 10.1111/cobi.12702.

Hazen, E.L., B. Abrahms, S. Brodie, G. Carroll, M.G. Jacox, M.S. Savoca et al. 2019. Marine top predators as climate and ecosystem sentinels. Front. Ecol. Environ. 17: 565–574. doi: 10.1002/fee.2125.

Heerah, K., M.P. Dias, K. Delord, S. Oppel, C. Barbraud, H. Weimerskirch et al. 2019. Important areas and conservation sites for a community of globally threatened marine predators of the Southern Indian Ocean. Biol. Conserv. 234: 192–201. doi: 10.1016/j.biocon.2019.03.037.

Hindell, M.A., R.R. Reisinger, Y. Ropert-Coudert, L.A. Hückstädt, P.N. Trathan, H. Bornemann et al. 2020. Tracking of marine predators to protect Southern Ocean ecosystems. Nature 580: 87–92. doi: 10.1038/s41586-020-2126-y.

Hislop, C. and J. Jabour. 2015. Quality counts: High Seas Marine Protected Areas in the Southern Ocean, in: Ocean Yearbook Online. pp. 166–191. doi: 10.1163/22116001-02901009.

Humphries, G.R.W., R. Naveen, M. Schwaller, C. Che-Castaldo, P. McDowall, M. Schrimpf et al. 2017. Mapping Application for Penguin Populations and Projected Dynamics (MAPPPD): Data and tools for dynamic management and decision support. Polar Rec. 53: 160–166. doi: 10.1017/S0032247417000055.

IUCN. 2021. The IUCN Red List of threatened species. Version 2020–3. IUCN, Gland, Switzerland. Retrieved August 2020 from https://www.iucnredlist.org.

Jantke, K., K.R. Jones, J.R. Allan, A.L.M. Chauvenet, J.E.M. Watson and H.P. Possingham. 2018. Poor ecological representation by an expensive reserve system: Evaluating 35 years of marine protected area expansion. Conserv. Lett. 11: 1–8. Doi: 10.1111/conl.12584.

Jenkins, C.N. and K.S. Van Houtan. 2016. Global and regional priorities for marine biodiversity protection. Biol. Conserv. 204: 333–339. Doi: 10.1016/j.biocon.2016.10.005.

Klein, C.J., C.J. Brown, B.S. Halpern, D.B. Segan, J. McGowan, M. Beger et al. 2015. Shortfalls in the global protected area network at representing marine biodiversity. Sci. Rep. 5: 1–7. Doi: 10.1038/srep17539.

Kroodsma, D.A., J. Mayorga, T. Hochberg, N.A. Miller, K. Boerder, F. Ferretti et al. 2018. Tracking the global footprint of fisheries. Science 359: 904–908. Doi: 10.1126/science.aao5646.

Krüger, L., J.A. Ramos, J.C. Xavier, D. Grémillet, J. González-Solís, Y. Kolbeinsson et al. 2017a. Identification of candidate pelagic marine protected areas through a seabird seasonal-, multispecific- and extinction risk-based approach. Anim. Conserv. 20: 409–424. Doi: 10.1111/acv.12339.

Krüger, L., V.H. Paiva, M.V. Petry, R.C. Montone and J.A. Ramos. 2017b. Population estimate of Trindade Petrel *Pterodroma arminjoniana* by the use of Predictive Nest Habitat Modelling. Bird Conserv. Int: 197–207. Doi: 10.1017/S0959270916000289.

Krüger, L., J. Pereira, I. Ramírez, J.A. Ramos and V.H. Paiva. 2018a. How the future climate will modulate the non-breeding distribution of a vulnerable gadfly petrel? Mar. Ecol. Prog. Ser. 599: 253–266. Doi: 10.3354/meps12637.

Krüger, L., J.A. Ramos, J.C. Xavier, D. Grémillet, J. González-Solís, M.V. Petry et al. 2018b. Projected distributions of Southern Ocean albatrosses, petrels and fisheries as a consequence of climatic change. Ecography 41: 195–208. Doi: 10.1111/ecog.02590.

Krüger, L., M.F. Huerta, F. Santa Cruz and C.A. Cárdenas. 2021. Antarctic krill fishery effects over penguin population under adverse climate conditions: Implications for the management of fishing practices. Ambio 50: 560 – 571. doi: 10.1007/s13280-020-01386-w.

Kuletz, K.J., M.C. Ferguson, B. Hurley, A.E. Gall, E.A. Labunski and T.C. Morgan. 2015. Seasonal spatial patterns in seabird and marine mammal distribution in the eastern Chukchi and western Beaufort seas: Identifying biologically important pelagic areas. Prog. Oceanogr. 136: 175–200. Doi: 10.1016/j.pocean.2015.05.012.

Laffoley, D., J.M. Baxter, J.C. Day, L. Wenzel, P. Bueno and K. Zischka. 2018. Marine protected areas. World Seas An Environ. Eval. Vol. III Ecol. Issues Environ. Impacts 2019: 549–569. Doi: 10.1016/B978-0-12-805052-1.00027-9.

Lambert, C., A. Virgili, E. Pettex, J. Delavenne, V. Toison, A. Blanck et al. 2017. Habitat modelling predictions highlight seasonal relevance of Marine Protected Areas for marine megafauna. Deep. Res. Part II Top. Stud. Oceanogr. 141: 262–274. Doi: 10.1016/j.dsr2.2017.03.016.

Lascelles, B.G., G.M. Langham, R.A. Ronconi and J.B. Reid. 2012. From hotspots to site protection: Identifying Marine Protected Areas for seabirds around the globe. Biol. Conserv. 156: 5–14. Doi: 10.1016/j.biocon.2011.12.008.

Lascelles, B.G., P.R. Taylor, M.G.R. Miller, M.P. Dias, S. Oppel, L. Torres et al. 2016. Applying global criteria to tracking data to define important areas for marine conservation. Divers. Distrib. 22: 422–431. Doi: 10.1111/ddi.12411.

Lenton, T.M., J. Rockström, O. Gaffney, S. Rahmstorf, K. Richardson, W. Steffen et al. 2019. Too risky to bet against. Nature 575: 593–595.

Levin, S.A. and J. Lubchenco. 2008. Resilience, robustness, and marine ecosystem-based management. Bioscience 58: 27–32. Doi: 10.1641/B580107.

Maréchaux, I., A.S.L. Rodrigues and A. Charpentier. 2017. The value of coarse species range maps to inform local biodiversity conservation in a global context. Ecography 40: 1166–1176. Doi: 10.1111/ecog.02598.

Marine Conservation Institute. 2021. MPAtlas. Marine Conservation Institute, Seattle, USA. Retrieved April 2021 from www.mpatlas.org.

Markus, T. and P.P.S. Sánchez. 2018. Managing and regulating underwater noise pollution. Handb. Mar. Environ. Prot. 971–995. doi: 10.1007/978-3-319-60156-4_52.

Martín, B., A. Onrubia, J. González-Arias and J.A. Vicente-Vírseda. 2020. Citizen science for predicting spatio-temporal patterns in seabird abundance during migration. PLoS One 15: e0236631. doi: 10.1371/journal.pone.0236631.

Mason, C., R. Alderman, J. McGowan, H.P. Possingham, A.J. Hobday, M. Sumner et al. 2018. Telemetry reveals existing marine protected areas are worse than random for protecting the foraging habitat of threatened shy albatross (*Thalassarche cauta*). Divers. Distrib. 24: 1744–1755. doi: 10.1111/ddi.12830.

McGowan, J., R.J. Smith, M. Di Marco, R.H. Clarke and H.P. Possingham. 2018. An evaluation of marine important bird and biodiversity areas in the context of spatial conservation prioritization. Conserv. Lett. 11: 1–8. doi: 10.1111/conl.12399.

McKinley, D.C., A.J. Miller-Rushing, H.L. Ballard, R. Bonney, H. Brown, S.C. Cook-Patton et al. 2017. Citizen science can improve conservation science, natural resource management, and environmental protection. Biol. Conserv. 208: 15–28. doi: 10.1016/j.biocon.2016.05.015.

Melo-Merino, S.M., H. Reyes-Bonilla and A. Lira-Noriega. 2020. Ecological niche models and species distribution models in marine environments: A literature review and spatial analysis of evidence. Ecol. Modell. 415: 108837. doi: 10.1016/j.ecolmodel.2019.108837.

Miller, M.G.R., N. Carlile, J.S. Phillips, F. McDuie and B.C. Congdon. 2018. Importance of tropical tuna for seabird foraging over a marine productivity gradient. Mar. Ecol. Prog. Ser. 586: 233–249. doi: 10.3354/meps12376.

Morato, T., D. Varkey, C. Damaso, M. Machete, M. Santos, R. Prieto et al. 2008. Evidence of a seamount effect on aggregating visitors. Mar. Ecol. Prog. Ser. 357: 23–32. doi: 10.3354/meps07269.

Moudrý, V. and R. Devillers. 2020. Quality and usability challenges of global marine biodiversity databases: An example for marine mammal data. Ecol. Inform. 56: 101051. doi: 10.1016/j.ecoinf.2020.101051.

Myers, S.S. 2017. Planetary health: protecting human health on a rapidly changing planet. Lancet 390: 2860–2868. doi: 10.1016/S0140-6736(17)32846-5.

O'Hara, C.C., J.C. Villaseñor-Derbez, G.M. Ralph and B.S. Halpern. 2019. Mapping status and conservation of global at-risk marine biodiversity. Conserv. Lett. 12: 1–9. doi: 10.1111/conl.12651.

Oliveira, P., F.M. Medina, M. Nogales and P.L. Geraldes. 2022. Eradication and control of invasive mammal species as a seabird conservation tool. pp. 224–242. *In*: J.A. Ramos and L. Pereira [eds.]. Seabird Biodiversity and Human Activities. CRC Press, Boca Raton, FL.

Oppel, S., A. Meirinho, I. Ramírez, B. Gardner, A.F. O'Connell, P.I. Miller et al. 2012. Comparison of five modelling techniques to predict the spatial distribution and abundance of seabirds. Biol. Conserv. 156: 94–104. doi: 10.1016/j.biocon.2011.11.013.

Oppel, S., M. Bolton, A.P.B. Carneiro, M.P. Dias, J.A. Green, J.F. Masello et al. 2018. Spatial scales of marine conservation management for breeding seabirds. Mar. Policy 98: 37–46. doi: 10.1016/j.marpol.2018.08.024.

Palumbi, S.R., P.A. Sandifer, J.D. Allan, M.W. Beck, D.G. Fautin, M.J. Fogarty et al. 2009. Managing for ocean biodiversity to sustain marine ecosystem services. Front. Ecol. Environ. 7: 204–211. doi: 10.1890/070135.

Patterson, T.A., R.J. Sharples, B. Raymond, D.C. Welsford, V. Andrews-Goff, M.A. Lea et al. 2016. Foraging distribution overlap and marine reserve usage amongst sub-Antarctic predators inferred from a multi-species satellite tagging experiment. Ecol. Indic. 70: 531–544. doi: 10.1016/j.ecolind.2016.05.049.

Pereira, J.M., L. Krüger, N. Oliveira, A. Meirinho, A. Silva, J.A. Ramos et al. 2018. Using a multi-model ensemble forecasting approach to identify key marine protected areas for seabirds in the Portuguese coast. Ocean Coast. Manag. 153: 98–107. doi: 10.1016/j.ocecoaman.2017.12.014.

Pereira, J.M., V.H. Paiva, L. Kruger and S.C. Votier. 2022. Tracking seabirds for conservation and marine spatial planning. pp. 59–74. *In*: J.A. Ramos and L. Pereira [eds.]. Seabird Biodiversity and Human Activities. CRC Press, Boca Raton, FL.

Péron, C., M. Authier, C. Barbraud, K. Delord, D. Besson and H. Weimerskirch. 2010. Interdecadal changes in at-sea distribution and abundance of subantarctic seabirds along a latitudinal gradient in the Southern Indian Ocean. Glob. Chang. Biol. 16: 1895–1909. doi: 10.1111/j.1365-2486.2010.02169.x.

Piatt, J.F., W.J. Sydeman and F. Wiese. 2007. Introduction: A modern role for seabirds as indicators. Mar. Ecol. Prog. Ser. 352: 199–204. doi: 10.3354/meps07070.

Poloczanska, E.S., M.T. Burrows, C.J. Brown, J. García-Molinos, B.S. Halpern, O. Hoegh-Guldberg et al. 2016. Responses of marine organisms to climate change across oceans. Front. Mar. Sci. 3: 62. doi: 10.3389/fmars.2016.00062.

Ramirez, I., M. Tarzia, M.P. Dias, I.J. Burfield, J.A. Ramos, S. Garthe et al. 2017. How well is the EU protecting its seabirds? Progress in implementing the Birds Directive at sea. Mar. Policy 81: 179–184. doi: 10.1016/j.marpol.2017.03.034.

Ratcliffe, N., M. Bell, T. Pelembe, D. Boyle, R. Benjamin, R. White et al. 2010. The eradication of feral cats from Ascension Island and its subsequent recolonization by seabirds. Oryx 44: 20–29. doi: 10.1017/S003060530999069X.

Raymond, B., M.A. Lea, T. Patterson, V. Andrews-Goff, R. Sharples, J.B. Charrassin et al. 2015. Important marine habitat off east Antarctica revealed by two decades of multi-species predator tracking. Ecography 38: 121–129. doi: 10.1111/ecog.01021.

Rayner, M.J., M.N. Clout, R.K. Stamp, M.J. Imber, D.H. Brunton and M.E. Hauber. 2007. Predictive habitat modelling for the population census of a burrowing seabird: A study of the endangered Cook's petrel. Biol. Conserv. 138: 235–247. doi: 10.1016/j.biocon.2007.04.021.

Ripple, W.J., C. Wolf, T.M. Newsome, M. Hoffmann, A.J. Wirsing and D.J. McCauley. 2017. Extinction risk is most acute for the world's largest and smallest vertebrates. Proc. Natl. Acad. Sci. 114: 10678–10683. doi: 10.1073/pnas.1702078114.

Roberts, C.M., B.C. O'Leary, D.J. McCauley, P.M. Cury, C.M. Duarte, J. Lubchenco et al. 2017. Marine reserves can mitigate and promote adaptation to climate change. Proc. Natl. Acad. Sci. 114: 6167–6175. doi: 10.1073/pnas.1701262114.

Rogers, A.D., O. Alvheim, E. Bemanaja, D. Benivary, P. Boersch-Supan, T.G. Bornman et al. 2017. Pelagic communities of the South West Indian Ocean seamounts: R/V Dr Fridtjof Nansen Cruise 2009-410. Deep Sea Res. Part II Top. Stud. Oceanogr. 136: 5–35. doi: 10.1016/j.dsr2.2016.12.010.

Rogers, A.D. 2018. The Biology of Seamounts: 25 Years on. Adv. Mar. Biol. 79: 137–224. doi: 10.1016/bs.amb.2018.06.001.

Rolland, V., H. Weimerskirch and C. Barbraud. 2010. Relative influence of fisheries and climate on the demography of four albatross species. Glob. Chang. Biol. 16: 1910–1922. doi: 10.1111/j.1365-2486.2009.02070.x.

Ronconi, R.A., B.G. Lascelles, G.M. Langham, J.B. Reid and D. Oro. 2012. The role of seabirds in Marine Protected Area identification, delineation, and monitoring: Introduction and synthesis. Biol. Conserv. 156: 1–4. doi: 10.1016/j.biocon.2012.02.016.

Sala, E. and K. Rechberger. 2018. Protecting half the ocean. pp. 239–261. *In*: R.M. Desai, H. Kato, H. Kharas and J. McArthur [eds.]. From Summits to Solutions. Innovations in Implementing the Sustainable Development Goals. Brookings Institution Press, Washington, DC, USA.

Sala, E. and S. Giakoumi. 2018. No-take marine reserves are the most effective protected areas in the ocean. ICES J. Mar. Sci. 75: 1166–1168. doi: 10.1093/icesjms/fsx059.

Sala, E., J. Lubchenco, K. Grorud-Colvert, C. Novelli, C. Roberts and U.R. Sumaila. 2018. Assessing real progress towards effective ocean protection. Mar. Policy 91: 11–13. doi: 10.1016/j.marpol.2018.02.004.

Scales, K.L., G.S. Schorr, E.L. Hazen, S.J. Bograd, P.I. Miller, R.D. Andrews et al. 2017. Should I stay or should I go? Modelling year-round habitat suitability and drivers of residency for fin whales in the California Current. Divers. Distrib. 23: 1204–1215. doi: 10.1111/ddi.12611.

Sergio, F., I. Newton, L. Marchesi and P. Pedrini. 2006. Ecologically justified charisma: Preservation of top predators delivers biodiversity conservation. J. Appl. Ecol. 43: 1049–1055. doi: 10.1111/j.1365-2664.2006.01218.x.

Sergio, F., T. Caro, D. Brown, B. Clucas, J. Hunter, J. Ketchum et al. 2008. Top predators as conservation tools: Ecological rationale, assumptions, and efficacy. Annu. Rev. Ecol. Evol. Syst. 39: 1–19. doi: 10.1146/annurev.ecolsys.39.110707.173545.

Serratosa, J., K.D. Hyrenbach, D. Miranda-Urbina, M. Portflitt-Toro, N. Luna and G. Luna-Jorquera. 2020. Environmental drivers of seabird at-sea distribution in the eastern south pacific ocean: Assemblage composition across a longitudinal productivity gradient. Front. Mar. Sci. 6: 1–13. doi: 10.3389/fmars.2019.00838.

Somveille, M., M.P. Dias, H. Weimerskirch and T.E. Davies. 2020. Projected migrations of Southern Indian Ocean albatrosses as a response to climate change. Ecography 43: 1683–1691. doi 10.1111/ecog.05066.

Spatz, D.R., K.M. Newton, R. Heinz, B. Tershy, N.D. Holmes, S.H.M. Butchart et al. 2014. The biogeography of threatened seabirds and island conservation opportunities. Cons. Biol. 28: 1282–1290. doi: 10.1111/cobi.12279.

Spatz, D.R, N.D. Holmes, B.G. Reguero, S.H.M. Butchart, B.R. Tershy and D.A. Croll. 2017. Managing invasive mammals to conserve globally threatened seabirds in a changing climate. Cons. Letters 10: 736–747. doi: 10.1111/conl.12373.

Steffen, W., K. Richardson, J. Rockström, S.E. Cornell, I. Fetzer, E.M. Bennett et al. 2015. Planetary boundaries: Guiding human development on a changing planet. Science 347: 1259855. doi: 10.1126/science.1259855.

Sundblad, G., U. Bergström and A. Sandström. 2011. Ecological coherence of marine protected area networks: A spatial assessment using species distribution models. J. Appl. Ecol. 48: 112–120. doi: 10.1111/j.1365-2664.2010.01892.x.

Sylvester, Z.T. and C.M. Brooks. 2019. Protecting Antarctica through Co-production of actionable science: Lessons from the CCAMLR marine protected area process. Mar. Policy 111: 103720. doi: 10.1016/j.marpol.2019.103720.

Thaxter, C.B., B. Lascelles, K. Sugar, A.S.C.P. Cook, S. Roos, M. Bolton et al. 2012. Seabird foraging ranges as a preliminary tool for identifying candidate Marine Protected Areas. Biol. Conserv. 156: 53–61. doi: 10.1016/j.biocon.2011.12.009.

Tickler, D., J.J. Meeuwig, M.L. Palomares, D. Pauly and D. Zeller. 2018. Far from home: Distance patterns of global fishing fleets. Sci. Adv. 4: 4–10. doi: 10.1126/sciadv.aar3279.

Toropova, C., I. Meliane, D. Laffoley, E. Matthews and M. Spalding. 2010. Global Ocean Protection. UNEP-WCMC and IUCN-WCPA, Cambridge.

Tuck, G.N., R.B. Thomson, C. Barbraud, K. Delord, M. Louzao, M. Herrera et al. 2015. An integrated assessment model of seabird population dynamics: Can individual heterogeneity in susceptibility to fishing explain abundance trends in Crozet wandering albatross? J. Appl. Ecol. 52: 950–959. doi: 10.1111/1365-2664.12462.

UNEP-WCMC and IUCN. 2021. Protected Planet: The World Database on Protected Areas (WDPA), UNEP-WCMC and IUCN, Cambridge, UK. Retrieved April 2021 from www.protectedplanet.net.

Urbanski, J.A., L. Stempniewicz, J.M. Węsławski, K. Dragańska-Deja, A. Wochna, M. Goc et al. 2017. Subglacial discharges create fluctuating foraging hotspots for sea birds in tidewater glacier bays. Sci. Rep. 7: 1–12. doi: 10.1038/srep43999.

van der Boog, C.G., M.F. de Jong, M. Scheidat, M.F. Leopold, S.C.V. Geelhoed, K. Schulz et al. 2019. Hydrographic and Biological Survey of a Surface-Intensified Anticyclonic Eddy in the Caribbean Sea. J. Geophys. Res. Ocean. 124: 6235–6251. doi: 10.1029/2018JC014877.

van Sebille, E., C. Wilcox, L. Lebreton, N. Maximenko, B.D. Hardesty, J.A. van Franeker et al. 2015. A global inventory of small floating plastic debris. Environ. Res. Lett. 10: 124006. doi: 10.1088/1748-9326/10/12/124006.

Ventura F., J.P. Granadeiro, O. Padget and P. Catry. 2020. Gadfly petrels use knowledge of the windscape, not memorized foraging patches, to optimize foraging trips on ocean-wide scales. Proc. R. Soc. B 287: 20191775. doi: 10.1098/rspb.2019.1775.

Waggitt, J.J., P.G.H. Evans, J. Andrade, A.N. Banks, O. Boisseau, M. Bolton et al. 2020. Distribution maps of cetacean and seabird populations in the North-East Atlantic. J. Appl. Ecol. 57: 253–269. doi: 10.1111/1365-2664.13525.

Waldron, A., V. Adams, J. Allan, A. Arnell, G. Asner, G. Atkinson et al. 2020. Protecting 30% of the planet for nature: Costs, benefits and economic implications. Working paper analysing the economic implications of the proposed 30% target for areal protection in the draft post-2020 Global Biodiversity Framework, Cambridge, UK. ttps://www.conservation.cam.ac.uk/files/waldron_report_30_by_30_publish.pdf.

Waliczky, Z., L.D.C. Fishpool, S.H.M. Butchart, D. Thomas, M.F. Heath, C. Hazin et al. 2019. Important Bird and Biodiversity Areas (IBAs): Their impact on conservation policy, advocacy and action. Bird Conserv. Int. 29: 199–215. doi: 10.1017/S0959270918000175.

Ward, E.J., K.N. Marshall, T. Ross, A. Sedgley, T. Hass, S.F. Pearson et al. 2015. Using citizen-science data to identify local hotspots of seabird occurrence. PeerJ 3: e704. doi: 10.7717/peerj.704.

Watson, R.A. and A. Tidd. 2018. Mapping nearly a century and a half of global marine fishing: 1869–2015. Mar. Policy 93: 171–177. doi: 10.1016/j.marpol.2018.04.023.

Watters, G.M., J.T. Hinke and C.S. Reiss. 2020. Long-term observations from Antarctica demonstrate that mismatched scales of fisheries management and predator-prey interaction lead to erroneous conclusions about precaution. Sci. Rep. 10: 2314. doi: 10.1038/s41598-020-59223-9.

Webb, T.J. and B.L. Mindel. 2015. Global patterns of extinction risk in marine and non-marine systems. Curr. Biol. 25: 506–511. doi: 10.1016/j.cub.2014.12.023.

Weber, M.M., R.D. Stevens, J.A.F. Diniz-Filho and C.E.V. Grelle. 2017. Is there a correlation between abundance and environmental suitability derived from ecological niche modelling? A meta-analysis. Ecography 40: 817–828. doi: 10.1111/ecog.02125.

Weidberg, N., A. Ospina-Alvarez, J. Bonicelli, M. Barahona, C.M. Aiken, B.R. Broitman et al. 2020. Spatial shifts in productivity of the coastal ocean over the past two decades induced by migration of the Pacific Anticyclone and Bakun's effect in the Humboldt Upwelling Ecosystem. Glob. Planet. Change 193: 103259. doi: 10.1016/j.gloplacha.2020.103259.

Weimerskirch, H., M. Louzao, S. de Grissac and K. Delord. 2012. Changes in wind pattern alter albatross distribution and life-history traits. Science 335: 211–214. doi: 10.1126/science.1210270.

Welch, H., R.L. Pressey and A.E. Reside. 2018. Using temporally explicit habitat suitability models to assess threats to mobile species and evaluate the effectiveness of marine protected areas. J. Nat. Conserv. 41: 106–115. doi: 10.1016/j.jnc.2017.12.003.

Wilson, K.L., D.P. Tittensor, B. Worm and H.K. Lotze. 2020. Incorporating climate change adaptation into marine protected area planning. Glob. Chang. Biol. 26: 3251–3267. doi: 10.1111/gcb.15094.

Wing, S.R. and L. Jack. 2013. Marine reserve networks conserve biodiversity by stabilizing communities and maintaining food web structure. Ecosphere 4: 1–14. doi: http://dx.doi.org/10.1890/ES13-00257.1.

Yamamoto, T., N. Kokubun, D.M. Kikuchi, N. Sato, A. Takahashi, A.P. Will et al. 2016. Differential responses of seabirds to environmental variability over 2 years in the continental shelf and oceanic habitats of southeastern Bering Sea. Biogeosciences 13: 2405–2414. doi: 10.5194/bg-13-2405-2016.

Young, H.S., S.M. Maxwell, M.G. Conners and S.A. Shaffer. 2015. Pelagic marine protected areas protect foraging habitat for multiple breeding seabirds in the central Pacific. Biol. Conserv. 181: 226–235. doi: 10.1016/j.biocon.2014.10.027.

Zhang, B., E.J. Matchinski, B. Chen, X. Ye, L. Jing and K. Lee. 2019. Marine oil spills-oil pollution, sources and effects. pp. 391–409. *In*: C. Sheppard [ed.]. World Seas: An Environmental Evaluation (Second Edition) Volume III: Ecological Issues and Environmental Impacts. Academic Press. doi: 10.1016/B978-0-12-805052-1.00024-3.

Index

I

L

M

N

O

P

R

S

T

U

W